Optical Wireless Communications

Optical Wireless Communications

System and Channel Modelling with MATLAB®

Z. Ghassemlooy

W. Popoola

S. Rajbhandari

CRC Press
Taylor & Francis Group
Boca Raton London New York

CRC Press is an imprint of the
Taylor & Francis Group, an **informa** business

CRC Press
Taylor & Francis Group
6000 Broken Sound Parkway NW, Suite 300
Boca Raton, FL 33487-2742

First issued in paperback 2017

© 2013 by Taylor & Francis Group, LLC
CRC Press is an imprint of Taylor & Francis Group, an Informa business

No claim to original U.S. Government works

Version Date: 20120615

ISBN 13: 978-1-4398-5188-3 (hbk)
ISBN 13: 978-1-138-07480-4 (pbk)

Library of Congress Cataloging-in-Publication Data

Ghassemlooy, Zabih.
    Optical wireless communications : system and channel modelling with MATLAB / Z. Ghassemlooy, W. Popoola, S. Rajbhandari.
        p. cm.
    Includes bibliographical references and index.
    ISBN 978-1-4398-5188-3 (hardback)
    1. Optical communications. 2. Signal processing. 3. Free space optical interconnects. 4. Signals and signaling--Mathematical models. 5. MATLAB. I. Popoola, W. II. Rajbhandari, S. III. Title.

TK5103.592.F73G46 2012
621.382'7028553--dc23   2012006800

Visit the Taylor & Francis Web site at
http://www.taylorandfrancis.com

and the CRC Press Web site at
http://www.crcpress.com

Dedicated to Azar, Odunayo and Kanchan

Contents

Preface

In recent years, we have been witnessing a growing demand by the end users for bandwidth in mobile communications to support broadband wireless services such as, high-definition TV, mobile videophones, video conferencing, high-speed Internet access to name a few. With the increasing demand for high-quality multimedia services, the quest for higher bandwidth is expected to grow higher in the next decade. In access networks, a number of technologies currently in use to address the end-users' communication needs include copper-based technologies, hybrid coaxial and optical fibre cables, fibre-to-the-home, broadband RF/microwave wireless technologies. As the global demand for bandwidth continues to accelerate, it is becoming exceedingly clear that copper/coaxial cables and RF cellular/microwave technologies with such limitations as limited bandwidth, congested spectrum, security issues, expensive licensing and high cost of installation and accessibility to all, cannot meet the upcoming needs. In some countries the network operators are starting to deploy new optical fibre-based access networks where the bandwidth available to customers is being increased dramatically. Nevertheless, it is often thought that optical fibre-based networks offer unlimited bandwidth; however, in reality, available architectural choices, compatibility of devices and components, performance constraints of networking equipment and deployment of the complete system do result in limited capacity being offered to the end users.

Optical wireless communications (OWC) is an innovative technology that has been around for the last three decades and is gaining more attention as the demand for capacity continues to increase. OWC is one of the most promising alternative technologies for indoor and outdoor applications. It offers flexible networking solutions that provide cost-effective, highly secure high-speed license-free wireless broadband connectivity for a number of applications, including voice and data, video and entertainment, enterprise connectivity, disaster recovery, illumination and data communications, surveillance and many others. Due to the unique properties of the optical signal, one can precisely define a footprint and hence can accommodate a number of devices within a small periphery; thus offering a perfect OWC system. OWCs, also referred to as free-space optical communication systems for outdoor applications, will play a significant role as a complementary technology to the RF systems in future information superhighways.

Having seen the developments of OWC systems in the last two decades, we feel that there is a need for a new textbook on this topic. Textbooks already published in this area are either on indoor or outdoor optical wireless systems. Therefore, there is a need for this new book that covers the theory of the technology in a concise but comprehensive manner suitable for undergraduate- and graduate-level courses as well as researchers and professional engineers. This book broadly covers the four important aspects of OWC systems: (a) the fundamental principles of OWC, (b) devices and systems, (c) modulation techniques and (d) channel models and system performance analysis. The book also covers different challenges encountered in OWCs as well as outlining possible solutions and current research trends. A major attraction of the book is the presentation of MATLAB® simulations and the inclusion of MATLAB codes to help readers understand the topic under discussion and to be able to carry out extensive simulations.

The book is structured into eight self-contained chapters. To facilitate a logical progression of materials and to enable a relatively easy access, each chapter/topic is started with the appropriate amount of background discussion and supported with adequate theoretical analysis. Additional supporting materials are included in end of chapter appendices. Starting with a bit of history, Chapter 1 presents an up-to-date review of OWC systems for indoor and outdoor applications, the current state of play as well as the future directions for OWC. The wireless access technologies, benefits and limitations, link configurations, eye safety, application areas and challenges of the OWC systems are all covered in this chapter. There are a number of light sources and photodetectors (PIN and/or avalanche photodiodes) that could be used for OWC systems. LEDs and low-power laser diodes are mainly used in short-range indoor applications. For long-range outdoor applications laser diodes are mostly used. Chapter 2 discusses the types of light sources, their structures and optical characteristics, as well as the process of optical detection. Different types of noise encountered in optical detection and the statistics of the optical detection process are also discussed in Chapter 2.

To design efficient optical communication systems, it is imperative that the characteristics of the channel are well understood. Characterization of a communication channel is performed by its channel impulse response, which is then used to analyse and offer solutions to the effects of channel distortions. A number of propagation models (ceiling bounce, Hayasaka–Ito and spherical) for line-of-sight and non-line-of-sight indoor applications are studied in Chapter 3. The artificial light interference that affects the indoor OWC link performance is also outlined in this chapter. The atmospheric outdoor channel is a very complex and dynamic environment that can affect the characteristics of the propagating optical beam, thus resulting in optical losses and turbulence-induced amplitude and phase fluctuation. There are a number of models to characterize the statistical nature of the atmospheric channel, and these are dealt with in Chapter 3. A practical test bed for investigating the atmospheric effect on the free space optics link, as well as measured data, is also covered in Chapter 3.

Most practical OWC systems currently in use are based on the intensity modulation/direct detection scheme. For the outdoor environment and atmospheric conditions in particular, heavy fog is the major problem, as the intensity of light beam reduces considerably in thick fog. Increasing the level of transmit power is one option to improve the situation. However, eye safety regulations will limit the amount of transmit optical power. For indoor applications the eye safety limit on transmit optical power is even more stringent. In Chapter 4, a number of modulation techniques that are most popular in terms of power efficiency and bandwidth efficiency for both indoor and outdoor OWC applications are discussed. The emphasis is more on digital modulation techniques including pulse position modulation (PPM), on-and-off keying (OOK), digital pulse interval modulation (DPIM) and so on. The spectral properties, the error probability and the power and bandwidth requirements of a number of modulation schemes are presented. Advance modulation techniques such as subcarrier intensity modulation and polarization shift keying are also covered in this chapter. In indoor scenarios, the additional periodic and deterministic form of noise that degrades the system performance is due to the presence of background artificial light sources. Diffuse indoor links suffer from the multipath-induced inter-symbol interference, thus limiting the maximum achievable data rates. The perfor-mance of the OOK, PPM and DPIM-based systems in the presence of the artificial light interference and intersymbol interference is investigated in Chapter 5. To improve the link performance, possible mitigation techniques using high-pass filter-ing, equalization, wavelet transform and the neural network are also outlined in this chapter.

Atmospheric turbulence is known to cause signal fading in the channel and with the existing mathematical models for describing the fading have already been introduced in Chapter 3, Chapter 6 analyses the effect of atmospheric turbulence on the different modulation techniques used in outdoor free space optics. The emphasis in this chapter will, however, be on the effect of atmospheric turbulence-induced fading on OOK, PPM and the phase shift keying premodulated subcarrier intensity modulation. The primary challenges attributed to the outdoor OWC (i.e., FSO) communications include building sway, scattering/absorption-induced atten-uation and scintillation-induced link fading. For building swaying, one needs accu-rate pointing and tracking mechanisms, photodetector arrays and/or wide beam profiles. For FSO systems the phase and irradiance fluctuation suffered by the tra-versing beam makes the optical coherent detection less attractive, simply because it is sensitive to both signal amplitude and phase fluctuations. Thus, the reasons for adopting direct detection in terrestrial FSO links. The available options for mitigat-ing the effect of channel fading include but are not limited to increased power, frequency diversity, spatial and temporal diversity schemes including the multiple-input multiple-output (MIMO) scheme. In Chapter 7, we look into the time and spatial diversities as well as the subcarrier intensity modulation to mitigate the channel fading in FSO links. Link performance using equal gain combining, opti-mal combining or equivalently maximum ratio combining and selection combining diversity schemes in log-normal atmospheric channels employing a range of diver-sity techniques is also outlined in this chapter.

Finally, Chapter 8 is dedicated to visible light communications (VLC), a subject that over the last few years has witnessed an increased level of research activities. Putting it simply, VLC is the idea of using light-emitting diodes for both illumination and data communications. The main drivers for this technology include the increas-ing popularity of the solid state lighting and longer lifetime of high-brightness light-emitting diodes compared to other sources of artificial light. The dual functionality offered by visible light devices (i.e., lighting and data communication) has created a whole range of interesting applications including home networking, car-to-car com-munication, high-speed communication in aeroplane cabins, in-train data communi-cation, traffic light management and communications to name a few. The levels of power efficiency and reliability offered by light-emitting diodes are superior com-pared to the traditional incandescent light sources used for lighting. This chapter gives an overview of the VLC technology, highlighting the fundamental theoretical background, devices available, modulation and diming techniques, and system performance analysis. Multiple-input multiple-output and cellular VLC systems are also covered in this chapter.

MATLAB codes are included in each chapter to enable the reader to carry out simulations. Recent, relevant and up-to-date references that provide a guide for further reading are also included at the end of each chapter. A complete list of common abbreviations used in the text is also provided. Throughout the book, SI units are used.

Many thanks are extended to the authors of all journals and conference papers, articles and books that have been consulted in writing this book. Special thanks to those authors, publishers and companies for kindly granting permission for the reproduction of their figures. We would also like to extend our gratitude to all our past and current PhD students for their immense contributions to knowledge in the area of OWCs. Their contributions have undoubtedly enriched the content of this book. Finally, we remain extremely grateful to our families and friends who have continued to be supportive and provide the needed encouragement. In particular, our very special thanks go to our wives Azar, Odunayo and Kanchan for their continuous patience and unconditional support that has enabled us to finally complete this challenging task. Their support has been fantastic.

MATLAB® is a registered trademark of The MathWorks, Inc. For product information, please contact:

The MathWorks, Inc.
3 Apple Hill Drive
Natick, MA 01760-2098 USA
Tel: 508 647 7000
Fax: 508-647-7001
E-mail: info@mathworks.com
Web: www.mathworks.com

Authors

Professor Z. Ghassemlooy (CEng, Fellow of IET, senior member of IEEE) received his BSc (Hons.) in electrical and electronics engineering from the Manchester Metropolitan University in 1981, and his MSc and PhD in optical communications from the University of Manchester Institute of Science and Technology (UMIST) in 1984 and 1987, respectively with scholarships from the Engineering and Physical Science Research Council, UK. From 1986 to 1987, he worked as a demonstrator at UMIST and from 1987 to 1988, he was a postdoctoral research fellow at the City University, London. In 1988, he joined Sheffield Hallam University as a lecturer, becoming a reader in 1995, and a professor in optical communications in 1997. He was the group leader for communication engineering and digital signal processing, and also head of Optical Communications Research Group until 2004. In 2004, he moved to the University of Northumbria at Newcastle upon Tyne, UK as an associate dean for research in the School of Computing, Engineering and Information Sciences. He also heads the Northumbria Communications Research Laboratories within the School. In 2001, he received the Tan Chin Tuan Fellowship in engineering from the Nanyang Technological University in Singapore to work on photonic technology. He has been a visiting professor at number of international institutions. He is the editor-in-chief of *The Mediterranean Journals of Computers and Networks, and Electronics and Communications*. He currently serves on the editorial committees of a number of international journals. He is the founder and the chairman of the IEEE, IET International Symposium on Communication Systems, Network and Digital Signal Processing, the chairman of NOC 2011, and a member of technical committees of number of international conferences. He is a college member of the Engineering and Physical Science Research Council, UK (2003–2009) and (2009–present), and has served on a number of international research and advisory committees including a panel member of the Romanian Research Assessment Exercise 2011. His research interests are on photonics switching, optical wireless and wired communications, visible light communications and mobile communications. He has received a number of research grants from UK Research Councils, European Union, Industry and the UK government. He has supervised more than 33 PhD students and has published over 410 papers and presented several keynotes and invited talks. He is a coeditor of an IET book on analogue optical fibre communications, the proceedings of the *CSNDSP* 2006, 2008, 2010, *CSDSP'98, NOC 2011,* and the *First International Workshop on Materials for Optoelectronics 1995.* He is the guest coeditor of a number of special issues: *The Mediterranean Journal of Electronics and Communications* on 'Free Space Optics–RF', July 2006, the *IEE Proceedings J.* 1994, and 2000, *IET Proceedings Circuits, Devices and Systems*, special issue on the best papers for *CSNDSP Conference*, Vol. 2(1), 2008, *IEE Proceedings Circuits,*

Devices and Systems, special issue on the best papers for *CSNDSP Conference*, Vol. 153(4), 2006, *International Journal of Communications Systems* 2000, *Journal of Communications* 2009, *Ubiquitous Computing and Communication Journal—Selected Papers from CSNDSP 2008 Conference*, 2009, *International Journal of Communications—Special Issue on Optical Wireless Communications*, 2009. From 2004 to 2006 he was the IEEE UK/IR Communications chapter secretary, the vice-chairman during 2004–2008, the chairman during 2008–2011, and the chairman of the IET Northumbria Network from October 2011–present.

 **Dr. W. Popoola** had his national diploma in electrical engineering from The Federal Polytechnic, Ilaro, Nigeria and later graduated with a first class honours degree in electronic and electrical engineering from Obafemi Awolowo University, Nigeria. Thereafter he worked briefly, during the national service year, as a teaching assistant at Nnamdi Azikiwe University, Nigeria between 2003 and 2004. He later proceeded to Northumbria University at Newcastle upon Tyne, UK for his MSc in optoelectronic and communication systems where he graduated with distinction in 2006. He was awarded his PhD in 2009 at the Northumbria University for his research work in free-space optical communications. During his PhD studies he was awarded the 'Xcel Best Engineering and Technology Student of the year 2009'.

Dr. Popoola has published about 30 scholarly articles in the area of optical wireless communications. One of his papers was ranked second most downloaded paper in terms of the number of full-text downloads within the IEEE Xplore database in 2008, from the hundreds of papers published by IET Optoelectronics since 1980. He has also coauthored two invited book chapters. He is currently a researcher with the Institute for Digital Communications, University of Edinburgh, UK working on visible light communications. His team recently demonstrated 130 Mbit/s data streaming over a single off-the-shelf white LED. His research interests include optical communications, wireless communications and digital signal processing. He is an associate member of the Institute of Physics.

 **Dr. S. Rajbhandari** obtained his bachelor degree in electronics and communication engineering from the Institute of Engineering, Pulchowk Campus (Tribhuvan University), Nepal in 2004. He obtained an MSc in optoelectronic and communication systems with distinction in 2006 and was awarded the P. O. Byrne prize for most innovative project. He then joined the Optical Communications Research Group (OCRG) at Northumbria University as a PhD candidate and was awarded a PhD degree in 2010. His PhD thesis was on mitigating channel effect on indoor optical wireless communications using wavelet transform and a neural network. Since 2009 he has been with the OCRG at Northumbria University working as a postdoctorate researcher.

Dr. Rajbhandari has published more than 50 scholarly articles in the area of optical wireless communications. He has served as a local organizing committee member for CSNDSP 2010 and publication chair for NOC/OC&I 2011 as well as reviewer for several leading publications including the *IEEE/OSA Journal of Lightwave Technology, Selected Areas on Communications, Photonics Technology Letters, Communications Letters, IET Communications,* and several international journals and conferences. He has also served as a coeditor for the *Proceedings of the NOC/OC&I 2011.* His research interests lie in the area of optical wireless communications, modulation techniques, equalization, artificial intelligence and wavelet transforms. He is a member of IEEE and an associate member of the Institute of Physics.

List of Figures

List of Tables

Abbreviations

3G	3rd generation
4G	4th generation
ADC	Analogue-to-digital converter
ADSL	Asymmetric digital subscriber line
AFC	Automatic frequency control
AIM	Analogue intensity modulation
ALI	Ambient light interference
AM	Amplitude modulation
ANN	Artificial neural network
ANSI	American National Standards Institute
APD	Avalanche photodiode detector
ASK	Amplitude shift keying
AWGN	Additive white Gaussian noise
BAM	Bit angle modulation
BER	Bit error rate
BLL	Beer–Lambert law
BPL	Broadband over power line
BPoLSK	Binary polarization shift keying
BPSK	Binary phase-shift keying
CC	Convolutional codes
CDMA	Code division multiple access
CDRH	Center for Devices and Radiological Health
CENELEC	European Committee for Electrotechnical Standardization
CNR	Carrier-to-noise ratio
CPoLSK	Circular polarization shift keying
CWT	Continuous wavelet transform
DAC	Digital-to-analogue converter
DAPPM	Differential amplitude pulse position modulation
DC	Direct current
DCPoLSK	Differential circle polarization shift keying
DD	Direct detection
DFB	Distributed feedback laser
DFE	Decision feedback equalizer
DH-PIM	Dual-header pulse interval
DMT	Discrete multitone modulation
DPIM	Digital pulse internal modulation
DPLL	Digital phase-locked loop
DPPM	Differential pulse position modulation
DPSK	Differential phase shift keying
DSSS	Direct sequence spread spectrum
DTRIC	Dielectric totally internally reflecting concentrator

DVB	Digital video broadcast
DVB-S2	Digital Video Broadcast Satellite Standard
DWT	Discrete wavelet transform
eV	Electron volt
EGC	Equal gain combining
EPM	Edge position modulation
ERP	Emitter radiation pattern
FEC	Forward error correction
FFT	Fast Fourier transform
FIR	Finite impulse response
FLI	Fluorescent light interference
FOV	Field of view
FPL	Fabry–Perot laser
FSK	Frequency shift keying
GaAs	Gallium arsenide
Ge	Germanium
GS	Guard slot
HDD	Hard decision decoding
HPF	High-pass filter
H–V	Hufnagel–Valley model
IEC	International Electrotechnical Commission
IF	Intermediate frequency
IFFT	Inverse fast Fourier transform
IM	Intensity modulation
IMD	Intermodulation distortion
IMP	Intermodulation product
InP	Indium phosphide
IR	Infrared
IrDA	Infrared Data Association
ISI	Intersymbol interference
ISM	International Scientific and Medical
LAN	Local area networks
LCD	Liquid crystal display
LD	Laser diode
LDPC	Low-density parity check
LED	Light-emitting diode
LIA	Laser Institute of America
$LiNbO_3$	Lithium niobate
LO	Local oscillator
LOS	Line of sight
LPF	Low-pass filter
LSD	Light shaping diffuser
MF	Matched filter
MIMO	Multiple input multiple output
MISO	Multiple input single output
ML	Maximum likelihood

MLCD	Mars laser communication demonstration
MMSE	Minimum mean square error
MMW	Millimetre wave
MPE	Maximum permissible exposures
MPPM	Multiple pulse position modulation
MRC	Maximum ratio combining
MSC	Mobile switching centres
MSD	Multiple spot diffusing
MSLD	Maximum likelihood sequence detection
MVR	Meteorological visual range
MZI	Mach–Zehnder interferometry
MZM	Mach–Zehnder modulator
NEC	Nippon Electric Company
NEP	Noise-equivalent power
NLOS	Nonline of sight
NOPR	Normalized optical power requirement
NRZ	Non-return to zero
OBPF	Optical band pass filter
OFDM	Orthogonal Frequency Division Multiplexing
OIM	Optical impulse modulation
OLED	Organic light-emitting diode
OLO	Optical local oscillator
OMIMO	Optical multiple input multiple output
OOK	On and off keying
OPP	Optical power penalty
OPPM	Overlapping pulse position modulation
OSNR	Optical signal-to-noise ratio
OWC	Optical wireless communications
PAM	Pulse amplitude modulation
PAPR	Peak-to-average power ratio
PBS	Polarization beam splitter
PD	Photodetector
PDAs	Personal digital assistants
PDF	Probability density function
PER	Packet error rate
PIM	Pulse interval modulation
PMOPR	Peak-to-mean optical power ratio
PoLSK	Optical polarization shift keying
PPM	Pulse position modulation
PPM+	Pulse position modulation plus
PRBS	Pseudo random binary signal
PSD	Power spectral density
PSK	Phase shift keying
PTM	Pulse time modulation
PWM	Pulse width modulation
QAM	Quadrature amplitude modulation

QCL	Quantum cascaded laser
QPSK	Quadrature phase shift keying
RF	Radio frequency
RGB	Red, green and blue
RIN	Laser relative intensity noise
RMS	Root means square
Rx	Receiver
RZ	Return to zero
SCM	Sub carrier multiplexing
SDD	Soft decision decoding
SEC	Switch-and-examine combining
SelC	Selection combining
SER	Slot error rate
Si	Silicon
SI	Scintillation Index
SILEX	Semiconductor-laser Intersatellite Link Experiment
SIM	Subcarrier intensity modulation
SIMO	Single input multiple output
SLD	Superluminescent diodes
SNR	Signal-to-noise ratio
SOP	State of polarization
SSC	Switch-and-stay combining
STBC	Space–time block code
STDD	Subcarrier time delay diversity
STFT	Short-term Fourier transform
TCM	Trellis coded modulation
TDL	Time delay line
TL	Training length
TPC	Turbo product code
Tx	Transmitter
UT	User terminal
UV	Ultraviolet
UWB	Ultrawideband
VCSEL	Vertical cavity surface emitting laser
VFIr	Very fast infrared
VL	Visible light
VLC	Visible light communication
VLCC	Visible Light Communications Consortium
WiMax	Worldwide Interoperability for Microwave Access
WDM	Wavelength division multiplexing
WPAN	Wireless personal area networks
ZFE	Zero forcing equalizer

1 Introduction
Optical Wireless Communication Systems

We are seeing a growing demand for bandwidth in mobile communication, as the number of users is increasing significantly. The next-generation wireless communication systems therefore should be able to offer higher capacity to support various broadband wireless services such as the high-definition TV (HDTV—4–20 Mbps), computer network applications (up to 100 Mbps), mobile videophones, video conferencing, high-speed Internet access, and so on. In access networks, the technologies currently in use include the copper and coaxial cables, wireless Internet access, broadband radio frequency (RF)/microwave and optical fibre. These technologies, in particular copper/coaxial cables and RF based, have limitations such as a congested spectrum, a lower data rate, an expensive licensing, security issues and a high cost of installation and accessibility to all. Optical wireless communications (OWC) is an age-long technology that entails the transmission of information-laden optical radiation through the free-space channel. OWC technology is one of the most promising alternative schemes for addressing the 'last mile' bottleneck in the emerging broadband access markets. OWC offers a flexible networking solution that delivers the truly broadband services. Only the OWC technology provides the essential combination of virtues vital to bring the high-speed traffic to the optical fibre backbone. That is offering a license-free spectrum with almost an unlimited data rate, a low cost of development and ease and speediness of installation. This chapter gives an overview of the OWC systems for indoor and outdoor applications as a complementary technology to the existing schemes. The rest of the chapter is organized as follows. Section 1.1 outlines the wireless access technologies, followed by a brief history of OWC in Section 1.2. The comparative study of the optical and RF spectrum is summarized in Section 1.3. The link configuration and applications of the OWC are given in Sections 1.4 and 1.5. The eye safety issues are covered in the last section.

1.1 WIRELESS ACCESS SCHEMES

In rural areas, the installation of buried optical fibre cable to provide a high-speed network is not an economic proposition. In addition, the migration of mobile services to applications requiring broadband services, together with the move to larger numbers of small nano- and microcells of range up to 1 km will put increasing demands on mobile network infrastructure, particularly the 'backhaul' connections from mobile base stations to base stations and to mobile switching centres (MSC). Current systems and base station separations of the order of approximately 10–20 km employ the microwave radio technology with a data rate of 2 Mbps per user to provide data links or 'mobile backhaul network'. This approach cannot deal with higher data rates, thus leading to the data throughput 'backhaul bottleneck'. The third- and fourth-generation (3G and 4G) wireless communication systems will provide omnipresent connectivity, ranging from high-mobility cellular systems to the fixed and low-mobility indoor environments. Today's RF-based networks may be able to support one or perhaps two high-capacity users per cell which is highly wasteful with cell sizes of ~100 m accommodating tens of users. Multiple high-capacity users could only be serviced by deploying a similar number of systems, all within the same locale. This would create a situation where the multiple cells almost completely overlap, which then raises concerns with regard to the interference, the carrier reuse, and so on. It will require the development of very high-capacity short-range links, which could be achieved with the direct connection of the base station to the trunk network via an optical fibre. Table 1.1 shows the possible access network technologies mainly aimed for the last mile connections.

Public telephone networks still use copper cable-based technology together with the old time division multiplex-based network infrastructure with limited bandwidth (e.g., 1.5 Mbps (T-1) or 2.024 Mbps (E-1)) per end user. Digital subscriber line (DSL) technology provides high-speed Internet access to homes and small businesses, but its penetration rates have been throttled by sluggish deployment. Cable-based modem technologies offered as a package of cable TV channels have had more success in noncommercial markets, but lower data rate and link security are major issues. Wireless Internet access is still slow, although this is changing quite rapidly. Broadband RF/microwave (or millimetre-wave) technologies offer excellent mobility but still have severe limitations due to high attenuation (16 dB/km at 60 GHz), and thus reduced

TABLE 1.1

Access Technologies for the Last Mile Link

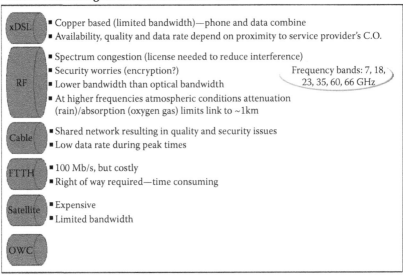

transmission link. The RF spectrum is congested, costly and offer maximum data rates which are low compared to the optical fibre systems. The satellite technology offers limited bandwidth and is very expensive. Fibre optic technology does offer the best of everything, but does not reach everyone, particularly the end users and the rural areas. Fibre to the homes (FTTH) can be offered using different architectures. Time division multiplexed passive optical networks (TDM-PONs) offer shared bandwidth to multiple users; however, many network service provider and operators have been upgrading to wavelength division multiplexing (WDM)-PON. WDM-PON offers a dedicated high bandwidth to every user over a shared PON architecture. However, for future development, there are a number of problems such as (i) extra space in the physical infrastructure to accommodate additional equipments, (ii) operating environment in the customer premises, for example, temperature variation at homes/offices could affect the system performance and (iii) practicality of installing fibres.

As for the fixed wireless access schemes, the options are as follows:

1. Worldwide interoperability for microwave access (WiMax), which is based on the IEEE 802.16d standard for fixed broadband wireless access with theoretical data rates up to 120 Mbps over a line-of-sight (LOS) link range of 50 km.
2. Broadband over power line (BPL) using the existing comprehensive wired network.
3. Ultrawideband (UWB) technology, which bypasses the spectrum regulation, offers good propagation characteristics over a range less than a few tens of metres. With the introduction of microcells with reduced distances between transmitter and user, up to 1 km, higher data rates and mobile broadband services could be offered to a large number of end users. This will require the development of very high-capacity short-range links connecting the base station to the MSC, which in turn could be connected to the main trunk network via optical fibre cables.

In terms of mobility in indoor and outdoor environments, there is absolutely no doubt that RF communication will be the preferred technology. In such environments, the data rates available to individual users could possibly be limited and can be met with additional allocation of available spectrum. In indoor environments with both fixed and mobile terminals, the circumstances are less apparent. For fixed terminals, 10 Mbps Ethernet-wired local area networks (LANS) are predominantly used, with 100 Mbps and 1 Gbps currently being developed. Mobile connections are provided via the existing cellular networks. The emergence of portable computing devices such as laptops, palmtops and personal digital assistants (PDAs) has fuelled the demand for mobile connectivity and hence led to the development of RF wireless LANs and wireless personal area networks (WPAN). Wireless LANs and WPANS offer users increased mobility and flexibility compared with traditional wired networks, and may be classified as either infrastructure wireless or ad hoc wireless networks [1]. It also allows users to maintain network connectivity while roaming anywhere within the coverage area of the network. This configuration requires the use of access points, or base stations, which are connected to the wired LAN and act as interfaces to the wireless devices. In contrast, ad hoc wireless LANs are simple peer-to-peer networks in which each client only has access to the resources of the other clients on the network and not a central server. Ad hoc wireless LANs require no administration or preconfiguration, and are created on demand for only as long as the network is needed. The term 'wireless' is synonymous with radio, and there are numerous radio LAN products in the market today. The RF wireless LANs available at present use unregulated 'free' spectrum regions, offering 1–2 Mbps at 2.4 GHz international scientific and medical (ISM). However, the available bandwidth is limited to 83.5 MHz, and therefore must be shared with numerous other products in the market such as cordless telephones and baby monitors. The next generation of radio LAN products that operate in the 5 GHz band offering 20 Mbps has been allocated exclusively for use by wireless LAN products. Consequently, this allows systems to be optimized in terms of data rate and efficiency, free from the constraints associated with coexisting with other products. There are currently two competing standards in this band, IEEE 802.11a and HiperLAN2, both of which specify maximum data rates of 54 Mbps [2,3]. As the popularity of wireless LANs and WPAN increases to carry enormous amounts of data for multimedia and other services within the limited frequency spectrum they will face spectrum congestion, thus, leading to service degradation and therefore the reduced viability of the technology in the long run. Possible solutions would be to move up the frequency spectrum, exploiting the unlicensed 60 GHz band, currently being proposed by the Wireless Gigabit Alliance [4]. It has a system bandwidth of 7 GHz, offering almost 7 Gbps data rate. The 60 GHz band is also being considered within the IEEE 802.11ad framework [5]. However, at 60 GHz, the path loss is very high, and therefore, these are not suitable as an alternative technology to conventional wireless LANs. They are more attractive for very short-range indoor applications with very high data rates.

Figure 1.1 shows the power per user for the core, metro edge, fixed access and wireless access. The fixed access network power usage is expected to remain flat for

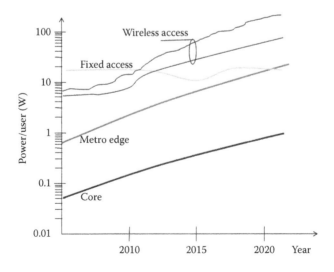

FIGURE 1.1   Power per user for different access technologies.

the next 10 years. Wireless RF access power could grow by a factor of 100 in the next 10 years. By 2020, it is predicted that the wireless RF access power consumption would dominate the global network.

Optical fibre with enormous transmission capacity of about 4 Tbps at 1.55 µm wavelength together with the erbium-doped fibre amplifiers (EDFA) can provide the best option. However, deployment cost is still high for the end users (i.e., the 'last mile' and rural areas). A hybrid millimetre wave and optical fibre could be one option to achieve the requirements of broadband wireless system. Therefore, an integration of optical and RF wireless network is an excellent cost-effective means for transmitting various wideband applications. Millimetre-wave wireless over optical fibre (or better known as radio over fibre) systems are capable of offering data rates in excess of 500 Mbps to both fixed and mobile users [6–8]. However, the overall offered transmission capacity is still limited due to the low carrier frequency.

Along with radio, the term 'wireless' is also applicable to systems which utilize other regions of the electromagnetic spectrum, such as infrared (IR) also known as OWC. The enormous growth of number of information terminals and portable devices in indoor as well as outdoor environments has accelerated the research and development in the OWC technology. As a complementary access technology to the RF techniques, OWC was first proposed as a medium for short-range wireless communication more than two decades ago [9]. Most practical OWC systems use light-emitting diodes (LEDs) or laser diodes (LDs) as transmitter and PIN photodiode or avalanche photodiode (APD) as receiver. Intensity modulation (IM) with direct detection (DD) is widely used at data rates below 2.5 Gbps, whereas for higher data rates, external modulation is employed.

There is wide diversity in the field of OWC applications, including a very short-range (mm range) optical interconnect within integrated circuits, high-volume ubiquitous consumer electronic products, outdoor intrabuilding links (a few kilometre range), intersatellite links (4500 km) and so on. OWC systems offer a number of unique advantages over its RF counterpart, such as [10–12]:

- Abundance of unregulated bandwidth (200 THz in the 700–1500 nm range).
- No utilization tariffs.
- No multipath fading when IM and DD is used.
- Highly secure connectivity. It requires a matching transceiver carefully aligned to complete the transmission.
- The optical beams transmitted are narrow and invisible, making them harder to find and even harder to intercept.
- Higher capacity per unit volume (bps/m^3) due to neighbouring cells sharing the same frequency.
- Small, light, compact smaller size components and relatively low cost.
- Well-defined cell boundaries and no interchannel interference.
- Use one wavelength to cover a large number of cell, therefore no frequency reuse problem as in RF.
- No need to dig up roads and is easily installed.
- Minimal absorption effects at 800–890 nm and 1550 nm.
- Health-friendly (no RF radiation hazards).

- Lower power consumption.
- Immunity to the electromagnetic interference.

On the other hand, OWC links with an inherent low probability of intercept and antijamming characteristics is among the most secure of all wide-area connectivity solutions. Unlike many RF systems that radiate signals in all directions, thus making the signal available to all within the receiving range, the OWC (in particular outdoor free-space optical (FSO) links) transceivers use a highly directional and cone-shaped laser beam normally installed high above the street level with an LOS propagation path. Therefore, the interception of a laser beam is extraordinarily difficult and anyone tapping into the systems can easily be detected as the intercept equipment must be placed within the very narrow optical foot print. Even if a portion of the beam is intercepted, an anomalous power loss at the receiver could cause an alarm via the management software. To protect the overshoot energy against being intercepted at the receiver part, a window or a wall can be set up directly behind the receiver [11,13]. Based on these features, OWC communication systems developed for voice, video and broadband data communications are used by security organizations such as government and military [12].

OWC operating at the near-IR region of 750–950 nm has most of the physical properties of the visible light, except that it is at the lower part of the optical frequency spectrum making it invisible to the human eye. However, the eye is very sensitive to this wavelength range and therefore must be protected by limiting the transmission intensity. The eye is also affected by the ambient light sources (sun, fluorescence, etc.) (see Figure 1.2). At the higher wavelength region of 1550 nm, the eye is less sensitive to light and, therefore, the eye safety requirement is more relaxed

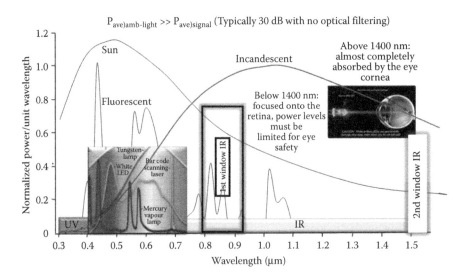

FIGURE 1.2   (**See colour insert.**) Normalized power/unit wavelength for optical wireless spectrum and ambient light sources.

and the interference due to ambient light sources is considerably reduced. This wavelength is also compatible with the third window backbone optical fibre communication network. IR light, similar to visible light, will not pass through opaque barriers and will reflect off the walls, ceiling and most other objects in a room.

1.2   A BRIEF HISTORY OF OWC

OWC or better still FSO communication is an age-old technology that entails the transmission of information-laden optical radiation through the air from one point to the other. Around 800 BC, ancients Greeks and Romans used fire beacons for signalling and by 150 BC, the American Indians were using smoke signals for the same purpose of signalling. Other optical signalling techniques such as the semaphore were used by the French sea navigators in the 1790s; but what can be termed as the first optical communication in an unguided channel was the photophone experiment by Alexander Graham Bell in 1880 [9]. In his experiment, Bell modulated the sun radiation with voice signal and transmitted it over a distance of about 200 m. The receiver was made of a parabolic mirror with a selenium cell at its focal point. However, the experiment did not go very well because of the crudity of the devices used and the intermittent nature of the sun radiation.

The fortune of OWC changed in the 1960s with the discovery of optical sources, most importantly, the laser. A flurry of FSO demonstrations was recorded in the early 1960s into the 1970s. Some of these included the spectacular transmission of television signal over a 30-mile (48-km) distance using GaAs LED by researchers working in the MIT Lincolns Laboratory in 1962; a record 118 miles (190 km) transmission of voice-modulated He–Ne laser between Panamint Ridge and San Gabriel Mountain, USA in May 1963, and the first TV-over-laser demonstration in March 1963 by a group of researchers working in the North American Aviation. The first laser link to handle commercial traffic was built in Japan by Nippon Electric Company (NEC) around 1970. The link was a full duplex 0.6328 μm He–Ne laser FSO between Yokohama and Tamagawa, a distance of 14 km [10].

From this time onward, OWC has continued to be researched and used chiefly by the military for covert communications. The technology has also been heavily researched for deep space applications by NASA and ESA with programmes such as the then Mars Laser Communication Demonstration (MLCD) and the Semiconductor-Laser Intersatellite Link Experiment (SILEX), respectively. Although deep space OWC lies outside the scope this book, it is worth mentioning that over the last decade, near Earth FSO were successfully demonstrated in space between satellites at data rates of up to 10 Gbps [11]. In spite of early knowledge of the necessary techniques to build an operational OWC systems, the usefulness and practicality of an OWC system were until recently questionable for many reasons [10]: first, existing communications systems were adequate to handle the demands of the time; second, considerable research and development were required to improve the reliability of components to assure reliable system operation; third, a system in the atmosphere would always be subject to interruption in the presence of heavy fog; fourth, the use of the system in space, where atmospheric effects could be neglected, required accurate pointing and tracking optical systems which were not available then. In view of

TABLE 1.2
History of OWC Systems

Date	Systems/Devices/Standards
800 BC	Fire beacons—by the ancient Greeks and Romans
150 BC	Smoke signal—by the American Indians
1790s	Optical telegraph—by Claude Chappe, France
1880	Photophone—by Alexander Graham Bell, USA
1960	Laser
1970s	FSO mainly used in secure military applications
1979	Indoor OWM systems—F. R. Gfeller and G. Bapst
1993	Open standard for IR data communications—The Infrared Data Association (IrDA)
2003	The Visible Light Communications Consortium (VLCC)–Japan
2008	Global standards for home networking (infrared and VLC technologies)—Home Gigabit Access (OMEGA) Project—EU
2009	IEEE802.15.7—Standard on VLC

these problems, it is not surprising that until now FSO had to endure a slow penetration into the communication networks.

With the rapid development and maturity of optoelectronic devices, OWC has now witnessed a rebirth. Also, the increasing demand for more bandwidth in the face of new and emerging applications implies that the old practice of relying on just one access technology to connect with the end users has to give way. These forces coupled with the recorded success of its application in military applications have rejuvenated interest in its civil applications within the access network. Several successful field trials have been recorded in the last few years in various parts of the world which have further encouraged investments in the field. This has now culminated into the increased commercialization and deployment of OWC in today's communication infrastructures. Full duplex outdoor OWC systems running at 1.25 Gbps between two static nodes are now common sights in today's market just like systems that operate reliably in all weather conditions over a range of up to 3.5 km. In 2008, the first 10 Gbps outdoor OWC system was introduced in the market, making it the highest-speed commercially available wireless technology [12]. Efforts are continuing to further increase the capacity via integrated FSO/fibre communication systems and wavelength division multiplexed (WDM) FSO systems which are currently at experimental stages [13]. Table 1.2 gives a concise history of OWC.

1.3 OWC/RADIO COMPARISON

The comparison between radio and IR for indoor wireless communications is shown in Table 1.3. There is very little doubt that RF-based technology will provide mobility both indoor and outdoor over small and large coverage areas. However, the data rates available to end users are limited. Wired LANS are predominantly 10 Mbps Ethernet, with 100 Mbps and 1 Gbps standards being developed, which are used to

TABLE 1.3

Properties of Radio and Infrared Wireless

Property	Radio	Infrared	Implication for Infrared
Bandwidth regulated	Yes	No	• Approval not required • Worldwide compatibility
Passes through walls	Yes	No	• Inherently secure • Carrier reuse in adjacent rooms
Rain	Yes	No	Reduced range
Snow	No	Yes	Reduced range
Multipath fading	Yes	No	Simple link design
Multipath dispersion	Yes	Yes	Problematic at high data rates
Path loss	High	High	Short range
Dominant noise	Other users	Background light	Short range
Average power proportional to	$\int |f(t)|^2\, dt$	$\int f(t)\, dt$	$f(t)$ is the input signal with high peak-average ratio

connect fixed terminals. Mobile connections are available using the traditional cellular networks. The RF wireless LANs uses the unregulated spectrum bands, with networks operating at 2.4 GHz ISM (data rate of 1–2 Mbps), 5.7 GHz (data rate of ~20 Mbps), and the proposed 17 and 60 GHz where available data rates are much higher. The RF technology is excellent at providing coverage at lower data rates (i.e., lower carrier frequencies). This is due to the diffraction and scattering of RF waves and the sensitivity of the receivers. RF channels are also robust to blocking and shadowing and can provide full coverage between rooms. At higher carrier frequencies (i.e., offering much larger data rates), the RF propagation becomes more LOS and, therefore, the problems encountered are similar to that using light. At these frequencies, the RF components are expensive, and the key advantages of the RF technology (i.e., the mobility, coverage and receiver sensitivity) become less clear compared to the OWC systems.

In contrast, OWC systems (indoor and outdoor) covering a wide unlicensed spectral range of 700–10,000 nm has the potential to offer a cost-effective protocol-free link at data rates exceeding 2.5 Gbps per wavelength up to 5 km range (see Figure 1.3). For indoor applications through multiple user-sized cells, and because of the intrinsically abrupt boundary of these cells, interference would be negligible and carrier reuse would not be an issue in order to increase the capacity. Indeed, the optical wireless technology is a future proofed solution since additional capacity far beyond the capabilities of radio could be delivered to users as their needs increase with time. Perhaps the largest installed short-range wireless communication links are optical rather than RF. It is argued that OWC has a part to play in the wider 4G vision. In large open environments where individual users require greater than 100 Mbps or more, OWC is a more sensible solution because of its multiple user-sized cells, reduced interference and improved carrier reuse capabilities due to its intrinsically abrupt cells boundary. Optical wireless LAN has been utilized in large scale in a number of applications such as telemedicine, emergency situations, high-speed

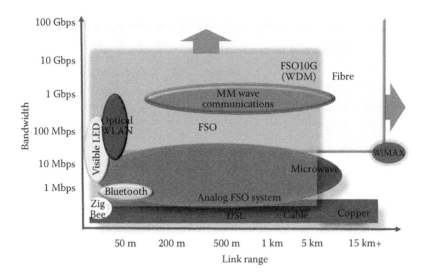

FIGURE 1.3 (**See colour insert.**) Bandwidth capabilities for a range of optical and RF technologies.

trains, laptops, PDAs, museums, and so on [14,15]. According to the IEEE 802.11 specifications, the IR physical layer can support data rate up to 2 Mbps with a potential to migration to higher data rates. The Infrared Data Association (IrDA) has standardized low cost and short range (1–8 m) IR data links supporting data rates ranging from 2.5 kbps to 16 Mbps at a wavelength between 850 nm and 900 nm [16]. Figure 1.4 shows the mobility of optical wireless systems versus bit rate.

To an extent, radio and IR may be viewed as complementary rather than competitive media. For example, if a wireless LAN is required to cover a large area,

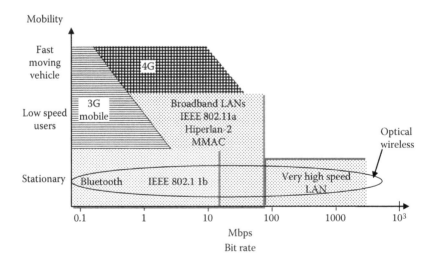

FIGURE 1.4 Mobility of optical wireless systems versus bit rate.

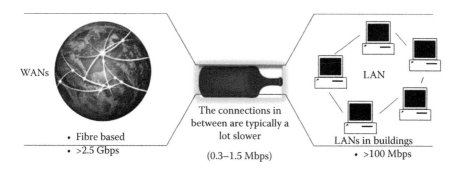

FIGURE 1.5   **(See colour insert.)** Access network bottleneck.

where users can roam freely and remain connected to the network at all times, then radio is the only cost-effective medium that can achieve this. If, however, a wireless LAN is required to cover a more modest area, but deliver advanced bandwidth-hungry multimedia network services such as video conferencing and video on demand, then IR is the only medium which truly has the bandwidth available to deliver this. For the last mile access network, the OWC technology could be employed to overcome the bandwidth bottleneck (see Figure 1.5). To take advantage of OWC high transmission rates and yet have a weather-resistant access, the hybrid OWC/MMW (millimetre wave) technology would be desirable in a number of outdoor applications where higher data rates at all weather conditions is desirable [17]. In fact, this could be one of the best solutions for future 4G communication systems.

1.4 LINK CONFIGURATION

Figure 1.6 shows a simple block diagram of the OWC system. Both LED and LD could be used as the source. To date, commercially available OWC systems have come close to delivering the high data rates over a link space up to 5 km at a marginally higher cost than RF technology [18,19], which are significantly faster than the latest radio LAN products that are currently available. Nevertheless, OWC is a challenging medium and there are numerous considerations which must be taken into account when designing high-speed OWC links. In indoor environment, light will reflect off the ceiling, walls and most other objects in a typical room or enclosure, but will not go through opaque barriers, whereas in outdoor environment, light will be scattered and absorbed due to the atmospheric conditions.

There are numerous ways through which an optical link can be physically configured. These are typically grouped into four system configurations:

1. Directed LOS
2. Nondirected LOS
3. Diffuse
4. Tracked

A number of receiver configurations as well as hybrid variations of the above configurations are also shown in Figure 1.7.

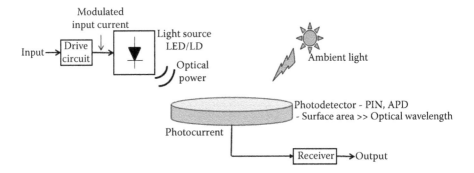

FIGURE 1.6 A system block diagram of OWC system.

Directed LOS—Typically used for point-to-point communication links in mainly outdoor and in some cases in indoor environment too. The optical is concentrated in a very narrow beam, thus exhibiting low power requirements as well as creating a high-power flux density at the photodetector. Furthermore, the LOS link offers the

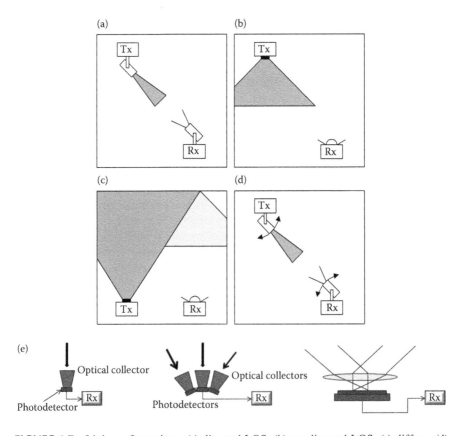

FIGURE 1.7 Link configurations: (a) directed LOS, (b) nondirected LOS, (c) diffuse, (d) tracked and (e) receiver configurations.

highest data rate (hundreds of Mbps and above) over a link span from a few metres to 5 km. Furthermore, directed LOS links do not suffer from multipath-induced signal distortion, and noise from the ambient light sources is also largely rejected when used with a narrow field-of-view (FOV) receiver [20]. The data rate is therefore limited by free-space path loss rather than the effects of multipath dispersion [20–22]. However, there are disadvantages. For indoor applications, the coverage area provided by a signal channel could be very small, so providing area coverage and roaming could become problematic. Directed LOS links cannot support mobile users because of the requirement for alignment of receiver and transmitter modules. In LOS OWC-based LANs catering for single users, the transmitter/receiver base station is usually mounted to the room ceiling. However, in situations where there is a need to offer services to multiple users within a relatively large room, the possible option would be to adopt a cellular topology, where the narrow transmit optical beam is replaced with a source of a wide optical foot print (see Figure 1.8) [23]. In cellular OWC links, the optical signal is broadcast to all mobile users within the cell, and the communication between mobile users is established via a base station located at the centre of each cell. Cellular OWC topology offer reduced power efficiency as well as mobility.

Additionally directed LOS links (mainly for outdoor applications) must be accurately 'pointed' or aligned before use and require an uninterrupted LOS path making them susceptible to beam blocking. In addition to this, by their very nature, they are more suited to point-to-point links rather than point-to-multipoint broadcast-type links, thus reducing their flexibility. Directed LOS is the most well-known link topology, and has been used for many years in low bit rate, simplex remote control applications for domestic electrical equipment, such as televisions and audio equipment. In recent years, we have seen a growing interest in application of LOS links for a number of outdoor applications: university campuses, last miles access networks, cellular communication backhaul, intersatellite communications, satellite uplink/downlink, deep space probes to ground, disaster recovery, fibre communication back-up, video

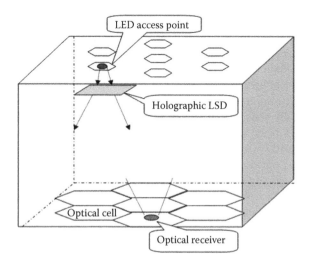

FIGURE 1.8 Cellular OWC system.

conferencing, links in difficult terrains, wide and local area networks, ground-to-ground (short and long distance terrestrial links), deep space probes to ground, and ground-to-air/air-to-ground terminal (UAV, HAP, etc.) and temporary links [24]. The LOS OWC link offers a number of advantages, including higher data rates easily exceeding 100 Gbit/s, using WDM techniques, and security as well as a small terminal size, a light weight, minimal aperture sizes and a low-power consumption.

In hybrid LOS systems, either the transmitter or the receiver has a wide FOV, while the other element has a narrow FOV. A typical hybrid LOS system uses ceiling-mounted transmitters that illuminate the wide area. Like in directed LOS links, the transmitter has a narrow angle emission and the receiver has a narrow FOV for directed non-LOS links. However, to overcome any obstacles between them, the transmitter is aimed at a reflective surface so that the first reflection can reach the receiver. In addition to overcoming the barrier, the information signal is received after a single reflection, which minimizes the multipath dispersion [11]. However, the alignment is problematic due to a highly directional transmitter and receiver. The hybrid system incorporates either a wide beam transmitter or a narrow FOV receiver or vice versa. This relaxes the need for strict alignment between the transmitter and the receiver. Though blocking probability can be significantly reduced using the link design, the system is affected by multipath propagation.

Nondirected LOS—For indoor applications, the nondirected LOS, considered to be the most flexible configuration, uses wide beam transmitters, wide FOV receivers and scatter from surfaces within the room to achieve a broader coverage area, thus offering optical 'ether' similar to the RF offering data rate in excess of 150 Mbps. Non-directed links are suitable for point to multipoint broadcast applications. They offer robustness to shadowing and blockage and require no alignment and tracking. They overcome blocking problem by relying on reflections from surfaces of objects within rooms, so that a high proportion of the transmitted light is detected at the photodetector from a number of different directions. However, they incur a high optical path loss (thus higher transmit power) and they must also contend with multipath-induced dispersion. While multipath propagation does not result in multipath fading in indoor IR systems, since detector sizes are huge in comparison with the wavelength, it does give rise to inter-symbol interference (ISI), thus limiting the data rate to around a few Mbps in a typical size room. In addition to this, OWC links ought to be able to operate in environments with intense ambient light levels, thus degrading the link performance.

Diffuse configuration—Also known as the nondirected non-LOS proposed in Ref. [9], it typically consists of a transmitter that points directly towards the ceiling emitting a wide IR beam. The diffuse OWC indoor topology is the most convenient for LAN ad hoc networks since it does not require careful alignment of the transmitter and receiver modules, nor does it require a LOS path to be maintained and is almost immune to blockage of the transmission path. In addition to this, it is also extremely flexible, and can be used for both infrastructure and ad hoc networks [9,20]. Unfortunately, diffuse links experience high path loss, typically 50–70 dB for a horizontal separation of 5 m [25]. The path loss is increased further if a temporary obstruction, such as a person, obscures the receiver such that the main signal path is blocked, a situation referred to as shadowing [25]. In addition, a photodetector with a wide FOV normally collects signals that have undergone one or more reflections from ceiling, walls and room objects. Such reflections attenuate the signal with typical reflection coefficients between 0.4 and 0.9 [9]. The received signal can also suffer from severe multipath dispersion, where the transmitted pulses spread out in time over alternative routes of differing lengths, thus limiting the maximum unequalized bit rate R_b achievable with a room volume of $10 \times 10 \times 3$ m^3 to typically around 16 Mbps [9]. Dispersion-induced ISI incurs a power penalty and thus bit-error rate (BER) degradation. Thus, as a result of these factors, relatively high optical transmit powers are required. For example, at high data rates, the power requirements of pulse-position modulation (PPM) are higher than the on-off keying (OOK). However, the average optical power emitted by an IR transceiver is limited by eye safety regulations and electrical power consumption in portable (battery-powered) devices. Therefore, the use of power-efficient modulation techniques is desirable. In addition, in diffuse systems, the entire room needs to be illuminated by a single or multiple transmitters. This is naturally realizable by diffused light propagation after a few reflections but requiring a relatively huge transmitted optical power. For instance, 475 mW of transmit power for a diffuse 50-Mbps link at a horizontal link separation of ~5 between transmitter and for receiver both directed to the ceiling [14]. Table 1.4 shows the key features of the directed and diffuse OWC links.

Replacing the nonimaging receiver in both nondirected and non-LOS links by a single imaging light concentrator and an array of photodetector can further improve the link performance by the way of reducing the multipath distortion and the ambient light noise. Performance of diffuse systems can be substantially improved by the use of multibeam transmitter together with a multiple element narrow FOV angle-diversity receiver. In such a configuration (see Figure 1.9), the light power is projected in the form of multiple narrow beams of equal optical intensity over a regular grid of small spots on to a ceiling. In this configuration, each diffusing spot would be considered a source offering a LOS link to a narrow FOV receiver, thus offering reduced path loss leading to the increased link signal-to-noise ratio (SNR) by ~20 dB [26–30]. Angle-diversity receivers consist of multiple receiver elements effectively pointing in different directions either because of their physical orientation or due to imaging concentrators [26]. The photocurrent generated by each receiver element can be processed in a number of ways, including the use of digital wavelet-AI techniques. This type of receiver can reduce the effects of ambient noise and multipath distortion due to the fact that these unwanted contributions are usually received from different directions to that of the desired signal [27]. The performance of such schemes has been analysed in Refs. [28–42]. Other enhanced optical designs such as the recently proposed dielectric totally internally reflecting concentrator (DTRIC) optical antenna are also used to combat the ISI [43]. Such methods can be used to partner other techniques to provide a combined improvement as in Ref. [44], where

TABLE 1.4
Features of Directed and Diffuse Links

	Capacity	Bandwidth	Power Efficiency	Mobility	Reliable Access
Directed link	High	High	Good	Yes with tracking	Yes with tracking
Diffuse link	Low	Low	Average	Yes	Yes

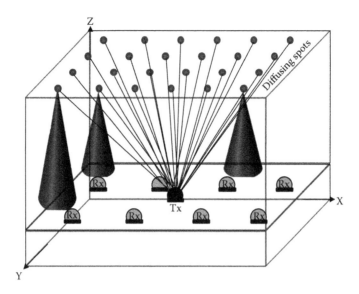

FIGURE 1.9 Multispot diffusing configuration.

angle diversity is employed in conjunction with code division multiple access (CDMA).

Tracked systems—Here, the transmitter (Tx)/receiver (Rx) base station is mounted on the ceiling and the mobile stations (MSs) are located at the table-top height (see Figure 1.8). For the LOS link, where the Tx optical beam for both the up and down links are focussed onto the Rx, less power is needed than the diffuse link. The combination of a narrow Tx beam and a receiver with a smaller FOV results in reduced multipath-induced ISI and the ambient light interference. In contrast, the diffuse IR link offers improved mobility since it uses reflections from the ceiling, walls and object within the room to transmit data from Tx to MS Rx, thus requiring no alignment. In such configuration, full connectivity between Tx and Rx can be maintained even when the LOS path is blocked. However, the diffuse link suffers from multipath-induced ISI, and hence, reduced data rate. As shown in Figure 1.10, this scheme offers high power efficiency and potentially high bit rates (1 Gbps using mechanical steerable optics [45]) of the directed LOS links with the increased coverage area enjoyed by nondirected LOS links.

Unfortunately, mechanical steerable optics are expensive to realize and electronic tracking schemes have been proposed in Ref. [46] followed by a 100 Mbps practical system operating at a wavelength of 1550 nm in Ref. [47] and a 155 Mbps link in Ref. [48]. Solid-state systems have been proposed which are conceptually similar to angle-diversity techniques. Solid-state tracked system, using multielement transmitter and receiver arrays along with a lens arrangement, has been proposed [20,47]. Using this arrangement, steering is merely a matter of selecting the appropriate array element. More recently, a single-channel imaging receiver has been proposed [27], which could be described as a hybrid tracked and angle-diversity system. Optical multiple input multiple output (OMIMO) systems [49] improve channel quality due to spatial diversity, or system speed using multiplexing. Note that, along with diffuse links using multibeam

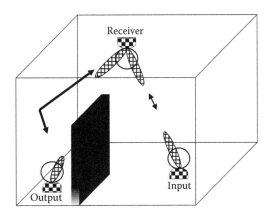

FIGURE 1.10   Hybrid diffuse and tracked LOS links.

transmitters and angle-diversity receivers, tracked systems offer the potential to implement space-division multiplexing, whereby multiple users can communicate without suffering a loss of per-user capacity, since each user is located in a different cell [50].

OWC links employing one-dimensional tracking systems using arrays of lasers and photodetectors, angle-diversity and imaging receivers have been developed to provide high data rates, mobility and coverage area using LOS channels in indoor environment. LOS links employing a receiver with a narrow FOV are more efficient offering a high optical signal gain and a low optical noise [20]. For outdoor applications with LOS link, the main challenges are the atmospheric conditions that limit the link range and thus the link availability. In LOS optical wireless LANs, the narrow transmit beam is now replaced by a wider light foot-print to cover a large area within a room (see Figure 1.11), thus offering optical 'cellular' capability but at the cost of reduced power efficiency, since more transmit power is required to ensure the adequate optical power flux density at the receiving end. In optical cellular configurations, optical signal is delivered to all the terminals within the cell and communications between subscribers could be accomplished in a star-network topology. One advantage of the optical cellular system is that the same wavelength could

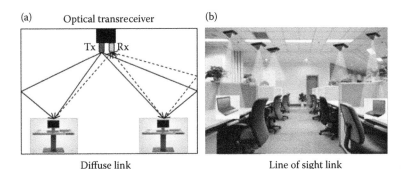

FIGURE 1.11   (**See colour insert.**) Optical wireless LANs: (a) diffuse and (b) line of sight.

be used to cover a large number of cells within the same area, thus overcoming the need for wavelength reuse (or frequency reuse as in RF cellular systems).

A hybrid cellular and NLOS tracked systems (see Figure 1.10) would offer additional flexibility such as

1. Tracking and pointing functions, thus ensuring 100% link availability at all times
2. Power efficiency, by switching off the sources that do not illuminate a user terminal [26,27]
3. Increased system capacity, where each source in the emitter array transmits different data streams
4. Increased transmission bandwidth due to the detector array low capacitance
5. Reduced ambient light due to narrow FOV at the transmitter and detectors

In such a system, two-dimensional arrays of transmitter and photodetectors are used, where the sources in transmitter array emit normally to the plane of the array. Using optical systems, a light beam is deflected in a specific angle depending on the spatial position of the source within the array. Thus, a cell is split into microcells, each illuminated by a single light source of the array. At the receiver, depending on the angle of arrival of the optical signal, the light is focussed onto a particular detector in the photodetector array. To increase the data rate, spectrally efficient modulation formats have been proposed (see Chapter 4). Multielement optical transmitters and receivers, and quasi-diffuse OWC links employing discrete and imaging receivers as well as a combination of coding techniques have also been investigated to improve the optical power efficiency [26,28,41,51].

In hybrid LOS systems, either the transmitter or the receiver has a wide FOV, while the other element has a narrow FOV. A typical hybrid LOS system uses ceiling-mounted transmitters that illuminate the wide area. Like in directed LOS links, the transmitter has a narrow angle emission and the receiver has a narrow FOV for directed non-LOS links. However, to overcome any obstacles between them, the transmitter is aimed at a reflective surface so that the first reflection can reach the receiver. In addition to overcoming barrier, the information signal is received after a single refection, which minimizes the multipath dispersion [9]. However, the alignment is problematic due to highly directional transmitter and receiver. The hybrid system incorporates either a wide beam transmitter or a narrow FOV receiver or vice versa. This relaxes the need for strict alignment between the transmitter and the receiver. Though the blocking probability can be significantly reduced using the link design, the system is affected by multipath propagation.

For outdoor environment, FSO links are capable of offering similar performance of optical fibre communication over a link range spanning from of a few hundred metres up to a few kilometres. The optical transmitter power is up to 100 mW with a sufficient power margin at the receiver to deal with the attenuation due to the atmospheric and meteorological conditions such as fog, snow, rain, drizzle, turbulence and thermal expansion of structures. Atmospheric and meteorological conditions lead to fading of the received signal more than the RF waves. Fog is the most difficult to cope with, due to a very high attenuation, compared to rain and snow. For the link length

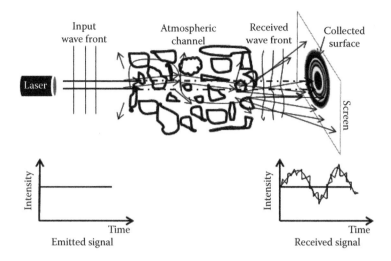

FIGURE 1.12   The effects of scintillation.

less than a few hundred metres, temperature induced scintillation will effect the laser beam propagation [44] (i.e., defocussing and deflecting the beam, see Figure 1.12).

1.5   OWC APPLICATION AREAS

The outdoor OWC systems are very attractive for various applications within the access and the metro networks. It can conveniently complement other technologies, such as wired and wireless RF communications, fibre-to-the-X technologies and hybrid fibre coaxial, among others, in making the huge bandwidth that resides in an optical fibre backbone available to the end users. The point that most end users are within a short distance from the backbone—one mile or less—makes FSO very attractive as a data bridge between the backbone and the end users. Among other emerging areas of application, terrestrial FSO has been found suitable for use in the following areas:

Last mile access—FSO is used to bridge the bandwidth gap (last mile bottle-neck) that exists between the end-users and the fibre optic backbone. Links ranging from 50 m up to a few km are readily available in the market with data rates covering 1 Mbps to 10 Gbps.

Optical fibre back-up link—Used to provide back-up against loss of data or communication breakdown in the event of damage or unavailability of the main optical fibre link. Link range could be up to 10 km with data rates up to 10 Gbps.

Cellular communication backhaul—Can be used to backhaul traffic between base stations and switching centres in the third/fourth-generation (3G/4G) networks, as well as transporting IS-95 CDMA signals from macro- and microcell sites to the base stations.

Disaster recovery/temporary links—The technology finds application where a temporary link is needed, be it for a conference or ad hoc connectivity in the event of a collapse of an existing communication network.

Multicampus communication network—FSO has found application in inter-connecting campus networks and providing back-up links at fast-Ethernet or gigabit-Ethernet speeds. Systems with data rates up to a few gigabits per second covering a link span of 1–2 km are already available in the market.

Difficult terrains—FSO is an attractive data bridge in instances such as across a river, a very busy street, rail tracks or where right of way is not available or too expensive to pursue.

High-definition television—In view of the huge bandwidth requirement of high-definition cameras and television signals, FSO is increasingly being used in the broadcast industry to transport live signals from high-definition cameras in remote locations to a central office.

There are commercially available systems operating at data rates exceeding 1 Gbps over a link range of 500–5000 m, offering a range of services from gigabit Ethernet to 270 Mbps serial digital TV transmission (at 270 Mbps) for outside broadcast applications.

Figure 1.13 illustrates a block diagram of the outdoor OWC system. For such systems, there are a number of challenges that may affect the OWC link performance.

1. The ambient light sources—with the spectra well within the receiver bandwidth of silicon photodetectors, thus resulting in the background noise [52].
2. Blocking and shadowing—the blockage or partial obstruction of optical rays due to multipath propagation in indoor environment, which also causes pulse spreading, thus leading to ISI [53]. To reduce the impact of ISI, there are a number of mitigation techniques, including diversity detection and emission [26,37], equalization both at the transmitter and receiver [54] and adaptive threshold detection [55].
3. Alignment and tracking—by employing highly directional and narrow beams of light, the changes in mispointing of the transmit beam as well as error due to the tracking of the receiver will introduce signal fading. One-dimensional tracking systems using arrays of lasers and photodetectors, and angle-diversity and imaging receivers have been developed to provide the coverage area using LOS channels.
4. Adverse atmospheric weather—can have serious impact on the outdoor OWC link availability and performance. The performance of outdoor laser-based OWC systems is highly dependent on the operating transmission windows [56]. Shorter wavelengths are most widely used because of the availability of low-cost devices and components. Previous study based on available measured weather statistic has shown that weather effects can be mitigated by switching from IR bands (700–1000 and 1550 nm) to far-IR [56,57]. Thus, quantum cascaded laser (QCL) operating at the midwave and long-wave IR band (3–20 µm) can offer a solution to the atmospheric-induced optical losses [58–61].

In recent years, we have seen the emergence of visible light technology for both illumination and communications. These devices are mechanically robust with a high energy efficiency offering simultaneous illumination and IM at a data rate in

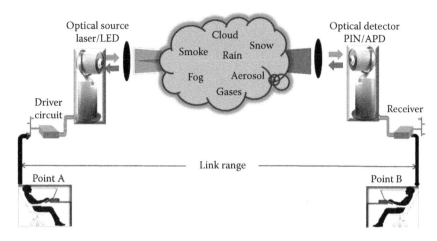

FIGURE 1.13   A system block diagram of an outdoor OWC link.

excess of 100 Mbps [62–64]. Such devices are opening up new applications in indoor environment such as mobile-to-mobile links, building and industrial automation (see Chapter 8).

1.6 SAFETY AND REGULATIONS

IR light sources pose a potential safety hazard to human if operated incorrectly. The optical beams can cause injury to both the skin and the eye, but the damage to the eye is far more significant because of the eye's ability to focus and concentrate optical energy. The eye can focus light covering the wavelengths around 0.4–1.4 μm on to the retina; other wavelengths tend to be absorbed by the front part of the eye (the cornea) before the energy is focused. Figure 1.14 shows the absorption of the eye at different wavelengths. At 700–1000 nm spectral range, where optical sources and detectors are low cost, the eye safety regulations are particularly stringent. The maximum measured radiant intensity with a maximum permissible exposure (MPE) at a wavelength of 900 nm is ~143 mW/sr [35]. However, at longer wavelengths of 1500 nm and above, that is, the third transmission window in optical fibre backbone networks, the eye safety regulations are much less strict, but the devices available are relatively costly. Therefore, when designing optical communication systems, efforts must be made to ensure that the optical radiation is safe and does not cause any damage to the people that might come into contact with it.

There are a number of international standard bodies which provide guidelines on safety of optical beams, notable among which are [65]

Center for Devices and Radiological Health (CDRH)—An agency within the U.S. Food and Drug Administration (FDA). It establishes regulatory standards for lasers and laser equipment that are enforceable by law (21 CFR 1040).
International Electrotechnical Commission (IEC)—Publishes international standards related to all electrical equipment, including lasers and laser equipment (IEC60825-1). These standards are not automatically enforceable

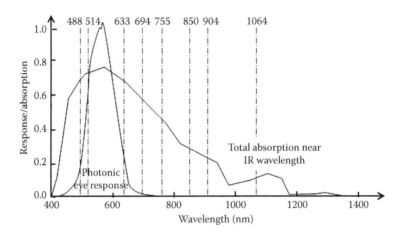

FIGURE 1.14 Response/absorption of the human eye at various wavelengths.

by law, and the decision to adopt and enforce IEC standards is at the discretion of individual countries.

American National Standards Institute (ANSI)—Publishes standards for laser use (ANSI Z136.1). ANSI standards are not enforceable by law but do form the basis for the U.S. Occupational Safety and Health Administration (OSHA) legal standards, as well as comparable legal standards that have been adopted by various state regulatory agencies.

European Committee for Electrotechnical Standardization (CENELEC)—An organization that establishes electrotechnical standards based on recommendations made by 19 European member nations. CENELEC standards are not directly enforceable by law but, as with IEC standards, are often integrated into the legal requirements developed by individual countries.

Laser Institute of America (LIA)—LIA is an organization that promotes the safe use of lasers, provides laser safety information, and sponsors laser conferences, symposia, publications and training courses.

Each of these organizations has developed ways of classifying lasers; the specific criteria vary slightly from one body to the other but the IEC classifications will be considered in this section. Lasers are generally divided into four groups, Class 1 to Class 4, with Class 1 being the least powerful and Class 4 being the most powerful. Each class is defined by the accessible emission limits (AEL) metric, which depends on the wavelength of the optical source, the geometry of the emitter and the intensity of the source [66].

Table 1.5 presents the main characteristics and requirements for the classification system as specified by the revised IEC 60825-1 standard [22,67]. In addition, classes 2 and higher must have the triangular warning label and other labels are required in specific cases indicating laser emission, laser apertures, skin hazards and invisible wavelengths. For outdoor LOS OC links, generally high-power lasers, Class 3B band, are used in order to accomplish good power budget. Therefore, to ensure eye safety, stringent regulations such as ANSI Z-136 and IEC 825 series have been established [67].

TABLE 1.5

Classification of Lasers according to the IEC 60825-1 Standard

Category	Description
Class 1	Low-power device emitting radiation at a wavelength in the band 302.5–4000 nm. Device intrinsically without danger from its technical design under all reasonably foreseeable usage conditions, including vision using optical instruments (binoculars, microscope, monocular).
Class 1M	Same as Class 1 but there is possibility of danger when viewed with optical instruments such as binoculars, telescope, etc. Class 1M lasers produce large-diameter beams, or beams that are divergent.
Class 2	Low-power device emitting visible radiation (in the band 400–700 nm). Eye protection is normally ensured by the defence reflexes including the palpebral reflex (closing of the eyelid). The palpebral reflex provides effective protection under all reasonably foreseeable usage conditions, including vision using optical instruments (binoculars, microscope, monocular).
Class 2M	Low-power device emitting visible radiation (in the band 400–700 nm). Eye protection is normally ensured by the defence reflexes including the palpebral reflex (closing of the eyelid). The palpebral reflex provides an effective protection under all reasonably foreseeable usage conditions, with the exception of vision using optical instruments (binoculars, microscope, monocular).
Class 3R	Average-power device emitting a radiation in the band 302.5–4000 nm. Direct vision is potentially dangerous. Generally located on the rooftops.
Class 3B	Average-power device emitting a radiation in the band 302.5–4000 nm. Direct vision of the beam is always dangerous. Medical checks and specific training required before installation or maintenance is carried out. Generally located on the rooftops.
Class 4	High-power device there is always danger to the eye and for the skin, fire risk exists. Must be equipped with a key switch and a safety interlock. Medical checks and specific training required before installation or maintenance is carried out.

The AEL values at two wavelength mostly used for FSO are presented in Table 1.6. Lasers classified as Class 1 are most desirable for OWC systems since their radiation are safe under all conditions and circumstances. Figure 1.15 shows the eye safety limits for the two most popular wavelengths of 900 and 1550 nm adopted in OWC systems. For collimated light source, the 1550 nm wavelength is preferable.

Class 1 lasers require no warning labels and can be used without any special safety precautions. As shown in Table 1.5, the power available to Class 1 lasers is limited. Most commercial terrestrial FSO links operating at up to 1.5 Gbps use Class 1M lasers; these are inherently safe except when viewed with optical instruments, such as binoculars. On certain instances, higher-class lasers are used for FSO, and the safety of these systems is maintained by installing the optical beams on rooftops with safety labels and warning or on towers to prevent inadvertent interruption [68].

As shown in Table 1.6, a Class 1 laser system operating at 1550 nm is allowed to transmit approximately 50 times more power than a system operating in the shorter IR wavelength range, such as 850 nm, when both have the same-size aperture lens. It should however be noted that no wavelength is inherently dangerous or eye-safe; it

---

TABLE 1.6

Accessible Emission Limits for Two Wavelengths, 850 and 1550 nm

	Average Optical Power Output (mW)	
Class	λ = 850 nm	λ = 1550 nm
1	<0.22	<10
2	Category reserved for the range 400–700 nm—same AEL as for Class 1	
3R	0.22–2.2	10–50
3B	2.2–500	50–500
4	>500	>500

Source: Adapted from O. Bouchet et al. *Free-Space Optics: Propagation and Communication*, London: ISTE Ltd, 2006.

---

is the output power that determines the laser classification. It is therefore possible to design eye-safe FSO systems that operate at any wavelength of choice. It is also important to understand that the regulation addresses the power density in front of the transmit aperture rather than the absolute power created by an LD inside the equipment. For example, the LD inside the FSO equipment can actually be Class 3B even though the system itself is considered to be a Class 1 or 1M laser product if the

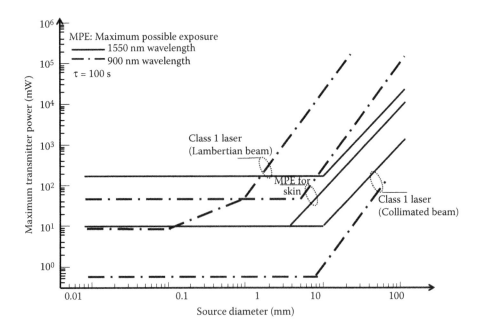

FIGURE 1.15 The eye safety limits for 900 and 1550 nm wavelengths.

TABLE 1.7

Example of MPE Values (W/m²) of the Eye (Cornea) at 850 nm and 1550 nm Wavelengths

Exposure duration (s)	1	2	4	10	100	1000	10,000
MPE (W/m²) at 850 nm	36	30	25	20	11	6.5	3.6
MPE (W/m²) at 1550 nm	5600	3300	1900	1000	1000	1000	1000

Source: Adapted from O. Bouchet et al. *Free-Space Optics: Propagation and Communication*, London: ISTE Ltd, 2006.

light is launched from a large-diameter lens that spreads out the radiation over a large area before it enters the space in front of the aperture [65]. In order to maintain the Class 1 or 1M safety classifications, it is possible to use higher-power laser with increased lens aperture or to use multiple large transmission apertures [65].

LED sources do not produce a concentrated light beam and therefore cannot be focussed on to the retina. This makes the use of LEDs ideal for indoor applications. The penalty, however, is reduced bandwidth compared to the LDs [69]. Recent developments in vertical-cavity surface-emitting lasers (VCSEL), offering a safer peak wavelength at 1550 nm [70–73], is becoming an attractive option for outdoor and even indoor applications. This is due to their well-controlled, narrow beam properties, high data rate, excellent reliability, low power consumption and the possibility of the array configuration.

1.6.1 Maximum Permissible Exposures

To provide a better protection to the eye or skin of anyone who might come in contact with the radiation of laser equipments, the MPE limits have been stipulated by standard organizations. It is the highest radiation power or energy, measured in W/cm² or J/cm², that is considered safe with a negligible probability of causing damage. The MPE is measured at the cornea of the human eye or at the skin, for a given wavelength and exposure time. It is usually about 10% of the dose that has a 50% chance of creating damage under worst-case conditions [74]. In Table 1.7, the MPE values for the eye at two commonly used wavelengths for FSO are presented; the values for the skin are much lower since the skin is usually less sensitive to laser radiation.

As shown in Table 1.7, the MPE is higher for brief exposure durations than for high exposures times. The MPE for the eye is much higher at 1550 nm than at 850 nm; this is related to the laser radiation absorption at the level of the various eye components. This difference in MPE values can be explained by the fact that at 850 nm, approximately 50% of the signal can reach the retina whereas at 1550 nm, the signal is almost completely absorbed by the cornea and aqueous humour [22].

1.7 OWC CHALLENGES

In Table 1.8, the major challenges encountered in the design and implementation of OWC systems are outlined.

TABLE 1.8

Challenges in OWC Systems

	Causes	Effects	Solutions	Indoor/FSO
Intersymbol interference (ISI)	Multipath propagation Depends on: • Room size • Reflection coefficient within the room	• Quality of the transmission • Multipath distortion or dispersion • Reduce data rates	• Equalization and predistortion equalization[40] • Forward error correction (FEC) • Spread spectrum techniques • Multiple-subcarrier modulation (MSM) • More bandwidth efficient than single-carrier • OFDM, MSM • Line strip spot diffusing transmitter (indoor) • Diversity detection and emission (indoor)[75] • Adaptive threshold detection[41] • Multibeam transmitter • FOV controlling	In/FSO Indoor
Safety	Laser radiation	Damage to eyes and skin	Power efficient modulation PPM, DPIM and so on Use LED, holographic diffuser	In/FSO Indoor
Wavelengths	Source @ 800–900 nm	Damaging the eye	Use 1550 nm	FSO
Noise	Dark current noise	Limit performance of communication systems	Optical and electrical filtering	In/FSO
	Shot noise	BER deterioration	Preamplification	In/FSO
	Background noise		FEC	In/FSO
	Thermal noise		Low noise amplifier	In/FSO
	Relative intensity noise			In/FSO
	Excess noise			In/FSO
	ASE (only if optical amplifier is used)	APD-amplification process	Optical filter	In/FSO

continued

Turbulence	Random refractive index variation	Irradiance fluctuation (scintillation) Image dancing Phase fluctuation Beam spreading Polarization fluctuation	• Coding, for example, LDPC, FEC, MIMO • Diversity reception (temporal and spatial) • Adaptive optics • Robust modulation techniques • Adaptive optics • Coherent detection not used due to phase	FSO
Fog	From 0.22–272 dB/km of loss	• Mie scattering • Photon absorption	• Increase transmit optical power • Hybrid FSO/RF • Diversity • More efficient modulations	FSO
Reflection index	Different materials	Higher losses due to reflected surfaces		Indoor
Rain and snow	Heavy rainfall (15 cm/h) —>20–30 dB/km of loss Light snow—>3 dB/km power loss Blizzard—>60 dB/km power loss	Photon absorption	Increase transmit optical power	FSO
Pointing stability and swaying buildings		• Loss of signal • Multipath induced distortions • Low power due to beam divergence and spreading • Short term loss of signal	• Spatial diversity • Mesh architectures: using diverse routes • Ring topology: User's n/w become nodes at least one hop away from the ring • Fixed tracking (short buildings) • Active tracking (tall buildings)	FSO

TABLE 1.8 (continued)
Challenges in OWC Systems

	Causes	Effects	Solutions	Indoor/FSO
Blocking	Furniture	Link loss	Diffuse link	Indoor
	Moving objects		Cellular system	In/FSO
	Walls		Multibeam	In/FSO
	Birds		Hybrid FOS/RF	In/FSO
Roaming		Higher link availability at low speed	• Tracking	Indoor (LOS)
			• Using a narrow beam transmitter and a narrow FOV receiver [76]	
Bit error	• Noise	Bit error rate	• FEC	In/FSO
	• Multipath		• Block code	
			• Convolutional Code	
			• Turbo Code	
Aerosols gases and smoke		• Mie scattering	• Increase transmit power	FSO
		• Photon absorption	• Diversity techniques	
		• Rayleigh scattering	• Hybrid FSO-RF	
Nonuniform power distribution			• Spot diffusing transmitter and fly-eye receivers, study power budget, ambient light interference	Indoor (diffuse)
			• Direction diversity in the IR receiver—uses a cluster of arrow FOV detectors oriented in different directions to improve the sensitivity of rotation	

- Holographic diffusers
- Angle diversity—using multiple narrow beam transmitter and multiple nonimaging receivers. High ambient light rejection and reduced multipath distortion due to narrow FOV detector
- Imaging diversity reception—A single imaging lens and a photodetector segmented into multiple pixels
- Reduced ambient light
- Multibeam transmitter and imaging diversity receivers
- Multispot diffusing—combine the advantages of LOS and diffuse systems using holographic spot array generators. Decreased delay spread and also reduction in the transmitter power requirements
- Line strip spot diffusing transmitter—A multibeam transmitter located on the floor produces multiple diffusing spots on the middle of the ceiling in the form of a line strip. Reduces ISI and improved SNR
- Multitransmitter broadcast system—A number of transmitters placed in a grid near the ceiling of the room
- A square grid located near the room ceiling

REFERENCES

1. F. Halsall, *Data Communications, Computer Networks and Open Systems*, 4th ed. Harlow, UK: Addison Wesley, 1995.
2. Institute of Electrical and Electronics Engineers, IEEE Standard 802.11a-1999.
3. M. Johnsson, HiperLAN/2—The broadband radio transmission technology operating in the 5 GHz frequency band, *HiperLAN/2 Global Forum*, 1999. Available at: http://www.hiperlan2.com/site/specific/whitepaper.pdf
4. WiGig specification, whitepaper, http://wirelessgigabitalliance.org/specifications/, Last visited November 2011.
5. Very high throughput in 60 GHz, http://www.ieee802.org/11/Reports/tgad update.htm, Last visited November 2011.
6. L. Noel, D. Wake, D. G. Moodie, D. D. Marcenac, L. D. Westbrook and D. Nesset, Novel techniques for high-capacity 60 GHz fiber-radio transmission systems, *IEEE Transactions on Microwave Theory and Techniques*, 45, 1416–1423, 1997.
7. H. Ogawa, D. Polifko and S. Banba, Millimeter-wave fiber optics systems for personal radio communication, *IEEE Transactions on Microwave Theory and Techniques*, 40, 2285–2293, 1992.
8. A. J. Cooper, "Fibre/radio" for the provision of cordless/mobile telephony services in the access network, *Electronics Letters*, 26, 2054–2056, 1990.
9. F. R. Gfeller and U. Bapst, Wireless in-house data communication via diffuse infrared radiation, *Proceedings of the IEEE*, 67, 1474–1486, 1979.
10. J. R. Barry, *Wireless Infrared Communications*, Boston: Kluwer Academic Publishers, 1994.
11. R. Ramirez-Iniguez, S. M. Idrus and Z. Sun, *Optical Wireless Communications: IR for Wireless Connectivity*, Boca Raton: CRC Press, 2008.
12. E. Ciaramella, Y. Arimoto, G. Contestabile, M. Presi, A. D'Errico, V. Guarino and M. Matsumoto, 1.28 terabit/s (32×40 Gbit/s) WDM transmission system for free space optical communications, *IEEE Journal on Selected Areas in Communications*, 27, 1639–1645, 2009.
13. W. O. Popoola and Z. Ghassemlooy, BPSK subcarrier intensity modulated free-space optical communications in atmospheric turbulence, *Journal of Lightwave Technology*, 27, 967–973, 2009.
14. J. M. Kahn and J. R. Barry, Wireless infrared communications, *Proceedings of IEEE*, 85, 265–298, 1997.
15. A. J. C. Moreira, R. T. Valadas and A. M. d. O. Duarte, Optical interference produced by artificial light, *Wireless Networks*, 3, 131–140, 1997.
16. Infrared Data Association Releases IrDA Global Market Report 2007. Available: http://www.irda.org.
17. S. Bloom and W. S. Hartley, The last-mile solution: Hybrid FSO radio, *AirFiber, Inc. White paper*, http://www.freespaceoptic.com/WhitePapers/Hybrid_FSO.pdf, Last visited July 2007.
18. J. Bellon, M. J. N. Sibley, D. R. Wisely and S. D. Greaves, Hub architecture for infra-red wireless networks in office environments, *IEE Proceedings—Optoelectronics*, 146, 78–82, 1999.
19. V. Jungnickel, T. Haustein, A. Forck and C. von Helmolt, 155 Mbit/s wireless transmission with imaging infrared receiver, *Electronics Letters*, 37, 314–315, 2001.
20. A. M. Street, P. N. Stavrinou, D. C. Obrien and D. J. Edwards, Indoor optical wireless systems—A review, *Optical and Quantum Electronics*, 29, 349–378, 1997.
21. P. S. Peter, L. E. Philip, T. D. Kieran, R. W. David, M. Paul and W. David, Optical wireless: A prognosis, *Proc. SPIE*, PA, USA, 1995, pp. 212–225.

22. O. Bouchet, H. Sizun, C. Boisrobert, F. De Fornel and P. Favennec, *Free-Space Optics: Propagation and Communication*, London: ISTE Ltd, 2006.

23. M. J. McCullagh and D. R. Wisely, 155 Mbit/s optical wireless link using a bootstrapped silicon APD receiver, *Electronics Letters*, 30, 430–432, 1994.

24. A. K. Majumdar and J. C. Ricklin, *Free-Space Laser Communications: Principles and Advances*, New York: Springer, 2008.

25. J. M. Kahn, W. J. Krause and J. B. Carruthers, Experimental characterization of non-directed indoor infrared channels, *IEEE Transactions on Communications*, 43, 1613–1623, 1995.

26. O. Bouchet, M. El Tabach, M. Wolf, D. C. O'Brien, G. E. Faulkner, J. W. Walewski, S. Randel, M. Franke, S. Nerreter, K. D. Langer, J. Grubor and T. Kamalakis, Hybrid wireless optics (HWO): Building the next-generation home network, *Communication Systems, Networks and Digital Signal Processing, 2008. CNSDSP 2008. 6th International Symposium on*, Graz, Austria, 2008, pp. 283–287.

27. M. A. Naboulsi, H. Sizun and F. d. Fornel, Wavelength selection for the free space optical telecommunication technology, *SPIE*, 5465, 168–179, 2004.

28. J. M. Kahn, P. Djahani, A. G. Weisbin, K. T. Beh, A. P. Tang and R. You, Imaging diversity receivers for high-speed infrared wireless communication, *IEEE Communications Magazine*, 36, 88–94, 1998.

29. W. Binbin, B. Marchant and M. Kavehrad, Dispersion analysis of 1.55 um free-space optical communications through a heavy fog medium, *IEEE Global Telecommunications Conference*, Washington, DC, 2007, pp. 527–531.

30. A. Sivabalan and J. John, Improved power distribution in diffuse indoor optical wireless systems employing multiple transmitter configurations, *Optical and Quantum Electronics*, 38, 711–725, 2006.

31. V. Kvicera, M. Grabner and J. Vasicek, Assessing availability performances of free space optical links from airport visibility data, *7th International Symposium on Communication Systems Networks and Digital Signal Processing (CSNDSP)*, Newcastle, UK, pp. 562–565, 2010.

32. A. Prokes, Atmospheric effects on availability of free space optics systems, *Opt. Eng.*, 48, 066001–066010, 2009.

33. World Meteorological Organisation, *Guide to Meteorological Instruments and Methods of Observation*, Geneva, Switzerland, 2006.

34. M. Grabner and V. Kvicera, On the relation between atmospheric visibility and optical wave attenuation, *16th IST Mobile and Wireless Communications Summit*, Budapest, Hungary, pp. 1–5, 2007.

35. R. T. Valadas, A. M. R. Tavares and A. M. Duarte, Angle diversity to combat the ambient noise in indoor optical wireless communication systems, *International Journal of Wireless Information Networks*, 4, 275–288, 1997.

36. K. L. Sterckx, J. M. H. Elmirghani and R. A. Cryan, Pyramidal fly-eye detection antenna for optical wireless systems, *IEE Colloquium on Optical Wireless Communications (Ref. No. 1999/128)*, London, UK, pp. 5/1–5/6, 1999.

37. M. Grabner and V. Kvicera, Case study of fog attenuation on 830 nm and 1550 nm free-space optical links, *Proceedings of the Fourth European Conference on Antennas and Propagation (EuCAP)*, Barcelona, Spain, pp. 1–4, 2010.

38. F. Nadeem, V. Kvicera, M. S. Awan, E. Leitgeb, S. Muhammad and G. Kandus, Weather effects on hybrid FSO/RF communication link, *IEEE Journal on Selected Areas in Communications*, 27, 1687–1697, 2009.

39. D. P. Greenwood, Bandwidth specification for adaptive optics systems, *J. Opt. Soc. Am.*, 67, 390–393, 1977.

40. M.A. Al-Habash, L.C. Andrews, and R. L. Phillips, Mathematical model for the irradiance probability density function of a laser beam propagating through turbulent media, *Optical Engineering*, 40, 1554–1562, 2001.
41. S. Rajbhandari, Z. Ghassemlooy, J. Perez, H. Le Minh, M. Ijaz, E. Leitgeb, G. Kandus and V. Kvicera, On the study of the FSO link performance under controlled turbulence and fog atmospheric conditions, *Proceedings of the 2011 11th International Conference on Telecommunications (ConTEL)*, Graz, Austria, pp. 223–226, 2011.
42. S. Rajbhandari, Application of wavelets and artificial neural network for indoor optical wireless communication systems, PhD thesis, Northumbria University, Newcastle upon Tyne, UK, 2010.
43. M. Abramowitz and I. S. Stegun, *Handbook of Mathematical Functions with Formulars, Graphs and Mathematical Tables*, Dover Publication Inc., New York, 1977.
44. R. Ramirez-Iniguez and R. J. Green, Optical antenna design for indoor optical wireless communication systems, *International Journal of Communication Systems—Special Issue on Indoor Optical Wireless Communication Systems and Networks*, 18, 229–245, 2005.
45. R. Barakat, Sums of independent lognormally distributed random variables, *J. Opt. Soc. Am.*, 66, 211–216, 1976.
46. R. D. Wisely, A 1 Gbit/s optical wireless tracked architecture for ATM delivery, *IEE Colloquium on Optical Free Space Communication Links*, London, UK, 1996, pp. 14/1–14/7.
47. S. M. Haas and J. H. Shapiro, Capacity of wireless optical communications, *IEEE Journal on Selected Areas in Communications*, 21, 1346–1357, 2003.
48. E. Jakeman and P. Pusey, A model for non-Rayleigh sea echo, *IEEE Transactions on Antennas and Propagation*, 24, 806–814, 1976.
49. M. Uysal and J. Li, BER performance of coded free-space optical links over strong turbulence channels, *IEEE 59th Vehicular Technology Conference, VTC 2004-Spring*, Tamsui, Taiwan, Vol. 1, pp. 352–356, 2004.
50. K. Kiasaleh, Performance of coherent DPSK free-space optical communication systems in K-distributed turbulence, *IEEE Transactions on Communications*, 54, 604–607, 2006.
51. B. Braua and D. Barua, Channel capacity of MIMO FSO under strong turbulance conditions, *International Journal of Electrical & Computer Sciences*, 11, 1–5, 2011.
52. A. C. Boucouvalas, Indoor ambient light noise and its effect on wireless optical links, *IEE Proceedings—Optoelectronics*, 143, 334–338, 1996.
53. J. R. Barry, J. M. Kahn, W. J. Krause, E. A. Lee and D. G. Messerschmitt, Simulation of multipath impulse response for indoor wireless optical channels, *IEEE Journal on Selected Areas in Communications*, 11, 367–379, 1993.
54. G. W. Marsh and J. M. Kahn, 50-Mb/s diffuse infrared free-space link using on-off keying with decision-feedback equalization, *IEEE Photonics Technology Letters*, 6, 1268–1270, 1994.
55. C. C. Motlagh, V. Ahmadi and Z. Ghassemlooy, Performance of free space optical communication using M-array receivers at atmospheric condition, *17th ICEE*, Tehran, Iran, May 12–14, 2009.
56. N. S. Kopeika and J. Bordogna, Background noise in optical communication systems, *Proceedings of the IEEE*, 58, 1571–1577, 1970.
57. I. I. Kim and E. Korevaar, Availability of free space optics and hybrid FSO/RF systems, *Proceedings of SPIE: Optical Wireless Communications IV*, 4530, 84–95, 2001.
58. E. J. Lee and V. W. S. Chan, Part 1: Optical communication over the clear turbulent atmospheric channel using diversity, *IEEE Journal on Selected Areas in Communications*, 22, 1896–1906, 2004.
59. E. Telatar, Capacity of multi-antenna Gaussian channels, *European Transactions on Telecommunications*, 10, 585–595, 1999.

60. G. J. Foschini, Layered space-time architecture for wireless communication in a fading environment when using multi-element antennas, *Bell Labs Technical Journal*, 1, 41–59, 1996.

61. S. G. Wilson, M. Brandt-Pearce, C. Qianling and M. Baedke, Optical repetition MIMO transmission with multipulse PPM, *IEEE Journal on Selected Areas in Communications*, 23, 1901–1910, 2005.

62. I. Hen, MIMO architecture for wireless communication, *Intel. Technology Journal*, 10, 157–166, 2006.

63. F. Xu, A. Khalighi, P. Caussé and S. Bourennane, Channel coding and time-diversity for optical wireless links, *Opt. Express*, 17, 872–887, 2009.

64. J. Chen, Y. Ai and Y. Tan, Improved free space optical communications performance by using time diversity, *Chin. Opt. Lett.*, 6, 797–799, 2008.

65. S. Bloom, E. Korevaar, J. Schuster and H. Willebrand, Understanding the performance of free-space optics *Journal of Optical Networking*, 2, 178–200, 2003.

66. S. Hranilovic, *Wireless Optical Communication Systems*, New York: Springer-Verlag, 2004.

67. *Safety of Laser Products—Part 1: Equipment Classification and Requirements*, International Electrotechnical Commission, IEC 60825-1 ed2.0, 2007.

68. D. Heatley, D. R. Wisely and P. Cochrane, Optical wireless: The story so far, *IEEE Communications Magazine*, 36, 72–82, 1998.

69. M. Czaputa, T. Javornik, E. Leitgeb, G. Kandus and Z. Ghassemlooy, Investigation of punctured LDPC codes and time-diversity on free-space optical links, *Proceedings of the 2011 11th International Conference on Telecommunications (ConTEL)*, Graz, Austria, 2011, pp. 359–362.

70. W. Zixiong, Z. Wen-De, F. Songnian and L. Chinlon, Performance comparison of different modulation formats over free-space optical (FSO) turbulence links with space diversity reception technique, *IEEE Photonics Journal*, 1, 277–285, 2009.

71. F. S. Vetelino, C. Young, L. Andrews and J. Recolons, Aperture averaging effects on the probability density of irradiance fluctuations in moderate-to-strong turbulence, *Appl. Opt.*, 46, 2099–2108, 2007.

72. J. H. Churnside, Aperture averaging of optical scintillations in the turbulent atmosphere, *Appl. Opt.*, 30, 1982–1994, 1991.

73. L. C. Andrews, R. L. Phillips and C. Y. Hopen, Aperture averaging of optical scintillations: power fluctuations and the temporal spectrum, *Waves in Random Media*, 10, 53–70, 2000.

74. American National Standards Institute, ANSI Z136.1—2000, American National Standard for Safe Use of Lasers. LIA, Orlando, FL, 2000.

75. A. P. Tang, J. M. Kahn and K.-P. Ho, Wireless infrared communication links using multi-beam transmittersand imaging receivers, *IEEE International Conference on Communications*, Dallas, USA, 1996, pp. 180–186.

76. P. Djahani and J. M. Kahn, Analysis of infrared wireless links employing multibeam transmitters and imaging diversity receivers, *IEEE Transactions on Communications*, 48, 2077–2088, 2000.

2 Optical Sources and Detectors

There are a number of light sources and photodetectors that could be for OWC systems. The most commonly used light sources used are the incoherent sources-light emitting diodes (LEDs) and coherent sources—laser diodes (LD). LEDs are mainly used for indoor applications. However, for short link (e.g., up to a kilometre) and moderate data rates, it is also possible to use LEDs in place of LDs. Lasers, because of their highly directional beam profile, are mostly employed for outdoor applications. Particularly for long transmission links, it is crucial to direct the energy of the information to be transmitted precisely in the form of a well-collimated laser beam. This is to limit the often still very large channel power loss between the transmitter and the receiver. In order to limit the beam divergence, ideally, one should use a diffraction-limited light source together with a relatively large high-quality optical telescope. At the receiving end, it is also advantageous to use a high-directionality telescope not only to collect as much of the transmitted power as possible but also to reduce the background ambient light, which introduces noise and thus reduces the performance of the link. As for detectors, both the PIN and the APD photodetectors could readily be used. This chapter discusses the types of lights sources, their structures and their optical characteristics. The process of optical direct detection as well as coherent detection is also covered in this chapter. Different types of noise encountered in optical detection will be introduced and the statistics of the optical detection process is also discussed.

2.1 LIGHT SOURCES

For optical communication systems, light sources adopted must have the appropriate wavelength, linewidth, numerical aperture, high radiance with a small emitting surface area, a long life, a high reliability and a high modulation bandwidth. There are a number of light sources available but the most commonly used source in optical communications are LEDs and LDs, both of which rely on the electronic excitation of semiconductor materials for their operation [1]. The optical radiation of these luminescent devices, LED and LD, excludes any thermal radiation due to the temperature of the material as is the case in incandescent devices. Both LD and LED light sources offer small size, low forward voltage and drive current, excellent brightness in the visible wavelengths and with the option of emission at a single wavelength or range of wavelengths. Which light source to choose mainly depends on the particular applications and their key features, including optical power versus current characteristics, speed and the beam profile. Both devices supply similar power (about 10–50 mW) [2].

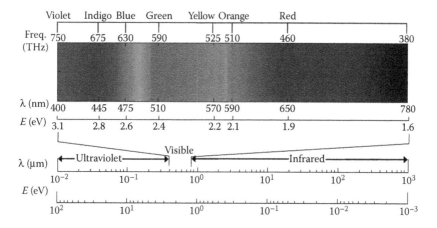

FIGURE 2.1 (**See colour insert.**) Wavelength and energy of ultraviolet, visible and infrared portion of the electromagnetic spectrum.

LEDs/LDs can be fabricated to emit light across a wide range of wavelengths (colours) from the visible to the infrared (IR) parts of the electromagnetic spectrum. These wavelengths and their corresponding energies are shown in Figure 2.1. The visual range of the human eye only extends from 400 nm to 700 nm. All these wavelengths are of great interest in OWC. The IR region of the electromagnetic spectrum can be classified into the following segments:

- *Near infrared* (IR-A (0.780–1.4 μm))—commonly used in optical fibre communications
- *Short-wavelength infrared* (IR-B (1.4–3 μm))—1530–1560 nm range is the dominant spectral region for long-distance telecommunications, including optical fibre and free-space optics
- *Mid-wavelength infrared* (IR-C (3–5 μm))
- *Long-wavelength infrared* (IR-C (8–15 μm))
- *Far infrared* (15–1000 μm)

The first and a part of the second segment are adopted in both optical fibre and optical wireless communications. The green part of the visible spectrum, 495–570 nm, is of particular interest in underwater OWC because of a low attenuation window within this band; ultraviolet (UV) radiations are currently being explored for indoor applications while IR has long been widely used for optical wireless applications, including TV remote controls, file sharing between phones and other personal devices as well as point-to-point outdoor FSO links.

The IR band can be further classified into [3]:

- *Near infrared* (0.7–1.0 μm)—region closest in wavelength to the radiation detectable by the human eye. The boundary between IR and visible light is not accurately defined. The human eye is noticeably less sensitive to light above 700 nm wavelength.

- *Short-wave infrared* (1.0–3 μm).
- *Midwave infrared* (3–5 μm).
- *Long-wave infrared* (8–12 or 7–14 μm).
- *Very-long-wave infrared* (12 to about 30 μm).

There is another classification for optical communications that is based on availability of light sources, optical fibre attenuation/dispersion, free-space channel losses and detector availability [4]:

- First window: 800–900 nm—widely used in early optical fibre communications.
- Second window: 1260–1360 nm—used in long-haul optical fibre communications.
- Third window (S-band): 1460–1530 nm.
- Third window (C-band): 1530–1565 nm—the dominant band for long-distance optical fibre communication networks. Transmissions at 1550 nm do not pass through the corneal filter, and cannot harm the sensitive retina. This means that at these wavelength bands, the emitted optical power could be allowed to reach values up to 10 mW [5] for OWC links [6,7].
- Fourth window (L-band): 1565–1625 nm—for future photonic networks.
- Fourth window (U-band): 1625–1675 nm—for future photonic networks.

In general, the generation of light is due to the transition of an electron from an excited state to a lower energy state. The energy difference due to the transition of the electron leads to a radiative or a nonradiative process. The radiative processes will lead to light generation, thus optical sources, whereas the nonradiative process typically leads to the creation of heat. In both types of light source devices, the recombination of carrier is used to provide a photon flux. In solids, the photon interacts with an electron in the following three distinct ways [1]:

1. The photon transfers its energy to an electron in the filled valence band. The electron is then excited to an empty state in the conduction band. This is the photon–electron interaction that is associated with the solar cell.
2. An electron in a filled conduction band can spontaneously return to the empty state in the valence band releasing a *photon* in the process. This is termed as the spontaneous radiative recombination and it is the process associated with LEDs.
3. Another interaction is when a photon is used to stimulate the radiative recombination process. The incident photon causes the emission of a second photon which is of the same phase, that is, the two photons are coherent and this is the type of emission on which the operation of a laser is based.

In a radiative recombination process, it is usually the case to assume that the emitted photon has the same energy as the energy band-gap of the material. However, at temperatures above the absolute zero, the extra thermal energy causes the electrons in the conduction band to reside just above the band edge and holes to sit just below the band edge in the valence band. Hence, the photon energy in a radiative recombination will be slightly higher than the band energy [1].

The frequency and the wavelength of the emitted or absorbed photon are related to the difference in energy E given by

$$E = E_2 - E_1 = hf = \frac{hc}{\lambda} \qquad (2.1)$$

where E_2 and E_1 are the two energetic states, $h = 6.626 \times 10^{-34}$ Js is the Planck's constant, f is the frequency, c is the speed of light 3×10^8 m/s and λ is the wavelength of the absorbed or emitted light.

2.2 LIGHT-EMITTING DIODE

The LED is a semiconductor p–n junction device that gives off spontaneous optical radiation when subjected to electronic excitation. The electronic excitation is achieved by applying a forward bias voltage across the p–n junction. This excitation energizes electrons within the material into an 'excited' state which is unstable. When the energized electrons return to the stable state, they release energy in the process and this energy is given off in the form of photons. The radiated photons could be in the UV, visible or IR part of the electromagnetic spectrum depending on the energy band-gap of the semiconductor material. In LEDs, the conversion process is fairly efficient, thus resulting in very little heat compared to incandescent lights. In the working of an LED, the electronic excitation causes electron(s) in the conduction band to spontaneously return to the valence band. This process is often referred to as the radiative recombination. It is so called because the electron returning to the valence band gives off its energy as photon as shown in Figure 2.2. In effect, the energy

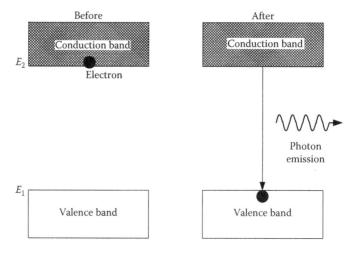

FIGURE 2.2   Spontaneous emission of photon.

of the emitted photon is equal to the energy difference between the conduction and the valence bands, that is, the band-gap energy. Nonradiative recombination occurs when the falling electron only gives out phonons (heat) and not photons.

In the case of spontaneous radiative recombination process just described, the rate of photon emission and its wavelength are given by [1]

$$I(E = hv) \propto \sqrt{E - E_g} \exp\left(-\frac{E}{kT}\right) \tag{2.2a}$$

$$\lambda = \frac{hc}{E(\text{eV})} \ \mu\text{m} \tag{2.2b}$$

where E_g is the band-gap energy of the semiconductor material, v is the radiation frequency, E is the energy level, k is the Boltzmann's constant and T is the absolute temperature. This expression is plotted in Figure 2.3. From this figure, it can be seen that the spontaneous emission has a threshold energy E_g and a half-power width of $1.8kT$ which translates into a wavelength spectra width given by [1]

$$\Delta\lambda = \frac{1.8kT}{hc} \tag{2.3}$$

where c is the speed of light. Since the operation of an LED is spontaneous emission based (i.e., transition of electrons from many energy levels of conduction and valence bands), the spectral width ($\Delta\lambda$) is rather wide. Additionally, transition, and therefore photon radiation, occurs randomly with no phase correlation between different photons thus leading to what is called an *incoherent* light source. A typical LED operating at a wavelength of 850 nm has $\Delta\lambda$ of about 60 nm and about 170 nm at an operating wavelength of 1300 nm [1].

An increase in temperature causes a decrease in the band-gap energy of a material. And since in a radiative recombination process, the emitted photon energy is

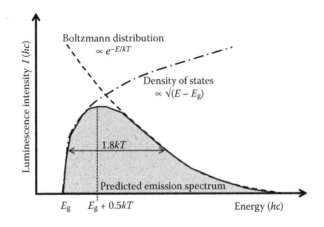

FIGURE 2.3 Theoretical spectrum of spontaneous emission.

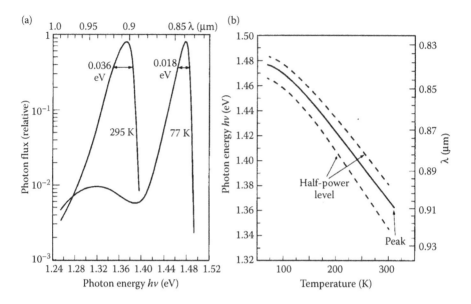

FIGURE 2.4 (a) GaAs diode emission spectrum at 300 and 77 K and (b) dependence of emission peak and half-power width as a function of temperature.

strongly tied to the band-gap energy, the peak photon energy thus decreases as the temperature increases. This is illustrated in Figure 2.4; the figure depicts the emission spectra for GaAs p–n junction at 77 and 300 K. In Figure 2.4, the peak photon energy and half-power points of the p–n junction are shown as a function of temperature.

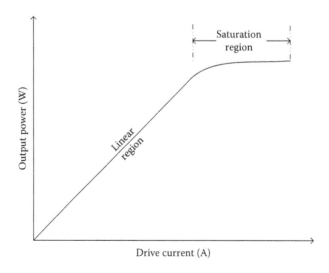

FIGURE 2.5 An illustration of the radiated optical power against driving current of an LED.

TABLE 2.1

Common LED Materials and Their Optical Radiation Wavelengths

LED Material/Substrate	Peak Wavelength or Range (nm)
AlGaN/GaN	230–350
InGaN/GaN	360–525
ZnTe/ZnSe	459
SiC	470
GaP	470
$GaAs_{0.15}P_{0.85}$	589
AlGaInP/GaAs	625–700
$GaAs_{0.35}P_{0.65}$/GaAs	632
$GaAs_{0.6}P_{0.4}$/GaAs	650
GaAsP/GaAs	700
$Ga_{1-x}Al_xAs$/GaAs	650–900
GaAs	910–1020
InGaAsP/InP	600–1600

Since photons are radiated in arbitrary directions, very few of them create light in the desired direction, a factor that reduces the LED output power. Therefore, this results in a low current-to-light conversion efficiency. Figure 2.5 illustrates the relationship between the radiated optical power and the driving current passing through the p–n junction. This clearly shows that the radiated optical power increases as the driving current is increased (i.e., a linear characteristic). The response becomes nonlinear for larger current (optical power level) values, thus introducing distortions.

As mentioned earlier, the peak wavelength (or colour) of the spontaneous emission depends very much on the band-gap energy of the p–n junction which in turn depends on the semiconductor material(s) making up the junction. Table 2.1 presents common LED materials and their respective peak radiation wavelength or range of wavelengths.

2.2.1 LED STRUCTURE

The basic structure of an LED is a p–n junction in which the extrinsic material types dictate the radiation wavelength(s). If the structure of the p–n junction is such that the semiconductor materials making up the junction are dissimilar, with different band-gap energies, a heterostructure device is formed; otherwise it is a homostructure. In the structure of an LED, there is nothing like a resonant cavity or gain medium; hence, its radiation is not going to be as intense as that of a laser diode that is discussed in Section 2.3. There are various structures of an LED depending on the application and more importantly how the light generated at the p–n junction is radiated out of the device. To illustrate the basic structure of an LED, only the planar, dome and edge-emitting LEDs will be discussed. A comprehensive discussion of all possible structures and their detailed analyses is beyond the scope of this book; interested readers are however referred to Refs. [1,2].

2.2.2 PLANAR AND DOME LED

The planar LED has the simplest structure and it is fabricated by either liquid- or vapour-phase epitaxial process over the whole surface of a GaAs substrate [8]. The planar LED structure shown in Figure 2.6 emits light from all surfaces and the emission is therefore termed as Lambertian. In Lambertian emission, the optical power radiated from a unit area into a unit solid angle (otherwise called the surface irradiance) is constant. In terms of intensity, the maximum intensity, I_o, in a Lambertian radiation is perpendicular to the planar surface but reduces on the sides with the viewing angle θ according to expression (2.4). The Lambertian intensity distribution is illustrated in Figure 2.7.

$$I_{li}(\theta) = I_o \cos\theta \tag{2.4}$$

The structure of a typical dome LED is shown in Figure 2.8. In this structure, a hemisphere of n-type GaAs is formed around a diffused p-type region. The diameter of the dome is chosen to maximize the external quantum efficiency of the device. The geometry of the dome is such that it is much larger than the recombination area; this gives a greater effective emission area and thus reduces the radiance. Because the dome structure does not suffer as much internal reflection as the planar LED, it therefore has a higher external quantum efficiency [8].

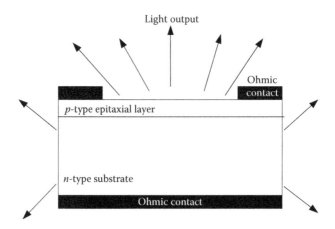

FIGURE 2.6 Planar LED structure showing light emission on all surfaces.

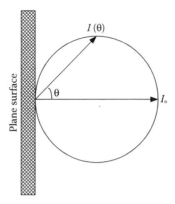

FIGURE 2.7   The Lambertian intensity distribution.

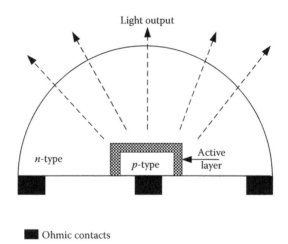

FIGURE 2.8   The basic structure of a dome (hemispherical) LED.

2.2.3   EDGE-EMITTING LED

As discussed above, the planar structure of an LED emits light in all directions but there are very many instances in optical communications when this is not desirable and the light will be preferred confined. The edge-emitting LED does just this by confining the light in a thin (50–100 μm) narrow stripe in the plane of the $p–n$ junction. This is illustrated in the figure of a double heterostructure AlGaAs edge-emitting LED depicted in Figure 2.9. The confinement is achieved by deliberately making the active layer of higher refractive index than the surrounding materials. This then creates a waveguide (through the process of total internal reflection) for the light generated at the junction to travel to both ends of the device. The waveguiding results in higher efficiency for the device and also narrows the beam divergence to a half power width of about 30 degrees in the plane perpendicular to the

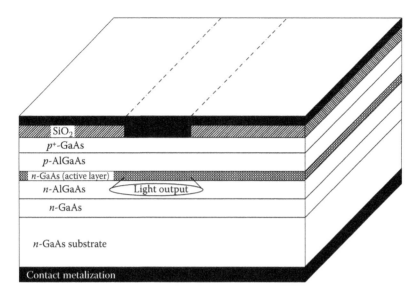

FIGURE 2.9 The structure of a double heterojunction AlGaAs edge-emitting LED.

$p–n$ junction. However, the radiation in the junction plane is Lambertian with a half-power width of about 120° since there is no such waveguiding in the plane of the junction [2,8]. The emerging optical beam is therefore elliptical in shape.

The confined light is channelled to both ends of the device by the waveguiding, to ensure that the light is only radiated from one face; the emitting edge is coated with an antireflection while a reflector is situated at the other end.

2.2.4 LED Efficiencies

2.2.4.1 Internal Quantum Efficiency

This term relates to the conversion of carriers into photons within the device. It can therefore be expressed as the ratio of the number of internally emitted photons to the number of carrier passing through the $p–n$ junction. That is

$$\eta_{in} = \frac{\text{No. of photons emitted internally}}{\text{No. of carriers passing junction}} \tag{2.5}$$

The internal quantum efficiency can also be related to the fraction of injected carriers that recombine radiatively to the total recombination rate and this is directly linked to the carrier lifetimes by the equation

$$\eta_{in} = \frac{R_r}{R_r + R_{nr}} = \frac{\tau_r}{\tau_r + \tau_{nr}} = \left(1 + \frac{\tau_{nr}}{\tau_r}\right)^{-1} \tag{2.6}$$

where R_r and R_{nr} represent the radiative and nonradiative recombination rates, respectively, while τ_r and τ_{nr} stand for the radiative and nonradiate lifetimes in that order.

2.2.4.2   External Quantum Efficiency

This quantum efficiency is different from the internal quantum efficiency described above in that it relates to the amount of photons emitted externally by the device. As such, it is defined as the ratio of the radiative recombination to nonradiative recombination and by the absorption of the generated light by the semiconductor material.

$$\eta_{ex} = \frac{\text{No. of photons emitted externally}}{\text{No. of carriers passing junction}} = \eta_{in}\eta_t \qquad (2.7)$$

The optical efficiency given by η_t denotes the ratio of photons that get emitted externally from the LED to that emitted internally. This ratio is a function of the optics in and around the device. For a typical planar semiconductor LED device, the optical efficiency could be as low as 2%. A simple expression for the estimation of the optical efficiency is given by

$$\eta_t = \frac{1}{2}(1 - \cos\theta_c) \approx \frac{1}{4}\frac{n_o}{n_s} \qquad (2.8)$$

where θ_c is the critical angle inside the device while n_o and n_s are, respectively, the refractive indexes of the LED's (semiconductor) material and that of the medium which the device is emitting light into, that is, the ambient.

The amount of photons emitted externally by the optical source and indeed the external quantum efficiency depends on loss mechanisms within the device. These losses could be from a combination of

1. The total internal reflection of photons incident to the surface at angles greater than the critical angle—critical angle loss.
2. Absorption loss—This accounts for the loss due to photon absorption by the LED material. For LEDs made using the opaque GaAs substrate, up to 85% of the total generated photons could be absorbed and for transparent substrate like the GaP with isoelectronic centres, the value is much lower, around 25% [1].
3. Fresnel loss—Fresnel reflection, which is always present at the interface of materials with different index of refraction, causes some of the generated photon to be reflected back to the semiconductor material instead of being emitted externally.

2.2.4.3   Power Efficiency

This is simply the ratio of the optical power output to the electrical power input to the LED. That is

$$\begin{aligned}\eta_{\mathrm{p}} &= \frac{\text{Optical output power}}{\text{Electrical input power}}\\[2mm]&= \frac{\text{No. of photons emitted externally} \times hv}{\text{No. of carriers passing junction} \times q \times V}\end{aligned} \tag{2.9}$$

Assuming that the electrical bias qV is approximately equal to the energy band-gap, hv, $\eta_{\mathrm{p}} \approx \eta_{\mathrm{ex}}$.

2.2.4.4 Luminous Efficiency

This metric is often used to characterize LEDs radiating within the visible spectrum. The luminous efficiency basically normalizes the power efficiency defined above by a factor that is related to the eye sensitivity shown in Figure 2.10. It is hence the ratio of the luminous flux to the input electrical power. The luminous flux (in lumens) given by Equation 2.10 describes the 'weighted' total emitted light. The luminous flux is total emitted flux weighted or scaled appropriately to reflect the varying sensitivity of the human eye to different wavelengths of light.

$$\text{Luminous flux} = 683 \int V(\lambda)P_{\mathrm{op}}(\lambda)\,\mathrm{d}\lambda \tag{2.10}$$

$V(\lambda)$ is the relative eye sensitivity shown in Figure 2.10. Its value is normalized to unity at the peak wavelength of 555 nm and $P_{\mathrm{op}}(\lambda)$ is the radiation power spectrum of the LED.

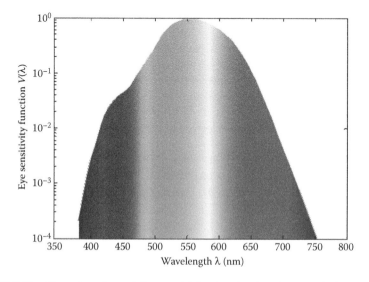

FIGURE 2.10 (**See colour insert.**) Eye sensitivity function based on the 1978 CIE data.

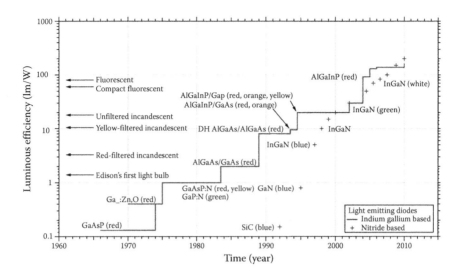

FIGURE 2.11 LED luminous efficiency.

The luminous efficiency is therefore given by

$$\eta_{lu} = \frac{\text{Luminous flux}}{\text{Electrical input power}} = \frac{683 \int V(\lambda) P_{op}(\lambda) d\lambda}{VI} \, lm/W \qquad (2.11)$$

Figure 2.11 shows the progression of the LED luminous efficiencies with time, including that of conventional lighting [1].

2.2.4.5 LED Modulation Bandwidth

The amount of modulation bandwidth and the frequency response of an LED depend on (1) the injected current, (2) the junction capacitance and (3) the parasitic capacitance. The capacitance values are nearly invariant while the response increases with increasing current [9]. As such, the effects of the aforementioned factor can be reduced by superimposing the AC signal on a constant DC bias. If the DC power is given as P_o and that at frequency, ω, by $P(\omega)$, then the relative optical power output at any given frequency is given as [10]

$$\frac{P(\omega)}{P_o} = \frac{1}{\sqrt{1 + (\omega\tau)^2}} \qquad (2.12)$$

The optical 3-dB bandwidth is obtained by setting Equation 2.12 to 0.5 while the equivalent electrical bandwidth is obtained by equating the same expression to 0.707 as illustrated in Figure 2.12.

From Equation 2.12, it follows therefore that the inherent modulation bandwidth of an LED is essentially limited by the minority carrier lifetime τ. There are two ways

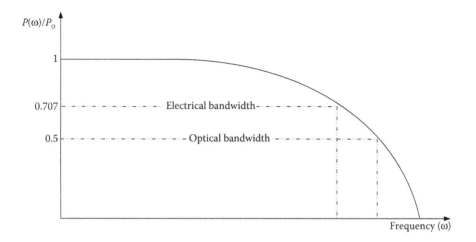

FIGURE 2.12   An illustration of the optical and electrical bandwidth.

to reduce the effective carrier lifetime and increase the modulation bandwidth. One is to increase the doping level in the recombination region. It should be noted that the lifetime has both radiative and nonradiative components to it and are combined according to

$$\frac{1}{\tau} = \frac{1}{\tau_r} + \frac{1}{\tau_{nr}}$$ (2.13)

So, increasing the doping level increases the internal quantum efficiency for only as long as τ_r is reduced. At very high doping levels, however, an excessive number of nonradiative centres is formed; hence, the increase in modulation bandwidth due to increased doping levels is accompanied by a decrease in quantum efficiency. The second approach is to increase the carrier density. Lasers have considerably higher modulation bandwidth because the radiative lifetime is further shortened by the stimulated emission process.

2.3   THE LASER

2.3.1   Operating Principle of a Laser

Laser, which is the acronym for light amplification by stimulated emitted radiation, is a device that amplifies (generated) light, as the name suggests, but they are rarely used for this purpose in practice. Lasers are used mainly as an optical oscillator with light bouncing back and forth in an optical cavity. One end of the cavity is made to have almost 100% reflection while the other is significantly less to allow the emission of monochromatic light (not exactly single wavelength but a very narrow band of wavelengths, typically 0.1–5 nm). In the operation of a laser, certain vital conditions are required to be met before lasing can take place. These prerequisite lasing concepts are discussed below and together they explain the principle behind the operation of a laser. In laser, the conversion process is fairly efficient compared to LEDs and it has a high output power. In comparison, for an LED to radiate 1 mW of output power, up to 150 mA of forward current is required, whereas for a laser diode to radiate the same power only 10 mA or less of current is needed.

2.3.2 STIMULATED EMISSION

In order to describe the basic principle of stimulated emission, a two-level atomic system will be considered. In this simple system, an atom in the upper level E_2 can fall into level E_1, as shown in Figure 2.13, in a random manner giving off a photon with energy $(E_2 - E_1)$ in the process called spontaneous emission. Another possibility, one which is of greater interest here, is when an external source with energy $hv = E_2 - E_1$ is incident on the atom in level E_2. This forces the atom in the upper level E_2 to transit to the lower level E_1. The change in energy involved in this process is then given off or emitted as a photon that has the same phase and frequency as the incident (exciting) photon (see Figure 2.13). And so the stimulated radiation is *coherent*. This describes stimulated emission, a process that underpins the working of a laser. The impinging photon (or external source of energy) stimulates the emission of another photon (note that $E_p = hv = hc/\lambda$). In other words, an external stimulus is required to achieve stimulated emission; this can be in the form of an optical excitation as just described or in the form of carrier injection through forward bias in the case of laser diodes. This property of the laser ensures that its spectral width will be very narrow compared to LEDs. In fact, it is quite common for laser diodes to have 1 nm or less spectral width at both 1300 nm and 1550 nm.

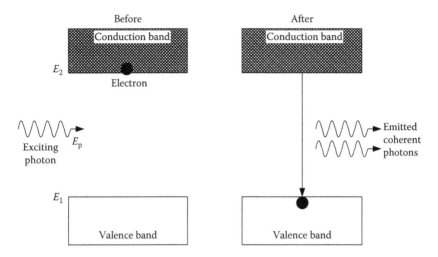

FIGURE 2.13   Stimulated emission.

2.3.2.1 Population Inversion

In the two-energy-level atomic system under consideration, under thermal equilibrium, the number of atoms N_1 in the lower level E_1 is greater than the number of atoms N_2 in the upper level E_2 as stipulated by the Boltzmann statistics given by

$$\frac{N_1}{N_2} = \frac{g_1}{g_2}\exp\left(\frac{E_2 - E_1}{kT}\right) \tag{2.14}$$

To attain optical amplification necessary for lasing to take place, we require a nonequilibrium distribution of atoms such that the population of the E_1 level is lower than that of the upper level E_2, that is, $N_2 > N_1$. This condition is called population inversion. By using an external excitation, otherwise called 'pumping', atoms from the lower level are excited into the upper level through the process of stimulated absorption. A two-level atomic system is not the best in terms of lasing action as the probability of absorption and stimulated emission are equal, providing at best equal populations in the two levels E_1 and E_2. A practical laser will have one or more meta-stable levels in between. An example is the four-level He–Ne laser illustrated in Figure 2.14. To attain population inversion, atoms are pumped from the ground-state level E_0 into level E_3. The atoms there then decay very rapidly into a metastable level E_2 because they are unstable in level E_3. This increases the population at level E_2 thereby creating population inversion between E_2 and E_1. Lasing can then take place between E_2 and E_1.

2.3.3 OPTICAL FEEDBACK AND LASER OSCILLATION

Light amplification in the laser occurs when a photon collides with an atom in an excited energy level and causes the stimulated emission of a second photon; these are

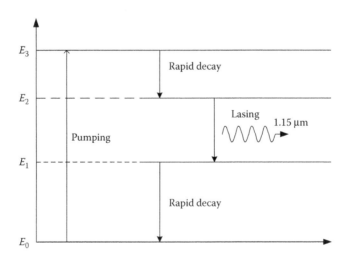

FIGURE 2.14 An illustration of lasing action based on a four-level He–Ne laser.

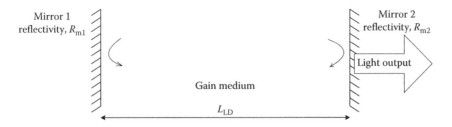

FIGURE 2.15 An illustration of optical feedback and laser oscillation.

then fed back to release two more photons [2]. To sustain the process and maintain coherence, two mirrors (plane or curved) are placed at either end of the amplifying medium as shown in Figure 2.15. The optical signal is thus fed back many times while receiving amplification as it passes through the gain medium. A stable output is obtained at saturation when the optical gain is exactly matched by the losses incurred in the amplifying medium.

The output of a laser is not exactly monochromatic but very narrow spectral band of the order of 1 nm; its centre frequency is equal to the mean energy difference of the stimulated emission transition divided by Planck's constant. While population inversion is necessary for lasing, it is not a sufficient condition. A threshold or minimum gain given by Equation 2.15 within the amplifying medium must be attained before lasing can occur.

$$g_{th} = \alpha + \frac{1}{2L_{LD}} \ln \frac{1}{R_{m1}R_{m2}}$$ (2.15)

In Equation 2.15, the loss coefficient of the gain medium in dB/km is represented by α, and the fraction of incident power reflected by the mirrors by R_{m1} and R_{m2}.

In laser diodes, the forward bias voltage brings about population inversion and the stimulated emission condition necessary for lasing. Optical feedback is achieved by polishing the end faces of the $p-n$ junction to act as mirrors while the sides are deliberately roughened to prevent any unwanted radiations.

2.3.4 BASIC SEMICONDUCTOR LASER STRUCTURE

The structure of a semiconductor laser diode resembles that of an LED, except of course, that an optical feedback is deliberately introduced through the inclusion of polished mirrors at either end of the structure as illustrated in Figure 2.15. The optical feedback effect is significant in crossing the threshold for lasing to occur and also to maintain the lasing action when driving the laser diode at a current that is sufficient to produce population inversion (i.e., above the threshold point); otherwise, its effect is negligible. At low drive currents—below the threshold level—the emission of the laser diode is spontaneous just as in LED where the gain is less than the loss (see Figure 2.16). However, as the drive current increases, more electron–hole pairs are produced leading to spontaneous emission of many more photons.

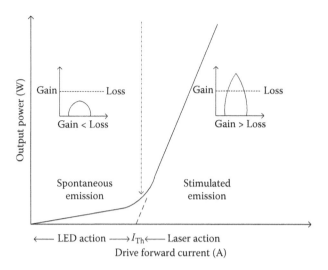

FIGURE 2.16   Laser output power against drive current plot.

The increasing spontaneous emission in return makes it more likely for any of these photons to encounter and stimulate further emissions. This continues until the drive current is high enough to produce population inversion which then results in a cascade of stimulated emission as the laser crosses the threshold with a gain larger than the loss (see Figure 2.16) [11]. The electrical power needed to reach the threshold current winds up as heat in the laser as well as a fraction of the above-threshold current that is not converted into light. The heat generated does not just go down as wasted power; it degrades the laser performance and shortens its lifetime. This explains why lower threshold lasers tend to have longer lifetimes [11]. A typical VCSEL diode, for example, the OPV314, with 1.5 mW output power, has about 3 mA of threshold current and 12 mA driving current.

As shown in Figure 2.16, the laser response is linear above the threshold level. Single-mode lasers exhibit a linear response above the threshold, whereas in multi-mode laser, we sometimes observe a slightly nonlinear response above the threshold level due to mode-hopping. The linearity of the light sources (LED and LD) is particularly important for analogue systems.

When an electron–hole recombines, the energy it releases in the process, and the wavelength in the case of radiative recombination, depend on the energy gap between the conduction band occupied by the free electrons and the valence band where the hole is. As previously mentioned, the energy band-gap depends on the atomic composition for the semiconductor material(s). So, changing the composition of the band-gap is a good way of manipulating the radiation wavelength(s). Table 2.2 shows some common laser diode materials and their corresponding radiation wavelengths. It should be mentioned that in order for the atoms in adjacent layers of a laser to line up properly, or to be structurally compatible, the materials making the different layers must have nearly identical lattice constants. This is absolutely necessary to avoid strains (or defects) developing in the layers due to lattice mismatch.

TABLE 2.2
Laser Diode Materials and Their Corresponding Radiation Wavelengths

Laser Material/Substrate	Wavelength
AlGaN	350–400 nm
GaAlN	375–440 nm
ZnSSe	447–480 nm
ZnCdSe	490–525 nm
AlGaInP/GaAs	620–680 nm
$Ga_{0.5}In_{0.5}P$/GaAs	670–680 nm
GaAlAs/GaAs	750–900 nm
GaAs/GaAs	904 nm
InGaAs/GaAs	915–1050 nm
InGaAsP/InP	1100–1650 nm
InGaAsSb	2000–5000 nm
PbCdS	2700–4200 nm
Quantum cascade	3–50 μm
PbSSe	4.2–8 μm
PbSnTe	6.5–30 μm
PbSnSe	8–30 μm

For LDs, both the internal and external efficiencies can be defined by Equations 2.5 and 2.7.

2.3.5 THE STRUCTURE OF COMMON LASER TYPES

2.3.5.1 Fabry–Perot Laser

The simplest laser structure is that of Fabry–Perot laser (FPL); the structure is essentially a resonating cavity of the type shown in Figure 2.15. The cavity mirrors are formed by the boundary between the high refractive index semiconductor crystal and the lower refractive index air. One facet is coated to reflect the entire light incident on it from within the cavity (gain medium) while the other facet is deliberately made to reflect less in order to emit the light generated within the cavity.

The gain curve of the FPL is quite broad; the implication of this is that the laser emits most of its light on one wavelength—the centre wavelength—but also emits at other wavelengths called the longitudinal modes. These modes or side bands are distributed and equally spaced on either side of the centre wavelength. When the intensity of the FPL is modulated, the centre wavelength does not stay fixed. It shifts and takes on different values at different times during the modulation process. This effect is called mode-hopping and it is undesirable from a communication standpoint. Mode-hopping occurs because the refractive index of the device varies with temperature. And the temperature itself varies with the driving current. So, modulating the laser by varying the driving current results in temperature changes which then leads to changes in the cavity's index of refraction and in turn the centre wavelength. The mode-hopping problem is one of the reasons why the FPL is only used for low-speed optical communications and other applications such as CD players where some variations in the centre wavelength can be tolerated.

2.3.5.2 Distributed Feedback Laser

To reduce the spectral width, we need to make a laser diode merely radiate in only one longitudinal mode. The distributed feedback (DFB) laser, which is a special type of edge-emitting lasers, is optimized for single-mode (single-frequency) operation. The single-mode operation is achieved by incorporating a periodic structure or a Bragg grating near the active layer as depicted in Figure 2.17.

The grating is placed so near the active layer that the oscillating transverse mode will interact with it [5]. The grating structure usually employed is the distributed Bragg diffraction grating. As for the meaning of the term 'DFB laser diode', the word *feedback* emphasizes that we have the means to return stimulated photons to an active medium. This is done by reflecting a portion of the light at each slope of the grating. The word *distributed* implies that reflection occurs not at a single point but at many points dispersed along the active region. The net result of this arrangement is that only one wave with the wavelength λ_o is radiated. The waves that satisfy the Bragg condition of Equation 2.16 are scattered off the successive corrugations and interfere constructively. The generated lightwave that satisfies this Bragg condition is then reflected back to the active region of the device. It is only the reflected lightwave that receives the feedback and optical amplification needed for lasing action.

$$\Lambda_g = \frac{m\lambda_g}{2n_{eff}} \tag{2.16}$$

In Equation 2.16, Λ_g is the grating period required for operations at a wavelength λ_g, n_{eff} is the effective refractive index of the waveguide and m is the integer order of the grating.

Since the radiation contains only a single longitudinal mode, therefore, as a result, the *spectral width* of the DFB laser is extremely narrow, which is ideal for wavelength

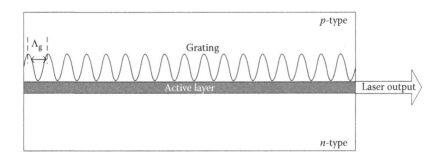

FIGURE 2.17   The structure of a DFB laser.

division multiplexing for backbone optical fibre communication networks in particular at 1550 nm. A modified version of using a Bragg grating as a reflector is better known as the distributed-Bragg-reflector laser diode, where the active region is located between two Bragg gratings that work as reflectors.

2.3.5.3 Vertical-Cavity Surface-Emitting Laser

All the laser diodes discussed above are based on edge-emitting devices, which are characterized by the significant length of their active region and their asymmetrical radiation pattern. However, we have seen the emergence of a new type of laser diode known as the VCSEL. A basic arrangement of a VCSEL is shown in Figure 2.18, where the resonant cavity is vertical as the name implies and perpendicular to the active layer. Above and below the active layer, in the vertical direction, are narrow band mirror layers and the light beam emerges from the surface of the wafer, through the substrate as outlined in Figure 2.18. The narrow band mirrors are made from alternating high and low refractive index layers, each of which is designed to a quarter wavelength of the laser's operating wavelength [5]. In terms

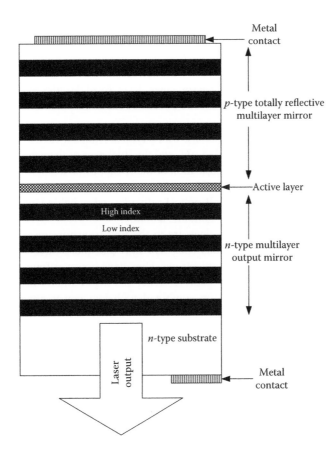

FIGURE 2.18   The basic structure of a VCSEL.

of materials, the GaAs-based VCSELs are very popular because they are relatively easy to make. The refractive index of GaAlAs is known to vary considerably with aluminium content, thereby providing the refractive index contrast needed for the multilayer mirrors. As a result, 850 nm VCSELs developed from this process are very widely used in optical communications. More recently, VCSELs have been developed using InGaAsP compounds that emit at the 1300 nm and 1550 nm windows [11]. Listed below are some of the features that distinguish VCSELs from other types of lasers:

1. Optical oscillation is perpendicular to the surface of the thin active layer, thus radiating a circular output beam in contrast to that radiated by edge-emitting lasers.
2. The short resonant cavity (on the order of 2 μm) of a VCSEL ensures single-mode operations (a spectral width of a gain curve of a few nanometres) and direct modulation for data rate well above one gigabit per second [11].
3. Lower radiated power and higher switching speed compared to edge-emitting lasers. VCSEL devices can radiate 3 mW of output power at 10 mA forward current with an intrinsic modulation bandwidth up to 200 GHz.
4. Comparatively low threshold current that translates into higher efficiency and longer lifespan.
5. A very stable lasing wavelength since it is fixed by the short (1~1.5 wavelength thick) Fabry–Perot cavity. VCSELs can also operate reliably at temperatures up to 80°C.
6. They have very small dimensions, with a typical resonant cavity and diameter of the active region of 1–5 μm, and the thickness of the active layer of 25 nm, thus allowing the fabrication of many diodes on a single substrate. This therefore makes the fabrication of monolithic two-dimensional (matrix) array of diodes for all-optical networks and switching, which is not possible with edge-emitting emitters, a reality.
7. Commercially available devices are in the first transmission window of optical fibre communications (i.e., 850 nm) and these have been used extensively for fibre communications. They have however found applications in areas such as the gigabit Ethernet and the computer mice (offering a high tracking precision combined with low power consumption) among others.

2.3.5.4 Superluminescent Diodes

Superluminescent diode (SLD) properties are intermediate between that of LEDs and LDs. These devices use a double heterostructure to confine the active region under conditions of high current density, thus ensuring population inversion so that the source is able to amplify light. However, there is no feedback in SLDs, and thus spontaneous emission radiation, which is a more powerful and more sharply confined beam than a regular LED. But an SLD's radiation is not as monochromatic, well directed, and coherent as an LD's radiation.

TABLE 2.3

A Comparison of an LED and a Semiconductor Laser Diode

Characteristics	LED	LD
Optical output power	Low power	High power
Optical spectral width	25–100 nm	0.01–5 nm
Modulation bandwidth	Tens of kHz to hundreds of MHz	Tens of kHz to tens of GHz
E/O conversion efficiency	10–20%	30–70%
Eye safety	Considered eye safe	Must be rendered eye safe
Directionality	Beam is broader and spreading	Beam is directional and is highly collimated
Reliability	High	Moderate
Coherence	Noncoherent	Coherent
Temperature dependence	Little temperature dependence	Very temperature dependent
Drive and control circuitry	Simple to use and control	Threshold and temperature compensation circuitry
Cost	Low	Moderate to high
Harmonic distortions	High	Less
Receiving filter	Wide—increase noise floor	Narrow—lower noise floor

2.3.6 COMPARISON OF LED AND LASER DIODES

Table 2.3 gives a succinct comparison between a semiconductor laser diode and an LED.

For outdoor application, the choice of wavelength will of course depend upon the atmospheric propagation conditions and characteristics, optical background noise, laser technology, photodetectors and optical filters.

2.4 PHOTODETECTORS

The photodetector is a square-law optoelectronic transducer that generates an electrical signal which is proportional to the square of the instantaneous optical field impinging on its surface. Thus, the signal generated by a photodetector is always proportional to the instantaneous (received) optical power. Since the optical signal is generally weak, having travelled through the communication channel, the photodetector must therefore meet stringent performance requirements such as high sensitivity within its operational range of wavelengths, a low noise level and an adequate bandwidth to accommodate the desired data rate. The effect of temperature fluctuations on the response of the photodetector is required to be minimal and the device must equally have a long operating life. The wavelengths at which the detector responds to light depend on the detector's material composition. Figure 2.19 shows the detector response curve for different materials.

The ratio of the number of electron–hole (e–h) pairs generated by a photodetector to the incident photons in a given time is termed the quantum efficiency η, which is

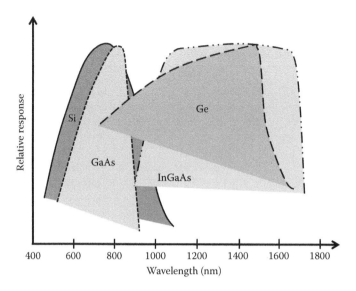

FIGURE 2.19 Relative response of photodetectors for different materials. (Adapted from W. V. Etten and J. V. D. Plaats, *Fundamentals of Optical Fiber Communications*, Prentice Hall International, 1991.)

defined and calculated as [5]

$$\eta_{qe} = \frac{\text{Electrons out}}{\text{Photons input}}$$

$$\eta_{qe} = (1 - R)\xi(1 - e^{-\alpha d}) \tag{2.17}$$

where R is the reflection coefficient at the air–semiconductor interface, ξ is the fraction of the e–h pairs that contributes to the photocurrent, α is the absorption coefficient and d is the distance where optical power is absorbed. The quantum efficiency is often used as a metric to characterize a photodetector.

> *Dark current*—Dark current is the current through the photodiode in the absence of light, when it is operated in the photoconductive mode. The dark current includes photocurrent generated by background radiation and the saturation current of the semiconductor junction. Dark current sets a floor on the minimum detectable signal, because a signal must produce more current than the dark current in order to be detected. Dark current depends on operating temperature, bias voltage and the type of detector. Dark current must be accounted for by calibration if a photodiode is used to make an accurate optical power measurement, and it is also a source of noise when a photodiode is used in an optical communication system.
>
> *Noise-equivalent power*—Noise-equivalent power (NEP) is the minimum input optical power to generate photocurrent, equal to the root mean square (RMS) noise current in a 1 Hz bandwidth. This measures the minimum detectable

signal more directly because it compares noise directly to the optical power. NEP depends on the frequency of the modulated signal, the bandwidth over which noise is measured, the detector area and the operating temperature. Consider an incoming optical radiation with an average power P_r impinging on a photodetector over a period T. Then the electric current (quantity of charge per unit time) generated by the detector is given by [2]

$$\langle i \rangle = \frac{\eta_{qe} q \lambda P_r}{hc} = R P_r \qquad (2.18)$$

where q is the electronic charge and R is the photodetector responsivity defined as the photocurrent generated per unit incident optical power. With λ in μm, R is expressed as

$$R = \frac{\lambda q \eta_{qe}}{hc} = \frac{\lambda}{1.24} \eta_{qe} \qquad (2.19)$$

The speed of response and the bandwidth of a photodetector depend on

1. The *transit time* of the photon-generated carriers through the depletion region
2. The *electrical frequency response* determined by the RC time constant, which depends on the diode's capacitance
3. The *slow diffusion* of carriers generated outside the depletion region

Similar to optical sources adopted for OWC systems, there are a number of criteria that define some of the most important performance and compatibility requirements for detectors.

- Large detection surface area—thus a large collection aperture and wider detection FOV. For high-speed applications, photodetector arrays with small surface area are the best possible option
- High sensitivity and responsivity at the operating wavelength
- Low noise, thus high SNR
- Fast response time
- Low cost, small size and high reliability

There are four types of photodetectors that could be used in optical receivers: (i) PIN photodiodes (with no internal gain, which is compensated by a larger bandwidth), (ii) APDs, (iii) photoconductors and (iv) metal–semiconductor metal photodiodes. The last three have internal gain. Both PIN and APD are the most popular and widely used detectors for OWC systems, meeting all the requirements mentioned above.

2.4.1 PIN PHOTODETECTOR

The PIN photodetector consists of p- and n-type semiconductor materials separated by a very lightly n-doped intrinsic region [12]. In normal operating conditions, a sufficiently large reverse bias voltage is applied across the device as shown in the

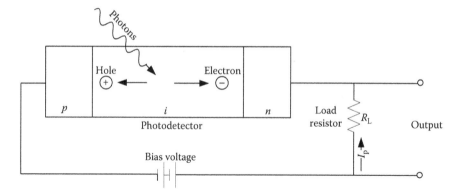

FIGURE 2.20 PIN photodetector schematic diagram.

schematic of Figure 2.20. The reverse bias ensures that the intrinsic region is depleted of any charge carriers.

For the device to convert an incident photon into an electron/electric current, the energy of the incoming photon must not be less than the band-gap energy of the semiconductor material. The incident photon uses its energy to excite an electron from the valence band to the conduction band, thereby generating a free electron–hole pair in the process. Normally, the incident light is concentrated on the depleted intrinsic region. The high electric field present in this depleted region causes the generated charge carriers to separate and be collected across the reverse biased junction. This gives rise to a current flow in an external circuit as shown in Figure 2.20; there is one electron flowing for every carrier pair generated.

The semiconductor material of the photodetector determines over what wavelength range the device can be used (see Figure 2.19). The upper cut-off wavelength λ_c, in micrometres (μm) is generally given by Equation 2.20, where E_g is the energy band-gap of the semiconductor material in electron-volt (eV).

$$\lambda_c\,(\mu m) = \frac{hc}{E_g} = \frac{1.24}{E_g} \tag{2.20}$$

The operating wavelength ranges for different photodetector materials are summarized in Table 2.4.

The responsivity of a PIN photodetector is always less than unity and a graph showing typical responsivity values for different PIN photodetectors is depicted in Figure 2.21.

PIN photodetectors are capable of operating at very high bit rates exceeding 100 Gbps [13–15]. A high-efficiency waveguide InGaAs PIN photodiode with a bandwidth greater than 40 GHz at a responsivity of 0.55 A/W (external quantum efficiency of 44%) has been reported in Ref. [16], whereas InP/InGaAs uni-travelling-carrier photodiode with 310 GHz bandwidth at 1550 nm wavelength is demonstrated in Ref. [17]. However, the commercially available devices only offer bandwidth up to 20 GHz; this is mainly due to limitations of the packaging.

TABLE 2.4

Operating Wavelength Ranges for Different Photodetector Materials

Material	Energy Gap (eV)	Cut-Off Wavelength (nm)	Wavelength Band (nm)
Silicon	1.17	1060	400–1060
Germanium	0.775	1600	600–1600
GaAs	1.424	870	650–870
InGaAs	0.73	1700	900–1700
InGaAsP	0.75–1.35	1650–920	800–1650

Source: Adapted from G. Keiser, *Optical Fiber Communications*, 3rd ed. New York: McGraw-Hill, 2000.

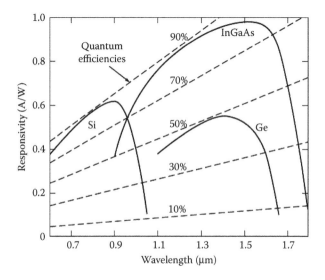

FIGURE 2.21 Responsivity and quantum efficiency as a function of wavelength for PIN photodetectors. (Reproduced from G. Keiser, *Optical Fiber Communications*, 3rd ed. New York: McGraw-Hill, 2000. With permission.)

2.4.2 APD PHOTODETECTOR

The APD is different from the PIN photodetector in that it provides an inherent current gain through the process called repeated electron ionization. This culminates in increased sensitivity since the photocurrent is now multiplied before encountering the thermal noise associated with the receiver circuit. Hence, the expression for the responsivity of an APD includes a multiplication (or gain) factor given by

$$M = I_T/I_p \tag{2.21}$$

where I_T is the average value of the total output current and $I_p = RP_r$ is the primary unmultiplied photocurrent (i.e., PIN diode case).

TABLE 2.5

Typical Performance Characteristics of Photodetectors

Parameter	Silicon		Germanium		InGaAs	
	PIN	**APD**	**PIN**	**APD**	**PIN**	**APD**
Wavelength range (nm)	400–1100		800–1800		900–1700	
Peak (nm)	900	830	1550	1300	1300 (1550)	1300 (1550)
Responsivity R (A/W)	0.6	77–130	0.65–0.7	3–28	0.63–0.8 (0.75–0.97)	
Quantum efficiency (%)	65–90	77	50–55	55–75	60–70	60–70
Gain	1	150–250	1	5–40	1	10–30
Excess noise factor	—	0.3–0.5	—	0.95–1	—	0.7
Bias voltage (−V)	45–100	220	6–10	20–35	5	<30
Dark current (nA)	1–10	0.1–1.0	50–500	10–500	1–20	1–5
Capacitance (pF)	1.2–3	1.3–2	2–5	2–5	0.5–2	0.5
Rise time (ns)	0.5–1	0.1–2	0.1–0.5	0.5–0.8	0.06–0.5	0.1–0.5

Typical gain values lie in the range 50–300 [18]; thus, the responsivity value of an APD can be greater than unity. The APD offers a higher sensitivity than the PIN detector but the statistical nature of the ionization/avalanche process means that there is always a multiplication noise associated with the APD [19]. The avalanche process is also very temperature sensitive. These factors are very important and must always be taken into account whenever an APD is used in an optical communication system. Table 2.5 summarizes the typical performance characteristics of a range of photodetectors.

2.5   PHOTODETECTION TECHNIQUES

Photodetection is the process of converting information-bearing optical radiation into its equivalent electrical signal with the aim of recovering the transmitted information. At the transmitter, the information can be encoded on the frequency, phase or the intensity of the radiation from an optical source. This encoded radiation is then transmitted to the receiver via the free-space channel or the optical fibre. The receiver front-end devices (telescope and optical filter) focus the filtered radiation onto the photodetecting surface in the focal plane. There are two possible detection schemes widely adopted in optical communications: IM-DD and coherent schemes. IM-DD is the simplest and widely used. Coherent detection schemes offer the potential of restoring full information on optical carriers, namely the amplitude (in-phase (I) component) and phase (quadrature (Q) component) of the complex optical electric field and the state of polarization of the signal. However, such receivers are sensitive to the phase and the state of polarization of the received optical signal. Table 2.6 depicts the comparison between IM-DD and coherent schemes [20].

TABLE 2.6

Comparison between IM-DD and Coherent Schemes

	IM-DD	Coherent
Modulation parameters	Intensity	I and Q components
Detection method	Direct detection	Heterodyne or homodyne detection
Adaptive control	Not needed	Needed for the carrier phase and state of polarization

2.5.1 DIRECT DETECTION

In intensity-modulation direct detection, only one degree of freedom, the intensity of the light emitted from an LD or an LED, is employed to convey the information. In direct detection scheme, a local oscillator is not used in the detection process and for this type of receiver to recover the encoded information, it is essential that the transmitted information be associated with the intensity variation of the transmitted field [18]. Hence, this type of detection is also known as the envelope detection. For an instantaneous incident power $P(t)$, the instantaneous photodetector current $i(t)$ is given by

$$i(t) = \frac{\eta_{qe}q\lambda}{hc} MP(t) \qquad (2.22)$$

where M is the photodetector gain factor whose value is unity for the PIN photodetector. The block diagram of the direct detection receiver is illustrated in Figure 2.22.

2.5.2 COHERENT DETECTION

In coherent optical communications, the optical signal is modulated by the information using amplitude, phase and frequency of the lightwave. At the receiving end, an optical local oscillator (OLO) is used and by combining the OLO with the received signal, optical heterodyne or homodyne detection is performed [21]. The frequency of the local oscillator does not have to be the same as that of the incoming information-bearing radiation. This possibility is thus responsible for the two variants of coherent detection discussed below. In coherent heterodyne detection schemes, the OLO frequency is about several gigahertz different from the optical frequency of the received optical signal. The basic block diagram of a coherent receiver is shown

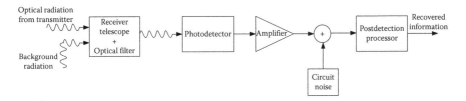

FIGURE 2.22 Block diagram of a direct detection optical receiver.

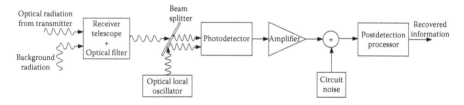

FIGURE 2.23   Block diagram of a coherent detection optical receiver.

in Figure 2.23. It is pertinent to clarify that the term 'coherent detection in optical detection' is not synonymous with coherent detection in RF parlance. In contrast to RF coherent detection, the output of the local oscillator in optical coherent detection is not required to have the same phase as the incoming radiation.

The electric fields of the received optical and the local oscillator signals are defined as

$$E_c(t) = A_c \exp[-i(\omega_0 t + \theta_c)] \tag{2.23}$$

$$E_L(t) = A_L \exp[-i(\omega_L t + \theta_L)] \tag{2.24}$$

where A_c and θ_c are the amplitude and phase of the carrier field, respectively, while the local oscillator amplitude and phase are A_L and θ_L in that order.

The generated photocurrent at the output of the photodetector, which operates as a square-law device, is given by

$$i_{IF-c}(t) = RM\left(E_c + E_L\right)^2 \tag{2.25}$$

Since the optical power is proportional to the intensity, the received optical power is defined by

$$P(t) = P_c + P_L + 2\sqrt{P_c P_L}\,\cos(\omega_{IF} t + \theta_c - \theta_L) \tag{2.26a}$$

where

$$P_c = KA_c^2, \quad P_L = KA_L^2 \quad \text{and} \quad \omega_{IF} = \omega_0 - \omega_L \tag{2.26b}$$

Depending on whether the intermediate frequency ω_{IF} is equal to zero or not, there are two different coherent detection schemes known as heterodyne and homodyne, as outlined below.

2.5.2.1   Heterodyne Detection

In a heterodyne detection optical receiver, the incoming radiation (carrier) is combined with a reference wave from the OLO (i.e., a laser) on the photodetector surface as shown in Figure 2.23. This optical mixing process produces another wave at the intermediate frequency ω_{IF} by the square-law characteristics of the photodetector [21]. This ω_{IF}, which is the difference between the incoming laser carrier and the reference OLO signal frequencies, passes through a band-pass filter to an electrical second detector (the postdetection processor in Figure 2.23) for the final demodulation. When the instantaneous field amplitudes $E_c(t)$ and $E_L(t)$ in Equations 2.23 and 2.24 combine on the photodetector surface, they produce an instantaneous signal whose intensity is given by

$$C(t) = \left(E_c(t) + E_L(t)\right)^2 \tag{2.27}$$

The time average of $C(t)$ multiplied by the responsivity gives the resultant instantaneous carrier and local oscillator current $i_p(t)$, at the photodetector output.

Hence

$$i_p(t) = R\left\{\left\langle A_L^2 \cos^2(\omega_L t + \theta_L)\right\rangle + \left\langle A_c^2 \cos^2(\omega_{co} t + \theta_c)\right\rangle + \left\langle A_L A_c \cos\left[(\omega_L - \omega_{co})t\right.\right.\right.$$
$$\left. + (\theta_L - \theta_c)\right]\rangle + \left\langle A_L A_c \cos\left[(\omega_L + \omega_{co})t + (\theta_L + \theta_c)\right]\right\rangle\right\} \tag{2.28}$$

The first two terms are time invariant, the third term is very slowly varying with respect to the short time over which the average is taken, while the fourth term is out of the IF band [21]. Equation 2.28 is now reduced to

$$i_p(t) = R\left\{\frac{A_L^2}{2} + \frac{A_c^2}{2} + \text{3rd term} + \text{4th term}\right\} \tag{2.29}$$

Therefore, the IF filter, which is an integral part of the postdetection processor of Figure 2.23, only allows the third term to go through while the others are suppressed, resulting in the following expression for the instantaneous current:

$$i_p(t) = RA_L A_c \cos[(\omega_L - \omega_{co})t + (\theta_L - \theta_c)] \tag{2.30}$$

The above result makes it possible to recover any information impressed on the carrier field amplitude, frequency or even phase. The following points can therefore be deduced from this that heterodyne detection offers

1. A relatively easy means of amplifying the photocurrent by simply increasing the local oscillator power.
2. Improved SNR. This is achieved by increasing the local oscillator power so much that its inherent shot noise dwarfs the thermal and the shot noise from other sources.

However, the frequency of an optical source is known to drift over time. Therefore, the ω_{IF} needs to be continually monitored at the input of the electrical detector and the local oscillator frequency varied accordingly to keep the IF centre frequency constant. Also, the optical source, particularly a laser, does suffer from phase noise which means that θ_c and θ_L in Equation 2.30 are not absolutely fixed; they fluctuate. These factors contribute to the challenges of implementing a coherent optical communication system.

2.5.2.2 Homodyne Detection

This is similar to the heterodyne detection process discussed above except that the OLO has the same frequency and phase as the incoming optical radiation/carrier, so that the modulated light signal can be directly demodulated at the photodetector into the baseband signal for further processing [21]. The resultant photocurrent thus contains the information signal at the baseband. Following the same step as in heterodyne detection, the instantaneous photocurrent is obtained thus

$$i_p(t) = R\left\{\frac{A_L^2}{2} + \frac{A_c^2}{2} + \left\langle A_L A_c \cos(\theta_L - \theta_c)\right\rangle + \left\langle A_L A_c \cos(2\omega_{co} t + \theta_L + \theta_c)\right\rangle\right\} \quad (2.31)$$

The third term is time invariant and the fourth term is suppressible via filtering to obtain

$$i_p(t) = R\left\{\frac{A_L^2}{2} + \frac{A_c^2}{2} + A_L A_c \cos(\theta_L - \theta_c)\right\} \quad (2.32)$$

By increasing the locally generated radiation power such that $A_c A_L \gg 0.5 A_c$ [21], expression (2.32) reduces to

$$i_p(t) = R[A_L A_c \cos(\theta_L - \theta_c)] \quad (2.33)$$

In coherent detection systems, the noise in the receiver is mainly dominated by OLO-induced shot noise [22]. The shot noise-limited receivers offer improved sensitivity, by up to 20 dB, compared to the IM-DD schemes [23]. In IM-DD systems, the optical carrier signals must be aligned at a large spacing in the optical wavelength domain. This is because of the bandwidth of the optical band-pass filter (OBPF), which is 2–3 nm. Therefore, the spacing between the optical carriers should be no less than several nanometres, which corresponds to hundreds to thousands of gigahertz. By using coherent schemes, the optical carrier signal could be aligned closely at a spacing 10 times or more than the data rate in the frequency domain; thus the possibility of employing frequency division multiplexing.

2.6 PHOTODETECTION NOISE

The noise sources as well as the frequency and distortion performance of a wireless optical link are decisive factors in determining the link performance. In line

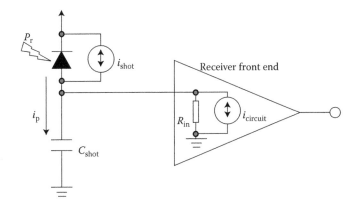

FIGURE 2.24 Diagram of a front-end photodiode detector along with channel impairments.

with nearly all communication systems, identification of the noise sources at the receiver front end, where the incoming signal contains the least power, is critical. The two primary sources of noise at the input of the receiver are due to shot noise from the received photocurrent and noise from the receiver electronics. Figure 2.24 illustrates a schematic of a receiver front end with photodetector as well as noise sources indicated. Photogenerated shot noise is a major noise source in the wireless optical link. It arises fundamentally due to the discrete nature of energy and charge in the photodiode. Thermal noise, due to the resistive element, is generated independently of the received optical signal and has a Gaussian distribution. Thermal noise is shaped by a transfer function of the receiver (i.e., the topology of the preamplifier). Thus, the circuit noise is also modelled as being Gaussian distributed.

The various sources of noise in optical communications are discussed below. For FSO systems, the noise from the background radiation can be significant, while in fibre optics communication, the background radiation noise is negligible.

2.6.1 PHOTON FLUCTUATION NOISE

For an ideal photodetector, the only significant noise that affects its performance is that associated with the quantum nature of light itself, the by-product of which is that the number of photons emitted by a coherent optical source in a given time is never constant. Although for a constant power optical source, the mean number of photons generated per second is constant, the actual number of photons per second follows the Poisson distribution. This results in photon fluctuation or quantum noise. Quantum noise (also termed photon noise) is a shot noise that is present in all photon detectors due to the random arrival rate of photons from the data carrying optical source and background radiation.

The quantum fluctuation is also important because it dominates over the thermal fluctuations within the photodetector, since $hf > \kappa T_e$ where h and f are the Planck's constant and the radiation frequency, respectively, while κ and T_e represent the

Boltzmann's constant and temperature, respectively. The quantum noise is a shot noise with variance

$$\sigma^2_{q-\text{pin}} = 2q\langle i\rangle B \qquad\qquad (2.34)$$

$$\sigma^2_{q-\text{adp}} = 2q\langle i\rangle BFM^2 \qquad\qquad (2.35)$$

where the bandwidth of the electrical filter that follows the photodetector is represented by B Hz. The shot noise is in fact proportional to $(n_{\text{photon}})^{0.5}$, where n_{photon} is the equivalent number of photons after degradation caused by imperfect conversion due to the quantum efficiency.

For coherent receivers, the shot noise due to the optical signal and the OLO is given by

$$\sigma^2_{q-\text{CS}} = 2q\langle i_s\rangle M^2 B_c \qquad\qquad (2.36)$$

$$\sigma^2_{q-\text{CLO}} = 2q\langle i_{lo}\rangle M^2 B_c \qquad\qquad (2.37)$$

where B_c is the bandwidth of the coherent receiver.

For the heterodyne case, the LO shot noise (2.37) is much larger than the signal shot noise (2.36); therefore, the signal shot noise can be ignored.

2.6.2 DARK CURRENT AND EXCESS NOISE

Since the photocurrent is proportional to the incident light power, the photon shot noise increases with respect to the increase of the incident power (in square root fashion). The lower end of this relation is limited by the noise from the dark current, which is present even when there is no input light. It is produced by the transition of electrons from the valence to the conduction band due to causes other than photon-induced excitation; its magnitude is closely related to the energy band-gap of the photodetector material(s). Large band-gap materials, such as silicon (Si), indium phosphide (InP) and gallium arsenide (GaAs), show very low values of mean dark current, $\langle i_d\rangle$, while for germanium (Ge), the value could be significant when they are operated at room temperature [24]. The dark current is a combination of bulk and surface leakage currents, carries no useful information and thereby constitutes a shot noise whose variances are given by the following

$$\sigma^2_{\text{db}} = 2qI_d M^2 FB \qquad\qquad (2.38)$$

$$\sigma^2_{\text{ds}} = 2qI_l B \qquad\qquad (2.39)$$

where I_d and I_l are the detector's primary unmultiplied dark and the surface leakage currents, respectively.

TABLE 2.7
Dark Current Values for Different Materials

	Dark Current (nA)	
Photodetector Material	**PIN**	**APD**
Silicon	1–10	0.1–1
Germanium	50–500	50–500
InGaAs	0.5–2.0	10–50 at gain = 10

Source: Adapted from G. Keiser, *Optical Communications Essentials*,
1st ed. New York: McGraw-Hill Professional, 2003.

For coherent receivers, the dark current noise is given by

$$\sigma_{D-C}^2 = 2qI_d M^2 B_c \tag{2.40}$$

The dark current consists of diffusion, tunnel, leakage currents and generation–recombination taking place in the space–charge region and is proportional to the volume of the depletion region [24]. Typical dark current values for some photodetector materials are shown in Table 2.7.

Photodetectors that employ internal avalanche gain mechanism to boost the signal above the thermal noise of amplifier stages on the receiver exhibit what is referred to as excess noise [26]. According to Ref. [26], the excess noise in an APD is due to the multiplication process in the high-field region of the detector where each primary electron hole can generate an additional electron through impact ionization of bound electrons. These additional carriers can then create still additional carriers in a cascading process. The excess noise increases with the gain. Since the gain exhibits the wavelength dependence, the excess noise differs according to the incident wavelength. Similarly, the signal-generated photocurrent is also amplified by the gain, thus illustrating that the best SNR is obtained at a certain gain.

If all primary carriers were to be multiplied equally in an APD, the mean-square current gain $\langle g^2 \rangle$ would be equal to the mean gain g, and the excess noise, defined as $F = \langle g^2 \rangle / g$ would be equal to 1. This is the case for PIN photodetectors [26]. However, due to the statistical nature of the avalanche process, F is always greater than 1 in an APD and other avalanche devices.

The amount of optical power incident on the surface of a photodetector that produces a signal at the output of the detector, which is just equal to the noise generated internally by the photodetector, is defined as the NEP. The NEP is roughly the minimum detectable input power of a photodiode. Another equivalent way of stating NEP is to take the ratio of the total noise current to the responsivity at a particular wavelength. For a photodetector in which the dark current is the dominant noise source, the expression for NEP is given by [5]

$$NEP = \frac{hc(2qi_d)^{0.5}}{\eta_{qe}q\lambda} \tag{2.41}$$

2.6.3 BACKGROUND RADIATION

This type of noise is due to the detection of photons generated by the environment. Two types of sources contribute to background radiation (ambient light) noise: localized point sources (e.g., the Sun) and extended sources (e.g., the sky). Background radiation from other celestial bodies such as stars and reflected background radiation are assumed to be too weak to be considered for a terrestrial FSO link; however, they contribute significantly to background noise in deep space FSO. The following are the irradiance (power per unit area) expressions for both the extended and localized background sources [21,27,28]

$$I_{sky} = N(\lambda)\Delta\lambda\pi\,\Omega^2/4 \tag{2.42}$$

$$I_{sun} = W(\lambda)\Delta\lambda \tag{2.43}$$

where $N(\lambda)$ and $W(\lambda)$ are the spectral radiance of the sky and spectral radiant emittance of the sun, respectively, $\Delta\lambda$ is the bandwidth of the OBPF that precedes the photodetector and Ω is the photodetector's field of view angle in radians. By carefully choosing a receiver with a very narrow FOV and $\Delta\lambda$, the impact of background noise can be greatly reduced. OBPF in the form of coatings on the receiver optics/telescope with $\Delta\lambda < 1$ nm are now readily available. Empirical values of $N(\lambda)$ and $W(\lambda)$ under different observation conditions are also available in the literature [18,21,27]. The background radiation is a shot noise with variance [21]

$$\sigma_{bg}^2 = 2qBR(I_{sky} + I_{sun}) \tag{2.44}$$

For a coherent receiver, use Equation 2.44 and change B to B_c. In most practical systems, the receiver SNR is limited by the background *shot noise* that is much stronger than the quantum noise and/or by thermal noise in the electronics following the photodetector.

2.6.4 THERMAL NOISE

Thermal noise, also known as the Johnson noise, occurs in all conducting materials. It is caused by the thermal fluctuation of electrons in any receiver circuit of equivalent resistance R_L and temperature T_e. The electrons are in constant motion, but they collide frequently with the atoms or molecules of the substance. Every free flight of an electron constitutes a minute current. The sum of all these currents taken over a long period of time must, of course, be equal to zero. The thermal noise is regarded as a 'white' noise. This is because the power spectral density (PSD) is independent of frequency. Moreover, the thermal noise obeys the Gaussian distribution with zero mean and variance for IM-DD and coherent receivers defined by [25]

$$\sigma_{th-D}^2 = \frac{4\kappa T_e B}{R_L} \tag{2.45}$$

$$\sigma_{\text{th-CS}}^2 = \frac{4\kappa T_e B_c}{R_L} \tag{2.46}$$

2.6.5 Intensity Noise

This is generated by the amplitude fluctuation of the optical signal, and is usually expressed in terms of relative intensity noise (RIN) given by

$$\sigma_{\text{in-D}}^2 = \eta_{\text{RIN}} (RMP_r)^2 B \tag{2.47}$$

$$\sigma_{\text{in-C}}^2 = C\eta_{\text{RIN}} (RMP_r)^2 B_c \tag{2.48}$$

where C is the common mode rejection ratio for the balanced receiver.

For a typical LD, η_{RIN} is about −160 to −150 dB/Hz.

For heterodyne receivers, the influence of RIN can be avoided by employing a balanced photodiode configuration as shown in Figure 2.25. In this scheme, the signals from the photodetectors are subtracted and thus result in cancellation of RIN [17]. Thus

$$I_{p1}(t) = 0.5R[P_c + P_L + 2\sqrt{P_c P_L} \cos(\omega_{\text{IF}} t + \theta_c - \theta_L)] \tag{2.49}$$

$$I_{p2}(t) = 0.5R[P_c + P_L - 2\sqrt{P_c P_L} \cos(\omega_{\text{IF}} t + \theta_c - \theta_L)] \tag{2.50}$$

$$\Delta I_p(t) = I_{p1} - I_{p2} = 2R\sqrt{P_c P_L} \cos(\omega_{\text{IF}} t + \theta_c - \theta_L)] \tag{2.51}$$

It is commonly used in applications where there is a requirement for higher SNR. It is very effective in being able to cancel the RIN or 'common mode noise' in laser diodes and has the ability to detect small signal fluctuations on a large DC signal.

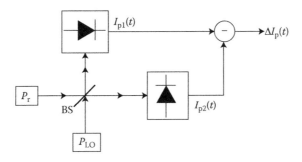

FIGURE 2.25   Balanced photodiode configuration, BS: beam splitter.

2.6.6 Signal-to-Noise Ratio

The SNR for IM-DD and coherent receivers is defined as

$$\text{SNR}_{\text{IM-DD}} = \frac{I_p^2}{N_T} = \frac{(RMP_r)^2}{\sigma_q^2 + \sigma_d^2 + \sigma_{bg}^2 + \sigma_{th}^2 + \sigma_{in}^2} \tag{2.52}$$

In the quantum limit (the ultimate limit), the SNR is defined as

$$\text{SNR}_{\text{IM-DD}} = \frac{RP_r}{2qB} \tag{2.53}$$

For a coherent receiver with a sufficiently large OLO power, the shot noise due to the OLO is the dominant term; thus

$$\text{SNR}_{\text{IM-C}} = \frac{RP_r}{2MB_c} \tag{2.54}$$

For PIN photodetector, $M = 1$.

2.7 OPTICAL DETECTION STATISTICS

According to the semiclassical approach, which treats an optical radiation as a wave and prescribes a probabilistic relation to account for its interaction with the atomic structure of the detector surface [18], the probability of a detector with an aperture area A_D, emitting n number of electrons from impinging photons during a period T obeys the Poisson distribution given as

$$p(n) = \frac{\langle n \rangle^n \exp(-\langle n \rangle)}{n!} \tag{2.55}$$

The mean count is related to the aperture area and the received irradiance $I(t,r)$ by the following expression

$$\langle n \rangle = \frac{\eta\lambda}{hc} \iint I(t,r)\,dt\,dr \tag{2.56}$$

The process of counting the number of electrons generated by the impinging photons is often referred to as photon or photoelectron counting.

Table 2.8 summarizes some basic statistical parameters of the Poisson random variable. It is worth noting that both the mean and variance of a Poisson random distribution are the same and are given by the mean count in this instance.

TABLE 2.8
Statistical Parameters of Poisson Random Distribution

Parameter	Definition	
Mean	$\displaystyle\sum_n np(n)$	$\langle n \rangle$
Mean-square value	$\displaystyle\sum_n n^2 p(n)$	$\langle n \rangle^2 + \langle n \rangle$
Characteristic function	$\displaystyle\sum_n e^{j\omega n} p(n)$	$\exp\left[(e^{j\omega}-1)\langle n\rangle\right]$
Moment generating function	$\displaystyle\sum_n (1-z)^n p(n)$	$\exp[-z\langle n\rangle]$
qth moment	$\displaystyle\sum_n n^q p(n)$	$\dfrac{\partial^q}{\partial z^q}[\exp(-z\langle n\rangle)]_{z=1}$

Source: Adapted from R. M. Gagliardi and S. Karp, *Optical Communications*, 2nd ed. New York: John Wiley & Sons Inc., 1995.

The mean current generated from the $\langle n \rangle$ electrons is $\langle i \rangle = q\langle n\rangle/T$ and its variance is given by

$$\sigma^2 = \left(\frac{q}{T}\right)^2 \sigma_n^2 \tag{2.57}$$

Since for a Poisson distribution the mean and variance are equal, and by choosing the electrical bandwidth of the postdetection filter as $1/2T$ (which is the Nyquist minimum bandwidth requirement), expression (2.40) becomes

$$\sigma^2 = \left(\frac{q}{T}\right)^2 \langle n\rangle = \left(\frac{q}{T}\right)^2 \frac{\langle i\rangle T}{q} = \frac{q}{T}\langle i\rangle = 2q\langle i\rangle B \tag{2.58}$$

This is the general expression for the variance of any shot noise process associated with photodetection. It should be mentioned that the photon counts from the desired incoming optical radiation, background radiation and that due to dark current are independent Poisson random variables. Hence, the probability density function (pdf) of the photoelectron emission due to all these processes occurring together is also a Poisson distribution whose mean is the sum of the means of the individual processes.

With large signal photoelectron counts, the generated signal current probability distribution can be approximated to be Gaussian [21]. That is

$$p(i) = \frac{1}{\sqrt{2\pi\sigma_{sh}^2}} \exp\left\{-\frac{(i - \langle i\rangle)^2}{2\sigma_{sh}^2}\right\} \tag{2.59}$$

By taking the additive white Gaussian noise (AWGN) into account, this expression is modified to

$$p(i) = \frac{1}{\sqrt{2\pi\left(\sigma_{sh}^2 + \sigma_{bg}^2\right)}} \exp\left\{-\frac{(i - \langle i \rangle)^2}{2\left(\sigma_{sh}^2 + \sigma_{bg}^2\right)}\right\} \tag{2.60}$$

Expressions (2.55), (2.59) and (2.60) are only valid for a nonvarying received optical field. However, if the incoming field is randomly varying, then the received irradiance and $\langle i \rangle$ also vary accordingly. Assuming the randomly varying field has a pdf given by $p(\langle i \rangle)$, it follows therefore that the generated photocurrent is now doubly stochastic and its pdf given by Equation 2.61 is obtained by averaging Equation 2.60 over the statistics of the varying field.

$$p(i) = \int_0^{\infty} \frac{1}{\sqrt{2\pi\left(\sigma_{sh}^2 + \sigma_{bg}^2\right)}} \exp\left\{-\frac{(i - \langle i \rangle)^2}{2\left(\sigma_{sh}^2 + \sigma_{bg}^2\right)}\right\} P(\langle i \rangle) \mathrm{d}\langle i \rangle \tag{2.61}$$

In FSO communications, atmospheric turbulence is the principal source of random fluctuations in the received optical radiation irradiance. This source of fluctuation is therefore the subject of discussion in the next chapter.

REFERENCES

1. S. M. Sze and K. K. Ng, *Physics of Semiconductor Devices*, 3rd ed. Hoboken, New Jersey: John Wiley & Sons Inc., 2007.
2. J. M. Senior, *Optical Fiber Communications Principles and Practice*, 3rd ed. Essex: Pearson Education Limited, 2009.
3. J. L. Miller, *Principles of Infrared Technology: A Practical Guide to the State of the Art*, New York: Springer, 1994.
4. R. Ramaswami, Optical fiber communication: From transmission to networking, *Communications Magazine, IEEE*, 40, 138–147, 2002.
5. C. C. Davis, *Lasers and Electro-Optics: Fundamentals and Engineering*, Cambridge, UK: Cambridge University Press, 1996.
6. D. C. O'Brien, G. E. Faulkner, K. Jim, E. B. Zyambo, D. J. Edwards, M. Whitehead, P. Stavrinou et al., High-speed integrated transceivers for optical wireless, *IEEE Communications Magazine*, 41, 58–62, 2003.
7. M. Czaputa, T. Javornik, E. Leitgeb, G. Kandus and Z. Ghassemlooy, Investigation of punctured LDPC codes and time-diversity on free-space optical links, *Proceedings of the 2011 11th International Conference on Telecommunications (ConTEL)*, Graz, Austria, 2011, pp. 359–362.
8. B. James (Ed.). *Proceedings of the NATO Advanced Study Institute on Unexploded Ordnance Detection and Mitigation II Ciocco, Italy, 20 July–2 August 2008, Book Series: NATO Science for Peace and Security Series, Subseries: NATO Science for Peace and Security, Series B: Physics and Biophysics*, Dordrecht, Netherlands: Springer, 2009.
9. W. V. Etten and J. V. D. Plaats, *Fundamentals of Optical Fiber Communications*, London: Prentice Hall International, 1991.

10. A. A. Bergh and J. A. Copeland, Optical sources for fiber transmission systems, *Proceedings of the IEEE*, 68, 1240–1247, 1980.

11. J. Hecht, *Understanding Fiber Optics*, 5th ed. Upper Saddle River, New Jersey: Prentice Hall, 2005.

12. G. Keiser, *Optical Fiber Communications*, 3rd ed. New York: McGraw-Hill, 2000.

13. K. Kishino, M. S. Unlu, J. I. Chyi, J. Reed, L. Arsenault and H. Morkoc, Resonant cavity-enhanced (RCE) photodetectors, *IEEE Journal of Quantum Electronics*, 27, 2025–2034, 1991.

14. W. Yih-Guei, K. S. Giboney, J. E. Bowers, M. J. W. Rodwell, P. Silvestre, P. Thiagarajan and G. Y. Robinson, 108 GHz GaInAs/InP p-i-n photodiodes with integrated bias tees and matched resistors, *IEEE Photonics Technology Letters*, 5, 1310–1312, 1993.

15. K. Kato, Ultrawide-band/high-frequency photodetectors, *IEEE Transactions on Microwave Theory and Techniques*, 47, 1265–1281, 1999.

16. K. Kato, S. Hata, A. Kozen, J. I. Yoshida and K. Kawano, High-efficiency waveguide InGaAs pin photodiode with bandwidth of over 40 GHz, *IEEE Photonics Technology Letters*, 3, 473–474, 1991.

17. H. Ito, T. Furuta, S. Kodama and T. Ishibashi, InP/InGaAs uni-travelling-carrier photodiode with 310 GHz bandwidth, *Electronics Letters*, 36, 1809–1810, 2000.

18. R. M. Gagliardi and S. Karp, *Optical Communications*, 2nd ed. New York: John Wiley & Sons Inc., 1995.

19. K. Shiba, T. Nakata, T. Takeuchi, T. Sasaki and K. Makita, 10 Gbit/s asymmetric waveguide APD with high sensitivity of −30 dBm, *Electronics Letters*, 42, 1177–1178, 2006.

20. K. Kikuchi, Coherent optical communications: Historical perspectives and future directions, *High Spectral Density Optical Communication Technologies*, M. Nakazawa, K. Kikuchi, and T. Miyazaki, eds., Berlin: Springer-Verlag, 2010, pp. 11–49.

21. W. K. Pratt, *Laser Communication Systems*, 1st ed. New York: John Wiley & Sons Inc., 1969.

22. T. Okoshi and K. Kikuchi, *Coherent Optical Fiber Communications*, Dordrecht, Holland: Kluwer Academic Publishers, 1988.

23. S. Rye, *Coherent Lightwave Communication Systems*, Boston, London: Artech House Publishers, 1995.

24. S. Betti, G. De Marchis and E. Iannone, *Coherent Optical Communication Systems*, 1st ed. Toronto, Canada: John Wiley & Sons Inc., 1995.

25. G. Keiser, *Optical Communications Essentials*, 1st ed. New York: McGraw-Hill Professional, 2003.

26. G. R. Osche, *Optical Detection Theory for Laser Applications*, Hoboken, New Jersey: John Wiley & Sons Inc., 2002.

27. N. S. Kopeika and J. Bordogna, Background noise in optical communication systems, *Proceedings of the IEEE*, 58, 1571–1577, 1970.

28. S. Karp, E. L. O'Neill and R. M. Gagliardi, Communication theory for the free-space optical channel, *Proceedings of the IEEE*, 58, 1626–1650, 1970.

3 Channel Modelling

To design, implement and operate efficient optical communication systems, it is imperative that the characteristics of the channel are well understood. Characterization of a communication channel is performed by its channel impulse response, which is then used to analyse and combat the effects of channel distortions. A considerable amount of work has been published on the channel characterization, covering both experimental measurement and computer modelling of indoor and outdoor systems. The power penalties directly associated with the channel may be separated into two factors, these being optical path loss and multipath dispersion. Two types of configurations are considered in an optical wireless channel as outlined in Chapter 1. For directed LOS and tracked configurations, reflections do not need to be taken into consideration, and consequently the path loss is easily calculated from knowledge of the transmitter beam divergence, receiver size and separation distance. However, a non-LOS configuration, also known as diffuse systems (mainly used in indoor environment), uses reflections off the room surfaces and furniture. These reflections could be seen as unwanted signals or multipath distortions which make the prediction of the path loss more complex. A number of propagation models (ceiling bounce, Hayasaka–Ito and spherical) for LOS and non-LOS are introduced in this chapter. The artificial light interference that affects the link performance is also outlined in this chapter. The atmospheric outdoor channel is a very complex and dynamic environment that can affect characteristics of the propagating optical beam, thus resulting in optical losses and turbulence-induced amplitude and phase fluctuation. There are a number of models to characterize the statistical nature of the atmospheric channel, and these will be discussed in this chapter. A practical test bed for investigating the atmospheric effect on the free space optics link, as well as measured data is also covered in this chapter.

3.1 INDOOR OPTICAL WIRELESS COMMUNICATION CHANNEL

As shown in Figure 1.7, there are a number of topologies that are commonly used for indoor applications. The configurations can be classified according to (1) the degree of directionality of transmitter and receiver and (2) the existence of the LOS path between the transmitter and the receiver [1–3]. Intensity modulation with direct detection (IM/DD) is the de-facto method of implementing optical wireless systems principally due to its reduced cost and complexity [1,4,5]. The drive current of an optical source is directly modulated by the modulating signal $m(t)$, which in turn varies the intensity of the optical source $x(t)$ (see Figure 3.1). The receiver employs a photodetector, with a response which is the integration of tens of thousands of very short wavelengths of the incident optical signal, that generates a photocurrent $y(t)$. This photocurrent is directly proportional to the instantaneous optical power incident on it, that is, proportional to the square of received electric field. An IM/DD-based optical wireless system has an equivalent baseband model that hides the high-frequency

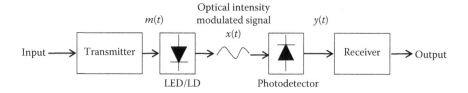

FIGURE 3.1 Block diagram of an optical intensity, direct detection communications channel.

FIGURE 3.2 Equivalent baseband model of an optical wireless system using IM/DD.

nature of the optical carrier [6]. The model is shown in Figure 3.2 where R is the photodetector responsivity, $h(t)$ is the baseband channel impulse response and $n(t)$ is the signal-independent shot noise, modelled throughout the book as the additive white Gaussian noise (AWGN) with a double-sided power spectral density (PSD) of $N_0/2$.

Non-LOS links, particularly in indoor applications, are subject to the effects of multipath propagation in the same way as RF systems and these effects are more pronounced. This type of link can suffer from severe multipath-induced performance penalties, as will be discussed in later chapters. Multipath propagation causes the electric field to suffer from severe amplitude fades on the scale of a wavelength. The detector would experience multipath fading if the detector size (i.e., the surface area) was proportional to one wavelength or less. Fortunately, OWC receivers use detectors with a surface area typically millions of square wavelengths. In addition, the total photocurrent generated is proportional to the integral of the optical power over the entire photodetector surface; this provides an inherent spatial diversity as shown in Figure 3.3 [1].

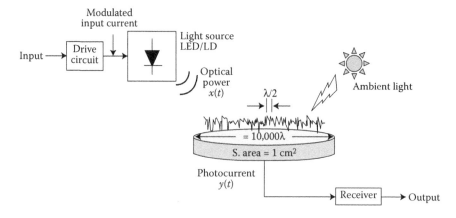

FIGURE 3.3 Equivalent baseband model of an optical wireless system using IM/DD.

Although indoor OWC links do not suffer from the effects of multipath fading, they do suffer from the effects of dispersion, which manifests itself in a practical sense as the intersymbol interference (IS). Dispersion is modelled as a linear baseband channel impulse response $h(t)$. The channel characteristic of an OWC link is fixed for a given position of transmitter, receiver and intervening reflecting objects. The channel characteristic only changes when these components are moved by distances of the order of centimetres [1]. Due to high bit rates, and the relatively slow movement of objects and people within a room, the channel will vary only on the time scale of many bit periods, and may therefore be considered as quasi-static [2].

The equivalent baseband model of an IM/DD optical wireless link can be summarized by the following equations

$$y(t) = Rx(t) \otimes h(t) + n(t)$$

$$= \int_{-\infty}^{\infty} Rx(\tau)h(t-\tau)d + n(t) \qquad (3.1)$$

where the symbol $\otimes$ denotes the convolution.

The impulse response $h(t)$ can be used to analyse or simulate the effects of multipath dispersion in indoor OWC channels. The channel impulse response was modelled by Gfeller and Bapst as follows [7]

$$h(t) = f(x) = \begin{cases} \dfrac{2t_0}{t^3 \sin^2(FOV)} & t_0 \leq t \leq \dfrac{t_0}{\cos(FOV)} \\ 0 & \text{elsewhere} \end{cases} \qquad (3.2)$$

where t_0 is the minimum delay.

While Equation 3.1 is simply a linear filter channel with AWGN, optical wireless systems differ from conventional electrical or radio systems since the instantaneous optical power is proportional to the generated electrical current. $x(t)$ represents the power rather than the amplitude signal. This imposes two constraints on the transmitted signal. Firstly, $x(t)$ must be nonnegative, that is

$$x(t) \geq 0 \qquad (3.3)$$

Secondly, the eye safety requirements limits the maximum optical transmit power that may be used. Generally, it is the average power requirement which is the most restrictive and hence, the average value of $x(t)$ must not exceed a specified maximum power value P_{max}, that is

$$P_{max} = \operatorname*{Lim}_{T \to \infty} \frac{1}{2T} \int_{-T}^{T} x(t) dt \qquad (3.4)$$

This is in contrast to the time-averaged value of the signal $|x(t)|^2$, which is the case for the conventional RF channels when $x(t)$ represents amplitude.

These differences have a profound effect on the system design. On conventional RF channels, the signal-to-noise ratio (SNR) is proportional to the average received power, whereas on optical wireless links, it is proportional to the square of the average received optical signal power given by

$$\text{SNR} = \frac{R^2 H^2(0) P_r^2}{R_b N_0} \tag{3.5}$$

where N_0 is the noise spectral density and $H(0)$ is the channel DC gain given by

$$H(0) = \int_{-\infty}^{\infty} h(t)\,dt \tag{3.6}$$

Thus, in optical systems relatively high optical transmit powers are required, and only a limited path loss can be tolerated. The fact that the average optical transmit power is limited suggests that modulation techniques possessing a high peak-to-mean power ratio are most favourable. This is generally achieved by trading off the power efficiency against the bandwidth efficiency. When shot noise is dominant, the SNR is also proportional to the photodetector area because the received electrical power and the variance of the shot noise are proportional to A_d^2 and A_d, respectively, where A_d is the receiver detector area. Thus, single element receivers favour the use of large-area detectors. However, as the detector area increases so does its capacitance, which has a limiting effect on receiver bandwidth, thus the transmission capacity. This is in direct conflict with the increased bandwidth requirement associated with power efficient modulation techniques, and hence, a trade-off exists between these two factors.

The optical wireless channel transfer function is defined by

$$H_{\text{OW}}(f) = H_{\text{los}} + H_{\text{diff}}(f) \tag{3.7}$$

where H_{los} is the contribution due to the LOS, which is basically independent on the modulation frequency, and it depends on the distance between transmitter and receiver and on their orientation with respect to the LOS, whereas H_{diff} is almost homogeneous and isotropic in most rooms.

In a directed link the power ratio between the LOS and the diffuse links can be increased by reducing the transmitter beam width and/or the receiver FOV. This can be quantified by the Rician factor

$$k_{\text{rf}} = \left(\frac{H_{\text{los}}}{H_{\text{diff}}} \right)^2 \tag{3.8}$$

As in radio communications it is defined via the electrical signal powers.

3.1.1 LOS PROPAGATION MODEL

In general, the indoor OWC system uses an LED as a source and large-area photode-tectors. The angular distribution of the radiation intensity pattern is modelled using a generalized Lambertian radiant intensity with the following distribution

$$R_0(\phi) = \begin{cases} \dfrac{(m_1+1)}{2\pi} \cos^{m_1}(\phi) & \text{for } \Phi \in [-\pi/2, \pi/2] \\ 0 & \text{for } \phi \geq \pi/2 \end{cases} \qquad (3.9)$$

where m_1 is the Lambert's mode number expressing directivity of the source beam, $\phi = 0$ is the angle of maximum radiated power. The order of Lambertian emission m_1 is related to the LED semiangle at half-power $\Phi_{1/2}$ by

$$m_1 = \frac{-\ln 2}{\ln(\cos \Phi_{1/2})} \qquad (3.10)$$

The radiant intensity is given by

$$S(\phi) = P_t \frac{(m_1+1)}{2\pi} \cos^{m_1}(\phi) \qquad (3.11)$$

The detector is modelled as an active area A_r collecting the radiation incident at angles ψ smaller than the detector FOV. The effective collection area of the detector is given by

$$A_{\text{eff}}(\psi) = \begin{cases} A_r \cos\psi & 0 \leq \psi \leq \pi/2 \\ 0 & \psi > \pi/2 \end{cases} \qquad (3.12)$$

Although ideally a large-area detector would be suitable for indoor OWC to col-lect as much power as possible, it would in practice cause a number of problems. Such as increased manufacture cost, increased junction capacitance and thus a decreased receiver bandwidth, and increased receiver noise. Hence, the use of non-imaging concentrator is a cost-effective solution in order to increase overall effective collection area. The optical gain of an ideal nonimaging concentrator having internal refractive index n is

$$g(\psi) = \begin{cases} \dfrac{n^2}{\sin^2 \Psi_c} & 0 \leq \psi \leq \Psi_c \\ 0 & \psi > \Psi_c \end{cases} \qquad (3.13)$$

where $\Psi_c \leq \pi/2$ is the FOV.

From the constant radiance theorem, the FOV of the receiver system is related to the collection area of lens A_{coll} and the photodetector area as [8]

$$A_{coll} \sin\left(\frac{\Psi_c}{2}\right) \le A_r \tag{3.14}$$

From Equation 3.14 it is apparent that the concentrator gain increases when the FOV is reduced.

The link length for indoor OWC is relatively short and hence attenuation due to the absorption and scattering is very low. Considering an OWC link with a Lambertian source, a receiver with an optical band-pass filter of transmission $T_s(\psi)$ and a non-imaging concentrator of gain $g(\psi)$, the DC gain for a receiver located at a distance of d and angle ϕ with respect to transmitter (see Figure 3.4) can be approximated as [1]

$$H_{los}(0) = \begin{cases} \dfrac{A_r(m_1+1)}{2\pi d^2} \cos^{m_1}(\phi) T_s(\psi) g(\psi) \cos\psi & 0 \le \psi \le \Psi_c \\ 0 & \text{elsewhere} \end{cases} \tag{3.15}$$

The received power therefore becomes

$$P_{r-los} = H_{los}(0) P_t \tag{3.16}$$

In the LOS link when transmitter and receiver are aligned the increase of the LOS signal is given by

$$H_{los}(m_1) = \frac{(m_1+1)}{2} H_{los} \tag{3.17}$$

where H_{los} refers to a Lambertian transmitter with $m_1 = 1$.

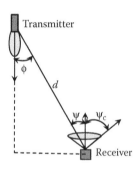

FIGURE 3.4   Geometry LOS propagation model.

In short-distance LOS links, multipath dispersion is seldom a problem and LOS links channel are often modelled as a linear attenuation and delay [6]. The optical LOS links are considered as nonfrequency selective and the path loss depends on the square of distance between the transmitter and the receiver. The impulse response can be expressed as

$$h_{\mathrm{los}}(t) = \frac{A_r(m_1+1)}{2\pi d^2}\cos^{m_1}(\phi)T_s(\psi)g(\psi)\cos\psi\,\delta\!\left(t-\frac{d}{c}\right) \qquad (3.18)$$

where c is the speed of the light in free space, $\delta(.)$ is the Dirac function and $\delta(t - d/c)$ represents the signal propagation delay. The expression assumes that $\phi < 90°$ and $\psi < \mathrm{FOV}$ and $d \gg \sqrt{A_r}$.

The MATLAB® codes for simulating the LOS channel gain are given in Program 3.1.

Program 3.1: MATLAB Codes to Calculate the LOS Channel Gain

```
theta=70;
% semi-angle at half power
m=-log10(2)/log10(cosd(theta));
%Lambertian order of emission
P_total=20;
%transmitted optical power by individual LED
Adet=1e-4;
%detector physical area of a PD

Ts=1;
%gain of an optical filter; ignore if no filter is used
index=1.5;
%refractive index of a lens at a PD; ignore if no lens is used
FOV=60*pi/180;
%FOV of a receiver
G_Con=(index^2)/sin(FOV);
%gain of an optical concentrator; ignore if no lens is used

lx=5; ly=5; lz=3;
% room dimension in metre
h=2.15;
%the distance between source and receiver plane

% [XT,YT]=meshgrid([-1.25 1.25],[-1.25 1.25]);
XT=0; YT=0;
% position of LED;
Nx=lx*10; Ny=ly*10;
% number of grid in the receiver plane
x=-lx/2:lx/Nx:lx/2;
y=-ly/2:ly/Ny:ly/2;
[XR,YR]=meshgrid(x,y);
% receiver plane grid
```

```
D1=sqrt((XR-XT(1,1)).^2+(YR-YT(1,1)).^2+h^2);
% distance vector from source 1
cosphi_A1=h./D1;
% angle vector

H_A1=(m+1)*Adet.*cosphi_A1.^(m+1)./(2*pi.*D1.^2);
% channel DC gain for source 1
P_rec=P_total.*H_A1.*Ts.*G_Con;
% received power from source 1;
P_rec_dBm=10*log10(P_rec);

meshc(x,y,P_rec_dBm);
xlabel('X (m)');
ylabel('Y (m)');
zlabel('Received power (dBm)');
axis([-lx/2 lx/2 -ly/2 ly/2 min(min(P_rec_dBm)) max(max(P_rec_
dBm))]);
```

3.1.2 NON-LOS PROPAGATION MODEL

For nondirected LOS and diffuse links, the optical path loss is more complex to predict since it is dependent on a multitude of factors, such as room dimensions, the reflectivity of the ceiling, walls and objects within the room, and the position and orientation of the transmitter and receiver, window size and place and other physical matters within a room. The received power is generally defined as

$$
\begin{aligned}
P_{r-nlos} &= (H_{los}(0) + H_{nlos}(0))P_t \\
&= \left(H_{los}(0) + \sum_{refl} H_{refl}(0) \right) P_t
\end{aligned}
\tag{3.19}
$$

where $H_{refl}(0)$ represents the reflected path.

The reflection characteristics of object surfaces within a room depend on several factors including, the transmission wavelength, surface material, the angle of incidence θ_i and roughness of the surface relative to the wavelength. The latter mainly determines the shape of the optical reflection pattern. Rayleigh criterion [7] is mostly adopted to determine the texture of a surface.

According to this criterion, a surface can be considered smooth if the maximum height of the surface irregularities conforms to the following

$$
h_{si} < \frac{\lambda}{8\sinh(\theta_i)}
\tag{3.20}
$$

To predict the path loss for nondirected LOS and diffuse links, it is necessary to analyse the distribution of optical power for a given set-up. There are different attempts to accurately characterize the indoor optical channel [6,9–29].

Gfeller and Bapst [7] have studied the power distribution for diffuse links, basing their model on single reflections only. They have shown that by using an optical source consisting of multiple elements oriented in different directions, a more uniform coverage can be obtained over a larger area, compared with a single wide-beam optical source. Lomba et al. [30] have addressed the optimization of the optical power distribution for diffuse and nondirected LOS links, and have proposed a specification for the emitter radiation pattern (ERP) of the IEEE 802.11 infrared physical layer standard [31,32]. For nondirected LOS links, as an alternative to adjusting the ERP, a grid of ceiling-mounted transmitters have been proposed in order to reduce the dynamic range of signal power [33–35].

Mathematically, the impulse response of the optical wireless channel is calculated by integrating the power of all the components arriving at the receiver after multipath propagation [10]. The received signal in the case of the non-LOS links consists of various components arriving from different path, the path length of these components differ in proportion to the room design, hence there is broadening of the pulse. The distribution of the channel gain in decibel (dB) for the LOS component follows a modified gamma distribution, and the channel gain in dB for LOS channels including all reflections follows a modified Rayleigh distribution for most transmitter–receiver distances [13]. The root mean square delay spread D_{rms} is a parameter which is commonly used to quantify the time-dispersive properties of multipath channels, and is defined as the square root of the second central moment of the magnitude squared of the channel impulse response [6,17], which can be calculated using the following expression

$$D_{rms} = \left[\frac{\int (t-\mu)^2 h^2(t)\,dt}{\int h^2(t)\,dt} \right]^{1/2} \tag{3.21}$$

where the mean delay spread μ is given by

$$\mu = \frac{\int t h^2(t)\,dt}{\int h^2(t)\,dt} \tag{3.22}$$

While determining the distribution of optical power throughout a room is adequate for basic power budget calculations, it does not allow the power penalty due to multipath propagation to be accurately predicted, since multiple reflections are not taken into consideration. Although the optical power associated with two or more reflections is relatively small, the signal arrives at the receiver much later than that undergoing only one reflection, and hence cannot be ignored when considering high-speed nondirected LOS and diffuse links. In order to generate an impulse response which includes higher order reflections, Barry et al. [16] developed a ray-tracing algorithm in which the path loss and time delay for every path containing a given number of reflections are calculated. The algorithm then sums together all contributions to give an overall impulse response. The authors considered empty rectangular rooms and assumed that the optical receiver was pointing vertically towards the ceiling. Given a particular single source S and receiver R_x in a room, the impulse response can be written as an infinite sum:

$$h_{\text{nlos}}(t,S,R_x) = \sum_{k=0}^{\infty} h_{\text{nlos}}^{(k)}(t,S,R_x) \tag{3.23}$$

where $h_{\text{nlos}}^{(k)}(t,S,R_x)$ is the impulse response due to light undergoing exactly k reflections.

For K multiple sources the above equation can be modified to

$$h_{\text{nlos}}(t,S,R_x) = \sum_{i}^{K} \sum_{k=0}^{\infty} h_{\text{nlos K}}^{(k)}(t,S,R_x) \tag{3.24}$$

The impulse response after k reflection $h^{(k)}(t,S,R_x)$ can be evaluated using a recursive algorithm as given by [11,16]

$$h_{\text{nlos}}(t,S,R_x) = \frac{m_1+1}{2\pi} \sum_{j=1}^{K} \rho_j \cos^{m_1}(\phi_j) \frac{\cos(\psi)}{d_{Sj}^2} \text{rect}\left(\frac{2\psi}{\pi}\right)$$
$$\times h_{\text{nlos}}^{(k-1)}\left(t - \frac{d_{Sj}}{c}, S, R_x\right) \Delta A. \tag{3.25}$$

where ΔA is the area of reflecting elements, κ is the total number of reflector elements in the room, ρ_j is the reflection coefficient of j, d_{Sj} is the distance from S to j, $h^{(k-1)}(t,S,R_x)$ is the impulse response of order $k-1$ between reflector j and R_x. Since $\|h^{(k)}(t,S,R_x)\| \to 0$, $k \to \infty$, the channel impulse response can be estimated considering only first $\Re$. Using recursive algorithm or iterative methods and considering only number of reflections up to 10, one can get excellent approximation of the channel.

To evaluate the first diffuse reflections, the surfaces of the room are divided by $\Re$ reflecting elements with an area of ΔA. The channel can be described by two components: (a) in the first component, each element of the surface with area of ΔA is considered as a receiver and (b) each element is then considered as a point source that re-emits the light-collected signal scaled by reflectivity ρ_j. Hence, the channel impulse response after one reflection in the ceiling (see Figure 3.5) can be approximated as

$$h_{\text{nlos}}^1(t,S,R_x) = \sum_{j=1}^{\Re} \frac{(m_1+1)\rho_j A_r \Delta A}{2\pi d_{Sj}^2 d_{R,j}^2} \cos^{m_1}(\phi_{Sj})$$
$$\times \cos(\psi_{Sj})\cos(\psi_{Rj})\delta\left(t - \frac{d_{Sj}+d_{Rj}}{c}\right). \tag{3.26}$$

Using the parameters given in Table 3.1, with the transmitter located at the ceiling, the orientation angles of transmitter and receiver is expressed by azimuth and

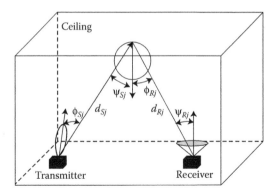

FIGURE 3.5 Geometry used to describe the single-refection propagation model.

elevation angle, the impulse responses of a LOS link and the first diffused reflection are shown in Figure 3.6. The corresponding MATLAB codes are given in Programs 3.2 and 3.3. The distribution of mean delay spread, RMS delay spread D_{rms} which is the root mean square of the channel time delay and maximum achievable data rate plots are illustrated in Figure 3.7. The channel impulse response in indoor optical wireless communication systems depends on the location of the receiver and varies by moving the receiver. The first reflection is the most important components which limits the data rates. The mean delay spread for the system under consideration varies from 10 ns to 16 ns with corresponding D_{rms} 0.3–0.6 ns. Hence, the maximum achievable data rates vary from 180 Mbps to 360 Mbps. Note that if the transmitter and receivers are located on the floor (as shown in Figure 3.5), the achievable data rates would significantly decrease.

TABLE 3.1
Simulation Parameters

	Parameters	Values
	Room size	$5 \times 5 \times 3$ m³
	$\rho_{North} = \rho_{South} = \rho_{West} = \rho_{East}$	0.8
Source	Location (x,y,z)	(2.5, 2.5, 3)
	Lambert's order	1
	Elevation	$-90°$
	Azimuth	$0°$
	Power	1 W
Receiver	Location (x,y,z)	(1.5, 1.5, 0)
	Active area (A_R)	1 cm²
	Half-angle FOV	$85°$
	Elevation	$90°$
	Azimuth	$0°$
	Δt	0.5 ns

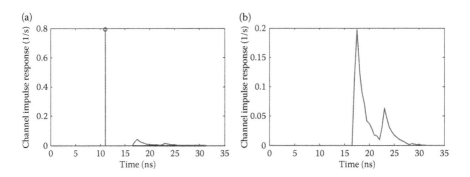

FIGURE 3.6 Impulse responses of the diffused links: (a) with a LOS path and (b) without LOS path.

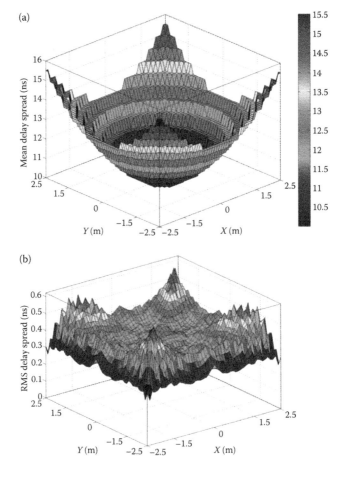

FIGURE 3.7 (**See colour insert.**) Channel delay spread: (a) mean delay spread, (b) RMS delay spread with LOS component, (c) RMS delay spread without LOS component and (d) maximum data rate distributions.

(c)

(d)

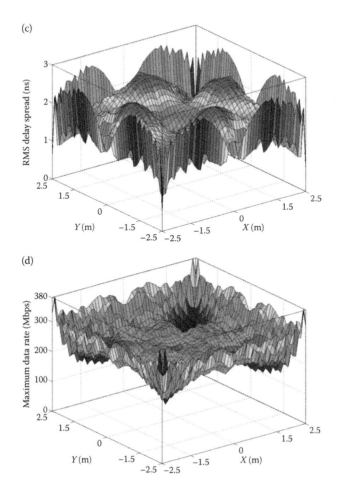

FIGURE 3.7 Continued.

Program 3.2: MATLAB Codes for Plotting the Optical Power Distribution in a Diffuse Channel

```
P_total=1;
% Total transmitted power
rho=0.8;
%reflection coefficient
lx=5; ly=5; lz=2.15;
% room dimension in metre
Nx=lx*3; Ny=ly*3; Nz=round(lz*3);
% number of grid in each surface
dA=lz*ly/(Ny*Nz);
% calculation grid area
```

```
x=linspace(-lx/2,lx/2,Nx);
y=linspace(-ly/2,ly/2,Ny);
z=linspace(-lz/2,lz/2,Nz);
[XR,YR,ZR]=meshgrid(x,y,-lz/2);

TP1=[0 0 lz/2];
% transmitter position

%%%%%%%%%%%%%%%%calculation for wall 1%%%%%%%%%%%%%%%%%%%%
for ii=1:Nx
  for jj=1:Ny
    RP=[x(ii) y(jj) -lz/2];
    % receiver position vector
    h1(ii,jj)=0;
    % reflection from North face
    for kk=1:Ny
      for ll=1:Nz
        WP1=[-lx/2 y(kk) z(ll)];
        % point of incidence in wall
        D1=sqrt(dot(TP1-WP1,TP1-WP1));
        % distance from transmitter to WP1
        cos_phi=abs(WP1(3)-TP1(3))/D1;
        cos_alpha=abs(TP1(1)-WP1(1))/D1;

        D2=sqrt(dot(WP1-RP,WP1-RP));
        % distance from WP1 to receiver
        cos_beta=abs(WP1(1)-RP(1))/D2;
        cos_psi=abs(WP1(3)-RP(3))/D2;

        if abs(acosd(cos_psi))<=FOV
          h1(ii,jj)=h1(ii,jj)+(m+1)*Adet*rho*dA*...
          cos_phi^m*cos_alpha*cos_beta*cos_psi/
          (2*pi^2*D1^2*D2^2);
        end
      end
    end
  end
end

% calculate channel gain (h2, h3 and h4) from other walls

P_rec_A1=(h1+h2+h3+h4)*P_total.*Ts.*G_Con;
```

Program 3.3: MATLAB Codes to Calculate the D_{rms} of a Diffuse Channel

```
%%
C=3e8*1e-9;
% time will be measured in ns in the program
theta=70;
% semi-angle at half power
```

```
m=-log10(2)/log10(cosd(theta));
% Lambertian order of emission
% P_LED=20;
% %transmitted optical power by individual LED
% nLED=60;
% number of LED array nLED*nLED
P_total=1;
% Total transmitted power
Adet=1e-4;
% detector physical area of a PD
rho=0.8;
% reflection coefficient

Ts=1;
% gain of an optical filter; ignore if no filter is used
index=1.5;
% refractive index of a lens at a PD; ignore if no lens is used
FOV=60;
% FOV of a receiver
G_Con=(index^2)/(sind(FOV).^2);
% gain of an optical concentrator; ignore if no lens is used

%%
%%%%%
lx=5; ly=5; lz=3-0.85;
% room dimension in metre

%[XT,YT,ZT]=meshgrid([-lx/4 lx/4],[-ly/4 ly/4],lz/2);
% position of Transmitter (LED);

Nx=lx*3; Ny=ly*3; Nz=round(lz*3);
% number of grid in each surface
dA=lz*ly/(Ny*Nz);
% calculation grid area
x=-lx/2:lx/Nx:lx/2;
y=-ly/2:ly/Ny:ly/2;
z=-lz/2:lz/Nz:lz/2;

%%
% first transmitter calculation
TP1=[0 0 lz/2];
% transmitter position
TPV=[0 0 -1];
% transmitter position vector
RPV=[0 0 1];
% receiver position vector

%%
WPV1=[1 0 0];
WPV2=[0 1 0];
WPV3=[-1 0 0];
```

```
WPV4=[0 -1 0];
% wall vectors
delta_t=1/2;
% time resolution in ns, use in the form of 1/2^m

for ii=1:Nx+1
  for jj=1:Ny+1
    RP=[x(ii) y(jj) -lz/2];
    t_vector=0:25/delta_t; % time vector in ns
    h_vector=zeros(1,length(t_vector));

    % receiver position vector
    % LOS channel gain
    D1=sqrt(dot(TP1-RP,TP1-RP));
    cosphi=lz/D1;
    tau0=D1/C;
    index=find(round(tau0/delta_t)==t_vector);
    if abs(acosd(cosphi))<=FOV
      h_vector(index)=h_vector(index)+(m+1)*Adet.*cosphi.
      ^(m+1)./(2*pi.*D1.^2);
    end
    %% reflection from first wall
    count=1;
    for kk=1:Ny+1
      for ll=1:Nz+1
        WP1=[-lx/2 y(kk) z(ll)];
        D1=sqrt(dot(TP1-WP1,TP1-WP1));
        cos_phi=abs(WP1(3)-TP1(3))/D1;
        cos_alpha=abs(TP1(1)-WP1(1))/D1;

        D2=sqrt(dot(WP1-RP,WP1-RP));
        cos_beta=abs(WP1(1)-RP(1))/D2;
        cos_psi=abs(WP1(3)-RP(3))/D2;
        tau1=(D1+D2)/C;
        index=find(round(tau1/delta_t)==t_vector);
        if abs(acosd(cos_psi))<=FOV
          h_vector(index)=h_vector(index)+(m+1)*Adet*rho*dA*...
          cos_phi^m*cos_alpha*cos_beta*cos_psi/(2*pi^2*D1^*D2^2);
        end
        count=count+1;
      end
    end

    %% Reflection from second wall
    count=1;
    for kk=1:Nx+1
      for ll=1:Nz+1
        WP2=[x(kk) -ly/2 z(ll)];
        D1=sqrt(dot(TP1-WP2,TP1-WP2));
        cos_phi= abs(WP2(3)-TP1(3))/D1;
        cos_alpha=abs(TP1(2)-WP2(2))/D1;
```

```
     D2=sqrt(dot(WP2-RP,WP2-RP));
     cos_beta=abs(WP2(2)-RP(2))/D2;
     cos_psi=abs(WP2(3)-RP(3))/D2;

     tau2=(D1+D2)/C;
     index=find(round(tau2/delta_t)==t_vector);
     if abs(acosd(cos_psi))<=FOV
       h_vector(index)=h_vector(index)+(m+1)*Adet*rho*dA*...
       cos_phi^m*cos_alpha*cos_beta*cos_psi/
       (2*pi^2*D1^2*D2^2);
     end
     count=count+1;
   end
end

%% Reflection from third wall
count=1;
for kk=1:Ny+1
    for ll=1:Nz+1
      WP3=[lx/2 y(kk) z(ll)];
      D1=sqrt(dot(TP1-WP3,TP1-WP3));
      cos_phi= abs(WP3(3)-TP1(3))/D1;
      cos_alpha=abs(TP1(1)-WP3(1))/D1;

      D2=sqrt(dot(WP3-RP,WP3-RP));
      cos_beta=abs(WP3(1)-RP(1))/D2;
      cos_psi=abs(WP3(3)-RP(3))/D2;

      tau3=(D1+D2)/C;
      index=find(round(tau3/delta_t)==t_vector);
      if abs(acosd(cos_psi))<=FOV
        h_vector(index)=h_vector(index)+(m+1)*Adet*rho*dA*...
        cos_phi^m*cos_alpha*cos_beta*cos_psi/
        (2*pi^2*D1^2*D2^2);
      end
      count=count+1;
    end
  end

  %% Reflection from fourth wall
  count=1;
  for kk=1:Nx+1
    for ll=1:Nz+1
      WP4=[x(kk) ly/2 z(ll)];
      D1=sqrt(dot(TP1-WP4,TP1-WP4));
      cos_phi= abs(WP4(3)-TP1(3))/D1;
      cos_alpha=abs(TP1(2)-WP4(2))/D1;

      D2=sqrt(dot(WP4-RP,WP4-RP));
      cos_beta=abs(WP4(2)-RP(2))/D2;
      cos_psi=abs(WP4(3)-RP(3))/D2;
```

```
        tau4=(D1+D2)/C;
        index=find(round(tau4/delta_t)==t_vector);
        if abs(acosd(cos_psi))<=FOV
          h_vector(index)=h_vector(index)+(m+1)*Adet*rho*dA*...
          cos_phi^m*cos_alpha*cos_beta*cos_psi/
          (2*pi^2*D1^2*D2^2);
        end
        count=count+1
      end
    end

    t_vector=t_vector*delta_t;
    mean_delay(ii,jj)=sum((h_vector).^2.*t_vector)/sum(h_
    vector.^2);
    Drms(ii,jj)=sqrt(sum((t_vector-mean_delay(ii,jj)).^2.
    *h_vector.^2)/sum(h_vector.^2));

  end
end

surf(x,y, Drms);
% surf(x,y,mean_delay);
% plot(t_vector*delta_t, h_vector)
save Drms_1LED mean_delay Drms x y z theta m P_total rho FOV
TP1;
axis([-lx/2 lx/2 -ly/2 ly/2 min(min(Drms)) max(max(Drms))]);
```

The model is applicable for a specific room configuration and does not take account of shadowing and furniture layout. Abtahi et al. [36] modified this work to consider the effects of furniture and people within the room, and also rooms of irregular shape. Pakravan and Kavehrad [29] used a neural network to speed up the algorithm developed by Barry, whereby only a fraction of the total number of points need to be calculated, from which the neural network learns the rest. A more efficient simulation model for the indoor channel is proposed by Lomba et al. [11,14] considering multiple reflection of the signal. The simulation efficiency is improved by 'time-delay agglutination' and 'time and space indexed tables' procedures. Iterative site-based model was suggested by Carruthers et al. [6,12] to simulate the effect of multipath reflections in the receiver. The study suggested that channels LOS paths must be modelled separately from those fully diffuse channels with no such path. A new approach was taken by López-Hernández et al. [27] to develop efficient simulation of the channel by slicing into time steps rather than into a number of reflections. The indoor optical channel is modelled as a parallel combination of LOS and diffuse paths and analytical model is introduced by Jungnickel et al. [9]. Most of the channel models described above are based on the Monte-Carlo ray-tracing algorithm and Lambert's model of reflection. However, the use of Phong's model can lead to different impulse response for the same channel with respect to Lambert's model when surfaces present a high specular component [21]. The impulse response can be modelled by decomposing the signal into primary and higher order reflections [10]. The gamma probability density function (pdf) is used for the matching function of the primary reflection impulse response. Impulse response of the higher order reflections is simulated using the spherical model. The analysis shows that the bandwidth characteristics are dominated by the response of the primary reflection.

Practical channel characterization was carried out by Kahn et al. [17], who measured channel frequency responses over the range 2–300 MHz using a swept frequency technique. From these measurements the authors computed impulse responses, path losses and RMS delay spreads. Both LOS and diffuse link configurations were considered, using different receiver locations in five different rooms, giving a total of ~100 different channels. Hashemi et al. [26,37] measured eight rooms at various positions and also took measurements for different orientations and rotations of the photodetector, giving a total of 160 frequency response profiles. The authors show that the channel response is sensitive not only to the position of the photodetector, but also to its orientation and rotation. Based on this knowledge, the authors proposed the angle diversity receiver structure.

A brief survey of different diffuse indoor optical wireless channel simulation model is given below.

3.1.3 CEILING BOUNCE MODEL

Although many models had been proposed and studied for simulating impulse response of indoor IR channel, by far the most used in simulation is the ceiling bounce model introduced by Carruthers and Kahn because of its excellent matching with the measured data and simplicity of the model. This method came up with a closed-form expression for the impulse response assuming the transmitter and the receiver to be colocated in planes parallel to the floor and directed towards the ceiling. In the ceiling bounce model, the multipath IR channel is characterized by only two parameters: the optical path loss and the RMS delay spread D_{rms}. The impulse response $h(t)$ of the ceiling bounce model is given by

$$h(t,a) = H(0)\frac{6a^6}{(t+a)^7}u(t) \tag{3.27}$$

where $u(t)$ is the unit step function, $a = 2H_c/c$, H_c is the height of the ceiling above the transmitter and receiver and c is the velocity of light. The parameter a is related to D_{rms} as

$$D_{rms}(h(t,a)) = \frac{a}{12}\sqrt{\frac{13}{11}} \tag{3.28}$$

The approximate the 3-dB cut-off frequency for the ceiling bounce model is

$$f_{3dB} = \frac{0.925}{4\pi D_{rms}} \tag{3.29}$$

For a more realistic channel model with multiple reflections, a needs to be modified. For the unshadowed and shadowed channels the expressions for a are given by

$$a(\text{unshad.}) = 12\sqrt{\frac{11}{13}}(2.1 - 5.0s + 20.8s^2)D_{\text{rms}}(h_1(t)) \tag{3.30}$$

$$a(\text{shad.}) = 12\sqrt{\frac{11}{13}}(2.0 + 9.4s)D_{\text{rms}}(h_1(t)) \tag{3.31}$$

where $h_1(t)$ is one bounce impulse response and s is defined as the ratio of the horizontal transmitter–receiver (TR) separation to the TR diagonal, the latter being defined as the length of the segment of the line between the transmitter and the receiver.

Another important parameter that is used for the study of performance at different date rates in a dispersive channel is the normalized delay spread D_T, which is a dimensionless parameter defined as

$$D_T = \frac{D_{\text{rms}}}{T_b} \tag{3.32}$$

A modified ceiling bounce model based on the statistical method of Perez-Jimenez et al. [38] and the Carruthers et al. approach is proposed in Ref. [20]. In this approach, the RMS delay spread for a diffuse indoor system depends upon the distance and transmission angle between transmitter and receiver, and mode number of the emitter radiation pattern, which is given by

$$D_{\text{rms}} = -2.37 + 0.007m + (0.8 - 0.002m)(d_{Sj} + d_{R_xj}) \tag{3.33}$$

3.1.4 HAYASAKA–ITO MODEL

In this model, the diffuse IR channel is analysed by the ray-tracing technique and the modified Monte Carlo method. Assuming a Lambertian model for the light sources, the impulse response of the channel is expressed as [17]

$$h(t) = h_1(t - T_1) + h_{\text{high}}(t - T_{\text{high}}) \tag{3.34}$$

where $h_1(t)$ is the impulse response of the primary reflection, $h_{\text{high}}(t)$ is that of the higher order reflections, T_1 and T_{high} are the starting times of the rise of the primary and higher order reflections.

Assuming no reflections from the side walls, the primary reflection follows a gamma probability density and hence the normalized impulse response $(h_1^0(t))$ given by

$$h_1^0(t) = \frac{\lambda^{-\alpha}}{\Gamma(\alpha)}t^{\alpha-1}\exp\left(-\frac{t}{\lambda}\right) \tag{3.35}$$

where $\Gamma(\alpha)$ is the gamma function, $\langle t \rangle = \alpha\lambda$, $\sigma^2 = \alpha\lambda^2$.

If $H_1^0(f)$ the Fourier transform of $h_1^0(t)$, then

$$| H_1^0(f) |^2 = \frac{1}{[1 + (2\pi\lambda f)^2]^\alpha} \tag{3.36}$$

Hence, the 3 dB cut-off frequency for primary reflection is given by

$$f_{3dB} = \frac{\sqrt{2^{1/\alpha} - 1}}{2\pi\lambda} \tag{3.37}$$

3.1.5 SPHERICAL MODEL

For the higher order reflections are model using spherical model [9] and is given by

$$h_{high} = \frac{\eta_{high}}{\tau} \exp\left(-\frac{t}{\tau_e}\right) \tag{3.38}$$

where the power efficiency of the diffuse signal η_{high}, the exponential decay time τ_e and the average transmission delay $\langle t \rangle$ for one reflection are given by

$$\eta_{high} = \frac{A_r}{A_{room}} \frac{\rho_{ceiling}\langle\rho\rangle}{1 - \langle\rho\rangle} \sin^2(\text{FOV})$$

$$\tau_e = \frac{-T_{ref}}{\ln\langle\rho\rangle} \tag{3.39}$$

$$\langle t \rangle = \frac{4V_{room}}{cAV_{room}}$$

Here V_{room} is the room volume, T_{ref} is the average transmission delay for a single ray reflectionary. The average reflectivity of walls $\langle\rho\rangle$ is defined as

$$\langle\rho\rangle = \frac{1}{A_{ro}} \sum_i A_i \rho_i \tag{3.40}$$

The frequency response of a diffused channel is

$$H_{diff} = \frac{\eta_{diff}}{1 + j2\pi f} \tag{3.41}$$

The 3-dB bandwidth for the high-order reflections is given by

$$f_{3dB} = \frac{1}{2\pi f \tau_e} \tag{3.42}$$

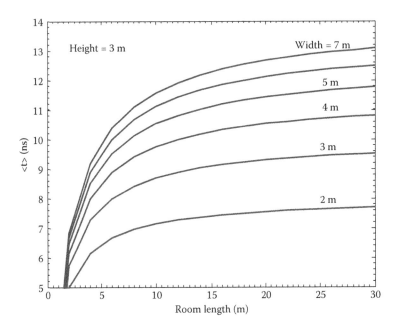

FIGURE 3.8 Average time between two reflections as a function of the room dimensions.

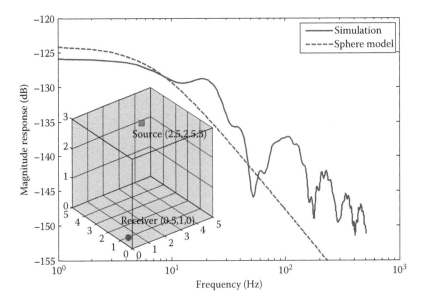

FIGURE 3.9 The frequency response of the diffused channel using the first diffused reflection and the sphere models.

Figure 3.8 shows the average time between two reflections as a function of the room dimensions. The frequency response of the diffused channel using the first diffused reflection and the sphere models is shown in Figure 3.9.

3.2 ARTIFICIAL LIGHT INTERFERENCE

Infrared transceivers operating in typical indoor environments are subject to intense ambient light, emanating from both natural and artificial sources. The main sources of ambient light are the sunlight, incandescent lamps and fluorescent lamps. The optical spectrum of the ambient light is shown in Figure 1.2. The spectra have been scaled to have equal maximum value, and the longer wavelength region of the fluorescent lamp spectrum has been amplified by a factor of 10 in order to make it clearly visible. The sunlight is typically the strongest source of noise and represents an unmodulated source of the ambient light with a very wide spectral width and a maximum PSD located at ~500 nm. Sunlight produces the highest levels of background current and is the major source of shot noise at the receiver photodiode. The background current due to artificial illumination is only few tens of μA, well below that produced by sunlight which could be as high as 5 mA [39].

The ambient light source (both natural and artificial) has an optical spectrum in both the operating line width of 830 nm and 1550 nm. The photocurrent generated by the ambient lights acts as a noise source in the receiver, which degrades the performance. The average combined power of the ambient light results in the photodetector generating a DC background photocurrent I_B that gives rise to shot noise $n(t)$ which is the dominant noise source in optical wireless systems [4]. The shot noise due to the ambient light is independent of the signal and can be modelled as a white Gaussian with one-sided PSD N_0 given by [39]

$$N_0 = 2qI_b \tag{3.43}$$

where q is the charge of an electron and I_b is the total DC photocurrent due to natural and artificial light sources.

Moreira et al. [40,41] carried out extensive measurements of a variety of ambient light sources and produced a model to describe the interference signal. Boucouvalas [42] also carried out similar measurements, which included a number of consumer products which use IR transmission. Along with experimental characterization of ambient light sources, a significant amount of work has been done on analysing the effect of ambient light interference (ALI) on the link performance [39,43–49].

All artificial ambient light sources are modulated, either by the mains frequency or, in the case of some fluorescent lamps, by a high-frequency switching signal. Incandescent lamps have a maximum PSD around 1 μm, and produce an interference signal which is a near perfect sinusoid with a frequency of 100 Hz [40]. Only the first few harmonics carry a significant amount of energy and interference effect can be effectively reduced by using an HPF [40,43,47]. The interference produced by fluorescent lamps driven by the conventional ballasts is distorted sinusoidal and extending up to 20 kHz. The interference can be effectively reduced using an HPF with a cut-off frequency of 4 kHz even at a low data rate of 1 Mbps and does not cause a significant

TABLE 3.2
Artificial Ambient Light Sources

Source	Details
Incandescent lamp	Bulb: Osram 60 W
Low-frequency fluorescent lamp	Ballast: Crompton C237 1 × 75 W
	Tube: Osram L 70 W/23
High-frequency fluorescent lamp	Ballast: Thorn G81016.4 1 × 70 W or 1 × 75 W (specified frequency = 35 kHz)
	Tube: Osram L 70 W/23

degradation in performance [43]. The interference produced by the fluorescent lamps driven by the electronic ballasts is the most serious source of performance degradations. The spectrum of the interference depends on the switching frequency. However, all models exhibit components at the switching frequency and at harmonics of that frequency. Harmonics of the switching frequency can extend into the megahertz range, and therefore present a much more serious impairment to optical wireless receivers. The interference can be modelled using a high-frequency component and low-frequency component and due to significant spectral component at high frequencies, even the HPF does not provide significant improvement in the performance for different modulation schemes even at higher data rates (>10 Mbps) [43,44].

Other noise sources like IR audio headphone transmitters and a TV remote control unit can be viewed as sources of the noise as the operating wavelengths of these equipments coincide with OW transceiver [42]. Measurements have been carried out to determine the time-domain waveforms and detected electrical spectra for the ambient light sources listed in Table 3.2.

The measurements were taken using a Thorlabs PDA55 amplified silicon detector, with a transimpedance of 15 kΩ and a 3 dB bandwidth ranging from 25 Hz to 7.9 MHz. For each measurement, the distance between the source and the detector was set such that the average received photocurrent was 100 μA. For the fluorescent lamps, all measurements were taken at the centre of the tube. Additionally, the effects of an RG780 optical long-pass filter were also investigated. The filter passes all wavelengths longer than 780 nm and when combined with the spectral response of the PDA55, results in an optical bandpass response ranging from 780 nm to ~1.1 μm.

3.2.1 INCANDESCENT LAMP

Incandescent lamps have a maximum power spectral density around 1 μm, and produce an interference signal which is a near perfect sinusoid with a frequency of 100 Hz. The slow response time of the filament means few harmonics are present. Figure 3.10 shows the time domain waveform and detected electrical spectrum for the incandescent bulb listed in Table 3.2. No optical filtering was used.

Only harmonics up to 400 Hz carry a significant amount of power, and beyond that, all harmonics are more than 60 dB below the fundamental. The average received

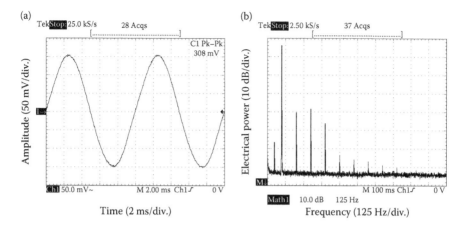

FIGURE 3.10   Incandescent bulb: (a) time domain waveform and (b) detected electrical.

TABLE 3.3

I_b and I_{pk-pk} for Incandescent Bulb with and without Optical Filtering

	Without Optical Filter	With Optical Filter	Reduction
I_b	100 μA	20.5 μA	79.5%
I_{pk-pk}	65.2 μA	12 μA	81.6%
I_b/I_{pk-pk}	1.53	1.71	

photocurrent I_b and peak-to-peak interference signal photocurrent I_{pk-pk} are given in Table 3.3, with and without optical filtering.

When optical filtering is used, I_b is reduced by 79.5%, and the peak-to-peak amplitude of the interference signal is reduced by 81.6%. The ratio of I_b/I_{pk-pk} is fairly similar both with and without optical filtering.

3.2.2   FLUORESCENT LAMP DRIVEN BY CONVENTIONAL BALLAST

Low-frequency fluorescent lamps are driven by the mains frequency. The interference signal is a distorted 100 Hz sinusoid, and the electrical spectrum contains harmonics into the tens of kilohertz. Figure 3.11 shows the time domain waveform and detected electrical spectrum for the low-frequency fluorescent lamp listed in Table 3.2. No optical filtering was used.

The average received photocurrent and peak-to-peak interference signal photocurrent, with and without optical filtering, are given in Table 3.4. Optical filtering gives a significant reduction in both the average background photocurrent and the peak-to-peak interference intensity. Since the reduction in I_b is greater than the reduction in I_{pk-pk}, with the optical filter in place, the peak-to-peak variation of the photocurrent is actually greater than the average background photocurrent.

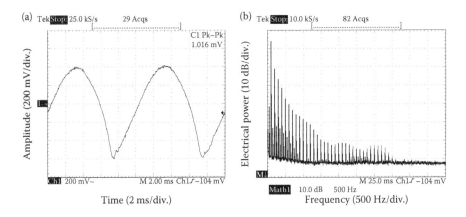

FIGURE 3.11 Low-frequency fluorescent lamp: (a) time-domain waveform and (b) detected electrical spectrum.

TABLE 3.4

I_b and I_{pk-pk} for Low-Frequency Fluorescent Lamp with and without Optical Filtering

	Without Optical Filter	With Optical Filter	Reduction
I_b	100 μA	5.4 μA	94.6%
I_{pk-pk}	67.7 μA	9.6 μA	85.8%
I_b/I_{pk-pk}	1.48	0.56	

3.2.3 FLUORESCENT LAMP MODEL

In recent years, fluorescent lamps have been introduced which are driven by high-frequency electronic ballasts. This type of lamp has a number of advantages over its low-frequency counterpart, such as reduced electrical power consumption for a given level of illumination and increased life expectancy of the tubes. The actual switching frequency used varies from one manufacturer to another, but is typically in the range 20–40 kHz. The detected electrical spectrum contains harmonics of the switching frequency and also harmonics of the mains frequency, similar to low-frequency fluorescent lamps. Harmonics of the switching frequency can extend into the megahertz range, and therefore present a much more serious impairment to optical wireless receivers. Figure 3.12 shows the time-domain waveform and detected electrical spectrum for the high-frequency fluorescent lamp listed in Table 3.2.

The average received photocurrent and peak-to-peak interference signal photocurrent, with and without optical filtering, are given in Table 3.5. Without optical filtering, for a given background photocurrent, the interference amplitude produced by the high-frequency fluorescent lamp is only about half that produced by the low-frequency fluorescent lamp. Optical filtering gives a similar reduction in I_b and I_{pk-pk} as it did for the low-frequency fluorescent lamp.

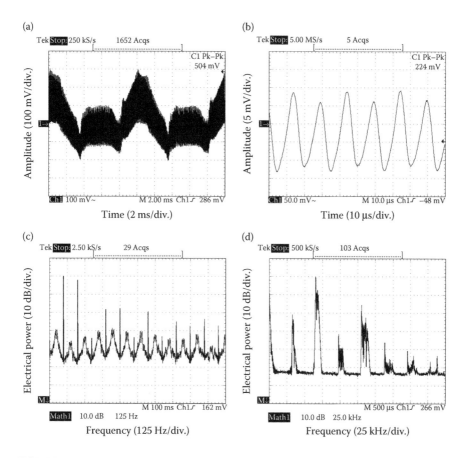

FIGURE 3.12 HF fluorescent lamp—time-domain waveform: (a) low-frequency component and (b) high-frequency component. Detected electrical spectrum: (c) low-frequency component and (d) high-frequency component.

TABLE 3.5

I_b and I_{pk-pk} for High-Frequency Fluorescent Lamp with and without Optical Filtering

	Without Optical Filter	With Optical Filter	Reduction
I_b	100 µA	3.9 µA	96.1%
I_{pk-pk}	33.6 µA	4.9 µA	85.4%
I_b/I_{pk-pk}	2.98	0.80	

On the basis of extensive measurements of a variety of fluorescent lamps driven by electronic ballasts, Moreira et al. produced a model to describe the interference signal [41]. All measurements were taken using an optical long-pass filter, and consequently, the values used in the model reflect this particular case. The interference signal is comprised of a low-frequency component, similar to that of a fluorescent lamp driven by a conventional ballast, and a high-frequency component, which is generated by the switching circuit of the electronic ballast. Hence, fluorescent lamps driven by electronic ballasts can be modelled using high-frequency m_{high} and low-frequency $m_{\text{low}}(t)$ components of the photocurrent at the receiver. Thus, the zero mean periodic photocurrent $m_{\text{fl}}(t)$ due to the fluorescent lamp is given as [40,41]

$$m_{\text{fl}}(t) = m_{\text{low}}(t) + m_{\text{high}}(t) \tag{3.44}$$

where

$$m_{\text{low}}(t) = \frac{I_b}{A_1} \sum_{i=1}^{20} \left[\Phi_i \cos(2\pi(100i - 50)t + \varphi_i) + \Psi_i \cos(2\pi 100 i t + \phi_i) \right] \tag{3.45}$$

where I_b is the average photocurrent generated by the fluorescent lamp, A_1 is the constant that relates the interference amplitude with I_b, φ_i and ϕ_i are the phase of the odd and even harmonics of 50 Hz, respectively, Φ_i and Ψ_i are the amplitudes of the odd and even harmonics of 50 Hz mains signal, respectively, given by

$$\Phi_i = 10^{(-13.1\ln(100i-50)+27.1)/20} \tag{3.46}$$

$$\Psi_i = 10^{(-20.8\ln(100i)+92.4)/20} \tag{3.47}$$

The high-frequency $m_{\text{high}}(t)$ component is given by

$$m_{\text{high}}(t) = \frac{I_b}{A_2} \sum_{j=1}^{22} \Gamma_j \cos\left(2\pi f_{\text{high}} j t + \theta_j\right) \tag{3.48}$$

Γ_j and θ_j are the amplitude and phase of the harmonics, f_{high} is the electronic ballast switching frequency, A_2 is the constant relating the interference amplitude to I_b.

TABLE 3.6
Low-Frequency Component Phase Values

i	φ_i (rad)	ϕ_i (rad)	i	φ_i (rad)	ϕ_i (rad)
1	4.65	0	11	1.26	6.00
2	2.86	0.08	12	1.29	6.17
3	5.43	6.00	13	1.28	5.69
4	3.90	5.31	14	0.63	5.37
5	2.00	2.27	15	6.06	4.00
6	5.98	5.70	16	5.49	3.69
7	2.38	2.07	17	4.45	1.86
8	4.35	3.44	18	3.24	1.38
9	5.87	5.01	19	2.07	5.91
10	0.70	6.01	20	0.87	4.88

TABLE 3.7

High-Frequency Component Amplitude and Phase Values

j	Γ_j (dB)	θ_j (rad)	j	Γ_j (dB)	θ_j (rad)
1	−22.2	5.09	12	−39.3	3.55
2	0	0	14	−42.7	4.15
4	−11.5	2.37	16	−46.4	1.64
6	−30.0	5.86	18	−48.1	4.51
8	−33.9	2.04	20	−53.1	3.55
10	−35.3	2.75	22	−54.9	1.78

The typical values for the parameters of $m_{low}(t)$ and $m_{high}(t)$ are given in Tables 3.6 and 3.7. For f_{high} of 37.5 kHz, there are 750 cycles of high-frequency component per cycle (20 μS) of the low-frequency component. With I_b set at 2 μA, one complete cycle of the low-frequency component and three complete cycles of high-frequency components of the interference photocurrent are shown in Figure 3.13.

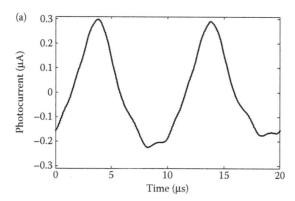

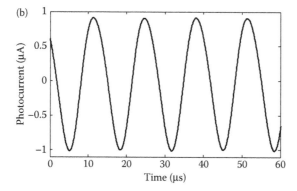

FIGURE 3.13 (a) Low-frequency interference component and (b) high-frequency interference component.

The MATLAB codes for simulating the FLI are given in Program 3.4.

Program 3.4: MATLAB Codes for the FLI Simulation Model

```
function i_elect=fl_model(Ib,samp_int,start_time,end_time)
% Generates a zero mean photocurrent, with maximum value i_max

global i_low i_high

t=start_time:samp_int:end_time;
no_of_points=length(t);

%***** low-frequency component *****
i=1:20;
p1=[4.65 2.86 5.43 3.90 2.00 5.98 2.38 4.35 5.87 0.70 1.26
1.29 1.28 0.63 6.06 5.49 4.45 3.24 2.07 0.87];
p2=[0.00 0.08 6.00 5.31 2.27 5.70 2.07 3.44 5.01 6.01 6.00
6.17 5.69 5.37 4.00 3.69 1.86 1.38 5.91 4.88];

b=10.^((-13.1*log((100*i)-50)+27.1)/20);
c=10.^((-20.8*log(100*i)+92.4)/20);

A3=5.9;

i_low=zeros(1,no_of_points);
for loop=1:no_of_points
  i_low(loop)=(Ib/A3)*sum(b.*cos(2*pi*(100*i-50)*
  t(loop)+p1)+ c.*cos(2*pi*100*i*t(loop)+p2));
end

%***** high-frequency component *****
j=[1 2 4 6 8 10 12 14 16 18 20 22];
d_db=[-22.2 0.00 -11.5 -30.0 -33.9 -35.3 -39.3 -42.7 -46.4
-48.1 -53.1 -54.9];
d=10.^(d_db/10); % convert from log to linear
thetaj=[5.09 0.00 2.37 5.86 2.04 2.75 3.55 4.15 1.64 4.51 3.55
1.78];

f_high=37500;
A4=2.1;

i_high=zeros(1,no_of_points); % preallocate array

for loop=1:no_of_points
  i_high(loop)=(Ib/A4)*sum(d.*cos(2*pi*f_high*j*
  t(loop)+thetaj));
end
i_elect=i_low+i_high;

end
```

3.3 OUTDOOR CHANNEL

The atmospheric channel is a very complex and dynamic environment that can affect the characteristics of the propagating optical beam (laser in most cases), thus resulting in optical losses and turbulence-induced amplitude and phase fluctuation. The properties of the atmospheric channel is random in nature; therefore its effects can be characterized by statistical means. There are a number of models in the literature to characterize the statistical nature of the atmospheric channel; these will be discussed in this section. Depending on the type of the model and its accuracy, one can estimate the received optical irradiance at the receiver.

3.3.1 ATMOSPHERIC CHANNEL LOSS

The loss mechanisms in an FSO are virtually identical to those in a line-of-sight RF (microwave and millimetre) channel but the fading level is higher than the RF signals. Optical signal propagating through a free space channel is very sensitive to the atmospheric conditions such as fog, rain and so forth. For optical radiation travelling through the atmospheric channel, the interaction between the photons and the molecular constituent of the atmosphere cause some of the photons to be extinguished while particulate constituents scatter the photons. These events ultimately result in a power transmission loss mechanism otherwise described by the Beer–Lambert law (BLL) [50]. BLL law describes the transmittance of an optical field through the atmosphere as function of the propagation distance. In addition, optical radiation traversing the atmosphere spreads out (beam divergence) due to the diffraction, causing the size of the received beam to be greater than the receiver aperture. These factors, combined with others herein discussed are responsible for the difference between the transmitted and the received optical powers.

Beam divergence can be minimized by employing a very narrow coherent laser source. Atmospheric absorption losses can be reduced by adopting wavelengths that lie in the low-loss 'transmission windows' in the visible or infrared bands. The concentrations of matter in the atmosphere, which result in the signal attenuation, vary spatially and temporally, and will depend on the current local weather conditions. For a terrestrial FSO link transmitting an optical signal through the atmosphere, the optical power reaching the receiver P_r is related to the transmitted power P_t by

$$P_r = P_t e^{-\tau_{od}} \tag{3.49}$$

where τ_{od} is optical depth. The fraction of the power transmitted in the optical link is defined as the transmittance and is given by

$$T = \frac{P_r}{P_t} = e^{-\tau_{od}} \tag{3.50}$$

The optical depth and optical atmospheric transmittance are related to the atmospheric attenuation coefficient and the transmission path length L by the Beer–Lambert's law as [51]

$$T(\lambda, L) = \frac{P_r}{P_t} = \exp\left[-\gamma_t(\lambda)L\right] \tag{3.51}$$

where $\gamma_t(\lambda)$ and $T(\lambda, L)$ represent the total attenuation/extinction coefficient (m^{-1}) and the transmittance of the atmosphere at wavelength λ, respectively. Note that γ is an optical source (i.e., laser) wavelength specific and depends on the path-integrated distribution of atmospheric constituents along the LOS and is defined as $\gamma = -10 \log_{10} T = 4.43 \, \tau_{od}$. Note that for τ_{od} of 0.7, the corresponding loss is 3 dB.

The attenuation of the optical signal in the atmosphere is due to scattering and absorption introduced by the molecular constituents (gases) and aerosols. The aerosol is made up of tiny particles of various shapes ranging from spherical to irregular shapes suspended in the atmosphere. Generally speaking, the atmospheric attenuation coefficient can be expressed as [52]

$$\gamma_t(\lambda) = \alpha_{ml}(\lambda) + \alpha_{al}(\lambda) + \beta_{ml}(\lambda) + \beta_{al}(\lambda) \tag{3.52}$$

where α_{ml} and α_{al} are respectively the absorption coefficient for the molecular and aerosol, and β_{ml} and β_{al} are the scattering coefficients for the molecular and aerosol, respectively.

Absorption and scattering of light by particles present in the atmosphere is a complex process involving Mie scattering and nonselective scattering by large size particles (such as fog, haze and rain) and Rayleigh scattering by smaller particles.

1. *Absorption*—This takes place when there is an interaction between the propagating photons and molecules (present in the atmosphere) along its path. A light photon is absorbed when the quantum state of a molecule is excited to a higher energy level. Some of the photons are extinguished and their energies converted into heat [53]. The absorption coefficient depends very much on the type of gas molecules and their concentration. Absorption is wavelength dependent and therefore selective. This leads to the atmosphere having transparent zones—range of wavelengths with minimal absorptions—referred to as the transmission windows as shown in Figure 3.14. Since it is not possible to change the physics of the atmosphere, the wavelengths adopted in FSO systems are basically chosen to coincide with the atmospheric transmission windows [54]. The attenuation coefficient is therefore dominated by scattering, and as such, $\gamma_t(\lambda) \cong \beta_{al}(\lambda)$.

2. *Scattering*—This results in angular redistribution of the optical field with and without wavelength modification. The scattering effect depends on the radius r_p of the particles (fog, aerosol) encountered during propagation. One way of describing this is to consider the size parameter $x_0 = 2\pi r_p/\lambda$. If

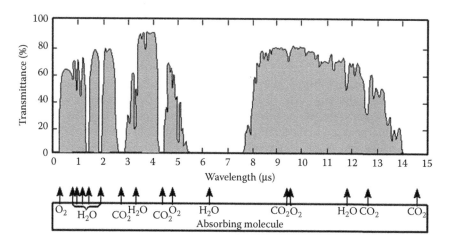

FIGURE 3.14 Atmospheric absorption transmittance over a sea level 1820 m horizontal path. (Adapted from H. Willebrand and B. S. Ghuman, *Free Space Optics: Enabling Optical Connectivity in Today's Network*. Indianapolis: SAMS Publishing, 2002.)

$x_0 \ll 1$, the scattering process is classified as Rayleigh scattering [55]; if $x_0 \approx 1$ it is Mie scattering and for $x_0 \gg 1$ [56], the scattering process can then be explained using the diffraction theory (geometric optics) [57]. Therefore, molecular and aerosol scatterings are due to the Rayleigh and Mie scattering, respectively. The scattering process for different scattering particles present in the atmosphere is summarized in Table 3.8. Scattering does not only degrade the optical beam propagating in the atmosphere, but it also induces noise in a free-space optical communication system by scattering the sky radiance into the receiver.

a. *Rayleigh Scattering*—A simplified expression describing the Rayleigh scattering is given by [58]

$$\beta_{Rayleigh}(\lambda) = 0.827 N_p A_p^3 \lambda^{-4} \qquad (3.53)$$

TABLE 3.8

Typical Atmospheric Scattering Particles with Their Radii and Scattering Process at $\lambda = 850$ nm

Type	Radius (μm)	Size Parameter x_0	Scattering Process
Air molecules	0.0001	0.00074	Rayleigh
Haze particle	0.01–1	0.074–7.4	Rayleigh–Mie
Fog droplet	1–20	7.4–147.8	Mie—Geometrical
Rain	100–10,000	740–74,000	Geometrical
Snow	1000–5000	7400–37,000	Geometrical
Hail	5000–50,000	37,000–370,000	Geometrical

where N_p is the number of particles per unit volume along the propagation path and A_p represents the cross-sectional area of scattering. The λ^{-4} proportionality of the Rayleigh scattering suggests that the shorter wavelengths get scattered more; this effect is what is responsible for the blue colour of the sky during the day.

b. *Mie Scattering*—Mie scattering occurs when the particle size (aerosol) is comparable to the beam size from submicrometre to a few tens of micrometres. In the atmosphere, aerosols differ in distribution, components and profile concentration, therefore influencing the interactions with the propagating optical beam in the forms of absorption and scattering. The largest concentration of aerosols is normally located at 1–2 km immediately above earth's surface, and is classified based on the following models:

   i. Maritime model: proximity or over the sea and ocean surfaces, and typically consist of salt particles in aggregation with water droplets.

   ii. Rural model: over land with aerosol composition consisting of dust and other particles mixed with water droplets. The aerosol composition, density and its particle size distribution will vary with land composition, vegetation, weather and seasonal climate variations.

   iii. Urban model: manmade aerosols, produced by industry.

   iv. Desert model: mainly airborne dust particles with concentration mainly depending on the wind speed.

The aerosol refraction index is described as

$$n_a = n_r - jn_i \tag{3.54}$$

where the real part n_r represents scattering and the imaginary part n_i represents absorption (see Figure 3.15).

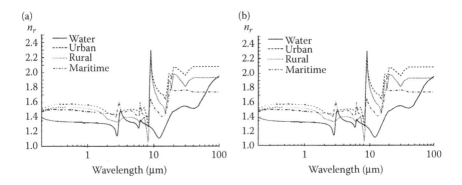

FIGURE 3.15 (a) Real and (b) complex parts of refraction for different aerosol models.

The absorption coefficient is defined as

$$\alpha_a = \frac{2\pi}{\lambda} n_i \tag{3.55}$$

Attenuation of optical beam due to the rain is rather low compared to the fog. Optical systems with a 25 dB or more link margin can operate under relatively heavy rain conditions. For rainfall, the scattering coefficient depends on the size of the raindrops, which is many times larger than the wavelength of the optical signal, and its distribution is defined by [52]

$$\beta_{rain} = 1.25 \times 10^{-16} \left(\Delta x / \Delta t\right)/a^2 \ (cm^{-1}) \tag{3.56}$$

where a is the parameter which characterizes the raindrop size distribution and $\Delta x / \Delta t$ is the rate of rainfall.

The attenuation due to the rain defined in terms of the precipitation rate R_{pr} is given as

$$\propto_{rain} = 1.076 \times R_{pr}^{0.67} \ (dB/km) \tag{3.57}$$

3.3.2 Fog and Visibility

Principally, fog particles reduce the visibility near the ground. The meteorological definition of fog is when the visibility drops to near 1 km. Different types of fog result in different levels of optical losses and this is mainly due to the distribution of the fog particles, size and the location. Two different types of fog found in the literature are the convection and maritime/advection fog. Convection or radiation fog is generated due to the ground cooling by radiation (see Figure 3.16). This fog appears when the air is sufficiently cool and becomes saturated. Principally, convection appears during the night and at the end of the day with the particle size distribution of 1–3 μm and the liquid water content varies between 0.01 and 0.1 g/m³. It affects shorter wavelengths (first and second transmission windows) and barely the 10 μm transmission window. A common visibility range for this type of fog is 500 m. Advection fog is created by the movement

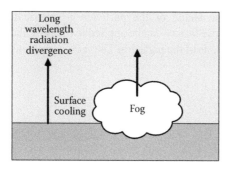 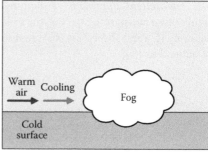

FIGURE 3.16 (See colour insert.) Schematic depiction of fog formation: (a) radiation and (b) advection.

TABLE 3.9

International Visibility Range and Attenuation Coefficient in the Visible Waveband for Various Weather Conditions

Atmospheric Conditions	International Visibility Code			Attenuation (dB/km)
	Weather Constituents (mm/h)		Visibility (m)	
Dense fog			50	315
Thick fog			200	75
Moderate fog			500	28.9
Light fog	Storm	100	770	18.3
Very light fog			1000	13.8
	Snow	Strong rain	25	1900	6.9
Light mist			2000	6.6
	Average rain	12.5	2800	4.6
Very light mist			4000	3.1
	Light rain	2.5	5900	2
			10,000	1.1
Clear air	Drizzle	0.25	18,100	0.6
			20,000	0.54
Very clear air			23,000	0.47
			50,000	0.19

Source: Adapted from M. S. Awan et al., *International Workshop on Satellite and Space Communications*, 2009, pp. 274–278.

of wet and warm air masses above colder maritime or terrestrial surfaces. It is characterized by liquid water content higher than 0.20 g/m^3 and a particle diameter close to 20 μm. A common visibility setting is 200 m for the advection fog [59]. Any visibility setting within the 0–1 km range is reasonable for the two types of fog [60–63]. Further, details on the relationship between the atmospheric weather constituents and the visibility are given in Table 3.9. The Mie-induced attenuation (mainly fog) in the FSO systems can be studied using both theoretical and empirical approach.

1. *Theoretical approach*   Assuming the shape of the particles is spherical, mainly those constituting the fog, the exact Mie theory can be applied to measure the scattering (C_s) and absorption (C_a) cross sections of the particles. These cross sections are defined as [64]

$$C_s = \frac{P_s}{I_0}$$

$$C_a = \frac{P_a}{I_0} \tag{3.58}$$

where P_s (W) is the electromagnetic energy scattered across the surface of an imaginary sphere cantered on the particle, P_a (W) is the electromagnetic energy absorbed across the surface of an imaginary sphere centred on the particle and I_o (W/m²) is the intensity of the incident radiation. In practice, the wavelengths adopted in FSO communication are at 0.69, 0.85 and 1.55 μm, where molecular absorption is negligible ($C_a \sim 0$). By knowing C_s, one can estimate the value of the normalized cross section or the scattering efficiency defined by

$$Q_s = \frac{C_s}{\pi r_p^2} \tag{3.59}$$

Q_s is a function of the size parameter $x_0 = 2\pi r_p/\lambda$, a unitless quantity, with a maximum value of 3.8 at $r_p/\lambda \approx 1$ (i.e., maximum scattering of the optical signal). However, the particle radius of the fog varies in the spatial domain in the atmosphere.

On the basis of the assumptions that the scattered light has the same wavelength as the incident light, only single scattering occurs while the multiple scattering effects are neglected, and that the particles are spherical in shape and are acting independently with a complex refractive index in space, the following expression for the Mie scattering was derived in [58]

$$\gamma_t(\lambda) \cong \beta_a(\lambda) = 10^5 \int_0^\infty Q_s\left(\frac{2\pi r_p}{\lambda}, n_r\right) \pi r^2 n(r_p) dr_p \tag{3.60}$$

where n_r is real part of the complex refractive index and $n(r_p)$ is the volume concentration, that is, the number of particles per unit volume per unit increment in radius. Here, $\beta_a(\lambda)$ is the absorption (specific attenuation) measured in dB/km and is defined as

$$\beta_a = C_1 \lambda^{-\delta} \tag{3.61}$$

where C_1 and δ are constant parameters related to the particle size distribution and the visibility, and λ is the operating wavelength. β_a can be calculated by summing up the attenuation effect of all the individual fog droplets present per unit volume per unit increase in radius. C_1 is related to the visual range (or visibility) V through Kim's model

$$C_1 = \frac{3.91}{V}(0.55)^\delta \tag{3.62}$$

where the visual range is 0.550 μm.

$n(r_p)$ is the most important parameter, and can be regarded as the key parameter to determine physical and optical properties of the particles (including fog). Since fog consists of an accumulation of small spheres of water particles with a particular drop size distribution, the effect of the individual drops must be summed over a unit volume, assuming that there is no interactivity between these spheres (i.e., multiple scattering is

TABLE 3.10

Particle Size Distribution for Advection and Maritime Fog

Fog Type	α	a	b	N (num/cm³)	W (g/m³)	r_m (μm)	V (m)
Dense advection fog	3	0.027	0.3	20	0.37	10	130
Moderate convection fog	6	607.5	3	200	0.02	2	450

Source: Adapted from I. I. Kim, B. McArthur and E. Korevaar, *SPIE Proceedings: Optical Wireless Communications III*, 4214, 26–37, 2001; M. A. Naboulsi, H. Sizun and F. Fornel, *SPIE*, 5465, 168–179, 2004.

negligible). Generally, this distribution is represented by analytical functions such as the log-normal distribution for aerosols and the modified gamma distribution for fog. The latter is used largely to model the various types of fog and clouds. The number of particles per unit volume and per unit increment of radius r_p is given by

$$n(r_p) = ar_p^\alpha \exp(-br_p) \tag{3.63}$$

α, *a*, *b* are parameters that characterize the distribution of the advection and convection type of fog given in Table 3.10.

Using Equation 3.61 the plot of the particle size distribution for the advection and the maritime fog are shown in Figure 3.17. The measured fog droplet size distribution falls within the variation of the mentioned models.

2. *Empirical approach*   The attenuation due to the fog depends on factors such as the location, particle size distribution, liquid water content and average particle diameter. As the particle concentration and size distribution vary from one location to another, this makes the task of predicting fog-induced attenuation a challenging one. Generally, the fog particle size distribution is not available and not reported in the standard meteorological data. Therefore, the fog-induced attenuation of the

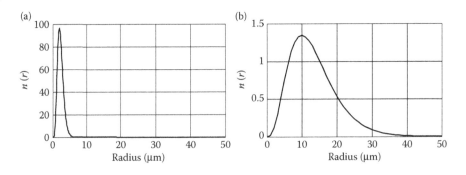

FIGURE 3.17 Particle size distribution versus particle radius (μm): (a) advection fog, (b) convection fog.

optical signal is only predicted using empirical models derived from the experimental observations. The empirical model uses the visibility V data in order to characterize the density of fog. The link visibility (i.e., the meteorological visual range, MVR) is used to measure the attenuation due to the fog [50,65–67]. The very fundamental law to measure the fog density indirectly on the basis of the visibility is the Koschmieder law. It defines the visibility as the distance to an object at which the visual contras/transmittance of an object drops to a certain value of the visual/ transmittance threshold T_{th} level of the original visual contrast (100%) along the propagation path. The meteorological visibility V (km) can therefore be expressed in terms of the atmospheric attenuation coefficient β_λ and T_{th} as

$$V = \frac{10\log(T_{th})}{\beta_\lambda} \quad (3.64)$$

T_{th} varies from 0.0077 to 0.06, where the smaller value defines a larger MVR for a certain atmosphere environment.

In 1924, Koschmieder defined a value for T_{th} of 2% following the Helmholtz theory. However, T_{th} is considered to be 5% for aeronautical requirements as the contrast of an object with respect to the surrounding area is much lower than that of an object against the horizon. Mostly used values of T_{th} are 2% and 5% [68]. The scattering coefficient of fog as earlier mentioned is a parameter that is dependent on the wavelength of the propagating optical beam and V. The following is a model developed by Kruse to link all these parameters together [69]

$$V(\text{km}) = \frac{10\log_{10} T_{th}}{\beta_\lambda} \left(\frac{\lambda}{\lambda_o}\right)^{-\delta} \quad (3.65)$$

Here, λ_o is the maximum spectrum of the solar band, δ is the particle size-related coefficient and its values are given as [69]

$$\delta = \begin{cases} 1.6 & \text{for } V > 50 \text{ km} \\ 1.3 & \text{for } 6 \text{ km} < V < 50 \text{ km} \\ 0.585V^{1/3} & \text{for } 0 \text{ km} < V < 6 \text{ km} \end{cases} \quad (3.67)$$

Figure 3.18a and b shows the atmospheric attenuation coefficient β_λ against the visibility for a range of wavelengths using the Kruse model for T_{th} of 2% and 5%, respectively. In Figure 3.18a for V of 500 m the attenuation difference at 0.69 and 1.55 μm is 10 dB increasing to 28 dB for the dense fog at V of 100 m. However, in Figure 3.18b with $T_{th} = 5\%$, the attenuation difference is 4 dB at V of 500 m increasing to 28 dB at V of 100 m. The Kruse model was originally developed for dense haze as reported in and is not very suitable to study the effect of fog for $V = 1$ km.

Further study based on the empirical data in region $V < 1$ km has indicated that β_λ is wavelength independent for $V < 0.5$ km [59]. This led to the Kim model which gives the values of δ as [56]

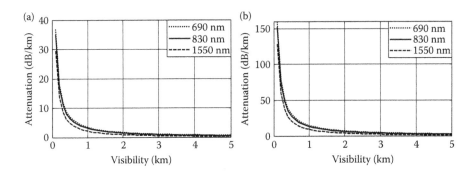

FIGURE 3.18 Attenuation versus the visibility using the Kruse model for (a) $T_{th} = 2\%$ and (b) $T_{th} = 5\%$.

$$\delta = \begin{cases} 1.6 & \text{for } V > 50 \text{ km} \\ 1.3 & \text{for } 6 \text{ km} < V < 50 \text{ km} \\ 0.16V + 0.34 & \text{for } 1 \text{ km} < V < 6 \text{ km} \\ V - 0.5 & \text{for } 0.5 \text{ km} < V < 1 \text{ km} \\ 0 & \text{for } V < 0.5 \text{ km} \end{cases} \tag{3.68}$$

The visibility range values under different weather conditions are presented in Table 3.11, while Figure 3.19 shows the attenuation coefficient values based on the Kim model against the visibility for a range of wavelengths. According to this model, in moderate-to-dense fog conditions, using higher wavelength for the FSO links offers no advantage over shorter ones since the attenuation coefficient is wavelength independent in such conditions. Further work on theoretical as well as practical investigation of this model for the low visibility is required to confirm that the attenuation is really independent of the wavelength for moderate-to-dense fog regimes as the Kim model was derived from calculation of exact Mie theory and by considering dense haze and fog.

TABLE 3.11

Weather Conditions and Their Visibility Range Values

Weather Condition	Visibility Range (m)
Thick fog	200
Moderate fog	500
Light fog	770–1000
Thin fog/heavy rain (25 mm/h)	1900–2000
Haze/medium rain (12.5 mm/h)	2800–40,000
Clear/drizzle (0.25 mm/h)	18,000–20,000
Very clear	23,000–50,000

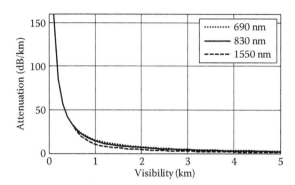

FIGURE 3.19 Kim model for visibility for T_{th} of 2% and a range of wavelengths.

The MATLAB codes for simulating the attenuation against visibility using the Kim's model are given in Program 3.5.

Program 3.5: MATLAB Codes for Simulation of Kim's Model

```
%****To evaluate attenuation coefficient as a function of
visibility
%***Using the Kim model

clear
clc
% Wavelength in nm
wavl=830;
% Visibility values in km
Visibility=0.1:0.1:50;

for i=1:length(Visibility)
  V=Visibility(i);
  if (V >= 50)
  q=1.6;
  elseif (V >=6) && (V< 50)
    q=1.3;
  elseif (V >=1) && (V< 6)
    q=0.16*V+0.34;
  elseif (V >=0.5) && (V< 1)
    q=V-0.5;
  else
    q=0;
  end

Att_coeff(i) = (3.91/V) * (wavl/550)^-q;
Att_coeff_dB_km(i) =10*Att_coeff(i)/log(10);
end
```

```
% Plot function
semilogy(Visibility,Att_coeff_dB_km)
xlabel('Visibility in km')
ylabel('Attenuation coefficient in dB/km')
```

Figure 3.20 shows the time series of attenuation results for an FSO link operating with a laser source of 785 nm wavelength in the city of Milan. This terrestrial FSO system goes into outage every time the measured attenuation exceeds the allowable dynamic range of 21 dB (~66 dB/km) set for atmospheric losses. From these data, the peak value of fog attenuation is estimated to be 154 dB/km at a corresponding visibility of ~114 m. The changes in specific attenuations were about ±8 dB/km averaged over a second scale [70]. This shows that optical attenuation in foggy environments poses a great challenge for a reliable operation of FSO links. In order to guarantee a reliable and 100% availability therefore, adequate link margin must be provided to account for the fog attenuation.

In Naboulsi proposed relations for the attenuation caused by the radiation and advection fog at a wavelength of 0.69–1.55 µm and a visibility range of 0.05–1 km. These are given by

$$\alpha_{con}(\lambda) = 10\log_e\left(\frac{0.11478\lambda + 3.8367}{V}\right) \tag{3.69}$$

$$\alpha_{adv}(\lambda) = 10\log_e\left(\frac{0.18126\lambda^2 + 0.13709\lambda + 3.7205}{V}\right) \tag{3.70}$$

This shows that the attenuation due to fog is higher at the 1.55 µm than 0.785 µm [59] which is in contradiction to the Kim's model for $V > 0.5$ km. Figure 3.21 shows the comparison of the attenuation due to fog using Kim and Naboulsi models for T_{th} of 2 and 5% at wavelengths of 830, 1550 and 2000 nm. The attenuation difference between the advection and the convection fog is very close to each other. The Naboulsi advection and convection fog models presents higher attenuation due to fog compared to the Kim model. For T_{th} of 2% the behaviour of the Kim model is very similar to the Naboulsi models (see Figure 3.21b). The Naboulsi models deviate from the Kim model for wavelengths higher than 900 nm (Figure 3.21c and d). The reason for this deviation is due to the experimental data collected from different locations.

Figure 3.22a–d shows the fog attenuation (dB/km) for the most important fog models for 0.5 µm $< \lambda <$ 1.55 µm and $V =$ 0.1, 0.25, 0.5 and 1 km. It shows the correlation for a range of 0.5 µm $< \lambda <$ 0.9 µm and for $V =$ 1 km. These models are also verified experimentally showing a good relation with the experimental data [71–73]. Kim's model underestimates the optical attenuation compared to the Naboulsi model, but the difference is small for 0.5 µm $< \lambda <$ 0.9 µm. Kim and Naboulsi convection models are best to estimate the attenuation of the optical signal in fog for the wavelength range of 0.5–0.9 µm and for $V =$ 1 km.

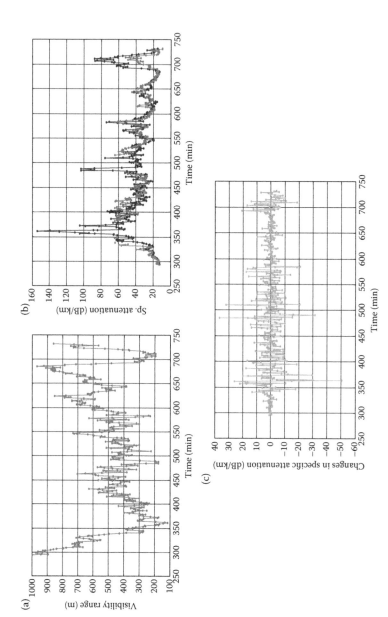

FIGURE 3.20 **(See colour insert.)** Time profiles of (a) visual range, (b) specific attenuation and (c) differences in specific attenuation during a fog event occurred in Milan on 11th January 2005. In (b) two profiles are shown: the measured laser attenuation (red curve) and the attenuation as estimated from visual range (blue curve). (Adapted from F. Nadeem et al., *1st International Conference on Wireless Communication, Vehicular Technology, Information Theory and Aerospace & Electronic Systems Technology,* 2009, pp. 565–570.)

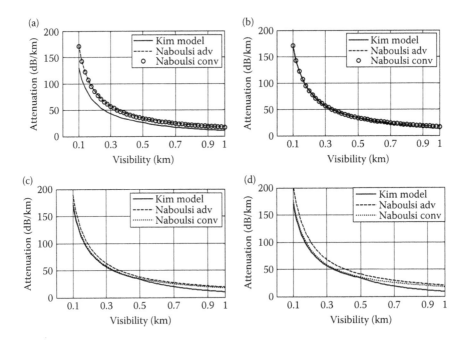

FIGURE 3.21 Attenuation versus visibility for Kim and Naboulsi models for fog at λ of 0.83 µm: (a) T_{th} of 5% and (b) T_{th} of 2%; and for T_{th} of 2% (c) 1550 nm and (d) 2000 nm.

The loss for rain and snow are defined as [74,75]

$$\delta_{rain} = \frac{2.9}{V}$$
$$\delta_{snow} = 20\delta_{rain}$$

(3.71)

The power loss due to the rain and snow are so low compared to that due to the Mie scattering, they still have to be accounted for in the link margin during the link budget analysis. A typical rainfall of 2.5 cm/h could result in an attenuation of ~6 dB/km [76] while a typical value for attenuation due to light snow to blizzard is 3–30 dB/km [52]. In early 2008 in Prague, Czech Republic, fog attenuation for an FSO link was measured and compared with the empirical fog attenuation models. This result is shown in Figure 3.23, with a visibility of less than 200 m, corresponding to thick fog, the attenuation at the location is ~200 dB/km. The empirical models provide a reasonable fit to the measured values with a maximum of about ±5 dB/km difference between any two empirical models.

3.3.3 BEAM DIVERGENCE

One of the main advantages of FSO systems is the ability to transmit a very narrow optical beam, thus offering enhanced security. But due to diffraction, the beam spreads out. This results in a situation in which the receiver aperture is only able to collect a fraction

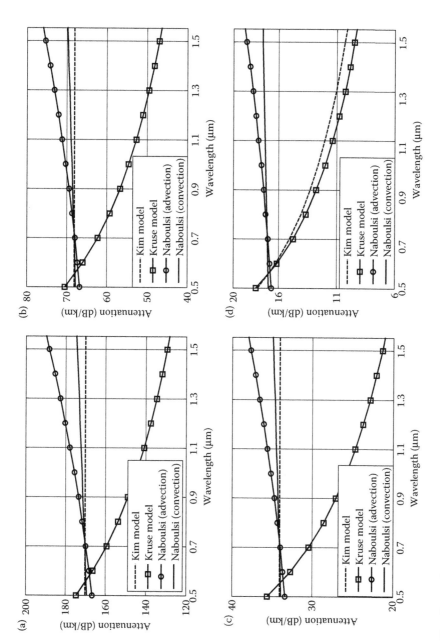

FIGURE 3.22 Comparison of different models for a range of wavelengths for different visibility values. (a) 0.1 km, (b) 0.25 km, (c) 0.5 km and (d) 1 km.

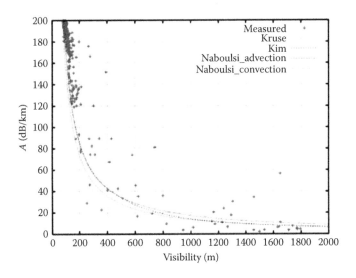

FIGURE 3.23 **(See colour insert.)** Measured attenuation coefficient as a function of visibility range at λ = 830 nm in early 2008, Prague, Czech Republic. (Adapted from M. Grabner and V. Kvicera, Experimental study of atmospheric visibility and optical wave attenuation for free-space optics communications, http://ursi-france.institut-telecom.fr/pages/pages_ursi/URSIGA08/papers/F06p5.pdf, last visited 2nd Sept. 2009.)

of the beam. The remaining uncollected beam then results in beam divergence loss. For relatively low data rate systems, a typical value for the beam divergence is 68 mrad and at higher data rates, 2 mrad beam divergence is commonly used. However, what really determines the size of the beam divergence are the optical channel and the pointing jitter. Considering the arrangement of an FSO communication link of Figure 3.24, and by invoking the thin lens approximation to the diffuse optical source whose irradiance is represented by I_s, the amount of optical power focused on the detector is derived as [78]

$$P_r = \frac{I_s A_t A_r}{L_p^2 A_s} \qquad (3.72)$$

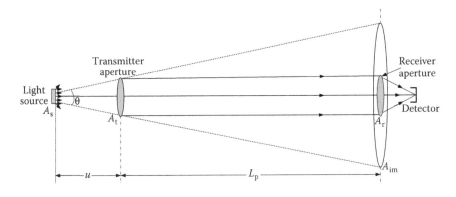

FIGURE 3.24 Beam divergence.

where A_t and A_r are the transmitter and receiver effective aperture areas, respectively while A_s is the area of the optical source and L_p is the path length. This clearly shows that a source with high radiance I_s/A_s and wide apertures are required in order to increase the received optical power.

For a nondiffuse, small source such as the laser, the size of the image formed at the receiver plane is no longer given by the thin lens approximation; it is determined by diffraction at the transmitter aperture. The diffraction pattern produced by a uniformly illuminated circular aperture of diameter d_t is known to consist of a set of concentric rings. The image size is said to be diffraction limited when the radius of the first intensity minimum or dark ring of the diffraction pattern becomes comparable in size with the diameter d_{im} of the normally focused image [78]. That is

$$d_{im} = \frac{L_p}{u} d_s < 1.22 \frac{\lambda L}{d_t} \tag{3.73}$$

Therefore, the source diameter is

$$d_s < 1.22 \frac{\lambda u}{d_t} \approx 1.22 \frac{\lambda f_1}{d_t} \tag{3.74}$$

where f_1 is the focal length, u is the distance from the source to the centre of the transmitter lens. The equation above shows that for diffraction to be the sole cause of beam divergence (diffraction limited), $d_s < 1.22 \lambda f_1/d_t$. A laser being inherently collimated and coherent normally produces a diffraction-limited image. The diffraction limited beam divergence angle in radians is thus given by $\theta_b \cong \lambda/d_t$. If the transmitter and receiver effective antenna gains are, respectively, given by

$$G_t = \frac{4\pi}{\Omega_b} \tag{3.75}$$

$$G_r = \frac{4\pi A_t}{\lambda^2} \tag{3.76}$$

And the free-space path loss is given by

$$L_{fs} = \left(\frac{\lambda}{4\pi L_p} \right)^2 \tag{3.77}$$

The received optical power then becomes

$$P_r = P_t L_{fs} G_t G_r = P_t \frac{4 A_r}{L_p^2 \Omega_b} \cong P_t \left(\frac{4}{\pi} \right)^2 \frac{A_t A_r}{L_p^2 \lambda^2} \tag{3.78}$$

where the radiation solid angle $\Omega_b \cong \pi\theta_b^2/4$. The diffraction limited beam spreading/geometric loss in dB is thus given by

$$L_{gl} = -10\left[\log\left(\frac{A_t A_r}{L_p^2 \lambda^2}\right) + 2\log\left(\frac{4}{\pi}\right)\right] \tag{3.79}$$

The same result given by Equation 3.79 can be obtained by substituting $A_{im} = \theta_b L_p$ for the image size in $P_r = P_t A_r/A_{im}$. In order to reduce the diffraction-limited beam divergence, a beam expander of the type shown in Figure 3.25, in which the diffracting aperture has been increased, can then be used. The beam expander reduces the beam divergence loss and increases the received signal power in the process.

For most practical sources, the beam divergence angle is usually greater than that dictated by diffraction and for a source with an angle of divergence θ_s, the beam size at a distance L_p away is $(d_t + \theta_s L_p)$. The fraction of the received power to the transmitted power is therefore given as

$$\frac{P_r}{P_t} = \frac{A_r}{A_{im}} = \frac{d_r^2}{\left(d_t + \theta_s L_p\right)^2} \tag{3.80}$$

where d_r is the receiving lens diameter.

The geometric loss in decibels is thus

$$L_{geom} = -20\log\left[\frac{d_r}{\left(d_t + \theta_s L_p\right)}\right] \tag{3.81}$$

The beam spreading loss for the diffraction limited source given by Equation 3.79 is, expectedly, lower than that of the nondiffraction limited case represented by Equation 3.81, this is because the image size is smaller by d_t in the diffraction limited case.

From the foregoing, it is clear that a source with a very narrow angle of divergence is preferable in terrestrial FSO systems. It should however be mentioned that, wide divergence angle sources are desirable in short-range FSO links to ease the alignment requirement, compensate for building sway and eliminate the need for

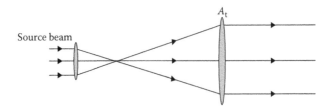

FIGURE 3.25 Typical beam expander.

active tracking systems at the expense of increased geometric loss apparently. A typical FSO transceiver has optical beam divergence in the range of 2–10 mrad and 0.05–1.0 mrad (equivalent to a beam spread of 2–10, and 5 cm to 1 m, respectively, at 1 km link range) for systems without and with tracking, respectively.

3.3.4 OPTICAL AND WINDOW LOSS

This class of power loss includes losses due to imperfect lenses and other optical elements used in the design of both the transmitter and receiver. It accounts for the reflection, absorption and scattering due to the lenses in the system [52]. The value of the optical loss L_{op}, can be obtained from the component manufacturer. It apparently depends on the characteristics of the equipments and quality of the lenses used. For FSO transceivers installed behind windows within a building, there exists an additional optical power loss due to the window glass attenuation. Although (glass) windows allow optical signals to pass through them, they contribute to the overall power loss of the signal. Uncoated glass windows usually attenuate 4% per surface, because of reflection. Coated windows display much higher losses and its magnitude is wavelength dependent.

3.3.5 POINTING LOSS

Additional power penalty is usually incurred due to lack of perfect alignment of the transmitter and receiver. The resulting power loss is catered for by including pointing/misalignment loss L_p in the link budget analysis. The transmitter is directly pointed at the receiver, the pointing loss is 0 dB. For short FSO links (<1 km), this might not be an issue but for longer link ranges, this can certainly not be neglected. Misalignments could result from building sway or strong wind effect on the FOS link head stand.

Including all the losses the received optical power then becomes

$$P_r = P_t \left(\eta_t \eta_A \frac{4\pi A_t}{\lambda_t^2} \right) L_{tp} L_{atm} L_{pol} L_{rp} \left(\frac{A_r}{4\pi L^2} \right) \eta_r \qquad (3.82)$$

where η_t is the transmit optics efficiency, η_A is the aperture illumination efficiency of the transmit antenna, λ_t is the transmit wavelength. L_{atm} and L_{pol} are the fractional loss due to absorption of the transmitting medium and mismatch of the transmit and receive antenna polarization patterns, respectively, the term $A_r/4\pi L_p^2$ is the fraction of power collected by A_r if the transmitter is an ideal isotropic radiator. η_r is the receiving optics collecting efficiency, L_{tp} and L_{rp} is the transmitter and receiver pointing losses, respectively.

For short links between two fixed terminals once the initial set-up is carried out there is no requirement for acquisition and tracking. However, for very tall buildings, where temperature and wind can cause swaying, the relative motion of buildings to each other should be considered when determining the transmit beam divergence angle and the FOV of the receiver. For links between mobile and fixed terminals, the

TABLE 3.12
Link Budget for the FSO Link

Parameters	No Tracking	With Tracking
Transmitter		
Power	30 dBm	30 dBm
Losses	3 dB	3 dB
Mis-pointing allowance	6 dB	3 dB
Channel losses	30 dB	14 dB
Receiver losses	3 dB	3 dB
Receiver sensitivity	−23 dB	−23 dB
Link safety margin—*for weather conditions*	5 dB	27 dB

knowledge of the exact position of the latter can be used by the former to relax the acquisition process. For the links between two mobile terminals there is no need to have a prior knowledge about the position of the counter terminal.

Table 3.12 shows a typical link budget for 2.5 Gbps for a 2 km link and 1550 nm wavelength [79].

3.3.6 THE ATMOSPHERIC TURBULENCE MODELS

Solar radiation absorbed by the Earth's surface causes air around the earth's surface to be warmer than that at higher altitude. This layer of warmer air becomes less dense and rises to mix turbulently with the surrounding cooler air causing the air temperature to fluctuate randomly [53]. Inhomogeneities caused by turbulence can be viewed as discrete cells, or eddies of different temperature, acting like refractive prisms of different sizes and indices of refraction. The interaction between the laser beam and the turbulent medium results in random phase and amplitude variations (scintillation) of the information-bearing optical beam which ultimately results in fading of the received optical power, thus leading to the system performance degradation. Of all the effects of atmospheric turbulence, only the modelling of the atmospheric turbulence-induced fluctuation of the received optical power (or irradiance) will be discussed in this chapter, since in the intensity modulated, direct detection FSO system under consideration, it is only the received power/irradiance that matters. Atmospheric turbulence is usually categorized in regimes depending on the magnitude of index of refraction variation and inhomogeneities. These regimes are a function of the distance travelled by the optical radiation through the atmosphere and are classified as weak, moderate, strong and saturation. Atmospheric turbulence results in signal fading thus impairing the FSO link performance severely.

Here we introduce models describing the pdf statistics of the irradiance fluctuation. Unfortunately, due to the extreme complexity involved in mathematically modelling of atmospheric turbulence, a universal model valid for all the turbulence regimes does not currently exist. As such, this chapter will review three most reported models for irradiance fluctuation in a turbulent channel. These are the log-normal, gamma–gamma and negative exponential models. Their respective ranges of validity, as reported in the literature [80–84], are in the weak, weak-to-strong and saturate regimes.

Atmospheric turbulence results in random fluctuation of the atmospheric refractive index n along the path of the optical field/radiation traversing the atmosphere. This refractive index fluctuation is the direct end product of random variations in atmospheric temperature from point to point [53]. These random temperature changes are a function of the atmospheric pressure, altitude and wind speed. The smallest and the largest of the turbulence eddies are termed as the inner scale l_0, and the outer scale L_0, of the turbulence, respectively. l_0 and L_0 are typically of the order of a few millimetres and several metres, respectively [82,85]. These weak lens-like eddies shown graphically in Figure 3.26 result in a randomized interference effect between different regions of the propagating beam causing the wavefront to be distorted in the process. Known effects of atmospheric turbulence include [81,82]: beam spreading, beam wander and beam scintillation. The beam wander effect caused by a large-scale turbulence, and diffraction effects are often assumed to be insignificant, in particular, when the receiver aperture diameter D is greater than the size of Fresnel zone $(L/k)^{0.5}$, where $k = 2\pi/\lambda$ is the optical wave number. Since the change of beam wander-induced fluctuation is very slow, it can be readily taken care of using the tracking schemes [81,82].

In an attempt to model the turbulent atmospheric channel, the widely accepted 'Taylor hypothesis' [80], which says that the turbulent eddies are fixed or frozen and can only move in their frozen form with the transverse component of the mean local wind will be followed. This hypothesis means that the temporal variations in the beam pattern or its statistical properties are caused by the component of the local wind that is perpendicular to the beam direction of propagation. In addition, the temporal coherence time τ_0 of atmospheric turbulence is reported to be of the order

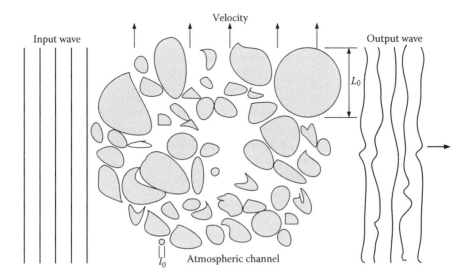

FIGURE 3.26 Atmospheric channel with turbulent eddies.

of milliseconds [84]. This value is very large compared to the duration of a typical data symbol, thus the turbulent atmospheric channel can be described as a 'slow fading channel' since it is static over the duration of a data symbol.

The relationship between the temperature of the atmosphere and its refractive index is given by Equation 3.83 [86] while for most engineering applications, the rate of change of the refractive index with respect to channel temperature is represented by Equation 3.8 [80]

$$n_{as} = 1 + 77.6(1 + 7.52 \times 10^{-3}\lambda^{-2})\frac{P_{as}}{T_e} \times 10^{-6} \tag{3.83}$$

$$\frac{-dn_{as}}{dT_e} = 7.8 \times 10^{-5}\frac{P_{as}}{T_e^2} \tag{3.84}$$

where P_{as} is the atmospheric pressure in millibars, T_e is the effective temperature in Kelvin and λ the wavelength in microns. Near the sea level, $-dn_{as}/dT_e \cong 10^{-6}\ K^{-1}$ [80]. The contribution of humidity to the refractive index fluctuation is not accounted for in Equation 3.83 because this is negligible at optical wavelengths [84].

The position and time-dependent index of refraction denoted by $n_{as}(r, t)$ can be expressed as the sum of its free-space (no turbulence) value n_{as_0}, and a turbulence-induced random fluctuation component $n_{as_1}(r,t)$. Thus

$$n_{as}(r,t) = n_{as_0} + n_{as_1}(r,t) \tag{3.85}$$

In accordance with the Taylor's 'frozen-flow' hypothesis, which implies that the temporal variations of the index of refraction of the channel are mainly due to the transverse component of the atmospheric wind, the randomly fluctuating part of Equation 3.85 can then be written as

$$n_{as_1}(r,t) = n_{as_1}(r - v_w t) \tag{3.86}$$

where $v_w(r)$ is the local wind velocity perpendicular to the field direction of travel.

In atmospheric turbulence, an important parameter for characterizing the amount of refractive index fluctuation is the index of refraction structure parameter C_n^2 introduced by Kolmogorov [87], which is a function of the wavelength, atmospheric altitude and temperature. A commonly used model to describe C_n^2 in terms of altitude is the Hufnagel–Valley (H–V) model given below as [82]

$$C_n^2(\hbar) = 0.00594(v_w/27)^2(10^{-5}\hbar)^{10}\exp(-\hbar/1000)$$
$$+ 2.7 \times 10^{-16}\exp(-\hbar/1500) + \hat{A}\exp(-\hbar/100) \tag{3.87a}$$

where $\hat{A}$ is taken as the nominal value of $C_n^2(0)$ at the ground level in $m^{-2/3}$ and $\hbar$ is the altitude in metres.

The most commonly used values for v_w is 21 m/s. The value of the index of refraction structure parameter varies with altitude, but for a horizontally propagating field it is usually assumed constant. For the link near the ground level C_n^2 is taken to be ~1.7×10^{-14} m$^{-2/3}$ and ~8.5×10^{-15} m$^{-2/3}$ for during daytime and at night, respectively. C_n^2 typically ranges from 10^{-12} m$^{-2/3}$ for the strong turbulence to 10^{-17} m$^{-2/3}$ for the weak turbulence with a typical average value being 10^{-15} m$^{-2/3}$ [83]. The strong turbulence is characterized by the way in which the statistical moments of velocity increments increase with the spatial separation L_s. Within the inertial subrange the refractive index structure is defined by the Kolmogorov two-thirds power law given as [82,88]

$$D_n(L_s) = C_n^2(\hbar)L_s^{2/3} \quad l_0 < L_s < L_0 \tag{3.87b}$$

A similar parameter for temperature variations is the temperature structure parameter and it is related to C_n^2 by [80]

$$C_n^2 = \left(\frac{dn_{as}}{dT_e}\right)^2 C_T^2 \tag{3.88}$$

In the spectral domain, the power spectral density of the refractive index fluctuation is related to C_n^2 by the following expression [87,89]

$$\Phi_n(K) = 0.033 C_n^2 K^{-11/3}; \quad 2\pi/L_0 \ll K \ll 2\pi/l_0 \tag{3.89}$$

where K is the spatial wave number. For a wider range of K, however, this expression has been modified by Tatarski and von Karman and reported in Refs. [83,90,91].

There are a number of other models as outlined below

1. Gurvich model [88]

    a. For $C_n^2|_{2.5m} > 10^{-13}$ m$^{-2/3}$ (strong turbulence):

$$C_n^2(\hbar) = \begin{cases} C_n^2|_{2.5m} (\hbar/2.5)^{-4/3} & 2.5\text{ m} \le \hbar \le 1000\text{ m} \\ C_n^2|_{1000m} \exp[-(\hbar - 1000)/9000] & 1000\text{ m} < \hbar \end{cases} \tag{3.90}$$

    b. For 10^{-13} m$^{-2/3} \ge C_n^2|_{2.5m} > 6.5 \times 10^{-15}$ m$^{-2/3}$

$$C_n^2(\hbar) = \begin{cases} C_n^2|_{2.5m} \left(\dfrac{\hbar}{2.5}\right)^{-\frac{2}{3}} & 2.5\text{ m} \le \hbar \le 50\text{ m} \\[2ex] C_n^2|_{50m} \hbar \left(\dfrac{\hbar}{2.5}\right)^{-\frac{4}{3}} & 50\text{ m} \le \hbar \le 1000\text{ m} \\[2ex] C_n^2|_{1000m} \exp[-(\hbar - 1000)/9000] & 1000\text{ m} < \hbar \end{cases} \tag{3.91}$$

c. For $16.5 \times 10^{-15} \text{m}^{-2/3} \geq C_n^2\big|_{2.5\text{m}} > 4.3 \times 10^{-16} \text{m}^{-2/3}$

$$C_n^2(\hbar) = \begin{cases} C_n^2\big|_{2.5\text{m}} (\hbar/2.5)^{-4/3} & 2.5\text{ m} \leq \hbar \leq 1000\text{ m} \\ C_n^2\big|_{1000\text{m}} \exp\left[-(\hbar - 1000)/9000\right] & 1000\text{ m} < \hbar \end{cases} \qquad (3.92)$$

d. For $4.3 \times 10^{-16} \text{m}^{-2/3} \geq C_n^2\big|_{2.5\text{m}}$ (weak turbulence):

$$C_n^2(\hbar) = \begin{cases} C_n^2\big|_{2.5\text{m}} & 2.5\text{ m} \leq \hbar \leq 1000\text{ m} \\ C_n^2\big|_{1000\text{m}} \exp\left[-(\hbar - 1000)/9000\right] & 1000\text{ m} < \hbar \end{cases} \qquad (3.93)$$

2. HV-Night model [62]—This is the modified HV which is given by

$$C_n^2(\hbar) = 1.9 \times 10^{-15} \exp(-\hbar/100) + 8.26 \times 10^{-54} \hbar^{10}$$
$$\times \exp(-\hbar/1000) + 3.02 \times 10^{-17} \exp(-\hbar/1500) \qquad (3.94)$$

3. Greenwood model [92]—A night time turbulence model which is mainly used for astronomical imaging from high grounds, which is given by

4.

$$C_n^2(\hbar) = [2.2 \times 10^{-15}(\hbar + 10)^{-1.3} + 4.3 \times 10^{-17}]\exp(-\hbar/4000) \qquad (3.95)$$

5. Submarine Laser Communication-Day model

$$C_n^2(\hbar) = \begin{cases} 0 & 0\text{ m} < \hbar < 19\text{ m} \\ 4.008 \times 10^{-13} \hbar^{-1.054} & 19\text{ m} < \hbar < 230\text{ m} \\ 1.3 \times 10^{-15} & 230\text{ m} < \hbar < 850\text{ m} \\ 6.352 \times 10^{-7} \hbar^{-2.966} & 850\text{ m} < \hbar < 7000\text{ m} \\ 6.352 \times 10^{-16} \hbar^{-0.6229} & 7000\text{ m} < \hbar < 20{,}000\text{ m} \end{cases} \qquad (3.96)$$

By their very nature, turbulent media are extremely difficult to describe mathematically, the difficulty, according to Ref. [84], is primarily due to the presence of nonlinear mixing of observable quantities which is fundamental to the process. In order to derive expressions for the statistical properties, namely the pdf and variance, of an optical beam travelling through the turbulent atmosphere; the following simplifying but valid assumptions will be used to reduce the mathematics to a manageable level [84]

1. The atmosphere is a nondissipative channel for the propagating wave. This assumption can be explained thus, in the event of absorption of the propagating wave or radiation by the atmosphere, the heat so generated is insignificant compared to diurnal contributions.

2. The process of scattering by the turbulent eddies does not result in loss of energy from the beam. Hence, the mean energy in the presence of turbulence is assumed equal to the mean energy in the absence of turbulence. This assumption is valid for plane and spherical waves. The plane wave is generally applicable to laser beams propagating over a long distance [51,84].

3.3.6.1 Log-Normal Turbulence Model

In describing the pdf of the irradiance fluctuation in a turbulent atmosphere, the beam is first represented by its constituent electric field $\vec{E}$. By employing Maxwell's electro-magnetic equations for the case of a spatially variant dielectric like the atmosphere, the following expression is derived [84]

$$\nabla^2 \vec{E} + k^2 n_{as}^2 \vec{E} + 2\nabla[\vec{E} \cdot \vec{\nabla} \ln(n_{as})] = 0 \tag{3.97}$$

where the wave number $k = 2\pi/\lambda$, and the vector gradient operator $\vec{\nabla} = (\partial/\partial x)\, i + (\partial/\partial y)\, j + (\partial/\partial z)\, k$ with i, j and k being the unit vectors along the x, y and z axes, respectively. The last term on the left-hand side of Equation 3.97 represents the turbulence-induced depolarization of the wave. In a weak atmospheric turbulence regime, which is characterized by single scattering event, the wave depolarization is negligible [80,93,94]. In fact, it has been shown both theoretically [95] and experimentally [96] that the depolarization is insignificant even for strong turbulence conditions. Equation 3.97 then reduces to

$$\nabla^2 \vec{E} + k^2 n_{as}^2 \vec{E} = 0 \tag{3.98}$$

The position vector will henceforth be denoted by r and $\vec{E}$ represented by $E(r)$ for convenience. In solving this last equation, Tatarski [90] in his approach introduced a Gaussian complex variable $\Psi(r)$ defined as the natural logarithm of the propagating field $E(r)$, and termed it as the Rytov transformation. That is,

$$\Psi(r) = \ln[E(r)] \tag{3.99}$$

The Rytov approach is also based on a fundamental assumption that the turbulence is weak and that it is characterized by single scattering process. By invoking the Rytov transformation (3.99), and equating the mean refractive index of the channel n_0, to unity, Equation 3.98 transforms to the following Riccati equation whose solution already exists

$$\nabla^2 \Psi + (\nabla\Psi)^2 + k^2(1 + n_{as_1})^2 = 0 \tag{3.100}$$

The next stage involves breaking $\Psi(r)$ down to its free-space form $\Psi_0(r)$, and its turbulence-induced departure form is represented by $\Psi_1(r)$. This is done via the smooth perturbing method [84], which in effect implies that $\Psi(r) = \Psi_0(r) + \Psi_1(r)$. Combining this with the Rytov change of variable (3.100) results in the following

$$\Psi_1(r) = \Psi(r) - \Psi_0(r) \tag{3.101}$$

$$\Psi_1(r) = \ln[E(r)] - \ln[E_0(r)] = \ln\left[\frac{E(r)}{E_0(r)}\right] \tag{3.102}$$

where the electric field and its free-space (without turbulence) form $E_0(r)$ are by definition given as

$$E(r) = A(r)\exp(i\phi(r)) \tag{3.103}$$

$$E_0(r) = A_0(r)\exp(i\phi_0(r)) \tag{3.104}$$

where $A(r)$ and $\phi(r)$, and $A_0(r)$ and $\phi_0(r)$ represent the amplitude and phase of the actual field with and without atmospheric turbulence, respectively.

These transformations can then be used to arrive at the solution of Equation 3.97 which describes the behaviour of a field in weak atmospheric turbulence. In finding the irradiance fluctuation statistical distribution, first combine Equations 3.99 and 3.100 to arrive at the turbulence-induced field amplitude fluctuation given below as

$$\Psi_1(r) = \ln\left[\frac{A(r)}{A_0(r)}\right] + i[\phi(r) - \phi_0(r)] = \mathcal{X} + i\mathcal{S} \tag{3.105}$$

Since $\Psi_1(r)$ is Gaussian, it follows therefore that, $\mathcal{X}$ is the Gaussian distributed log-amplitude fluctuation, and similarly $\mathcal{S}$ is the Gaussian distributed phase fluctuation of the field. By concentrating only on the field amplitude, however, the pdf of $\mathcal{X}$ is thus [83,84]

$$p(\mathcal{X}) = \frac{1}{\sqrt{2\pi\sigma_x^2}}\exp\left\{-\frac{(\mathcal{X} - E[\mathcal{X}])^2}{2\sigma_x^2}\right\} \tag{3.106}$$

where $E[\mathcal{X}]$ denotes the expectation of $\mathcal{X}$ and σ_x^2 is the log-amplitude variance, commonly referred to as the Rytov parameter. According to Ref. [80], the σ_x^2 which characterizes the extent of field amplitude fluctuation in atmospheric turbulence is related to the index of refraction structure parameter, the horizontal distance L_p, travelled by the optical field/radiation by the following equations

$$\sigma_x^2 = 0.56k^{7/6}\int_0^{L_p} C_n^2(x)(L_p - x)^{5/6}\,dx \quad \text{for a plane wave} \tag{3.107}$$

and

$$\sigma_x^2 = 0.563k^{7/6}\int_0^{L_p} C_n^2(x)(x/L)^{5/6}(L_p - x)^{5/6}dx \quad \text{for a spherical wave} \quad (3.108)$$

For a field propagating horizontally through the turbulent medium, as is the case in most terrestrial applications, the refractive index structure parameter C_n^2, is constant, and the log irradiance variance for a plane wave becomes

$$\sigma_l^2 = 1.23C_n^2k^{7/6}L_p^{11/6} \tag{3.109}$$

The field irradiance (intensity) in the turbulent medium is $I = |A(r)|^2$ while the intensity in free-space (no turbulence) is given by $I_0 = |A_0(r)|^2$, the log-intensity is then given by

$$l = \log_e \left|\frac{A(r)}{A_0(r)}\right|^2 = 2\chi \tag{3.110}$$

Hence

$$I = I_0 \exp(l) \tag{3.111}$$

To obtain the irradiance pdf, invoke the transformation of variable $p(I) = p(\chi)\left|\dfrac{d\chi}{dl}\right|$, to arrive at the log-normal distribution function given by

$$p(I) = \frac{1}{\sqrt{2\pi\sigma_l^2}}\frac{1}{I}\exp\left\{-\frac{(\ln(I/I_0) - E[l])^2}{2\sigma_l^2}\right\} \quad I \geq 0 \tag{3.112}$$

In the region of weak fluctuations, the statistics of the irradiance fluctuations have been experimentally found to obey the log-normal distribution. From Equation 3.106 the log-intensity variance $\sigma_l^2 = 4\sigma_x^2$ and the mean log intensity $E[l] = 2E[X]$. Based on the second assumption of Section 3.3.6, it follows that $E[\exp(l)] = E[I/I_0] = 1$ since there is no energy loss during the turbulence-induced scattering process and, as such, $E[I] = I_0$. The expectation, $E[l]$, is obtained by invoking the standard relation (3.113) which is valid for any real-valued Gaussian random variable [83].
An expression for $E[l]$ is then obtained as illustrated in the following steps

$$E[\exp(az)] = \exp(aE[z] + 0.5a^2\sigma_z^2) \tag{3.113}$$

$$1 = \exp(E[l] + 0.5\sigma_l^2) \tag{3.114}$$

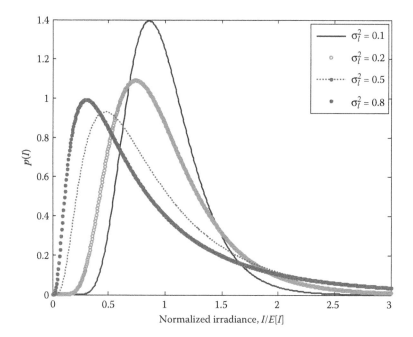

FIGURE 3.27 Log-normal pdf with $E[I] = 1$ for a range of log irradiance variance σ_l^2.

Hence

$$E[l] = -\frac{\sigma_l^2}{2} \qquad (3.115)$$

The log-normal pdf is plotted in Figure 3.27 for different values of log-irradiance variance σ_l^2. As the value of σ_l^2 increases, the distribution becomes more skewed with longer tails in the infinity direction. This denotes the extent of fluctuation of the irradiance as the channel inhomogeneity increases.

The MATLAB codes for the log-normal pdf are given in Program 3.6.

Program 3.6: MATLAB Codes for the Log-Normal pdf

```
%PDF of log normal distribution
clear
clc

% E[I] value
Io=1;

I=0:0.005:3;

%Log irradinace variance values
Var_l=[0.1,0.2,0.5,0.8];
```

```
for i=1:length(Var_1)
  for j=1:length(I)
    B=sqrt(2*pi*Var_1(i));
    C(j)=log(I(j)/Io)+(Var_1(i)/2);
    D=2*Var_1(i);
    pdf(i,j)=(1/I(j))*(1/B)*exp(-((C(j))^2)/D);
  end
end

%plot function
plot((I./Io),pdf)
xlabel('Normalised Irradiance, I/E[I]')
ylabel('p(I)')
```

After obtaining the pdf of the irradiance fluctuation, it is also paramount to derive an expression for the variance of the irradiance fluctuation σ_I^2, which characterizes the strength of irradiance fluctuation. This is carried out in the following steps:

$$\sigma_I^2 = E[I^2] - E[I]^2 = I_0^2\{E[\exp(2l)] - E[\exp(l)]^2\} \qquad (3.116)$$

By applying Equation 3.111 into Equation 3.115 and substituting for $E[l]$, the intensity variance is obtained as

$$\sigma_I^2 = I_0^2[\exp(\sigma_l^2) - 1] \qquad (3.117)$$

The normalized variance of intensity, often referred to as the scintillation index (S.I.), is thus

$$S.I. = \sigma_N^2 = \frac{\sigma_I^2}{I_0^2} = \exp(\sigma_l^2) - 1 \qquad (3.118)$$

3.3.6.2 Spatial Coherence in Weak Turbulence

When a coherent optical radiation propagates through a turbulent medium like the atmosphere, it experiences a decrease in its spatial coherence. The extent of this coherence degradation is, of course, a function of the atmospheric turbulence strength and the propagation distance. In effect, the turbulent channel breaks the coherent radiation up into various fragments whose diameters represent the reduced spatial coherence distance. Following from the Rytov approach used in modelling weak atmospheric turbulence, the spatial coherence of a field travelling through the atmosphere can be derived as [80]

$$\Gamma_x(\rho) = A^2 \exp[-(\rho/\rho_0)^{5/3}] \qquad (3.119)$$

where ρ_0 is the transverse coherence length of the field, this is the transverse distance at which the coherence of the field is reduced to e^{-1}. Equations 3.120 and 3.121 give the expressions for the transverse coherence distance ρ_0 for plane and spherical waves, respectively [80].

$$\rho_0 = \left[1.45k^2 \int\limits_0^{L_p} C_n^2(x)\mathrm{d}x \right]^{-3/5} \tag{3.120}$$

$$\rho_0 = \left[1.45k^2 \int\limits_0^{L_p} C_n^2(x)(x/L)^{5/3}\mathrm{d}x \right]^{-3/5} \tag{3.121}$$

The coherence distance is particularly useful in determining the size of the receiver aperture needed to collect the bulk of the propagating field through the process called aperture averaging, and also to determine the separation distance of detectors in a multiple receiver system. In order for the detectors in the array to receive uncorrelated signals, they must be spaced at a minimum distance ρ_0 apart. Figures 3.28 and 3.29 illustrate the coherence length for a horizontal link at two commonly used FSO wavelengths of 850 and 1550 nm, for typical values of C_n^2. The spatial coherence length is generally longer at longer wavelengths and decreases as both the propagation distance and the atmospheric turbulence increase.

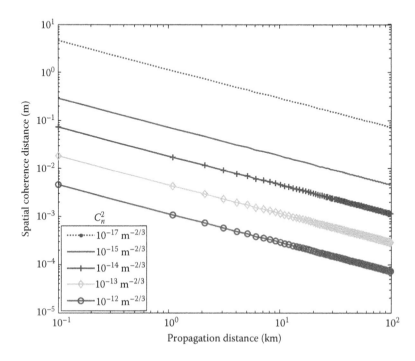

FIGURE 3.28   Plane wave transverse coherence length for $\lambda = 850$ nm and a range of C_n^2.

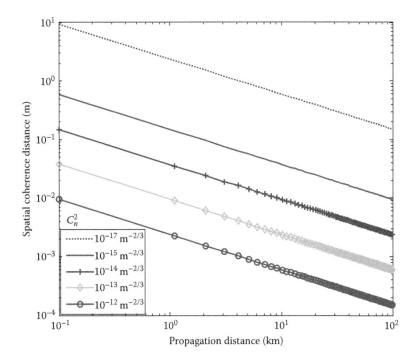

FIGURE 3.29   Plane wave transverse coherence length for $\lambda = 1550$ nm and a range of C_n^2.

3.3.6.3   Limit of Log-Normal Turbulence Model

So far, the Rytov approximation has been used to describe atmospheric turbulence and the log-normal turbulence model has been derived. The Rytov approximation predicts that the Rytov parameter increases without limit with the index of refraction structure parameter and/or the path length. However, based on experimental results as reported in Refs. [80,84], this prediction holds only in the weak turbulence regime when $\sigma_x^2 \leq 0.3$. As the turbulence strength increases beyond the weak regime, due to a combination of increased path length and/or increased C_n^2, the turbulent eddies result in multiple scatterings that are not accounted for by Rytov in his approximation [80,82]. On the basis of experiments reported in Refs. [80,84], the *S.I.* increases linearly with the Rytov parameter within the weak regime and continues to increase to a maximum value greater than unity. The regime in which *S.I.* attains its maximum value characterizes the highest strength of inhomogeneity or random focusing. The *S.I.* then starts to decrease due to self-interference as a result of multiple scattering and approaches unity as the Rytov parameter increases [84]. This observation is in contrast to the prediction of the Rytov approximation beyond the weak atmospheric regime. As the strength of atmospheric turbulence increases, multiple scattering effects must be considered and log-normal statistics display large deviations compared to the experimental data [82,97]. In the next section, two other irradiance fluctuation models that account for the multiple scatterings will be reviewed and their pdfs will be presented.

3.3.6.4 The Gamma–Gamma Turbulence Model

This model proposed by Andrews et al. [82] is based on the modulation process where the fluctuation of light radiation traversing a turbulent atmosphere is assumed to consist of small-scale (scattering) and large-scale (refraction) effects. The former includes contributions due to eddies/cells smaller than the Fresnel zone $R_F = (L_p/k)^{1/2}$ or the coherence radius ρ_0, whichever is smaller. Large-scale fluctuations on the other hand are generated by the turbulent eddies larger than that of the first Fresnel zone or the scattering disk $L/k\rho_0$, whichever is larger. The small-scale eddies are assumed to be modulated by the large-scale eddies. Consequently, the normalized received irradiance I is defined as the product of two statistically independent random processes I_x and I_y

$$I = I_x I_y \tag{3.122}$$

I_x and I_y arise from the large-scale and small-scale turbulent eddies, respectively, and are both proposed to obey the gamma distribution by Andrews et al. [81,82,98]. Their pdfs are thus given by

$$p(I_x) = \frac{\alpha(\alpha I_x)^{\alpha-1}}{\Gamma(\alpha)} \exp(-\alpha I_x) \quad I_x > 0; \quad \alpha > 0 \tag{3.123}$$

$$p(I_y) = \frac{\beta(\beta I_y)^{\beta-1}}{\Gamma(\beta)} \exp(-\beta I_y) \quad I_y > 0; \quad \beta > 0 \tag{3.124}$$

By fixing I_x and using the change of variable, $I_y = I/I_x$, the conditional pdf given by Equation 3.125 is obtained in which I_x is the (conditional) mean value of I.

$$p(I/I_x) = \frac{\beta(\beta I/I_x)^{\beta-1}}{I_x\Gamma(\beta)} \exp\left(\frac{-\beta I}{I_x}\right) \quad I > 0 \tag{3.125}$$

To obtain the unconditional irradiance distribution, the conditional probability $p(I/I_x)$ is averaged over the statistical distribution of I_x given by Equation 3.123 to obtain the following gamma–gamma irradiance distribution function.

$$\begin{aligned} p(I) &= \int_0^\infty p(I/I_x)p(I_x)\,dI_x \\ &= \frac{2(\alpha\beta)^{(\alpha+\beta)/2}}{\Gamma(\alpha)\Gamma(\beta)} I^{(\alpha+\beta/2)-1}K_{\alpha-\beta}\left(2\sqrt{\alpha\beta I}\right) \quad I > 0 \end{aligned} \tag{3.126}$$

where α and β, respectively, represent the effective number of large- and small-scale eddies of the scattering process. $K_n(\cdot)$ is the modified Bessel function of the 2nd kind of order n, and $\Gamma(\cdot)$ represents the gamma function. If the optical radiation at the receiver is assumed to be a plane wave, then the two parameters α and β that characterize the irradiance fluctuation pdf are related to the atmospheric conditions by [99]

$$\alpha = \left[\exp\left(\frac{0.49\sigma_l^2}{\left(1+1.11\sigma_l^{12/5}\right)^{7/6}} \right) - 1 \right]^{-1} \tag{3.127}$$

$$\beta = \left[\exp\left(\frac{0.51\sigma_l^2}{\left(1+0.69\sigma_l^{12/5}\right)^{5/6}} \right) - 1 \right]^{-1} \tag{3.128}$$

While the *S.I.* is given by

$$\sigma_N^2 = \exp\left[\frac{0.49\sigma_l^2}{\left(1+1.11\sigma_l^{12/5}\right)^{7/6}} + \frac{0.51\sigma_l^2}{\left(1+0.69\sigma_l^{12/5}\right)^{5/6}} \right] - 1 \tag{3.129}$$

A plot of this distribution, using Equation 3.126 is given in Figure 3.30 for three different turbulence regimes, namely weak, moderate and strong. The plot shows

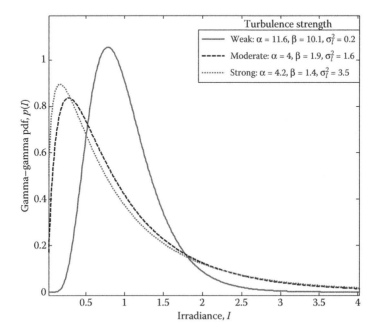

FIGURE 3.30 Gamma–gamma pdf for three different turbulence regimes, namely weak, moderate and strong.

that as the turbulence increases from the weak to strong regime, the distribution spreads out more, with an increase in the range of possible values of the irradiance.

The MATLAB codes for the gamma–gamma pdf are given in Program 3.7.

Program 3.7: MATLAB Codes for the Gamma–Gamma pdf

```
%The Gamma-gamma pdf
clear
clc
alpha=4.2;
a=alpha;
beta=1.4;
b=beta;
k=(a+b)/2;
k1=a*b;
K =2*(k1^k)/(gamma(a)*gamma(b));
I=0.01:0.01:5;
K1=I.^(k-1);
Z=2*sqrt(k1*I);
p=K.*K1.*besselk((a-b),Z);
plot(I,p)
xlabel('Irradiance, I')
ylabel('Gamma gamma pdf, p(I)')
```

The gamma–gamma turbulence model given by Equation 3.126 is valid for all turbulence scenarios from weak to strong and the values of α and β at any given regime can be obtained from Equations 3.127 and 3.128. Figure 3.31 shows the variation of *S.I.* as a function of the Rytov parameter based on Equation 3.129 this graph shows that as the Rytov parameter increases, the *S.I.* approaches a maximum value greater than 1, and then approaches unity as the turbulence-induced fading approaches the saturation regime. The values of α and β under different turbulence regimes are depicted in Figure 3.32.

The MATLAB codes for values of α and β under different turbulence regimes are given in Program 3.8.

Program 3.8: MATLAB Codes to Determine the Values of α
and β under Different Turbulence Regimes

```
%Plot of gamma-gamma model parameters alpha and beta against
log intensity
%variance (Rytov parameter)

clear
clc
```

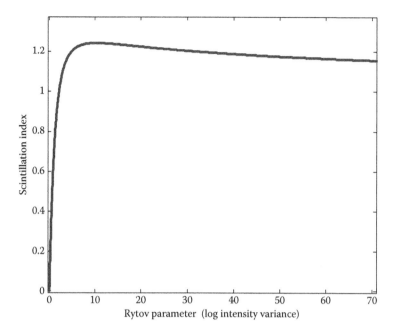

FIGURE 3.31 *S.I.* against log intensity variance for $C_n^2 = 10^{-15}$ m$^{-2/3}$ and $\lambda = 850$ nm.

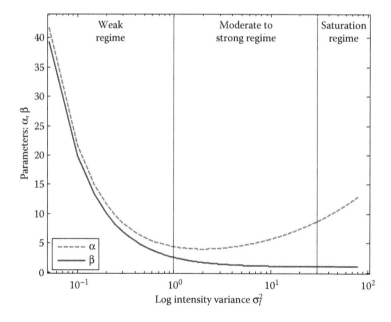

FIGURE 3.32 Values of α and β under different turbulence regimes: weak, moderate to strong and saturation.

```
R2=0:0.05:80; %Log intensity variance (Rytov variance)
for i=1:length(R2)
  R=R2(i);
  A=(0.49*R)/(1+1.11*R^(6/5))^(7/6);
  B=(0.51*R)/(1+0.69*R^(6/5))^(5/6);
  Sci_ind(i)=exp(A+B) - 1;
  alpha(i)=(exp(A) - 1)^-1;
  beta(i)=(exp(B) - 1)^-1;
end

%Plot function
semilogx(R2,alpha,R2,beta)
xlabel('Log intensity variance \sigma_l^2')
ylabel('Parameters: \alpha, \beta')
```

In the very weak turbulence regime, $\alpha \gg 1$ and $\beta \gg 1$ as shown in Figure 3.32, this means that the effective numbers of small and large-scale eddies are very large. But, as the irradiance fluctuations increase (beyond $\sigma_I^2 = 0.2$) and the focusing regime is approached, where α and β then decrease substantially as illustrated in Figure 3.32. Beyond the focusing (moderate-to-strong) regime and approaching the saturation regime, $\beta \to 1$. The implication of this according to, is that the effective number of small scale cells/eddies ultimately reduces to a value determined by the transverse spatial coherence radius of the optical wave. On the other hand, the effective number of discrete refractive scatterers α increases again with increasing turbulence and eventually becomes unbounded in the saturation regime as shown in Figure 3.32. Under these conditions, the gamma–gamma distribution approaches the negative exponential distribution below.

3.3.6.5 The Negative Exponential Turbulence Model

In the limit of strong irradiance fluctuations (i.e., in saturation regime and beyond) where the link length spans several kilometres, the number of independent scatterings becomes large [54,55]. This saturation regime is also called the fully developed speckle regime. The amplitude fluctuation of the field traversing the turbulent medium in this situation is generally believed and experimentally verified [100,101] to obey the Rayleigh distribution implying negative exponential statistics for the irradiance. That is

$$p(I) = \frac{1}{I_0} \exp\left(-\frac{I}{I_0}\right) \quad I_0 > 0 \tag{3.130}$$

where $E[I] = I_0$ is the mean received irradiance. During the saturation regime, the value of the S.I., S.I. $\to 1$. It is noteworthy that other turbulence models such as the log-normal-Rician [102] and the I–K [103] distributions, which are both valid from weak-to-strong turbulence regimes; the K-model [104,105], which is only valid for the strong regime, and the gamma–gamma turbulence models all reduce to the negative exponential in the limit of strong turbulence. The negative exponential pdf is shown in Figure 3.33.

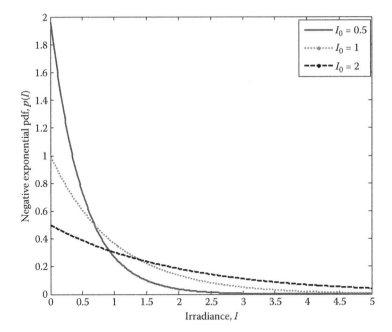

FIGURE 3.33 Negative exponential pdf for different values of I_0.

3.3.7 ATMOSPHERIC EFFECTS ON OWC TEST BED

In many cases, depending on the location, environment and so on, assessing the FSO link performance in real weather impairment conditions may take a very long time and will depend on the presence and duration of event such as fog, rain, snow and turbulence. Thus, there is need for an indoor atmospheric test bed where measurement could readily be carried under controlled environment. A typical FSO link set-up for data communication for a different modulation scheme is shown in Figure 3.34a. An arbitrary waveform generator (AWG) is used to produce different modulation formats and levels. The data signal is used to directly modulate a laser diode at a range of wavelength (600–1550 nm). To ensure the linearity of the system, the laser is properly biased and the peak-to-peak voltage of the input signal cannot exceed the specified dynamic range of the laser diode. The optical output from the laser experience different atmospheric effects including attenuation (due to absorption and scattering) and intensity fluctuation before it is collected at the receiver. The receiver front-end consists of an optical telescope (or lens) and a photodetector. The electrical signal at the output of the photodetector is amplified using a transimpedance amplifier and ideally followed by circuitry for clock and timing recovery and regeneration of the transmitted data. The motivation for developing the chamber is to simulate the atmospheric channel effects on optical signal traversing it in a controlled environment. Hence, in the

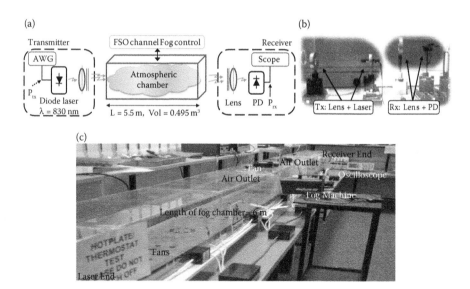

FIGURE 3.34 (**See colour insert.**) (a) Block diagram of the FSO experiment set-up, (b) the simulation chamber and (c) the laboratory chamber set up.

experimental set-up, data recovery circuitry is not used; rather the raw data at the receiver are analysed.

A closed glass chamber with dimension of $550 \times 30 \times 30$ cm^3 is used as the FSO channel (see Figure 3.34b and c). The chamber has multiple compartments (seven in total at this experiment) and the conditions (temperature and wind velocity) within a chamber can be controlled individually as necessary to mimic the atmospheric condition as far as possible. Two approaches are taken to generate the turbulence effect.

1. Two independent fans blow either hot air or cold air in the direction perpendicular to signal propagation to generate the temperature and wind speed variation (see Figure 3.34c). The cold air is set at about 20°C and hot air covers a temperature range of 20–80°C. Using a series of air vents, additional temperature control is achieved thus ensuring a constant temperature gradient between the source and the detector.
2. Each chamber has a heating element with a fan attached to it so that the turbulence effect can be generated and controlled.

The second approach gives more control over the channel parameter as the temperature variation can be adjusted as necessary within a resolution of 1°C. The descriptions that follow are based on an experimental carried out using the first approach.

Fog and smoke generators can be used to pump fog and/or smoke at a given rate to the chamber. The fog and/or smoke density within the chamber can be controlled by a combination of the amount of fog pumped in and the ventilation system. This set-up offers the advantage of repeatedly replicating atmospheric weather conditions in a controlled manner without the need for a long observation period in an outdoor environment. The path length through the chamber can be increased by using reflecting mirrors located on both ends of the chamber. The main set-up parameters are summarized in Table 3.13.

The receiver lens has 20 mm diameter and a focal length of 10 cm where the receiver is located. The PIN photodetector has a responsivity of 0.59 A/W at the peak wavelength ($\lambda = 830$ nm) with an active collection area of 1 mm^2. The electrical signal at the output of the photodetector is amplified using a trans impedance amplifier.

The atmospheric chamber is used to estimate the transmittance, T, based on the Beer–Lambert law [59], by comparing the average received optical intensity in the presence and absence of fog as

$$T = \frac{I(f)}{I(0)} = \exp(\beta_\lambda z) \tag{3.131}$$

where β_λ is the attenuation or the scattering coefficient due to fog, in units of km^{-1}, z is the propagation length and $I(f)$ and $I(0)$ are the average received optical intensities in the presence and absence of fog, respectively. In this case, the link visibility V parameter can then be obtained from the fog attenuation using Kim's model [56] with a distance shorter than 500 m, where β_λ is defined as

TABLE 3.13
Test Bed System Parameters

Transmitter Front End (Laser + AWG)		Receiver Front End (Photodiode + Lens)	
Operating wavelength	830 nm	Lens focal length	10 cm
Class	Class IIIb	Lens diameter	30 mm
Beam size at aperture	5 mm × 2 mm	Photodetector (PD) responsivity	0.59 A/W at 830 nm
Average optical output power	10 mW	PD active area	1 mm^2
Laser modulation depth, m	20%	PD half-angle view	±75°
Laser 3-dB bandwidth	50 MHz	PD rise and fall times	5 ns
Bit rate	12.5 Mbps		
PRBS length	2^{10}-1 bits		
Modulating signal amplitude	100 mVpp		
	Atmospheric Chamber		
Fog G. pumping rate	0.94 m^3/s	Dimension	550 × 30 × 30 cm^3
Total path length (with reflections)	12 m	Wind speed	1 m/s
Temperature range	20–80°C		

TABLE 3.14

System Characteristics for $\lambda = 830$ nm and $P_{tx} = -1.23$ dBm

Fog	Dense	Thick	Moderate	Light	Clear
V (m)	25–70	70–250	250–500	500–1000	>1000
T	<0.36	0.36–0.67	0.67–0.85	0.85–0.92	>0.92

$$\beta_\lambda = \frac{3.912}{V}\left(\frac{\lambda}{550}\right)^{-\delta} \tag{3.132}$$

where the constant δ depends on V. Equations 3.131 and 3.132 relate V and T are given in Table 3.14, thus enabling us to experimentally analyse the performance of the FSO link, Q-factor, in presence of fog.

As an example, let us consider the link margin M defined as

$$M = P_{tx} + S_{pd} - L_{fog} - L_{geo} - L_{dev} \tag{3.133}$$

where S_{pd} is the photodetector sensitivity (−36 dBm at 12.5 Mbps for bit error rate (BER) <10⁻⁶), P_{tx} is the total optical power transmitted, L_{geo} is the geometrical loss, L_{fog} is the loss due to the fog estimated by T and L_{dev} is the transceiver losses losses due to experimental issues. Using these values, the link margin for clear conditions is more than 6 dB, which is a typical value for existing commercial outdoor FSO communications with a range up to 2 km.

Using the chamber, it is also possible to experimentally determine the effect of turbulence and estimate the normalized variance of intensity of the received optical signal in a weak turbulence regime. To do this, the received signals in the presence of turbulence are compared with that of ideal channel (without turbulence), and estimated the *S.I.* The *S.I.* is calculated using

$$S.I = \frac{\langle I^2 \rangle}{\langle I \rangle^2} - 1 \tag{3.134}$$

where I denotes received optical irradiance and $\langle . \rangle$ denotes an ensemble average.

3.3.7.1 Demonstration of Scintillation Effect on Data Carrying Optical Radiation

To evaluate the strength of the turbulence generated using the above set-up, the intensity of laser beam is directly modulated by the pseudorandom binary signal (PRBS) of length 10 at different data rates. Note that the data are line encoded with the 4B5B encoding and hence the effective line rate is 1.25 times the data rate. The

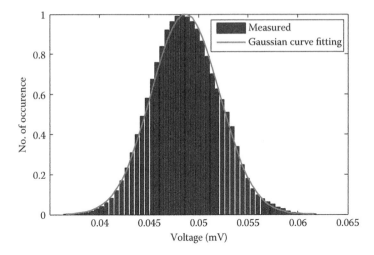

FIGURE 3.35   Received signal distribution without scintillation (mean = 0.0487, σ^2 = 1.2 E–5).

analogue peak-to-peak amplitude of the input signal is fixed to 500 mV, which corresponds to 3 dBm of transmit optical power for communications. Note that in order to avoid beam averaging at the receiver, no concentration lens is used at the transmitter. While carrying out the experiment without turbulence, the temperature across the turbulence simulation chamber is kept the same as the ambient temperature. With turbulence, the temperature at four different positions, T_1 to T_4 in Figure 3.34b, are constantly changing.

The histogram of the received signal for the level 1 sampled at the maximum eye-opening is given in Figures 3.35 and 3.36. Note that total number of occurrence is normalized, so the graph is a representation of the pdf of the received mean

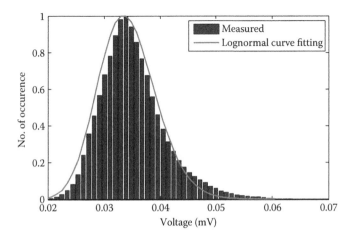

FIGURE 3.36   Received signal distribution with scintillation (mean = 0.0451, *S.I.* = 0.0164).

values. The MATLAB codes for generating curve fittings are given in Programs 3.5 and 3.6, and the corresponding eye diagrams are given in Figure 3.37. The variation observed in the received signal under no scintillation is down to the innate noise (shot and thermal) associated with the photodetection process. Figure 3.35 shows the mean value distribution with the simulated turbulence. In order to estimate the strength of the simulated scintillation effect, the *S.I.* is calculated using Equation 3.134. The histogram of the received signal and the lognormal curve fitting in the presence of the weak turbulence is illustrated in Figure 3.36. The *S.I.* value is around 0.0164. When the log intensity variance is in the range $0 < \sigma_l^2 \le 1$, the turbulence is generally classified to be in the weak regime (see Programs 3.9 through 3.11).

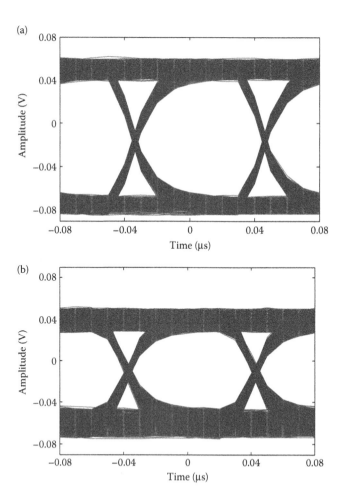

FIGURE 3.37 The measured eye diagram of received NRZ signal in the condition of (a) no turbulence, 12.5 Mbps and (b) weak–medium turbulence with $\sigma_l^2 = 0.0164$.

Program 3.9: MATLAB Codes to Estimate Mean and Standard Deviation for Practical Measurement

```
load OOK_NRZ_10M_500Ms.mat;

time=time-time(1);
data1=data1-mean(data1);
% removing DC content

Rb=10e6;
% bit rate
deltaT=(time(end)-time(1))/length(time);
% finding sampling rate;
Fs=1/deltaT;
% sampling rate
nsamp=floor(Fs/Rb);
% number of samples per bit

Fil_type=0;
% receiver filter
% 0 no filter; not recommended
% 1 sub-optimal low pass RC filter
% 2 butterworth low pass filter
% 3 matched filter

switch Fil_type
  case 0
    Rx_data=data1;

  case 1
    fc=0.75*Rb;
    t=deltaT:deltaT:20/Rb;
    lpf_impulse=exp(-t*2*pi*fc);
    Rx_data=filter(lpf_impulse,1,data1);

  case 2
    Wn=2*Rb/Fs;
    [b,a]=butter(4,Wn,'low');
    Rx_data=filter(b,a,data1);

  case 3
    MF_coef=ones(1,nsamp);
    Rx_data=filter(MF_coef,1,data1);
end

% finding a rising edge
for ii=1:10*nsamp
  if Rx_data(ii+1) > Rx_data(ii) && Rx_data(ii+1)*Rx_data(ii) < 0
    index=ii
    break
```

```
  end
end

eye_data=Rx_data(index:end);
eye_data=eye_data-mean(eye_data);

% H=commscope.eyediagram('SamplingFrequency', Fs, ...
% 'SamplesPerSymbol', nsamp, ...
% 'SymbolsPerTrace', 2, ...
% 'MinimumAmplitude', min(eye_data), ...
% 'MaximumAmplitude', max(eye_data), ...
% 'AmplitudeResolution', 0.00100, ...
% 'MeasurementDelay', 0, ...
% 'PlotType', '2D Color', ...
% 'PlotTimeOffset', 0, ...
% 'PlotPDFRange', [0 1], ...
% 'ColorScale', 'log', ...
% 'RefreshPlot', 'on');
% update(H, eye_data);
%
%
% **********sampling ***************************
rx_OOK=eye_data(nsamp-1:nsamp:end);
[m0 m1 s0 s1 Rytov_var]=approx_mean_var(rx_OOK);
Q_1point_sampling=(m1-m0)/(s0+s1);

% calculation of maximum Q-factor
for jj=1:nsamp
  temp=eye_data(jj:end);
  rx_OOK=temp(floor(nsamp/2):nsamp:end);
  [m0(jj) m1(jj) s0(jj) s1(jj) Rytov_var(jj)]=approx_mean_
var(rx_OOK);
  Q_factor(jj)=(m1(jj)-m0(jj))/(s0(jj)+s1(jj));
end
[Q_max index]=max(Q_factor);
%results=[Q_factor(index) m0(index) m1(index) s0(index)
s1(index) Rytov_var(jj)];

function [mean0 mean1 std0 std1]=approx_mean_var(data)
% function to approximate mean and standard deviation
% this method is valid only if Q-factor is large
% alternative method using first and second moment can be used
as well

min_val=min(data);
% minimum amplitude of received signal
max_val=max(data);
% maximum amplitude of received signal
temp=sort(data);

  for i=1:5
    % use iteration for close approximation
```

```
   mean_val=min_val+(max_val-min_val)/2;
   % approximate threshold value
   data0=temp(find(temp<=mean_val));
   % possible binary '0' data
   data1=temp(find(temp>mean_val));
   % possible binary '1' data
   mean0=mean(data0);
   % mean value for binary '0'
   mean1=mean(data1);
   % mean value for binary '1'
   std0=std(data0);
   % standard deviation for binary '0'
   std1=std(data1);
   % standard deviation for binary '1'
   max_val=mean1;
   min_val=mean0;
  end
end
```

Program 3.10: MATLAB Codes to Fit a Gaussian Distribution

```
% Gaussian curve fitting program
load data_WOT
th= m0_WOT+(m1_WOT-m0_WOT)/2;
% the threshold level
data_1=data_hist_WOT(find(data_hist_WOT>=th));
% received sequence greater than 0, possible 1 bit
data_0=data_hist_WOT(find(data_hist_WOT<th));
% received sequence less than 0, possible 0 bit

%********** for received bit '1' only**************
m1_WOT=mean(data_1);
% mean value for level 1
s1_WOT=std(data_1);
% standard deviation for level 1
[n,xout]=hist(data_1,50);
% histogram generation
y0=1/sqrt(2*pi*s1_WOT^2)*exp(-((xout-m1_WOT).^2/(2*s1_
WOT^2)));
% Gaussian curve fitting
y0=y0/max(y0);
% normalization
n=n/max(n);

Fig.
bar(xout,n);
hold on
plot(xout,y0,'r');
xlabel('Voltage (mV)');
ylabel('No. of Occurrence');
```

Program 3.11: MATLAB Codes to Fit the Log-Normal Distribution

```
load data_WT data_hist_WT Q_WT m0_WT m1_WT s0_WT s1_WT
load data_WOT data_hist_WOT Q_WOT m0_WOT m1_WOT s0_WOT s1_WOT

th= m0_WT+(m1_WT-m0_WT)/2;
data_1=data_hist_WT(find(data_hist_WT >=th));

% curve fitting lognormal Bit 1 *****************
[n,xout]=hist(data_1,50);
n=n/sum(n);
Rytov_var=(s1_WT/m1_WT)^2-(s1_WOT/m1_WOT)^2;
%Rytov_var=mean(data_1.^2)/mean(data_1)^2-1

I=xout(1):xout(2)-xout(1):1.2*xout(end);
I0=m1_WOT;
a=1/(sqrt(2*pi*Rytov_var));
b=(log(I/I0)-Rytov_var/2).^2;
y_WT=exp(-b/(2*Rytov_var))./(a*I);

y_awgn=1/sqrt(2*pi*s1_WOT^2)*exp(-((I-m1_WT).^2/(2*s1_
WOT^2)));

n=fliplr(n);
y0=y_WT;
y0=y0/max(y0);
n=n/max(n);
[d index]=find(n==max(n));
temp1=xout(index);
[d index]=find(y0==max(y0));
temp2=xout(index);
diff=temp2-temp1;
```

Figure 3.38 depicts the histogram of the received OOK–NRZ signal with/without turbulence for a range of Rytov variance. As can be seen without turbulence, bits '1' and '0' are clearly distinguishable from the signal distribution. The signal profiles for both bits are similar and are equally spaced on both sides of the zero mark. The decision threshold point can therefore be set at the zero mark. But in the presence of turbulence, the signal waveform becomes heavily distorted and the signal distributions are no longer distinguishable. It is still possible to set a threshold level with the *S.I.* of 0.0483. However, with the increase in the *S.I.*, there is an overlap in the signal level and therefore it becomes rather challenging to set the required threshold level (see Figure 3.38d). Compared with the case with no turbulence, the overlap indicates higher likelihood of detection error. From the foregoing, it can be said that in the absence of turbulence, the threshold level can be fixed at the zero signal level but this is no longer the case as soon as turbulence sets in. During turbulence, finding the exact decision threshold level that will minimize detection error will require the knowledge of the turbulence strength. What this means in essence is that, an adaptive threshold will be required for the type of data encoding being used.

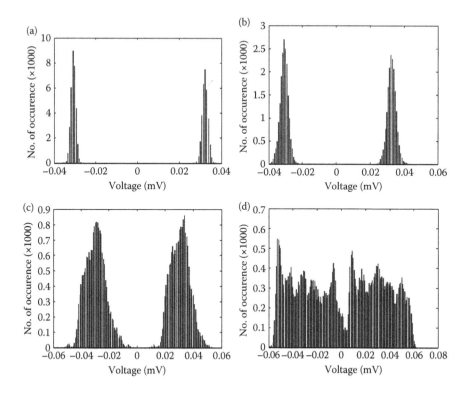

FIGURE 3.38 The histogram of the received OOK–NRZ signal: (a) without turbulence and with turbulence Rytov variance of (b) 0.00423, (c) 0.0483 and (d) 0.242.

Alternative to the histogram approach to investigate the link quality, the feature of signal can be analysed using the optical signal-to-noise-ratio (OSNR) or the Q-factor. The Q-factor for the binary modulation scheme is given by

$$Q = \frac{\upsilon_H - \upsilon_L}{\sigma_H + \sigma_L} \qquad (3.135)$$

where υ_H and υ_L are the mean received voltages and σ_H and σ_L are the standard deviation for 'high' and 'low' signal levels, respectively.

If SNR in the absence of the turbulence is known, the average SNR in the presence of turbulence with a *S.I.* of σ_I^2 can be expressed by

$$\langle \text{SNR} \rangle = \frac{\text{SNR}_0}{\sqrt{\sigma_I^2 \text{SNR}_0^2 + \frac{I_0}{\langle I \rangle}}} \qquad (3.136)$$

where SNR_0 is the OSNR in the absence of turbulence.

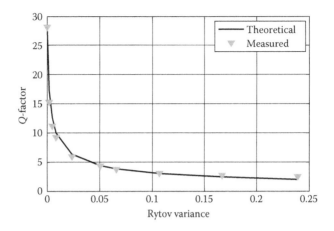

FIGURE 3.39   The predicted and measured Q-factor (OSNR) against Rytov variance for OOK–NRZ at 12.5 Mbps.

The predicted and measured Q-factor against Rytov variance for OOK-NRZ at 12.5 Mbps is given in Figure 3.39. Note that the effect of scintillation is independent of data rates and only depends on the SNR and the *S.I.* It can be observed that the SNR in the presence of the turbulence degrease exponentially with σ_l^2, meaning a small scintillation can significantly drop the system SNR. There are a number of methods to mitigate the effect of the turbulence in an FSO channel, which will be discussed in Chapter 7.

REFERENCES

1. J. M. Kahn and J. R. Barry, Wireless infrared communications, *Proceedings of IEEE*, 85, 265–298, 1997.
2. A. M. Street, P. N. Stavrinou, D. C. Obrien and D. J. Edwards, Indoor optical wireless systems—A review, *Optical and Quantum Electronics*, 29, 349–378, 1997.
3. G. W. Marsh and J. M. Kahn, Channel reuse strategies for indoor infrared wireless communications, *IEEE Transactions on Communications*, 45, 1280–1290, 1997.
4. J. R. Barry, *Wireless Infrared Communications*. Boston: Kluwer Academic Publishers, 1994.
5. S. Hranilovic, On the design of bandwidth efficient signalling for indoor wireless optical channels, *International Journal of Communication Systems*, 18, 205–228, 2005.
6. J. B. Carruthers and J. M. Kahn, Modeling of nondirected wireless Infrared channels, *IEEE Transaction on Communication*, 45, 1260–1268, 1997.
7. F. R. Gfeller and U. Bapst, Wireless in-house data communication via diffuse infrared radiation, *Proceedings of the IEEE*, 67, 1474–1486, 1979.
8. D. C. O'Brien, M. Katz, P. Wang, K. Kalliojarvi, S. Arnon, M. Matsumoto, R. Green and S. Jivkova, Short-range optical wireless communications, *Technologies for the Wireless Future: Wireless World Research Forum*, 2, 277–296, 2006.
9. V. Jungnickel, V. Pohl, S. Nonnig and C. v. Helmolt, A physical model of the wireless infrared communication channel, *IEEE Journal on Selected Areas in Communications*, 20, 631–640, 2002.
10. N. Hayasaka and T. Ito, Channel modeling of nondirected wireless infrared indoor diffuse link, *Electronics and Communications in Japan*, 90, 9–19, 2007.

11. C. R. Lomba, R. T. Valadas and A. M. Duarte, Efficient simulation of the impulse response of the indoor wireless optical channel, *International Journal of Communication Systems*, 13, 537–549, 2000.

12. J. B. Carruthers and P. Kannan, Iterative site-based modeling for wireless infrared channels, *IEEE Transactions on Antennas and Propagation*, 50, 759–765, 2002.

13. J. B. Carruthers and S. M. Carroll, Statistical impulse response models for indoor optical wireless channels, *International Journal of Communication Systems*, 18, 267–284, 2005.

14. C. R. Lomba, R. T. Valadas and A. M. Duarte, Propagation losses and impulse response of the indoor optical channel: A simulation package, *Mobile Communications Advanced Systems and Components*, Berlin: Springer, 1994, pp. 285–297.

15. J. B. Carruthers, S. M. Caroll and P. Kannan, Propagation modelling for indoor optical wireless communications using fast multi-receiver channel estimation, *IEE Proceedings-Optoelectronics*, 150, 473–481, 2003.

16. J. R. Barry, J. M. Kahn, W. J. Krause, E. A. Lee and D. G. Messerschmitt, Simulation of multipath impulse response for indoor wireless optical channels, *IEEE Journal on Selected Areas in Communications*, 11, 367–379, 1993.

17. J. M. Kahn, W. J. Krause and J. B. Carruthers, Experimental characterization of non-directed indoor infrared channels, *IEEE Transactions on Communications*, 43, 1613–1623, 1995.

18. X. Yang, M. Gong, H. Zhang, P. Yan, S. Zou, W. Jin, K. Zhang, F. Jiang and Y. Meng, Simulation of impulse response on IR wireless indoor channel with concentrator, *Proceedings of the SPIE*, 4873, 71–78, 2002.

19. R. P. Silvestre, P. J. Rafael, J. L. H. Francisco, B. G. H. Oswaldo, and J. A. A. Alejandro, Reflection model for calculation of the impulse response on IR-wireless indoor channels using ray-tracing algorithm, *Microwave and Optical Technology Letters*, 32, 296–300, 2002.

20. K. Smitha, A. Sivabalan and J. John, Estimation of channel impulse response using modified ceiling bounce model in non-directed indoor optical wireless systems, *Wireless Personal Communications*, 45, 1–10, 2008.

21. F. J. Lopez-Hernandez, R. Perez-Jimeniz and A. Santamaria, Monte Carlo calculation of impulse response on diffuse IR wireless indoor channels, *Electronics Letters*, 34, 1260–1262, 1998.

22. D. R. Biosca, P. Lopez and L. Jorge, Generalization of Monte Carlo ray-tracing algorithm for the calculation of the impulse response on indoor wireless infrared channels, *Universidad, Ciencia y Tecnología*, 9, 17–25, 2005.

23. A. Mihaescu and M. Otesteanu, Reduced size model method for diffuse optical indoor wireless channel characterization, *WSEAS Transactions on Communications*, 5, 155–160, 2006.

24. S. Arumugam and J. John, Effect of transmitter positions on received power and bandwidth in diffuse indoor optical wireless systems, *Optical and Quantum Electronics*, 37, 1–14, 2007.

25. O. Gonzalez, S. Rodriguez, R. Perez-Jimenez, B. R. Mendoza and A. Ayala, Error analysis of the simulated impulse response on indoor wireless optical channels using a Monte Carlo-based ray-tracing algorithm, *IEEE Transactions on Communications*, 53, 124–130, 2005.

26. H. Hashemi, G. Yun, M. Kavehrad and F. Behbahani, Frequency response measurements of the wireless indoor channel at infrared frequencies, *IEEE International Conference on Communications*, New Orleans, LA, USA, 1994, pp. 1511–1515.

27. F. J. Lopez-Hernandez and M. J. Betancor, DUSTIN: Algorithm for calculation of impulse response on IR wireless indoor channels, *Electronics Letters*, 33, 1804–1806, 1997.

28. V. Pohl, V. Jungnickel and C. Helmolt, A channel model for wireless infrared communication, *The 11th IEEE International Symposium on Personal, Indoor and Mobile Radio Communications, 2000*, 1, 297–303, 2000.

29. M. R. Pakravan and M. Kavehrad, Indoor wireless infrared channel characterization by measurements, *IEEE Transactions on Vehicular Technology*, 50, 1053–1073, 2001.
30. C. R. Lomba, R. T. Valadas and A. M. d. O. Duarte, Experimental characterisation and modelling of the reflection of infrared signals on indoor surfaces, *IEE Proceedings—Optoelectronics*, 145, 191–197, 1998.
31. Institute of Electrical and Electronics Engineers, IEEE Standard 802.11–1997, 1997.
32. R. T. Valadas, A. R. Tavares, A. M. d. Duarte, A. C. Moreira and C. T. Lomba, The infrared physical layer of the IEEE 802.11 standard for wireless local area networks, *IEEE Communications Magazine*, 36, 107–112, 1998.
33. J. J. G. Fernandes, P. A. Watson and J. C. Neves, Wireless LANs: Physical properties of infra-red systems vs. MMW systems, *IEEE Communications Magazine*, 32, 68–73, 1994.
34. G. N. Bakalidis, E. Glavas and P. Tsalides, Optical power distribution in wireless infrared LANs, *IEE Proceedings Communications*, 143, 93, 1996.
35. H. Yang and C. Lu, Infrared wireless LAN using multiple optical sources, *IEE Proceedings—Optoelectronics*, 147, 301–307, 2000.
36. D. Zwillinger, Differential PPM has a higher throughput than PPM for the band-limited and average-power-limited optical channel, *IEEE Transactions on Information Theory*, 34, 1269–1273, 1988.
37. H. Hashemi, Y. Gang, M. Kavehrad, F. Behbahani and P. A. Galko, Indoor propagation measurements at infrared frequencies for wireless local area networks applications, *IEEE Transactions on Vehicular Technology*, 43, 562–576, 1994.
38. R. Perez-Jimenez, J. Berges and M. J. Betancor, Statistical model for the impulse response on infrared indoor diffuse channels, *Electronics Letter*, 33, 1298–1300, 1997.
39. A. J. C. Moreira, A. M. R. Tavares, R. T. Valadas and A. M. Duarte, Modulation methods for wireless infrared transmission systems: performance under ambient light noise and interference, *Proceedings of SPIE Conference on Wireless Data Transmission*, Philadelphia, USA, 1995, pp. 226–237.
40. A. J. C. Moreira, R. T. Valadas and A. M. d. O. Duarte, Optical interference produced by artificial light, *Wireless Networks*, 3, 131–140, 1997.
41. A. J. C. Moreira, R. T. Valadas and A. M. d. O. Duarte, Characterisation and modelling of artificial light interference in optical wireless communication systems, *Proceedings of the 6th IEEE International Symposium on Personal, Indoor and Mobile Radio Communications*, Toronto, Canada, 1995.
42. A. C. Boucouvalas, Indoor ambient light noise and its effect on wireless optical links, *IEE Proceedings—Optoelectronics*, 143, 334–338, 1996.
43. A. J. C. Moreira, R. T. Valadas and A. M. d. O. Duarte, Performance of infrared transmission systems under ambient light interference, *IEE Proceedings-Optoelectronics*, 143, 339–346, 1996.
44. R. Narasimhan, M. D. Audeh and J. M. Kahn, Effect of electronic-ballast fluorescent lighting on wireless infrared links, *IEE Proceedings—Optoelectronics*, 143, 347–354, 1996.
45. A. M. R. Tavares, R. T. Valadas and A. M. d. O. Duarte, Performance of wireless infrared transmission systems considering both ambient light interference and inter-symbol interference due to multipath dispersion, *Conference on Optical Wireless Communications*, Boston, USA, 1998, pp. 82–93.
46. K. Samaras, A. M. Street, D. C. O'Brien and D. J. Edwards, Error rate evaluation of wireless infrared links, *Proceedings of IEEE International Conference on Communications*, Atlanta, USA, 1998, pp. 826–831.
47. A. J. C. Moreira, R. T. Valadas and A. M. Duarte, Reducing the effects of artificial light interference in wireless infrared transmission systems, *Proceedings of IEE Colloquium on Optical Free Space Communication Links*, London, UK, 1996, pp. 5/1–5/10.

48. K. K. Wong, T. O'Farrell and M. Kiatweerasakul, The performance of optical wireless OOK, 2-PPM and spread spectrum under the effects of multipath dispersion and artificial light interference, *International Journal of Communication Systems*, 13, 551–557, 2000.
49. K. K. Wong, T. O'Farrell and M. Kiatweerasakul, Infrared wireless communication using spread spectrum techniques, *IEE Proceedings—Optoelectronics*, 147, 308–314, 2000.
50. O. Bouchet, M. El Tabach, M. Wolf, D. C. O'Brien, G. E. Faulkner, J. W. Walewski, S. Randel et al. Hybrid wireless optics (HWO): Building the next-generation home network, *6th International Symposium on Communication Systems, Networks and Digital Signal Processing, CNSDSP*, Graz, Austria, 2008, pp. 283–287.
51. R. M. Gagliardi and S. Karp, *Optical Communications*, 2nd ed. New York: John Wiley, 1995.
52. H. Willebrand and B. S. Ghuman, *Free Space Optics: Enabling Optical Connectivity in Today's Network*. Indianapolis: SAMS Publishing, 2002.
53. W. K. Pratt, *Laser Communication Systems*, 1st ed. New York: John Wiley & Sons, Inc., 1969.
54. S. Bloom, E. Korevaar, J. Schuster and H. Willebrand, Understanding the performance of free-space optics, *Journal of Optical Networking*, 2, 178–200, 2003.
55. D. R. Bates, Rayleigh scattering by air, *Planetary Space Science*, 32, 785–790, 1984.
56. I. I. Kim, B. McArthur and E. Korevaar, Comparison of laser beam propagation at 785 nm and 1550 nm in fog and haze for optical wireless communications, *SPIE Proceedings: Optical Wireless Communications III*, 4214, 26–37, 2001.
57. H. Hemmati, Ed., *Deep Space Optical Communications*. Deep Space Communications and Navigation Series. California: Wiley-Interscience, 2006.
58. H. C. van de Hulst, *Light Scattering by Small Particles*. New York: Dover Publications 1981.
59. M. Al Naboulsi and H. Sizun, Fog attenuation prediction for optical and infrared waves, *Optical Engineering*, 23, 319–329, 2004.
60. J. C. Ricklin, S. M. Hammel, F. D. Eaton and S. L. Lachinova, Atmospheric channel effects on free space laser communication, *Journal of Optical and Fiber Communications Research*, 3, 111–158, 2006.
61. M. S. Awan, Marzuki, E. Leitgeb, B. Hillbrand, F. Nadeem and M. S. Khan, Cloud attenuations for free-space optical links, *International Workshop on Satellite and Space Communications, 2009*, Siena-Tuscany, Italy, 2009, pp. 274–278.
62. A. K. Majumdar and J. C. Ricklin, *Free-Space Laser Communications: Principles and Advances*. New York: Springer, 2008.
63. M. A. Naboulsi, H. Sizun and F. d. Fornel, Wavelength selection for the free space optical telecommunication technology, *SPIE*, 5465, 168–179, 2004.
64. W. Binbin, B. Marchant and M. Kavehrad, Dispersion analysis of 1.55 µm free-space optical communications through a heavy fog medium, *IEEE Global Telecommunications Conference*, Washington, 2007, pp. 527–531.
65. E. Leitgeb, M. Geghart and U. Birnbacher, Optical networks, last mile access and applications, *Journal of Optical and Fibre Communications Reports*, 2, 56–85, 2005.
66. V. Kvicera, M. Grabner and J. Vasicek, Assessing availability performances of free space optical links from airport visibility data, *7th International Symposium on Communication Systems Networks and Digital Signal Processing (CSNDSP)*, 2010, pp. 562–565.
67. A. Prokes, Atmospheric effects on availability of free space optics systems, *Opt. Eng.*, 48, 066001–066010, 2009.
68. *Guide to Meteorological Instruments and Methods of Observation*, World Meteorological Organisation, Geneva, Switzerland, 2006.
69. P. W. Kruse, L. D. McGlauchlin and R. B. McQuistan, *Elements of Infrared Technology: Generation, Transmission, and Detection*. New York: John Wiley & Sons, 1962.

70. F. Nadeem, B. Geiger, E. Leitgeb, M. S. Awan and G. Kandus, Evaluation of switch-over algorithms for hybrid FSO-WLAN systems, *1st International Conference on Wireless Communication, Vehicular Technology, Information Theory and Aerospace & Electronic Systems Technology*, Aalborg, Denmark, 2009, pp. 565–570.

71. M. Grabner and V. Kvicera, On the relation between atmospheric visibility and optical wave attenuation, *16th IST Mobile and Wireless Communications Summit*, 2007, pp. 1–5.

72. M. Grabner and V. Kvicera, Case study of fog attenuation on 830 nm and 1550 nm free-space optical links, *Proceedings of the Fourth European Conference on Antennas and Propagation (EuCAP)*, Barcelona, Spain, 2010, pp. 1–4.

73. F. Nadeem, V. Kvicera, M. S. Awan, E. Leitgeb, S. Muhammad and G. Kandus, Weather effects on hybrid FSO/RF communication link, *IEEE Journal on Selected Areas in Communications*, 27, 1687–1697, 2009.

74. D. Atlas, Shorter contribution optical extinction by rainfall, *Journal of Meteorology*, 10, 486–488, 1953.

75. H. W. O'Brien, Visibility and light attenuation in falling snow, *Journal of Applied Meteorology*, 9, 671–683, 1970.

76. I. I. Kim and E. Korevaar, Availability of free space optics and hybrid FSO/RF systems, *Proceedings of SPIE: Optical Wireless Communications IV*, 4530, 84–95, 2001.

77. M. Grabner and V. Kvicera, Experimental study of atmospheric visibility and optical wave attenuation for free-space optics communications, http://ursi-france.institut-telecom.fr/pages/pages_ursi/URSIGA08/papers/F06p5.pdf, last visited 2nd Sept. 2009.

78. J. Gowar, *Optical Communication Systems*, 2nd ed. New Jersey: Prentice-Hall, 1993.

79. J. Jesanathan and P. Inovo, Multi-gigabit-per-second optical wireless communications, *IEE Colloquium Optical Wireless Communications*, 818, 359–362, 1999.

80. S. Karp, R. M. Gagliardi, S. E. Moran and L. B. Stotts, *Optical Channels: Fibers, Clouds, Water and the Atmosphere*. New York: Plenum Press, 1988.

81. L. C. Andrews and R. L. Phillips, *Laser Beam Propagation through Random Media*, 2nd ed. Washington: SPIE Press, 2005.

82. L. C. Andrews, R. L. Phillips, and C. Y. Hopen, *Laser Beam Scintillation with Applications*. Bellingham: SPIE, 2001.

83. J. W. Goodman, *Statistical Optics*. New York: John Wiley, 1985.

84. G. R. Osche, *Optical Detection Theory for Laser Applications*. New Jersey: Wiley, 2002.

85. X. Zhu and J. M. Kahn, Free-space optical communication through atmospheric turbulence channels, *IEEE Transactions on Communications*, 50, 1293–1300, August 2002.

86. S. F. Clifford, Ed., *The Classical Theory of Wave Propagation in a Turbulent Medium* (*Laser Beam Propagation in the Atmosphere*). Berlin: Springer-Verlag, 1978.

87. A. Kolmogorov, Ed., *Turbulence. Classic Papers on Statistical Theory*. New York: Wiley-Interscience, 1961.

88. A. K. Majumdar and J. C. Ricklin, Effects of the atmospheric channel on free-space laser communications, *Free-Space Laser Communications V*, Vol. 5892, D. G. Voelz and J. C. Ricklin, Eds., New York: Springer, 2005, pp. 8920K-1–58920K-16.

89. A. Kolmogorov, The local structure of turbulence in incompressible viscous fluid for very large Reynold numbers, *Proceedings of Royal Society of London Series A-Mathematical and Physical*, 434, 9–13, 1991.

90. V. I. Tatarski, *Wave Propagation in a Turbulent Medium*. (Translated by R. A. Silverman). New York: McGraw-Hill, 1961.

91. A. Ishimaru, Ed., *The Beam Wave Case and Remote Sensing* (Topics in Applied Physics: Laser Beam Propagation in the Atmosphere). New York: Springer-Verlag, 1978.

92. D. P. Greenwood, Bandwidth specification for adaptive optics systems, *J. Opt. Soc. Am.*, 67, 390–393, 1977.

93. H. Hodara, Laser wave propagation through the atmosphere, *Proceedings of the IEEE*, 54, 368–375, 1966.

94. J. W. Strohbehn, Line-of-sight wave propagation through the turbulent atmosphere, *Proceedings of the IEEE*, 56, 1301–1318, 1968.

95. J. W. Strobehn and S. F. Clifford, Polarisation and angle of arrival fluctuations for a plane wave propagated through turbulent medium, *IEEE Transaction on Antennas Propagation*, AP-15, 416, 1967.

96. D. H. Hohn, Depolarisation of laser beam at 6328 Angstrom due to atmospheric transmission, *Applied Optics*, 8, 367, 1969.

97. M. Uysal, S. M. Navidpour and J. T. Li, Error rate performance of coded free-space optical links over strong turbulence channels, *IEEE Communication Letters*, 8, 635–637, 2004.

98. A.-H. M. A., A. L. C. and R. L. Phillips, Mathematical model for the irradiance probability density function of a laser beam propagating through turbulent media, *Optical Engineering*, 40, 1554–1562, 2001.

99. W. O. Popoola, Z. Ghassemlooy and E. Leitgeb, Free-space optical communication using subcarrier modulation in gamma–gamma atmospheric turbulence, *9th International Conference on Transparent Optical Networks (ICTON '07)*, Warsaw, Poland, Vol. 3, pp. 156–160, July 2007.

100. A. Garcia-Zambrana, Error rate performance for STBC in free-space optical communications through strong atmospheric turbulence, *IEEE Communication Letters*, 11, 390–392, 2007.

101. S. G. Wilson, M. Brandt-Pearce, Q. Cao and J. H. Leveque, Free-space optical MIMO transmission with Q-ary PPM, *IEEE Transactions on Communications*, 53, 1402–1412, 2005.

102. J. H. Churnside and S. F. Clifford, Log-normal Rician probability density function of optical scintillations in the turbulent atmosphere, *Journal of Optical Society of America*, 4, 1923–1930, 1987.

103. L. C. Andrews and R. L. Phillips, I–K distribution as a universal propagation model of laser beams in atmospheric turbulence, *Journal of Optical Society of America A*, 2, 160, 1985.

104. G. Parry and P. N. Pusey, K distributions in atmospheric propagation of laser light, *Journal of Optical Society of America*, 69, 796–798, 1979.

105. S. F. Clifford and R. J. Hill, Relation between irradiance and log-amplitude variance for optical scintillation described by the *K* distribution, *Journal of Optical Society of America*, 71, 112–114, 1981.

4 Modulation Techniques

Most practical OWC systems being currently deployed employ the IM/DD scheme for outdoor as well as indoor applications. Atmospheric conditions, in particular heavy fog, is the major problem, as the intensity of light propagating through a thick fog is reduced considerably. Therefore, intuitively, it appears that the best solution to high attenuation would be to pump more optical power or concentrate and focus more power into smaller areas. However, the eye safety introduces a limitation on the amount of optical power being transmitted. For indoor applications, the eye safety limit on transmit optical power is even more stringent. The optical channel differs significantly from the RF channels. Unlike RF systems where the amplitude, frequency and phase of the carrier signal are modulated, in optical systems, it is the intensity of the optical carrier that is modulated in most systems operating below 2.5 Gbps data rates. For data rates >2.5 Gbps, external modulation is normally adopted. Additionally, the use of photodetectors with a surface area many times larger than the optical wavelength facilitates the averaging of thousands of wavelength of the incident wave. In this chapter, a number of modulation techniques that are most popular, in terms of power efficiency and bandwidth efficiency, for both indoor and outdoor OWC applications are discussed. The analogue and digital baseband modulation techniques are discussed in Sections 4.2 and 4.3. The spectral properties, error probability as well as the power and bandwidth requirements are presented. Advanced modulation techniques such as the subcarrier intensity modulation and the polarization shift keying are presented in Sections 4.3 and 4.4.

4.1 INTRODUCTION

As discussed in Chapter 3, Section 3.1, the constraints (3.3) and (3.4) impose a limitation on the type of modulation schemes that could be adopted for OWC links. The conventional modulation techniques adopted in RF channels cannot be readily applied in optical channels. In fact, the constraint (3.4) prohibits the use of a large number of band-limited pulse shapes, including the sinc pulses and root-raised-cosine pulses to name a few. Figure 4.1 shows a number of modulation schemes that could be applied for optical channels. The transmission power employed in OWC configurations (mainly indoors) is limited by numerous factors, including eye safety, physical device limitations and power consumption. In outdoor FSO systems, to overcome attenuation due to fog, one could employ lasers with higher powers, but this is also limited by the eye-safety standards.

Both quadrature amplitude modulation (QAM) on discrete multitones (DMT) and multilevel pulse amplitude modulation (PAM) are spectrally efficient modulation schemes suitable for LED-based communications, but are less power efficient [1]. DMT is a baseband implementation of the more generalized orthogonal frequency division multiplexing (OFDM) [2] and is the most useful for channels with

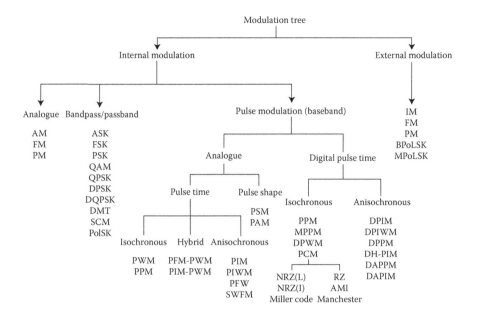

FIGURE 4.1   Modulation tree.

interference or strong low-frequency noise due to the artificial ambient light sources (e.g., fluorescent and incandescent) [3]. While L-PAM and L-QAM can provide higher bandwidth efficiency at the cost of reduced power efficiency, L-pulse time modulations (such as L-PPM and L-DPIM) can achieve higher power efficiency but at the expense of increased bandwidth requirement. Limitations on the optical power favour modulation schemes with a high peak-to-mean optical power ratio (PMOPR) such as the pulse position modulation (PPM) and digital pulse internal modulation (DPIM). OOK (the most widely used scheme in FSO systems) offers similar power requirement to the 2-PPM, whereas pass-band modulation schemes such as BPSK suffer from 1.8 dB power penalty. The bandwidth of high-data-rate systems is limited due to the capacitance constraints of large-area photodiodes and, therefore, a compromise between power and bandwidth requirements must be pursued. Selecting a modulation technique is one of the key technical decisions in the design of any communication system. Before selection can take place, it is necessary to define the criteria on which the various modulation techniques are to be assessed. For the optical wireless channel, these criteria are listed below in order of decreasing importance.

1. *Power efficiency:* In order to comply with the eye and skin safety regulations, the average optical power emitted by an optical wireless transceiver is limited [1,2]. Furthermore, in portable battery-powered equipment, it is desirable to keep the electrical power consumption to a minimum, which also imposes limitations on the optical transmit power. Consequently, the most important criterion when evaluating the modulation techniques suitable for indoor

optical wireless communication systems is the power efficiency. Each of these modulation schemes offers a certain optical average power, and therefore they are usually compared in terms of the average optical power required to achieve a desired BER performance or SNR. The power efficiency η_p of a modulation scheme is given by the average power required to achieve a given BER at a given data rate [3]. Mathematically, η_p is defined as [4]

$$\eta_p = \frac{E_{pulse}}{E_b} \tag{4.1}$$

where E_{pulse} is the energy per pulse and E_b is the average energy per bit.

2. *Bandwidth efficiency:* Although the optical carrier can be theoretically considered as having an 'unlimited bandwidth', the other constituents (photodetector area, channel capacity) in the system limit the amount of bandwidth that is practically available for a distortion-free communication system [2]. Also, the ensuing multipath propagation in diffuse link/nondirected LOS limits the available channel bandwidth [5]. This also makes the bandwidth efficiency a prime metric. The bandwidth efficiency η_B is defined as [4]

$$\eta_B = \frac{R_b}{B} \tag{4.2}$$

where R_b is the achievable bit rate and B is the bandwidth of the IR transceiver. The relationship between bandwidth and power efficiencies depends on the average duty cycle γ given by [4]

$$\eta_p = \frac{\eta_B}{\gamma} \tag{4.3}$$

When the shot noise is the dominant noise source, the received SNR is proportional to the photodetector surface area. Consequently, single-element receivers favour the use of large-area photodetectors. However, the high capacitance associated with large-area photodetectors has a limiting effect on the receiver bandwidth. In addition to this, for nondirected LOS and diffuse link configurations, the channel bandwidth is limited by multipath propagation. Therefore, it follows that modulation schemes which have a high bandwidth requirement are more susceptible to intersymbol interference (ISI), and consequently incur a greater power penalty. Thus, the second most important criterion when evaluating modulation techniques is the bandwidth efficiency.

3. *Transmission reliability:* A modulation technique should be able to offer a minimum acceptable error rate in adverse conditions as well as show resistance to the multipath-induced ISI and variations in the data signal DC component [6]. A long absence of '0 to 1' transition may be problematic as the clock recovery by a digital phase-locked loop (DPLL) might not be fea-

sible [7]. Moreover, multiple consecutive high pulses should be avoided, since the resulting signal would be distorted by the high-pass filter (HPF) in the receiver [8]. In addition, the modulation technique should be resistant to a number of factors such as the phase jitter due to variations of the signal power, pulse extensions due to diffusion component larger time constant and pulse distortion due to near-field signal clipping [4,6].

4. *Other considerations:* Optical wireless transceivers intended for mass-market applications are likely to have tight cost constraints imposed upon them. Consequently, it is highly desirable that the chosen modulation technique is rather simple to implement. Achieving excellent power efficiency and/or bandwidth efficiency is of little use if the scheme is so complex to implement that cost renders it unfeasible. Another consideration when evaluating modulation techniques is the ability to reject the interference emanating from artificial sources of ambient light. The simplest method to reduce the power level of the ambient light is to use electrical high-pass filtering. Consequently, it is desirable that the chosen modulation technique does not have a significant amount of its power located at DC and low frequencies, thereby reducing the effect of baseline wander and thus permitting the use of higher cut-on frequencies. In addition to this, if the chosen modulation technique is required to operate at medium to high data rates over non directed LOS or diffuse links, multipath dispersion becomes an issue. Consequently, it is also desirable that the scheme be resistant to ISI resulting from multipath propagation. For the intensity modulation direct detection indoor optical wireless channel discussed in this chapter, the candidate modulation techniques can be grouped into two general categories, these being baseband and subcarrier schemes.

4.2   ANALOGUE INTENSITY MODULATION

AIM techniques are simple and low cost compared to the digital modulation scheme that can be used in a number of applications, including cable TV distributions, high-definition TV transmission, radio-over-fibre, antenna remoting in cellular/personal communications and beam forming for phased-array radars. AIM can be implemented in two ways as shown in Figure 4.2. The simplest method is the IM/DD, where the intensity of the light source is modulated with the analogue signal (or a premodulated RF signal) and direct detection-based photodetector is used to recover the analogue signal. IM/DD technique is not suitable for high-frequency applications. This is so due to the limited bandwidth of the optical sources as well as their nonlinear characteristics. The alternative option is to use an external modulator (such as the Mach–Zehnder modulator [MZM]), to modulate the intensity of the continuous wave laser beam. At the receiver, a coherent receiver is used to recover the baseband signal.

Provided the source and the external modulator are linear, the free-space optical link acts only as an attenuator and is therefore transparent to the modulation format of the RF signal [9]. Thus, in such systems, amplitude modulation (AM) and multilevel

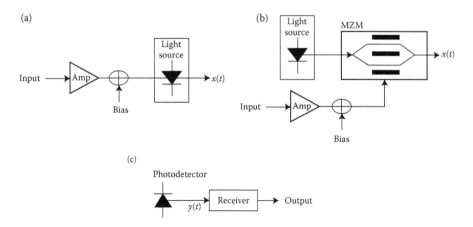

FIGURE 4.2 Optical transmission system block diagram: (a) intensity modulation, (b) external modulation and (c) receiver.

modulation formats such as QAM, pulse amplitude modulation (PAM), as well as subcarrier multiplexing (SCM) can be employed. Furthermore, millimetre wave signals approaching 100 GHz could be transmitted using external modulators but at the cost of power efficiency [10] and linearization requirements.

The main problem encountered with the AIM schemes is the optical source linearity. Both LEDs and LDs display acceptable linearity at low power levels, but at higher power levels they become progressively nonlinear, thus leading to harmonic and intermodulation distortions, particularly in frequency division multiplexed baseband transmission systems [11]. There are a number of techniques that could be employed to improve the distortion performance of AIM systems, including the predistortion, the feedback and feedforward. Predistortion of the drive waveform to take into account source nonlinearity is of limited use since it must be continually adjusted as the source ages. Negative feedback around the modulated optical source is more useful, but it requires fast amplifiers in order to restrict feedback loop delay to an acceptable value [12,13]. The most promising technique is feedforward compensation, where the characteristics of an identical source are used to modify the driving waveform to overcome its nonlinearity [14]. In this way, harmonic distortion can be reduced to around −70 dB at which point it is no longer a major consideration. A PIN diode is usually used instead of an APD at the receiver for better linearity performance. Although external modulators such as the MZM can support high-frequency RF signals, they require high drive voltages, which in turn lead to very costly drive amplifiers [12].

In AIM systems, the optical system bandwidth is the same as the baseband signal bandwidth, that is, $B_o = B_m$.

The average transmitted optical power in terms of IM signal $x(t)$ is given by

$$P_t = P_0(1 + x(t))$$
$$P_t = P_0(1 + m\cos\omega_m t)$$
$$(4.4)$$

where P_0 is the DC power and the modulation index $m = (i_p/I_b - I_{th})$; where i_p is the peak laser diode current above the DC bias current I_b, and I_{th} is the threshold current.

For FSO links with the receiver having an aperture diameter of D, the received optical power is defined as

$$P_r = \frac{\pi D^2}{8} I(0, L) \tag{4.5}$$

The performance of the optical wireless system defined in terms of SNR is given as

$$\text{SNR} = \frac{R^2 H^2(0) P_r^2}{\sigma_T^2} \tag{4.6}$$

where σ_T^2 is the total noise variance defined as

$$\sigma_T^2 = 2q(i_s + i_d + i_B) + \sigma_{th}^2 \tag{4.7}$$

where i_s is the shot noise, i_d is the dark current noise, i_B is the background noise and σ_{th} is the thermal noise.

Note that the SNR is a function of the square power; thus, an IM/DD-based system will naturally require high transmitted power and limited path loss. These differences have a profound effect on the system design. On conventional channels, the SNR is proportional (α) to the average received power P_r, whereas on OWC links, SNR α $[P_r]^2$, thus, implying the need for higher optical power requirement to deliver the same performance, as well as a limited path loss. The fact that P_t is limited due to the eye safety reasons suggests that modulation techniques offering a high peak-to-mean power ratio are favourable. This is generally achieved by trading off power efficiency against bandwidth efficiency. When the shot noise is the dominant noise source, the SNR is also proportional to the photodetector surface area A_{pd}. Thus, single-element receivers favour the use of large area detectors. However, as A_{pd} increases, so does its capacitance, thus limiting the system bandwidth. This is in direct conflict with the increased bandwidth requirement associated with power-efficient modulation techniques, and hence, there is a trade-off between these two factors, which designers must take into consideration.

$$\text{SNR} = \frac{(m/(1 + M)RGP_r)^2}{[2qG^{2+x}(I_d + RP_r) + \langle i_s \rangle^2 + (RGP_r)^2 \text{RIN}]B_m}$$
$$= m^2/(1 + m^2(\text{CNR})) \tag{4.8}$$

where R is the photodetector responsivity, I_d is the dark current, $\langle i_s \rangle^2$ is the receiver equivalent noise current variance, B_m is the baseband signal bandwidth, RIN is the laser relative intensity noise, G is the APD gain and CNR is the carrier to noise ratio. This method has no inherent SNR improvement over the CNR of the optical link.

Increasing m will result in SNR approaching CNR but usually at the cost of reduced system linearity.

4.3 DIGITAL BASEBAND MODULATION TECHNIQUES

4.3.1 BASEBAND MODULATIONS

For the signalling schemes in this class, the data have not been translated to a much higher carrier frequency prior to intensity modulation of the optical source. Thus, a significant portion of the signal power is restricted to the DC region. Baseband modulation techniques are so called because the spectrum of the modulated data is in the vicinity of DC. Baseband schemes that include, among others, OOK and the family of pulse time modulation (PTM) techniques, are more tolerant to the effects of the multipath channel. OOK is the simplest technique, in which the intensity of an optical source is directly modulated by the information sequence. In contrast, PTM techniques use the information sequence to vary some time-dependent property of a pulse train. Popular examples of such schemes include pulse width modulation (PWM), in which the width of the pulses convey the information, and PPM, in which the information is represented by the position of the pulses within fixed time frames [15,16]. In DPIM, the information is represented by the number of empty slots between pulses, potentially allowing higher data rates and improvements in power efficiency compared to OOK and PPM [17]. Dual-header pulse interval modulation (DH-PIM), which is a variation on PIM reduces the number of 'empty' slots, and therefore symbol length, by introducing a second pulse at the start of the information symbol. The technique offers a trade-off between the lower bandwidth requirement of the longer pulse and the subsequent higher average optical power requirement. At higher bit rates, the scheme is both bandwidth and power efficient compared to PPM [17]. Unlike the fixed symbol length of PPM, both DPIM and DH-PIM offer symbol synchronization due to the pulse always being at the start of the symbol. To reduce the performance degradation of pulse modulation schemes adopted in a highly dispersive indoor environment with a large delay spread, maximum likelihood sequence detection (MLSD) as well as decision feedback equalizer can be employed, but at the cost of higher system complexity.

This high bandwidth requirement problem in PPM has led to the development of several alternative modulation schemes, including the differential PPM [18,19], dicode PPM [20], multiple PPM [21], edge position modulation (EPM) [4] and hybrid modulation. The latter includes the differential amplitude pulse position modulation (DAPPM), which offers additional advantages over PPM in terms of the peak-to-average power ratio [22]. In EPM, the time is divided into discrete slots greater than the rise time and jitter of a pulse; however, the pulse width can be wider than one time slot, thus allowing more information to be transmitted than for a comparable PPM scheme. Multilevel modulation schemes such as PAM, L-PAM and QAM offered improved bandwidth efficiency by transmitting more information per symbol through the channel at the cost of lower power efficiency. However, compared to the binary schemes, they are more sensitive to the channel nonlinearities as well as the noise.

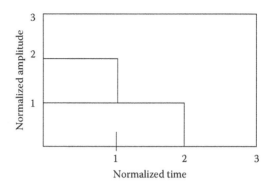

FIGURE 4.3 L-PAM time waveforms.

In *L*-PAM, a pulse is sent in each symbol interval, where the pulse amplitude takes on one of the *L* possible levels (see Figure 4.3). Its bandwidth efficiency is defined as

$$\eta_B = \frac{R_b}{B_m} = \log_2 L \tag{4.9}$$

Adopting the standard PAM constellation, the BER can be shown as [3]

$$P_{b-\text{PAM}} = \frac{2(L-1)}{L \log_2 L} Q\left(\frac{RP_r}{L-1} \sqrt{\log_2 L / \sigma_T^2 R_b} \right) \tag{4.10}$$

Binary direct sequence spread spectrum (DSSS) techniques have also been adopted to reduce the affect of multipath-induced ISI to achieve higher data rates [23].

4.3.2 On–Off Keying

OOK is the most reported modulation techniques for IM/DD in optical communication. This is apparently due to its simplicity. A bit one is simply represented by an optical pulse that occupies the entire or part of the bit duration while a bit zero is represented by the absence of an optical pulse. Both the return-to-zero (RZ) and non-return-to-zero (NRZ) schemes can be applied. In the NRZ scheme, a pulse with duration equal to the bit duration is transmitted to represent 1 while in the RZ scheme the pulse occupies only the partial duration of bit. Figure 4.4 shows the single mapping of OOK-NRZ and OOK-RZ with a duty cycle $\gamma = 0.5$ for average transmitted power of P_{avg}. Hence, the envelop for OOK-NRZ is given by

$$p(t) = \begin{cases} 2P_r & \text{for } t \in [0, T_b) \\ 0 & \text{elsewhere} \end{cases} \tag{4.11}$$

where P_r is the average power and T_b is the bit duration.

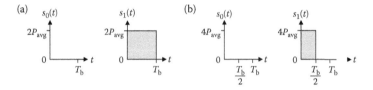

FIGURE 4.4 Transmitted waveforms for OOK: (a) NRZ and (b) RZ ($\gamma = 0.5$).

The simplicity of OOK has led to its use in commercial optical wireless systems such as IrDA, Fast IR links operating below 4 Mbit/s [24]. In these links, return-to-zero-inverted (RZI) signalling is used, in which a pulse represents a zero rather than a one. At bit rates up to and including 115.2 kbit/s, the pulse duration is nominally 3/16 of the bit duration. For data rates of 0.576 and 1.152 Mbit/s, the pulse duration is nominally 1/4 of the bit duration. PPM is also used at higher date rates (>4 Mbps).

The electrical power spectral densities (PSDs) of the OOK-NRZ and OOK-RZ ($\gamma = 0.5$) assuming independently and identically distributed (IID) one and zeros are given by [25]

$$S_{\text{OOK-NRZ}}(f) = (P_r R)^2 T_b \left(\frac{\sin \pi f T_b}{\pi f T_b} \right)^2 \left[1 + \frac{1}{T_b} \delta(f) \right] \tag{4.12}$$

$$S_{\text{OOK-RZ}(\gamma=0.5)}(f)(f) = (P_r R)^2 T_b \left(\frac{\sin(\pi f T_b/2)}{\pi f T_b/2} \right)^2 \left[1 + \frac{1}{T_b} \sum_{n=-\infty}^{\infty} \delta\left(f - \frac{n}{T_b} \right) \right] \tag{4.13}$$

where $\delta(\ )$ is the Dirac delta function.

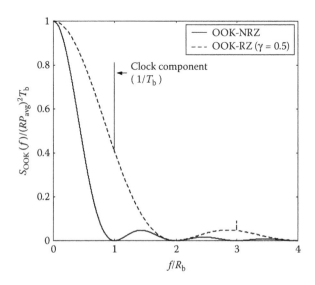

FIGURE 4.5 PSD of OOK-NRZ and OOK-RZ ($\gamma = 0.5$).

The PSDs of OOK-NRZ and OOK-RZ ($\gamma = 0.5$) are plotted in Figure 4.5 using the MATLAB® code in Programs 4.1–4.3. The power axis is normalized to the average electrical power multiplied by the bit duration $(P_r R)^2 T_b$ and the frequency axis is normalized to the bit rate R_b ($=1/T_b$). Both curves were plotted using the same average optical power P_r.

Program 4.1: MATLAB Codes to Calculate the Analytical PSD of OOK-NRZ

```
p_avg=1;
% average optical power
R=1;
% photodetecor sensitivity
Rb=1;
% normalized bit rate
Tb=1/Rb;
% bit duration
df=Rb/100;
% spectral resolution
f=0:df:5*Rb;
% frequency vector
x=f*Tb;
% normalized frequency

temp1=(sinc(x)).^2;
temp2=0;
a=2*R*p_avg;
% peak power is twice average power
p=(a^2*Tb).*temp1;
%p(1)=p(1)+(((a^2)*Tb))*(sinc(0)^2)*(1/Tb);
% delta function at DC

p=p/(((p_avg*R)^2)*Tb);
% power normalization by energy per bit
```

Program 4.2: MATLAB Codes to Calculate the Analytical PSD of OOK-RZ

```
p_avg=1;
% average optical power
R=1;
% photodetecor sensitivity
Rb=1;
% normalized bit rate
Tb=1/Rb;
% bit duration
df=Rb/100;
% spectral resolution
f=0:df:5*Rb;
% frequency vector
```

```
x=f*Tb/2;
% normalized frequency
term1=(sinc(x)).^2;

a=R*p_avg;
p=(((a^2)*Tb)).*term1;
%p(1)=p(1)+(((a^2)*Tb))*(sinc(0)^2)*(1/Tb)
% delta function at DC

p((Rb/df)+1)=p((Rb/df)+1)+(((a^2)*Tb))*(sinc(Rb*Tb/2)^2)*(1/Tb);
% delta function at f=Rb
p((2*Rb/df)+1)=p((2*Rb/df)+1)+(((a^2)*Tb))*(sinc(2*Rb
*Tb/2)^2)*(1/Tb);
% delta function at f=2Rb
p((3*Rb/df)+1)=p((3*Rb/df)+1)+(((a^2)*Tb))*(sinc(3*Rb
*Tb/2)^2)*(1/Tb);
% delta function at f=3Rb
p((4*Rb/df)+1)=p((4*Rb/df)+1)+(((a^2)*Tb))*(sinc(4*Rb
*Tb/2)^2)*(1/Tb);
% delta function at f=4Rb

p=p/(((p_avg*R)^2)*Tb);
```

Program 4.3: MATLAB Codes to Generate PSD of OOK
(NRZ and RZ) by Simulation

```
Rb=1;
Tb=1/Rb;
% bit duration
SigLen=1000;
% number of bits or symbols
fsamp=Rb*10;
% sampling frequency
nsamp=fsamp/Rb;
% number of samples per symbols

Tx_filter=ones(1,nsamp);
% transmitter filter for NRZ
%Tx_filter=[ones(1,nsamp/2) zeros(1,nsamp/2)];
% transmitter filter for RZ
bin_data=randint(1,SigLen);
%generating prbs of length SigLen
bin_signal=conv(Tx_filter,upsample(bin_data,nsamp));
% pulse shaping function
bin_signal=bin_signal(1:SigLen*nsamp);

% ************** psd of the signals ***********
Pxx=periodogram(bin_signal);
Hpsd=dspdata.psd(Pxx,'Fs',fsamp);
```

```
% Create PSD data object
figure;  plot(Hpsd); title('PSD of TX signal')
```

For baseband modulation techniques, the bandwidth requirement is generally defined as the span from DC to the first null in the PSD of the transmitted signal. As expected, OOK-RZ ($\gamma = 0.5$) has twice the bandwidth requirement of OOK-NRZ, since the pulses are only half as wide. Both OOK-NRZ and OOK-RZ ($\gamma = 0.5$) have discrete (impulse) terms at DC, with a weight P_r^2. OOK-RZ ($\gamma = 0.5$) also has discrete terms at odd multiples of the bit rate Figure 4.5. The impulse at $f = R_b$ can be used to recover the clock signal at the receiver. OOK-NRZ, on the other hand, has spectral nulls at multiples of the bit rate, and consequently requires the introduction of some nonlinearity in order to achieve clock recovery. Both OOK-NRZ and OOK-RZ have significant power contents at DC and low frequencies. This characteristic means that electrical high-pass filtering is not effective in reducing the interference produced by artificial sources of ambient light, since high cut-on frequencies cannot be used without introducing significant baseline wander. Comparing the areas under the two curves it is evident that, for a given average optical transmit power, OOK-RZ ($\gamma = 0.5$) has twice the average electrical power of OOK-NRZ.

The OOK-NRZ has power efficiency η_p of 2 and bandwidth efficiency η_B of 1. The OOK-RZ has the same power efficiency as OOK-NRZ; however, bandwidth efficiency depends on the duty cycle. The bandwidth efficiency for $\gamma = 1/4$ is 0.25. Furthermore, RZ does not support sample clock recovery at the receiver because it allows a long low signal without any 0 to 1 transition [4]. Therefore, bit stuffing is necessary which further decreases bandwidth efficiency.

4.3.3 Error Performance on Gaussian Channels

In error analysis for all modulation schemes, we have made a number of assumptions as outlined below:

1. The transmission link is a line of sight, and the channel imposes no multipath dispersion and no path loss.
2. The noise associated with the receiver is negligible and the dominant noise source is due to background shot noise, which is assumed as a white Gaussian.
3. There is no interference due to artificial light; this removes the need for a high-pass filter at the receiver and prevents baseline wander.
4. There is no bandwidth limitation imposed by the transmitter and receiver.

The ideal maximum likelihood (ML) receiver for OOK in the presence of additive white Gaussian noise (AWGN) consists of a continuous-time filter with an impulse response $r(t)$, which is matched to the transmitted pulse shape $p(t)$, followed by a sampler and threshold detector set midway between expected one and zero levels, as illustrated in Figure 4.6. The transmitter filter has a unit-amplitude rectangular impulse response $p(t)$, with a duration of one bit, T_b. The output of the transmitter

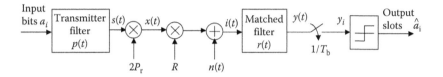

FIGURE 4.6 The block diagram of OOK system.

filter is scaled by the peak detected signal photocurrent $2P_r$, and P_r is the average received optical signal power. The signal-independent shot noise $n(t)$ is then added to the signal which is modelled as white and Gaussian, with a double-sided power spectral density $N_o/2$, given as [26]

$$\frac{N_o}{2} = qI_B \tag{4.14}$$

where q is the electron charge and I_B is the average photocurrent generated by the background light.

The detected signal at the input of the matched filter can be modelled as

$$i(t) = \begin{cases} I_p + n(t) & \text{for } a_i = 1 \\ n(t) & \text{for } a_i = 0 \end{cases} \tag{4.15}$$

where $n(t) \sim N(0, \sigma^2)$ is the additive white Gaussian noise due to ambient light with double-sided power spectral density $N_o/2$, zero mean and a variance of σ^2, and I_p is the peak photocurrent.

A digital symbol '1' is assumed to have been received if the received signal is above the threshold level and '0' otherwise.

The probability of error is therefore given as

$$P_e = p(0) \int_{i_{th}}^{\infty} p(i/0)\, di + p(1) \int_0^{i_{th}} p(1)\, di \tag{4.16}$$

where i_{th} is the threshold signal level, $p(0)$ and $p(1)$ are probabilities of 'zero' and 'one' and the marginal probabilities are defined as

$$p(i/0) = \frac{1}{\sqrt{2\pi\sigma^2}} \exp\left(\frac{-i^2}{2\sigma^2}\right) \tag{4.17}$$

$$p(i/1) = \frac{1}{\sqrt{2\pi\sigma^2}} \exp\left(\frac{-(i - I_p)^2}{2\sigma^2}\right) \tag{4.18}$$

For equiprobable symbols, $p(0) = p(1) = 0.5$; hence, the optimum threshold point is $i_{th} = 0.5\,I_p$ and the conditional probability of error reduces to

$$P_e = Q\left(\frac{i_{th}}{\sigma}\right) \qquad (4.19)$$

where $Q(\ )$ is Marcum's Q-function, which is the area under the Gaussian tail, given by

$$Q(x) = \frac{1}{\sqrt{2\pi}}\int\limits_{x}^{\infty} e^{-\alpha^2/2} d\alpha \qquad (4.20)$$

For the OOK-NRZ waveforms shown in Figure 4.4 neglecting any bandwidth limitations imposed by the transmitter or receiver, a unit-energy matched filter has a rectangular impulse response $r(t)$ with amplitude $1/\sqrt{T_b}$ and duration T_b. As illustrated in Figure 4.7, in the absence of noise, the peak output of this filter when a one transmitted is $I_p = \sqrt{E_b} = 2RP_r\sqrt{T_b}$, where R is the photodetector responsivity. When a zero is transmitted, the peak output of the matched filter is 0.

In the case of a matched filter, (4.15) can be replaced by

$$y_i = \begin{cases} E_p + n_i & \text{for } a_i = 1 \\ n_i & \text{for } a_i = 0 \end{cases} \qquad (4.21)$$

For a matched filter, the variance of the noise samples at the output of the filter is dependent only on the PSD of the noise input and the energy in the impulse response of the matched filter. Thus, if the input is AWGN with a double-sided PSD $N_0/2$, the variance of the noise at the output of the matched filter is given by [27]

$$\sigma^2 = \frac{N_0}{2}\int\limits_{t=0}^{T_b} r^2(t)\,dt \qquad (4.22)$$

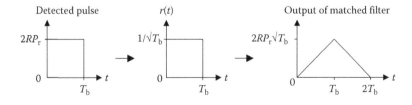

FIGURE 4.7 Matched filter output for detected OOK-NRZ pulse.

Hence, the standard deviation σ has the value

$$\sigma = \sqrt{\frac{N_0 E_p}{2}} \tag{4.23}$$

Therefore, incorporating (4.19), (4.21) and (4.23), that is, substituting E_p for I_p and $\sqrt{(N_0 E_p)/2}$ for σ, gives

$$P_{e\_bit\_OOK} = Q\left(\sqrt{\frac{E_b}{N_0}}\right) \tag{4.24}$$

where average energy per bit E_b is given by

$$E_b = \frac{E_p}{2} = 2(RP_r)^2 T_b \tag{4.25}$$

The ratio E_b/N_0 is usually referred to as the SNR per bit.

In the OOK-RZ formatting, the average energy per bit is increased by a factor of $1/\gamma$ which is given as

$$E_b = \frac{E_p}{2} = \left(2(RP_r)^2 T_b/\gamma\right) \tag{4.26}$$

Consequently, for a given value of P_r, OOK-RZ ($\gamma = 0.5$) has twice the ratio E_b/N_0 compared with OOK-NRZ. Therefore, in order to achieve the same error performance, OOK-RZ ($\gamma = 0.5$) requires 3 dB less electrical power or 1.5 dB less average optical power compared with OOK-NRZ. However, this improvement in power efficiency is achieved at the expense of doubling the bandwidth requirement. Similarly, setting $\gamma = 0.25$ results in a 3 dB reduction in average optical power requirement compared with OOK-NRZ, but requires four times the bandwidth.

The MATLAB codes for simulating the BER for OOK-NRZ using the matched filter based on one sample process are given in Programs 4.4 and 4.5.

Program 4.4: MATLAB Codes to Simulate BER of OOK-NRZ

```
clear;
clc;
close all

q=1.6e-19;
% Charge of Electron
Ib=202e-6;
% Background Noise Current+interfernce
N0=2*q*Ib;
% Noise Spectral Density, 2*q*Ib
```

```
R=1;
% Photodetector responsivity

Rb=1e6;
% Bit rate
Tb=1/Rb;
% bit duration
sig_length=1e5;
% number of bits
nsamp=10;
% samples per symbols
Tsamp=Tb/nsamp;
% sampling time

EbN0=1:12;
% signal-to-noise ratio in dB.
SNR=10.^(EbN0./10);
% signal-to-noise ratio

% ********** Simulation of probability of errors. ************
for i=1:length(SNR)
  P_avg(i)=sqrt(N0*Rb*SNR(i)/(2*R^2));
  % average transmitted optical power
  i_peak(i)=2*R*P_avg(i);
  % Peak Electrical amplitude
  Ep(i)=i_peak(i)^2*Tb;
  % Peak energy (Energy per bit is Ep/2)
  sgma(i)=sqrt(N0/2/Tsamp);
  % noise variance
  %sgma(i)=i_peak(i)/sqrt(2)*sqrt(nsamp/(2*SNR(i)));

  pt=ones(1,nsamp)*i_peak(i);
  % tranmitter filter
  rt=pt;
  % receiver filter matched to pt

  OOK=randint(1,sig_length);
  % random signal generation
  Tx_signal=rectpulse(OOK,nsamp)*i_peak(i);
  % Pulse shaping function (rectangular pulse)
  Rx_signal=R*Tx_signal+sgma(i)*randn(1,length(Tx_signal));
  % received signal (y=x+n)
  MF_out=conv(Rx_signal,rt)*Tsamp;
  % matched filter output
  MF_out_downsamp=MF_out(nsamp:nsamp:end);
  % sampling at end of bit period
  MF_out_downsamp=MF_out_downsamp(1:sig_length);
  % truncation

  Rx_th=zeros(1,sig_length);
  Rx_th(find(MF_out_downsamp>Ep(i)/2))=1;
  % thresholding
```

```
  [nerr ber(i)]=biterr(OOK,Rx_th);
  % bit error calculation
end

figure;
semilogy(EbN0,ber,'b');
hold on
semilogy(EbN0,qfunc(sqrt(10.^(EbN0/10))),'r-X','linewidth',2);
% theoretical ber, 'mx-');
grid on
legend('simulation','theory');
xlabel('Eb/No, dB');
ylabel('Bit Error Rate');
title('Bit error probability curve for OOK modulation')
```

Program 4.5: MATLAB Codes to Simulate BER of OOK-NRZ Using Matched Filter-Based Receiver

```
q=1.6e-19;
% Charge of Electron
Ib=202e-6;
% Background Noise Current+interfernce
N0=2*q*Ib;
% Noise Spectral Density, 2*q*Ib
Rb=1e6;
% bit rate.
Tb=1/Rb;
% bit duration
R=1;
% Receiver responsivity.
sig_length=1e5;
% No. of bits in the input OOK symbols.

snr_dB=0:9;
% signal-to-noise ratio in dB.
SNR=10.^(snr_dB./10);
% signal-to-noise ratio

for i=1:length(snr_dB)
  P_avg(i)=sqrt(N0*Rb*SNR(i)/(2*R^2));
  % average optical power
  i_peak(i)=2*R*P_avg(i);
  % peak photocurrent
  Ep(i)=i_peak(i)^2*Tb;
  % Peak Energy
  sgma(i)=sqrt(N0*Ep(i)/2);
  % sigma, standard deviation of noise after matched filter
  th=0.5*Ep(i);
  % threshold level
```

```
Tx=randint(1,sig_length);
% transmitted bit

for j=1 : sig_length;
  MF(j)=Tx(j)*Ep(i)+gngauss(sgma(i));
  %matched filter output
end

Rx=zeros(1,sig_length);
Rx(find(MF>th))=1;
%threshold detection
[No_of_Error(i) ber(i)]=biterr(Tx,Rx);
end
```

4.4 PULSE POSITION MODULATION

In LOS OWC links where the requirement for the bandwidth is not of a major concern, PPM with its significantly better power efficiency seems to be the most attractive option for a range of applications. PPM is an orthogonal modulation technique and a member of the pulse modulation family (see Figure 4.8). The PPM modulation technique improves on the power efficiency of OOK but at the expense of an increased bandwidth requirement and greater complexity. An L-PPM symbol consists of a pulse of constant power occupying one slot duration within L (= 2^M, where bit resolution $M > 0$ is an integer) possible time slots with the remaining slots being empty (see Figure 4.8). Information is encoded within the position of the pulse and the position of the pulse corresponds to the decimal value of the M-bit input data. In order to achieve the same throughput as OOK, PPM slot duration T_{s_PPM} is shorter than the OOK bit duration T_b by a factor L/M, that is

$$T_{s\_PPM} = \frac{T_b M}{L} \tag{4.27}$$

The transmit pulse shape for L-PPM is given by [11]

$$x(t)_{PPM} = \begin{cases} 1 & \text{for } t \in \left[(m-1)T_{s\_PPM}, mT_{s\_PPM}\right] \\ 0 & \text{elsewhere} \end{cases} \tag{4.28}$$

where $m \in \{1, 2, \ldots L\}$.

Hence, the PPM symbol sequence is given by

$$x(t)_{PPM} = LP_r \sum_{k=0}^{L-1} c_k p\left(t - \frac{kT_{symb}}{L}\right) \tag{4.29}$$

where $c_k \in \{c_0, c_1, c_2, \ldots, c_{L-1}\}$ is the PPM symbol sequence, $p(t)$ is the pulse shaping function of unity height and of duration T_{sysmb}/L, T_{symb} (= $T_b M$) is the symbol interval and LP_{avg} is the peak optical power of PPM symbol.

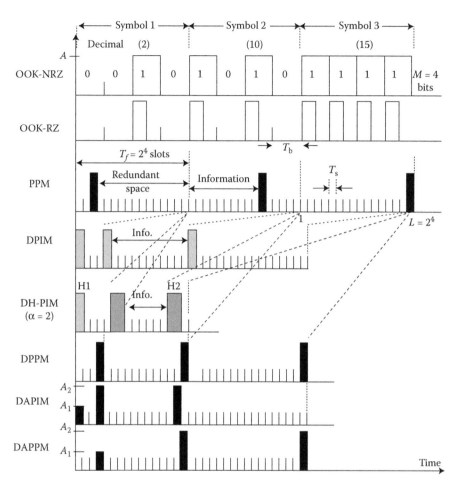

FIGURE 4.8   Time waveforms for OOK, PPM, DPI, DH-PIM, DPPM, DAPIM and DAPPM signals.

In PPM, all the signals are equidistant with

$$d_{\text{min-PPM}} = \min_{i \neq j} \int \left[x_i(t) - x_j(t) \right]^2 dt = 2LP^2 \log_2 \left(\frac{L}{R_b} \right) \quad (4.30)$$

The transmitted waveforms for 16-PPM and OOK are shown in Figure 4.8. Compared with OOK, employing PPM results in increased system complexity, since the receiver requires both slot and symbol synchronization in order to demodulate the signal. Nevertheless, primarily due to its power efficiency, PPM has been the most widely used modulation technique for OWC systems as well as deep space laser communications, and handheld devices where lower power consumption is one of the key factors [1]. The infrared physical layer section of the IEEE 802.11 standard on wireless LANs specifies 16-PPM for bit rates of 1 Mbit/s and 4-PPM for 2 Mbit/s (both schemes giving the same slot rate). In addition to this, IrDA serial data links operating at 4 Mbit/s specify 4-PPM.

The electrical power spectrum of L-PPM is given as [28]

$$S_{\text{PPM}}(f) = |P(f)|^2 \left[S_{\text{c,PPM}}(f) + S_{\text{d,PPM}}(f) \right] \tag{4.31}$$

where $P(f)$ is the Fourier transform of the pulse shape and $S_{\text{c,PPM}}(f)$ and $S_{\text{d,PPM}}(f)$ are the continuous and discrete components, respectively, which are given as

$$S_{\text{c,PPM}}(f) = \frac{1}{T_{\text{symb}}} \left[\left(1 - \frac{1}{L}\right) + \frac{2}{L} \sum_{k=1}^{L-1} \left(\frac{k}{L} - 1\right) \cos\left(\frac{k \, 2\pi f T_{\text{symb}}}{L}\right) \right] \tag{4.32}$$

$$S_{\text{d,PPM}}(f) = \frac{2\pi}{T_{\text{symb}}^2} \sum_{k=-\infty}^{\infty} \delta\left(f - \frac{kL}{T_{\text{symb}}}\right) \tag{4.33}$$

The PSD of PPM for $L = 4$, 8 and 16 is shown in Figure 4.9. The three curves were constructed using the same average optical power, using rectangular pulse shapes occupying the full slot duration. The power axis is normalized to the average electrical power multiplied by the duration. The frequency axis is normalized to the bit rate R_{b}. From Figure 4.9, it is easily seen that unlike OOK, the PSD of PPM falls to zero at DC for all values of L. This phenomenon provides an increased resistance to baseline wander over DC schemes and allows the use of higher cut on frequencies when mitigating high-pass filtering is employed to combat artificial light interference. As expected, by observing the positions of the first spectral nulls, it is clear that the bandwidth requirement increases as L increases. Furthermore, comparing the areas under the curves, it may also be observed that for a given average optical power, the detected

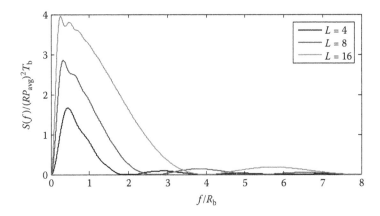

FIGURE 4.9 (**See colour insert.**) PSD of PPM for $L = 4$, 8 and 16.

electrical power increases as L increases. The MATLAB codes for generating the PPM symbols and PSDs are given in Programs 4.6 and 4.7, respectively.

Program 4.6: MATLAB Codes to Generate PPM

```
function PPM=generate_PPM(M,nsym)
% function to generate PPM
% 'M' bit resolution
% 'nsym': number of PPM symbol

PPM=[];
  for i= 1:nsym
    temp=randint(1,M);
    % random binary number
    dec_value=bi2de(temp,'left-msb');
    % converting to decimal value
    temp2=zeros(1,2^M);
    % zero sequence of length 2^M
    temp2(dec_value+1)=1;
    % placing a pulse accoring to decimal value,
    % note that in matlab index doesnot start from zero, so
    need to add 1;
    PPM=[PPM temp2];
    % PPM symbol
  end
end
```

Program 4.7: MATLAB Codes to Calculate the Analytical PSD of PPM

```
Rb=1;
% normalized bit rate
Tb=1/Rb;
% bit duration
M=4;
%Bit resolution
L=2^M;
% symbol length
p_avg=1;
% average optical power
R=1;
% photodetector responsivity
a=R*L*p_avg;
% electrical pulse amplitude
Ts=M/(L*Rb);
% slot duration
Rs=1/Ts;
% slot rate
```

```
df=Rs/1000;
% spectral resolution
f=0:df:8*Rb;
P_sq=(a*Ts)^2*(abs(sinc(f*Ts))).^2;
temp1=0;
for k=1:L-1
  temp1=temp1+(k/L-1).*cos(k*2*pi*f*Ts);
end
S_c=(1/(L*Ts))*(((L-1)/L)+(2/L)*temp1);
S=P_sq.*S_c;
S= S/(((p_avg*R)^2)*Tb);
```

4.4.1 ERROR PERFORMANCE ON GAUSSIAN CHANNELS

In isochronous schemes such as PPM, an error is confined to the symbol in which it occurs. Consequently, a single-slot error can affect a maximum of $\log_2 L$ bits. There are three types of error as shown in Figure 4.10, and outlined below:

- *Erasure error*: An erasure error occurs when a transmitted pulse in the kth is not detected due to the noise forcing the received pulse amplitude below the threshold level at the decision time. The probability of the erasure error can be defined as

$$P_{ee}(V < V_{th} : V_1) = \frac{1}{2}\mathrm{erfc}\left[(1-\alpha)\frac{A_i}{\sqrt{2}\sigma}\right] \qquad (4.34)$$

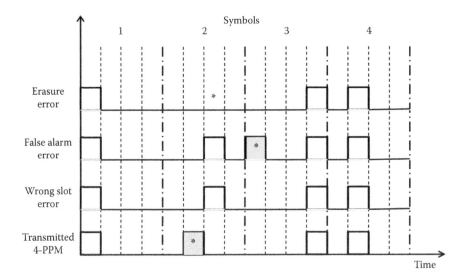

FIGURE 4.10 Error sources in PPM.

where σ is the noise deviation or RMS noise at the threshold detector, A is the pulse amplitude, V is the voltage level and $0 < \alpha < 1$.

- *False alarm error*: A false alarm error occurs when a transmitted zero is falsely detected as a one in the $(k + n)$th slot. It divides a symbol into two. The probability of the false alarm error can be defined as

$$P_{\text{fe}}(V_0 > V_{\text{th}} : V_0) = 0.5 \frac{T_s}{\tau} \text{erfc} \left[\frac{\alpha A_i}{\sqrt{2\sigma}} \right] \qquad (4.35)$$

- *Wrong slot error*: A wrong slot error occurs when a pulse is detected in a slot adjacent to the one in which it was transmitted. The probability of the wrong slot error can be defined as

$$P_{\text{wse}}(V > V_{\text{th}} : V_2 < V_{\text{th}}) = 0.5 \text{erfc} \left[\frac{V_{\text{th}} - V_2}{\sqrt{2\sigma}} \right] \qquad (4.36)$$

where V_2 is the noise-free voltage level at the sampling instant in the adjacent slot.

The block diagram of the matched filter-based receiver for the PPM scheme is given in Figure 4.11. The PPM encoder converts M-bit binary data sequence input bits $\{a_i; i = 1, \ldots, M\}$ into one of the possible L symbols. The PPM symbols are passed to a transmitter filter $p(t)$ which has a unit-amplitude rectangular impulse response with a duration of one slot $T_{\text{s\_PPM}}$. Since the duty cycle of PPM is fixed at L^{-1}, each pulse must have an amplitude of LP_r in order to maintain an average optical power of P_r, assuming that rectangular-shaped pulses are used. A unit-energy filter matched to the transmitted pulse shape $p(t)$ has a rectangular impulse response $r(t)$

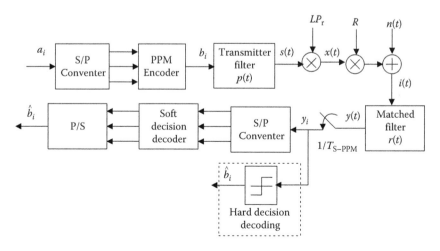

FIGURE 4.11 Block diagram of the matched filter-based receiver for PPM scheme with soft and hard decision decoding.

with amplitude $T_s^{-0.5}$ and duration T_s. In the absence of noise, the peak output of the matched filter when a pulse is transmitted is $\sqrt{E_s} = LRP_r\sqrt{T_s}$. When an empty slot is transmitted, the peak output of the matched filter is 0.

In the AWGN channel, two decoding algorithms can be used for the PPM signals:

1. Hard decision decoding using a threshold detector
2. Soft decision decoding using a maximum *a posteriori* (MAP) or maximum likelihood detector

Assuming that complete synchronization is maintained between the transmitter and receiver at all times, the hard decision decoding is carried out by comparing output of matched filter to a threshold level. A 'one' or a 'zero' is assigned depending on whether the matched filter output is above or below the threshold level at the sampling instant. For the 'soft' decision scheme, a block of L samples are passed to a soft decision detector, which assigns a one to the slot which contains the largest sample and zeros to the remaining slots. Since the relative amplitude of the slot within a symbol is of prime importance in soft decision decoding, soft decision decoding is optimal for systems in which AWGN is the dominant noise source and there is no channel distortion [27]. The soft decision decoding also offers more resilience to the artificial light interference, multipath distortion and turbulence compared to the hard decision decoding.

In the presence of AWGN with double-sided PSD $N_0/2$, the probability of slot error for the hard decoding may be derived as [25]

$$P_{se\text{-PPM-H}} = P(0)Q\left(\frac{\alpha_{T-opt}}{\sqrt{N_0/2}}\right) + P(1)Q\left(\frac{\sqrt{E_s} - \alpha_{T-opt}}{\sqrt{N_0/2}}\right) \qquad (4.37)$$

where α_{T-opt} is the optimum threshold level and $P(1)$ and $P(0)$ represent the probabilities of getting a pulse and an empty slot, respectively, as given by

$$P(0) = \frac{(L-1)}{L}; \quad P(1) = \frac{1}{L} \qquad (4.38)$$

Since the peak output of the matched filter when a pulse is transmitted is $\sqrt{E_s} = LRP_r\sqrt{T_s}$ and 0 for the empty slot, the slot error probability for hard decision is given by

$$P_{sle\text{-PPM-H}} = \frac{1}{L}Q\left(\frac{LRP_r\sqrt{T_s} - \alpha_{T-opt}}{\sqrt{N_0/2}}\right) + \frac{(L-1)}{L}Q\left(\frac{\alpha_{T-opt}}{\sqrt{N_0/2}}\right) \qquad (4.39)$$

Unlike OOK, the probabilities of receiving 'zero' and 'one' are not equal (probability of receiving zeros is $(L-1)$ times higher than receiving one). Hence, the optimum threshold level for hard decision decoding does not lie midway between one and zero levels. It is a complicated function of the signal and noise powers, and the order L. However, for low probability of error, the $\alpha_T = \sqrt{E}/2$ offers error probability very close to the optimum value and (4.39) can be simplified to

$$P_{\text{sle-PPM-H}} = Q\left(\sqrt{\frac{E_s}{2N_0}}\right) \tag{4.40}$$

Note that E_s is the energy of a symbol, which encodes $\log_2 L$ bits of data. Hence, the average energy per bit E_b is given by [27]

$$E_b = \frac{E_s}{\log_2 L} = L(RP_r)^2 T_b \tag{4.41}$$

Since each symbol contains L slots, the probability of slot error may be converted into a corresponding symbol error probability using

$$P_{\text{sye-PPM-H}} = 1 - (1 - P_{\text{sle-PPM-H}})^L \tag{4.42}$$

Assuming the data are IID, each symbol is equally likely and the probability of symbol error may be converted into a corresponding BER by [27]

$$P_{\text{be-PPM-H}} = \frac{L/2}{L-1} P_{\text{sye-PPM-H}} \tag{4.43}$$

For PPM soft decision decoding, in the presence of AWGN with double-sided PSD $N_0/2$, the probability of symbol error is given as [27]

$$P_{\text{sye-PPM-S}} = \frac{1}{\sqrt{2\pi}} \frac{1}{\sqrt{2\pi}} \int_{-\infty}^{\infty} \{1 - [1 - Q(y)]^{L-1}\} e^{-\left(y - \sqrt{2E_s/N_0}\right)^2/2} dy \tag{4.44}$$

The MATLAB codes for simulating slot error probabilities of PPM based on HDD and SDD are given in Program 4.8.

Program 4.8: MATLAB Codes to Simulate SER of PPM Based on HDD and SDD

```
M=3;
% bit resolutions
Lavg=2^M;
% Average symbol length
nsym=500;
```

```
% number of PPM symbols
Lsig=nsym*Lavg;
% length of PPM slots
Rb=1e6;
% Bit rate
Ts=M/(Lavg*Rb);
% slot duration
Tb=1/Rb;
% bit duration

EbN0=-10:5;
% Energy per slot
EsN0=EbN0+10*log10(M);
% Energy per symbol
SNR=10.^(EbN0./10);

for ii=1:length(EbN0)
  PPM= generate_PPM(M,nsym);
  MF_out=awgn(PPM,EsN0(ii)+3,'measured');

  %hard decision decoding
  Rx_PPM_th=zeros(1,Lsig);
  Rx_PPM_th(find(MF_out>0.5))=1;
  [No_of_Error(ii) ser_hdd(ii)]= biterr(Rx_PPM_th,PPM);

  % soft decision decoding
  PPM_SDD=[];
  start=1;
  finish=2^M;
  for k=1:nsym
    temp=MF_out(start:finish);
    m=max(temp);
    temp1=zeros(1,2^M);
    temp1(find(temp==m))=1;
    PPM_SDD=[PPM_SDD temp1];
    start=finish+1;
    finish=finish+2^M;
  end
[No_of_Error(ii) ser_sdd(ii)]=biterr(PPM_SDD,PPM);

end

% theoretical calculation
Pse_ppm_hard=qfunc(sqrt(M*2^M*0.5*SNR));
semilogy(EbN0,Pse_ppm_hard,'k-','linewidth',2);
Pse_ppm_soft=qfunc(sqrt(M*2^M*SNR));
semilogy(EbN0,Pse_ppm_soft,'r-','linewidth',2);
```

4.4.2 PPM VARIANTS

The biggest drawback of the PPM scheme is reduced bandwidth efficiency with increased power efficiency offered by larger constellation sizes. Additionally, in non-LOS configurations, the PPM is less effective at high data rates due to the

TABLE 4.1

Comparisons of Different Modulation Schemes

Modulation Type	SNR	Bandwidth Efficiency	Power Efficiency	Cost
AM	Low–moderate	High	Low–moderate	Low
FM/PM	Moderate	Moderate	Moderate	Moderate
Digital	High	Low	High	High
OOK-NRZ	Moderate	R_B	P	Low
OOK-RZ	Moderate	$2R_B$	$P–3$	Low
PPM	Low	$R_b L/\log_2 L$	$P–5\log10[(L/2)\log_2 L]$	Moderate

multipath-induced ISI, DPPM [18,19], dicode PPM [20], multiple PPM (MPPM) and overlapping PPM (OPPM) [21]. In order to gain improved bandwidth efficiency, a number of modified PPM schemes have been proposed, including (see Table 4.1) DPPM [18,19], dicode PPM [20], multiple PPM (MPPM) and overlapping PPM (OPPM) [21], EPM [29], PPM plus (PPM+) [4] and hybrid modulation. Tough PPM+, where a redundant low chip is inserted after each high chip, improves the reliability; it further decreases the bandwidth efficiency, whereas hybrid schemes, such as the differential amplitude pulse position modulation (DAPPM), offer the additional advantage of a peak-to-average power ratio [22]. In EPM, the time is divided into discrete slots greater than the rise time and jitter of a pulse; however, the pulse width can be wider than one time slot, thus allowing more information to be transmitted than for a comparable PPM scheme.

4.4.2.1 Multilevel PPM

In MPPM, each symbol of duration $T_{symb} = T_b \log_2 L$ is divided into n slots, each with a duration of T_{symb}/n. A pulse is transmitted in w slots, thereby giving $\binom{n}{w}$ possible symbols and hence, the potential to encode $\log_2 \binom{n}{w}$ bits, thus offering higher bandwidth efficiency compared to the standard PPM scheme [30–32]. As an example, $\binom{4}{2}$-MPPM has six possible symbols, as illustrated in Figure 4.12.

However, not all of the possible symbols are necessarily used. For example, a reduced symbol set may be chosen which has a large minimum Hamming distance. Park and Barry examined the performance of MPPM on the AWGN channel [32]. The authors found that with $w = 2$, MPPM outperforms PPM both in terms of bandwidth efficiency and power efficiency. Moving to $w = 8$ greatly increases the number of valid symbols, thereby giving an improvement in bandwidth efficiency at the expense of an increased power requirement. Park and Barry extended their analysis of MPPM by examining the effects of ISI [31]. For both unequalized detection and MLSD, the authors found that the power requirements of MPPM increase at a similar rate to those of PPM as the severity of ISI increases. As with PPM, the performance

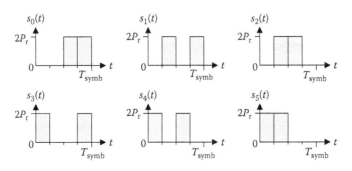

FIGURE 4.12 Valid symbols for $\binom{4}{2}$ MPPM.

of MPPM in the presence of ISI may be improved through the use of trellis coding. Park and Barry compared the performance of trellis-coded $\binom{17}{2}$ MPPM and trellis-coded 16-PPM over multipath channels, both using the same constraint length and MLSD [33]. The authors found that while PPM outperforms MPPM on channels which suffer only mildly from multipath dispersion, MPPM is the more power-efficient technique when the ISI is more severe.

To improve the performance of PPM and its variant-based OWC systems, forward error correction (FEC) schemes including Reed–Solomon (RS) codes [33], convolutional codes [34], trellis-coded modulation (TCM) [35] and turbo codes [36] have been proposed to reduce the power requirement. For FSO channels with no fog and under weak turbulence regime, the coding gains offered by the convolutional and RS codes are just sufficient. Under strong turbulence regime and in the presence of strong fog, more advanced FEC schemes, such as turbo [36] or low-density parity-check (LDPC) codes [37] are more suitable.

4.4.2.2 Differential PPM

The differential PPM (DPPM), which can be considered as inverted DPIM as shown in Figure 4.8, improves the power efficiency as well as the bandwidth efficiency or the throughput by removing all the empty slots that follow a pulse in a PPM symbol [19,38]. The average number of slots per symbol in DPPM $\bar{L}_{DPPM} = (L + 1)/2$ is almost half that of PPM (see Figure 4.8). This provides the possibility for improving the data throughput or bandwidth efficiency [19].

A block of $M = \log_2 L$ input bits is mapped to one of L distinct DPPM waveforms defined as [19]

$$x(t)_{DPPM} = P_p \sum_{k=-\infty}^{\infty} c_k p(t - kT_s) \tag{4.45}$$

where $c_k \in \{0, 1\}$ and P_p is the peak transmitted power, $p(t)$ is a unit-amplitude rectangular pulse shape with one time slot duration. The bandwidth requirement, spectral and error analysis of the DPPM is the same as that of the DPIM; the details can be found in Section 4.5.

Although every DPPM symbol ends with a pulse, thus displaying an inherent symbol synchronization capability at the receiver, a single slot error not only affects the corresponding symbol but also subsequent symbols, thus resulting in multiple symbol errors. There are two scenarios: (i) false alarm error—a pulse is detected in an empty slot following the pulse at the end of a symbol, thus resulting in the next symbol being demodulated in error, and (ii) erasure error—noise causing a pulse to be received as an empty slot, thus resulting in two symbols becoming one with the transmitted symbol being deleted and the next symbol being demodulated in error. Note that if error slots result in a run of empty slots longer than $(L - 1)$ slots, then the error is detected, otherwise it is not detected.

For the AWGN channel, for any given L, DPPM has a slightly higher power requirement but a much lower bandwidth requirement compared with PPM. A MAP detector scheme to overcome the ISI due to multipath propagation is suggested in Ref. [38] with significant performance improvement and more than 10 dB less power requirement than the hard decision. The performance of DPPM with concatenated coding in a diffuse channel is reported in Ref. [39] in which a combination of marker and Reed–Solomon codes is used to correct insertion/deletion errors.

4.4.2.3 Differential Amplitude Pulse Position Modulation

Combination of the PPM with the pulse amplitude modulation can be used to improve the data throughput, bandwidth capacity and peak-to-average power ratio (PAPR). A number of such variations had been suggested including DAPPM, which is a combination of PAM and DPPM, multiple pulse amplitude and position modulation [22,39]. In DAPPM, the symbol length and pulse amplitude are modulated according to the input data bit stream (see Figure 4.8). A block of $M = \log_2(A \times L)$ input bits is mapped to one of 2^M distinct waveforms with each having a pulse indicating the end of symbol, as in DPPM. The amplitude of the pulse chip and the length of a DAPPM symbol are selected from the sets $\{1, 2, \ldots, A\}$ and $\{1, 2, \ldots, L\}$, respectively. Note that DPPM can be considered as a special case of DAPPM with $A = 1$.

The transmitted DAPPM signal is defined as [22]

$$x(t)_{\text{DAPPM}} = \sum_{k=-\infty}^{\infty} \left(\frac{P_p}{A} \right) b_k p(t - kT_s) \qquad (4.46)$$

Assuming IID random data, each symbol is equally likely and the symbol length of DAPPM is given by

$$\bar{L}_{\text{DAPPM}} = \frac{L + 1}{2} \qquad (4.47)$$

4.5   PULSE INTERVAL MODULATION

A number of modulation schemes based on the PIM had been suggested and investigated [17,40,41]. The suggested modulation schemes either improve throughput or

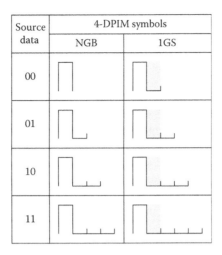

FIGURE 4.13 Mapping of source data to transmitted symbols for 4-DPIM(NGB) and 4-DPIM(1GS).

reduce power requirements adopting complex symbols pattern or by adopting multi-level amplitude. In PIM, information is encoded by inserting empty slots between two pulses. The PIM offers a reduced complexity compared to PPM due to its built-in symbol synchronization. The simplest method of PIM is the DPIM, which offers improved performance compared to PPM by removing the redundant space in PPM. Hence, DPIM is an anisochronous modulation technique, in which each block of $M\ (=\log_2 L)$ input data bits $\{d_i,\ i=1,\ 2,\ \ldots,\ M\}$ is mapped to one of L possible symbols $\{s(n),\ 0<n\le L\}$ of different length. A symbol is composed of a pulse of one slot duration followed by a series of empty slots, the number of which is dependent on the decimal value of the M-bit data stream being encoded. Consequently, the minimum and maximum symbol durations are T_s and LT_s, respectively, where T_s is the slot duration. In order to provide some immunity to the effects of ISI, a guard band consisting of one or more empty slots may be added to each symbol immediately following the pulse. Clearly, adding a single guard slot (GS) changes the minimum and maximum symbol durations to $2T_s$ and $(L+1)T_s$, respectively. The mapping of source data to transmitted symbols for 4-DPIM with no guard slot (NGS) and with a guard band consisting of one slot (1GS) is shown in Figure 4.13, and the MATLAB codes to generate DPIM symbols are given in Program 4.9. The DPIM can be used to achieve either higher bandwidth efficiency or power efficiency compared to PPM by varying the value of L. For a fixed average bit rate and fixed available bandwidth, improved average power efficiency can be achieved when using higher bit resolution (i.e., higher M) compared to PPM [42].

Program 4.9: MATLAB Codes to Generate DPIM Sequence

```
function DPIM=generate_DPIM(M,nsym,NGS)
% function to generate DPIM sequence
```

```
% M: bit resolution
% nsym: number of symbols
% NGS: number of guard slots (default value is zero)

if nargin == 2,
  NGS=0; %default number of guard slots
end

DPIM=[];
  for i= 1:nsym
    inpb(i,:)=randint(1,M);
  end

for i=1:nsym
  inpd=bi2de(inpb(i,:),'left-msb');
  % Converting binary to decimal number
  temp=[zeros(1,(inpd+NGS))];
  % inserting number of zeros in DPIM
  DPIM=[DPIM 1 temp];
  % inserting '1' at the start of each symbol
end
```

In DPIM, since the symbol duration is variable, the overall data rate is also variable. Therefore, the slot duration is chosen such that the mean symbol duration is equal to the time taken to transmit the same number of bits using fixed symbol (frame) length schemes such as OOK or PPM. This slot duration is given as

$$T_{s\_DPIM} = \frac{T_b M}{\bar{L}_{DPIM}} \tag{4.48}$$

where $\bar{L}_{DPIM}$ is the average symbol length of DPIM.

Assuming IID random data, each symbol is equally likely, and consequently, $\bar{L}_{DPIM}$ is given as

$$\bar{L}_{DPIM} = \frac{2^M + 1}{2} \quad \text{for DPIM(NGS)} \tag{4.49}$$

$$\bar{L}_{DPIM} = \frac{2^M + 3}{2} \quad \text{for DPIM(1GS)} \tag{4.50}$$

The DPIM symbol sequence can be expressed as

$$S_{DPIM}(t) = \bar{L}_{DPIM} P_r \sum_{k=\infty}^{\infty} c_k p(t - kT_{s\_DPIM} - \tau_n) \tag{4.51}$$

where $\{c_k\}$ is a random variable which represents the presence or absence of a pulse in the nth time slot and τ_n is the random jitter within a time slot at the threshold crossing at the receiver. Knowing that the data coded into DPIM is of a random nature, where the presence and absence of a pulse is presented by $c_k = 1$ and 0 with the probabilities of $P(c_k = 1) = (\bar{L}_{\text{DPIM}})^{-1}$ and $P(c_k = 0) = (\bar{L}_{\text{DPIM}} - 1)/\bar{L}_{\text{DPIM}}$.

The sequence $S_{\text{DPIM}}(t)$ is a cyclostationary process; its PSD may be calculated using

$$S_{\text{DPIM}}(f) = \frac{1}{T_{\text{s\_DPIM}}}|P(f)|^2\left[S_{\text{c,DPIM}}(f) + S_{\text{d,DPIM}}(f)\right] \tag{4.52}$$

where $P(f)$ is the Fourier transform of the pulse shape; $S_{\text{c,DPIM}}(f)$ and $S_{\text{d,DPIM}}(f)$ are the continuous and discrete components, respectively.

For unit-amplitude rectangular-shaped pulses with a duration of $T_{\text{s\_DPIM}}$, $P(f)$ is given by

$$P(f) = T_{\text{s\_DPIM}}\frac{\sin(\pi f T_{\text{s\_DPIM}})^2}{\pi f T_{\text{s\_DPIM}}} \tag{4.53}$$

$S_{\text{c,DPIM}}(f)$ and $S_{\text{d,DPIM}}(f)$ are given by [19,42]

$$S_{\text{c,DPIM}}(f) = \sum_{k=-5L}^{5L}\left(R_k - \frac{1}{\bar{L}_{\text{DPIM}}^2}\right)e^{-j2\pi kfT_{\text{s\_DPIM}}} \tag{4.54}$$

$$S_{\text{d,DPIM}}(f)(f) = \frac{2\pi}{T_{\text{s\_DPIM}}\bar{L}_{\text{DPIM}}^2}\sum_{k=-\infty}^{\infty}\delta\left(f - \frac{2\pi k}{T_{\text{s\_DPIM}}}\right) \tag{4.55}$$

where R_k is the slot autocorrelation function and can be approximated as follows for DPIM(NGB) and DPIM(1GS), respectively:

$$R_{k-\text{DPIM(NG)}} = \begin{cases} \dfrac{2}{(L+1)} & k = 0 \\[2mm] \dfrac{2}{L^k}(L+1)^{k-2} & 1 \le k \le L \\[2mm] \dfrac{1}{L}\sum_{i=1}^{L}R_{k-i} & k > L \end{cases} \quad \cdot R_k = \begin{cases} \dfrac{2}{(L+1)} & k = 0 \\[2mm] \dfrac{2}{L^k}(L+1)^{k-2} & 1 \le k \le L \\[2mm] \dfrac{1}{L}\sum_{i=1}^{L}R_{k-i} & k > L \end{cases} \tag{4.56}$$

$$R_{k-\mathrm{DPIM(1NG)}} =$$

$$\begin{cases} L_{\mathrm{avg}}^{-1} & k = 0 \\[2mm] 0 & k = 1 \\[2mm] \left(\dfrac{L_{\mathrm{avg}}^{-1}\, L^{-1}}{\sqrt{1+4L^{-1}}} \right) \left[\left(\dfrac{1+\sqrt{1+4L^{-1}}}{2} \right)^{k-1} - \left(\dfrac{1-\sqrt{1+4L^{-1}}}{2} \right)^{k-1} \right] & 2 \le k \le L+1 \\[4mm] \dfrac{1}{L} \displaystyle\sum_{i=1}^{L} R_{k-1-i} & k > L+1 \end{cases}$$

$$(4.57)$$

For reference, a full derivation of this expression may be found in Appendix 4.A. For both DPIM(NGB) and DPIM(1GS), R_k approaches $\bar{L}_{\mathrm{DPIM}}^{-2}$ as k increases, which is intuitive since two pulses far away from each other appear uncorrelated. It is found that for $k > 5L$, R_k may be approximated as $\bar{L}_{\mathrm{DPIM}}^{-2}$ with a good degree of accuracy. Since the mean value of the slot sequence is nonzero, $S_{\mathrm{DPIM}}(f)$ is composed of a continuous term, $S_{\mathrm{c,DPIM}}(f)$, and a discrete term, $S_{\mathrm{d,DPIM}}(f)$. When consideration is limited to rectangular-shaped pulses which occupy the full slot duration, the nulls of $|P(f)|^2$ cancel out the delta functions in $S_{\mathrm{d,DPIM}}(f)$, except at the DC region. Consequently, the discrete term is ignored in this analysis. The continuous component may be calculated from R_k using [19]

$$S_{\mathrm{c,DPIM}}(f) \cong \sum_{k=-5L}^{5L} \left(R_k - \bar{L}_{\mathrm{DPIM}}^{-2} \right) e^{j2\pi k f T_{\mathrm{s,DPIM}}} \qquad (4.58)$$

By substituting the appropriate R_k into (4.58), and then substituting for $P(f)$ and $S_{\mathrm{c,DPIM}}(f)$ in (4.53), the PSD of DPIM(NGB) and DPIM(1GS) using the MATLAB codes of Program 4.10 are plotted for $L = 4, 8, 16$ and 32, as shown in Figure 4.14. All curves were plotted for the same average optical power, using rectangular-shaped pulses occupying the full slot duration. The power axis is normalized to the average electrical power multiplied by the bit duration and the frequency axis is normalized to the bit rate R_b. The discrete terms at DC are not shown. Unlike PSD of PPM (Figure 4.9), DPIM has a DC content, although the power content at low frequencies is relatively small compared with OOK. Hence, DPIM will be more susceptible to the effects of baseline wander compared with PPM. Furthermore, if the areas under the curves are compared, for a given average optical power, the increase in detected electrical power as L increases is easily observed. Comparing the two DPIM schemes, for any given L, DPIM(NGB) has a slightly higher DC power component compared with DPIM(1GS), again suggesting a greater susceptibility to baseline wander. By observing the null positions, the slightly higher bandwidth requirement of DPIM(1GS) compared with DPIM(NGB) is also evident. Furthermore, if the areas under the curves are compared, for a given average

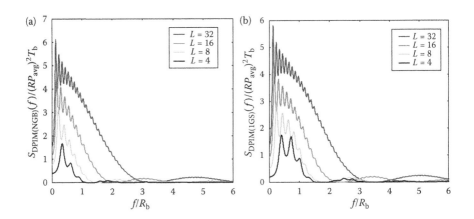

FIGURE 4.14   **(See colour insert.)** PSD of (a) DPIM(NGB) and (b) DPIM(1GS) for $L = 4, 8,$ 16 and 32.

optical power, the increase in detected electrical power as L increases is easily observed.

Program 4.10: MATLAB Codes to Calculate PSD of DPIM(0GS)

```
Rb=1;
% normalized bit rate
Tb=1/Rb;
% bit duration

for M=4
L=2^M;
% symbol length
Lavg=0.5*(2^M+1);
% average symbol length
p_avg=1;
% average optical power
R=1;
% photodetector responsivity
a=R*Lavg*p_avg;
% electrical pulse amplitude
Ts=M/(Lavg*Rb);
% slot duration
Rs=1/Ts;
% slot rate
df=Rs/100;
% spectral resolution
f=0:df:8*Rb;
x=f*Ts;

%***** Calculate ACF *****
 r(1)=2/(L+1); k=0;
```

```
for k=1:L
 r(k+1)=(2/(L^k))*((L+1)^(k-2));
end
for k=L+1:5*L
 temp=0;
 for i=1:L
  temp=temp+r(k-i+1);
 end
 r(k+1)=(1/L)*temp;
end

for k=(5*L)+1:1000
 r(k)=(1/Lavg)^2;
end
P_sq=(a*Ts)^2*(abs(sinc(f*Ts))).^2;

term2=0;
for ii=1:length(r)-1
 term2=term2+((r(ii+1)-((1/Lavg)^2))*cos(2*ii*pi*f*Ts));
end
p=(1/Ts)*P_sq.*((r(1)-1/Lavg^2)+2*term2);
% p=p/(((p_avg*R)^2)*Tb);
end
```

4.5.1 Error Performance on Gaussian Channels

In DPIM, however, since the pulses actually define the symbol boundaries, errors are not confined to the symbols in which they occur. To explain this further, consider the transmitted 8-DPIM(1GS) sequence shown in Figure 4.15a.

The erasure error (Figure 4.15b) combines two symbols into one longer-duration symbol. In the example given, data blocks 100 and 001 are combined into a single block 111. Only if the newly created symbol is longer than the maximum symbol duration will an error be detected. If the new symbol length is valid, the receiver will

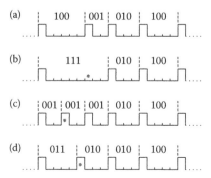

FIGURE 4.15 Types of error in DPIM: (a) transmitted 8-DPIM(1GS) signal, (b) erasure error, (c) false alarm error and (d) wrong slot error.

assume that this is the symbol which was transmitted and consequently, the remaining symbols in the packet will shift one position to the left with respect to the transmitted sequence. If no other errors occur in the packet, then the packet length will be short by $\log_2 L$ bits. The false alarm error (Figure 4.15c) has the effect of splitting one symbol into two shorter length symbols. In the example given, data block 100 is received as two blocks of 001. The error is only detected if one or both of the newly created symbols are shorter than the minimum symbol duration. If both new symbols are valid, then the receiver will assume that these two symbols were transmitted and consequently, the remaining symbols in the packet will shift one position to the right with respect to the transmitted sequence. If no further errors occur in the packet, then the packet will contain $\log_2 L$ extra bits. A wrong slot error occurs when a pulse is detected in a slot adjacent to the one in which it was transmitted (Figure 4.15d). This type of error may be thought of as an erasure error combined with a false alarm error. In the example given, adjacent data blocks 100 and 001 are demodulated as 011 and 010. Unless the error results in at least one of the newly created symbols being either shorter than the minimum symbol length or longer than the maximum symbol length, it will not be detected. A pulse detected in the wrong slot affects both symbols on either side of the pulse, but has no affect on the remaining symbols in the packet.

Thus, in the case of DPIM, since a single slot error has the potential to affect all the remaining bits in a packet, this makes the BER a meaningless measure of performance. Consequently, for the remainder of this chapter, the packet error rate (PER) is used when evaluating the error performance of modulation techniques. This is in accordance with the majority of network protocols, such as Ethernet and IEEE 802.11, which use packet-based error detection and automatic repeat request [43].

In DPPM, since symbol boundaries are not known prior to detection, the optimal soft decision decoding would require the use of MLSD, even in the absence of coding or ISI [18,19]. In practice, this means that for a packet containing n slots and w pulses, the receiver would have to compare the received sequence with every possible combination of w pulses in n slots. Even for very short packet lengths of say 64 bits, a 16-DPIM(NGB) packet would contain on average 136 slots of which 16 would be ones. There are $\binom{36}{16} \approx 10^{20}$ possible combinations of 16 pulses in 136 slots, and clearly, comparing the received sequence against each of these is unfeasible. Thus, the most likely method of detection for DPIM is hard decision decoding using a threshold detector.

The system block diagram of the matched filter-based receiver for DPIM system is shown in Figure 4.16. The DPIM encoder maps each block of $\log_2 L$ input bits to one of L possible symbols, each different in length. The symbols are passed to a transmitter filter, which has a unit-amplitude rectangular impulse response $p(t)$,

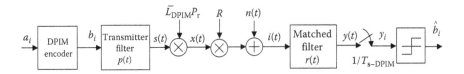

FIGURE 4.16   Block diagram of the matched filter-based receiver for the DPIM scheme.

with a duration of one slot T_{s_DPIM}. The output of the transmitter filter is scaled by the peak detected signal photocurrent $\bar{L}_{DPIM}P_r$. Shot noise $n(t)$ is then added to the signal. The receiver consists of a photodetector of responsivity R followed by a unit-energy matched filter with an impulse response $r(t)$, which is matched to $p(t)$. The filter output is sampled at the end of each slot period, and a one or zero is assigned depending on whether the signal is above or below the threshold level at the sampling instant.

In contrast to PPM, DPIM has a variable symbol length. As a result, an error in a symbol not only affects the symbol itself but also the preceding symbols. Therefore, it is difficult to localize an error to a particular symbol. Consequently, the error performance simply cannot be described in terms of BER. In such cases, the error performance is best described in terms of slot error rate (SER) P_{se} and packet error rate (PER) P_{pe}. The packet error rate can be derived from the SER as shown below:

$$P_{pe} = 1 - (1 - P_{sle})^{N_{pkt}\bar{L}_{DPIM}/M} \tag{4.59}$$

where P_{pe}, P_{sle} are packet and slot error rates, respectively, N_{pkt} is the average packet length and $\bar{L}_{DPIM}$ is the average symbol length.

For small slot error probability, the packet error probability can be approximated by

$$P_{pe} \approx \frac{P_{sle}N_{pkt}\bar{L}_{DPIM}}{M} \tag{4.60}$$

Packet lengths vary depending on the network protocol used. For example, the payload of an IEEE 802.11 packet may contain between 0 and 2500 bytes of data [44], whereas Ethernet packet payloads vary between 46 and 1500 bytes [43]. Figure 4.17 shows the relationship between PER and probability of slot error for 16-DPIM(NGB) using various packet lengths. Clearly, since longer packets contain more slots, they require a lower probability of slot error in order to achieve the same PER as shorter packets. Depending on the packet lengths in question, this reduction can be greater than an order of magnitude. Similarly, the probability of slot error required for a given PER decreases as L increases.

As illustrated in Figure 4.18, the functionality of a DPIM decoder is very simple, since it merely counts the number of empty slots between successive pulses. The falling edge of a detected pulse is used to initiate a counter, which operates at the slot rate. On the rising edge of the next detected pulse, the count value is equal to the $\log_2 L$ bits which have been encoded in that particular symbol. The count value is loaded into the data store and the counter is then reset, ready to begin counting again on the falling edge of the pulse. The data bits can then be read out serially from the data store as required, which could either be as soon as they are available or at the end of the current packet. If the DPIM symbols contain a guard band, this is handled simply by delaying the initiation of the counter by a predetermined number of slots. Due to its simple functionality, a DPIM decoder is also easily implemented in hardware, or expressed using a hardware description language such as VHDL.

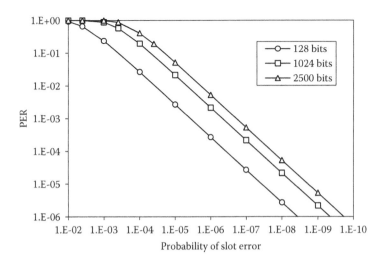

FIGURE 4.17 PER versus probability of slot error for 16-DPIM(NGB) using various packet lengths.

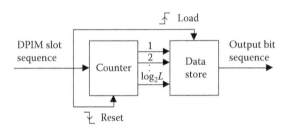

FIGURE 4.18 Block diagram of a DPIM demodulator.

Since DPIM symbol lengths are variable, the overall data rate is also variable. Within a network environment, packet buffers are generally used in both the transmitter and the receiver, and hence, the variable data rate does not pose a problem. If a fixed throughput is required, then one solution is to employ a dual-mapping technique, as suggested for DPPM in Ref. [19], whereby source bits are mapped to symbols either normally or in reverse fashion, whichever yields the shorter number of slots. A flag slot is added at the beginning of the packet to indicate the choice of mapping used, and empty slots are appended to the end of the packet until the mean packet duration is reached. If DPIM is required to operate in a non-packet-based real-time fixed-throughput application, one method of achieving this is to use a first-in first-out (FIFO) buffer in both the transmitter and the receiver, as illustrated in Figure 4.19.

By setting the slot rate such that DPIM yields the same average data rate as the fixed throughput required, and operating the FIFOs at nominally half full, the variable data rate of the DPIM sequence is effectively masked from the rest of the fixed rate system. The FIFOs must be sufficiently large to avoid underflow and overflow

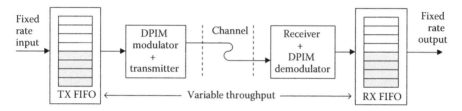

FIGURE 4.19   Using DPIM in a fixed-throughput system.

errors, and the minimum required size may be determined from the source data statistics. If a succession of symbols are transmitted which are shorter than the average symbol length, the DPIM data rate will be momentarily faster than the fixed data rate, causing the transmitter FIFO to empty and the receiver FIFO to fill. Conversely, if a number of longer-duration symbols are transmitted, the DPIM data rate will be momentarily slower than the fixed rate, causing the transmitter FIFO to fill and the receiver FIFO to empty.

4.5.1.1   DPIM with No Guard Band

Since the average duty cycle of a DPIM-encoded packet is $1/\bar{L}_{DPIM}$ it follows that each pulse must have an amplitude of $P_r\bar{L}_{DPIM}$ in order to maintain an average optical power of P_r, assuming that rectangular-shaped pulses are used. A unit-energy filter matched to the transmitted pulse shape $p(t)$ has a rectangular impulse response $r(t)$ with amplitude $1/\sqrt{T_s}$ and duration T_s. As illustrated in Figure 4.20, in the absence of noise, the peak output of this filter when a pulse is transmitted is $\sqrt{E_s} = \bar{L}_{DPIM}RP_rT_b$. When an empty slot is transmitted, the peak output of the matched filter is 0.

By adopting the similar approach taken for OOK and PPM, the slot error probability is given by

$$P_{\text{sle-DPIM(NGS)}} = P(0)Q\left(\frac{\alpha_{T-opt}}{\sqrt{N_0/2}}\right) + P(1)Q\left(\frac{\sqrt{E_s} - \alpha_{T-opt}}{\sqrt{N_0/2}}\right) \tag{4.61}$$

and

$$P(0) = \frac{(\bar{L}_{DPIM} - 1)}{\bar{L}_{DPIM}}; \quad P(1) = \frac{1}{\bar{L}_{DPIM}} \tag{4.62}$$

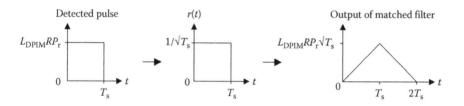

FIGURE 4.20   Matched filter output for the detected DPIM pulse.

For low error probability, a threshold level set midway between expected one and zero levels is very close to the optimum value. Hence, for $\alpha_T = \sqrt{E_s}/2$, (4.61) reduces to

$$P_{\text{sle-DPIM(NGS)}} = Q\left(\sqrt{\frac{E_s}{2N_0}}\right) \tag{4.63}$$

The energy of a symbol is given by

$$E_b = \frac{E_s}{\log_2 L} = \frac{(\bar{L}_{\text{DPIM}} R P_r)^2 T_{\text{s\_DPIM}}}{\log_2 L} \tag{4.64}$$

4.5.1.2 DPIM with One Guard Slot

When a guard band is employed, on detection of a pulse, the following slot(s) contained within the guard band are automatically assigned as zeros, regardless of whether or not the sampled output of the receiver filter is above or below the threshold level. In the case of DPIM(1GS), the probability of slot error for any given slot is dependent on the decision made for the previous slot. Thus, there are four possible scenarios which need to be considered:

1. *The previous slot was a one and was correctly detected*—If the previous slot was a one, the current slot is a guard slot and therefore must be a zero. Since the receiver detected the previous slot correctly, the current slot is automatically assigned a zero, and consequently it is not possible for an error to occur in the current slot. Thus

$$P_{\text{sle}} = 0 \tag{4.65}$$

2. *The previous slot was a one but was falsely detected as a zero*—If the previous slot was a one, the current slot is a guard slot and must therefore be a zero. However, since the previous slot was incorrectly detected as a zero, the receiver will not automatically assign a zero to the current slot. Therefore, a wrong decision could be made should a false alarm error occur in the current slot. The probability of this occurring is given by

$$P_{\text{sle-DPIM}} = \frac{1}{\bar{L}_{\text{DPIM}}} \cdot Q\left(\frac{\sqrt{E_s} - \alpha_T}{\sqrt{N_0/2}}\right) \cdot Q\left(\frac{\alpha_T}{\sqrt{N_0/2}}\right) \tag{4.66}$$

In this expression, the first term represents the probability that the previous slot was a one, the second term is the probability of that one being detected as a zero, and the third term is the probability that the current slot is falsely detected as a one. Clearly, when the probability of error is low, this scenario contributes very little to the overall average probability of slot error, since an erasure error and a false alarm error must occur in adjacent slots.

3. *The previous slot was a zero and was correctly detected*—If the previous slot was a zero, the current slot could be either a one or a zero. Since the previous slot was correctly detected, the current slot is not automatically assigned a zero. Thus, the expression for the probability of error for the current slot is given as

$$
P_{\text{sle-DPIM}} = \frac{\left(\overline{L}_{\text{DPIM}} - 1\right)}{\overline{L}_{\text{DPIM}}} \cdot \left[1 - Q\left(\frac{\alpha_T}{\sqrt{N_0/2}}\right)\right]
$$

$$
\cdot \left[\frac{\left(\overline{L}_{\text{DPIM}} - 1\right)}{\overline{L}_{\text{DPIM}}} \cdot Q\left(\frac{\alpha_T}{\sqrt{N_0/2}}\right) + \frac{1}{\overline{L}_{\text{DPIM}}} \cdot Q\left(\frac{\sqrt{E_s} - \alpha_T}{\sqrt{N_0/2}}\right)\right] \quad (4.67)
$$

In this expression, the first term represents the probability of a zero and the second term is the probability of that zero being correctly detected. The third term is simply the probability of error for the current slot, as given in (4.63) for DPIM(NGB).

4. *The previous slot was a zero but was falsely detected as a one*—If the previous slot was falsely detected as a one, the current slot is incorrectly assumed to be a guard slot and the receiver automatically assigns a zero to it. If the current slot should be a one, then an error will occur. The probability of this occurring is given by

$$
P_{\text{sle-DPIM}} = \frac{\left(\overline{L}_{\text{DPIM}} - 1\right)}{\overline{L}_{\text{DPIM}}} \cdot Q\left(\frac{\alpha_T}{\sqrt{N_0/2}}\right) \cdot \frac{1}{\overline{L}_{\text{DPIM}}} \quad (4.68)
$$

The first term represents the probability of a zero, the second term is the probability of that zero being falsely detected as a one, and the third term is the probability that the following slot is a one.

Thus, putting these possible scenarios together in one expression, the probability of slot error for DPIM(1GS) is given by

$$
P_{\text{sle-DPIM(1GS)}} = \left[\frac{1}{\overline{L}_{\text{DPIM}}} \cdot Q\left(\frac{\sqrt{E_s} - \alpha_T}{\sqrt{N_0/2}}\right) \cdot Q\left(\frac{\alpha_T}{\sqrt{N_0/2}}\right)\right]
$$

$$
+ \left[\frac{\left(\overline{L}_{\text{DPIM}} - 1\right)}{\overline{L}_{\text{DPIM}}} \cdot \left[1 - Q\left(\frac{\alpha_T}{\sqrt{N_0/2}}\right)\right]\right]
$$

$$
\cdot \left[\frac{\left(\overline{L}_{\text{DPIM}} - 1\right)}{\overline{L}_{\text{DPIM}}} \cdot Q\left(\frac{\alpha_T}{\sqrt{N_0/2}}\right) + \frac{1}{\overline{L}_{\text{DPIM}}} \cdot Q\left(\frac{\sqrt{E_s} - \alpha_T}{\sqrt{N_0/2}}\right)\right]\right]
$$

$$
+ \left[\frac{\left(\overline{L}_{\text{DPIM}} - 1\right)}{\overline{L}_{\text{DPIM}}} \cdot Q\left(\frac{\alpha_T}{\sqrt{N_0/2}}\right) \cdot \frac{1}{\overline{L}_{\text{DPIM}}}\right] \quad (4.69)
$$

4.5.2 OPTIMUM THRESHOLD LEVEL

Assume that OOK is used to transmit an information sequence comprising IID random data, and the transmitted signal is corrupted by signal-independent AWGN. If the received signal is passed through a matched filter and sampled at the optimum point, let s_0 denote the sample value of any given bit in the sequence. There are two conditional probability density functions for s_0, depending on whether a one or a zero was sent, as illustrated in Figure 4.21.

Since the AWGN is independent of the signal, the probability density functions are symmetrical, and are given by [25]

$$P(s_0 \mid 0) = \frac{1}{\sqrt{2\pi}\sigma} \cdot e^{-\frac{s_0^2}{2\sigma^2}} \tag{4.70}$$

$$P(s_0 \mid 1) = \frac{1}{\sqrt{2\pi}\sigma} \cdot e^{-\frac{(s_0 - E)^2}{2\sigma^2}} \tag{4.71}$$

where $P(s_0|0)$ is the probability of s_0 given that a zero was sent, $P(s_0|1)$ is the probability of s_0 given that a one was sent and σ is the standard deviation of the AWGN. A decision is made by comparing s_0 with the threshold level α_T, and assigning a one if $s_0 > \alpha_T$ and a zero otherwise. There are two possible ways in which errors can arise. If a one was sent and $s_0 > \alpha_T$, an erasure error occurs, as illustrated by the shaded area to the left of α in Figure 4.21. The probability of erasure error is given by [25]

$$P(ee \mid 1) = \int_{-\infty}^{\alpha_T} P(s_0 \mid 1)\, ds_0 \tag{4.72}$$

Similarly, if a zero was sent and $s_0 > \alpha_T$, a false alarm error occurs, as illustrated by the shaded area to the right of α_T in Figure 4.21. The probability of false alarm

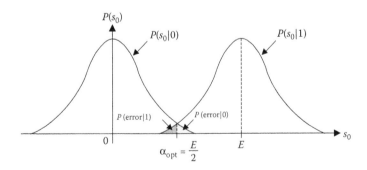

FIGURE 4.21 Conditional probability density functions of s_0 in the presence of signal-independent AWGN.

error is given by [25]

$$P(\text{fe} \mid 0) = \int_{\alpha_T}^{\infty} P(s_0 \mid 0) \, ds_0 \tag{4.73}$$

Thus, the overall probability of error is given by

$$P(\text{error}) = P(0) \cdot \int_{-\infty}^{\alpha_T} P(s_0 \mid 1) \, ds_0 + P(0) \cdot \int_{\alpha_T}^{\infty} P(s_0 \mid 0) \, ds_0 \tag{4.74}$$

The optimum threshold level which minimizes the probability of error may be found by differentiating (4.74) with respect to α_T, and then solving for the threshold level that set the derivative equal to zero [25,45]. In the case of OOK, since ones and zeros are equally likely and the probability distributions are identical, the optimum threshold level occurs at the point at which the two conditional probability density functions intersect, which is midway between expected one and zero levels [45,46]. Thus, for OOK, $\alpha_{T-\text{opt}} = E/2$.

With the exception of 2-DPIM(NGB), for all orders of DPIM(NGB) and DPIM(1GS), the probability of receiving a zero is greater than the probability of receiving a one. Therefore, the optimum threshold level does not lie midway between expected one and zero levels. Intuitively, since zeros are more likely, it is apparent that the probability of error can be improved by using a threshold level which is slightly higher than the midway value. This increases the probability of correctly detecting a zero, at the expense of increasing the probability of an erasure error. However, since zeros are more likely, an overall improvement in average error performance is achieved.

Generally, the cost of mistaking a zero for a one is the same as the cost of mistaking a one for a zero. In this case, if the probability densities are scaled by the *a priori* probabilities, then the optimum threshold level occurs where the probability densities intersect [46], as illustrated in Figure 4.22.

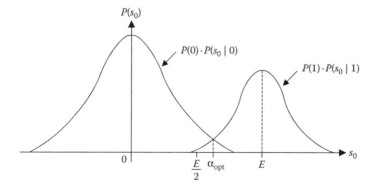

FIGURE 4.22   Scaled conditional probability density function of s_0 for $P(0) > P(1)$.

At the point of intersection

$$P(1) \cdot \frac{1}{\sqrt{2\pi}\sigma} \cdot e^{-\frac{(s_0-E)^2}{2\sigma^2}} = P(0) \cdot \frac{1}{\sqrt{2\pi}\sigma} \cdot e^{-\frac{s_0^2}{2\sigma^2}} \qquad (4.75)$$

Therefore, the optimum threshold level is equal to the value of s_0 which satisfies

$$f(s_0) = P(1) \cdot \frac{1}{\sqrt{2\pi}\sigma} \cdot e^{-\frac{(s_0-E)^2}{2\sigma^2}} - P(0) \cdot \frac{1}{\sqrt{2\pi}\sigma} \cdot e^{-\frac{s_0^2}{2\sigma^2}} = 0 \qquad (4.76)$$

In order to find the optimum threshold level from (4.76), the Newton–Raphson procedure for solving nonlinear equations may be used. The Newton–Raphson iteration procedure for obtaining an approximation to the root of $f(x)$ is given by [47]

$$x_{n+1} = x_n - \frac{f(x_n)}{f'(x_n)} \qquad (n = 0, 1, 2, \ldots) \qquad (4.77)$$

Thus, differentiating (4.76) with respect to s_0

$$f'(s_0) = P(1) \cdot \frac{(s_0 - E)}{\sigma^2} \cdot e^{-\frac{(s_0-E)^2}{2\sigma^2}} - P(0) \cdot \frac{s_0}{\sigma^2} \cdot e^{-\frac{s_0^2}{2\sigma^2}} = 0 \qquad (4.78)$$

Starting with an initial guess of a midway threshold level, that is, $\alpha = E/2$, the Newton–Raphson method finds the optimum threshold level to a good degree of accuracy after several iterations.

For various PERs, based on a packet length of 1 kbyte, the optimum threshold level was determined iteratively for various orders of DPIM(NGB) and DPIM(1GS). The method used to achieve this involves making an initial guess for P_{avg}, and then iteratively determining the optimum threshold level and hence, the minimum PER. This value is then compared with the target PER and, if necessary, P_{avg} is adjusted and the whole process repeated until the target PER is reached. The optimum threshold level versus PER for DPIM(NGB) and DPIM(1GS) are plotted in Figure 4.23a and b, respectively. In both figures, the optimum threshold level is normalized to the expected matched filter output when a one is transmitted.

From Figure 4.23a and b it is clear that as the probability of error falls, that is, as the SNR increases, the optimum threshold level tends towards the midway value for both DPIM(NGB) and DPIM(1GS). With the exception of moving from 2-DPIM(NGB) to 4-DPIM(NGB), it is also evident that increasing L moves the optimum threshold level further away from the midway value. For example, at a PER of 10^{-6}, the optimum threshold level for 4-DPIM(NGB) is 0.5% above the midway value, and this increases to ~3.3% when 32-DPIM(NGB) is used. This is due to the fact that increasing the number of bits per symbol increases $P(0)$ and decreases $P(1)$, thereby moving the point of intersection of the scaled conditional probability density

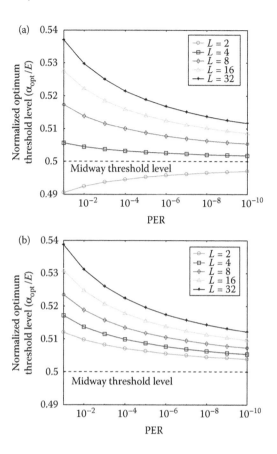

FIGURE 4.23   Normalized optimum threshold level versus PER: (a) DPIM(NGB) and (b) DPIM(1GS).

functions further away from the midway value. 2-DPIM(NGB) is the one case where the probability of a one is greater than the probability of a zero. Consequently, the optimum threshold level lies below the midway value. Increasing the order from 2-DPIM(NGB) to 4-DPIM(NGB) actually brings the *a priori* probabilities closer together, and consequently results in an optimum threshold level which is closer to the midway value. Comparing DPIM(NGB) and DPIM(1GS), for any given L and PER, the normalized optimum threshold level is slightly higher for DPIM(1GS). This is due to the fact that the presence of the guard slot increases $P(0)$ and reduces $P(1)$, which consequently increases the optimum threshold level slightly. The difference between the two schemes is more pronounced for small values of L, where adding a guard slot has a more significant effect on the *a priori* probabilities. As an example, for a PER of 10^{-6}, adding a guard slot increases the normalized optimum threshold level by ~1% for $L = 4$, while the increase is just ~0.2% for $L = 32$. Nevertheless, for an average PER of 10^{-6}, the optimum threshold level is within 3.5% of the midway value for all orders of DPIM(NGB) and DPIM(1GS) considered.

Program 4.11: MATLAB Codes to Simulate SER of DPIM in the AWGN Channel

```
M=4;
% bit resolutions
NGS=0;
% number of guard slots
Lavg=0.5*(2^M+1)+NGS;
% Average symbol length
nsym=1e4;
% number of PPM symbols

Rb=200e6;
% Bit rate
Tb=1/Rb;
% bit duration
Ts=M/(Lavg*Rb);
% slot duration

EbN0=-10:3;
% Energy per slot
EsN0=EbN0+10*log10(M);
% Energy per symbol
SNR=10.^(EbN0./10);

for ii=1:length(EbN0)
  DPIM=generate_DPIM(M,nsym,NGS);
  % generating DPIM sequence
  Lsig=length(DPIM);
  % actual packet length
  MF_out=awgn(DPIM,EsN0(ii)+3,'measured');
  % matched filter output assuming unit energy per bit
  Rx_DPIM_th=zeros(1,Lsig);
  Rx_DPIM_th(find(MF_out>0.5))=1;
  % Threshold detections
  [No_of_Error(ii) ser(ii)]=biterr(Rx_DPIM_th,DPIM);
end

semilogy(EbN0,ser,'ko','linewidth',2);
hold on
% theoretical calculation
Pse_DPIM=qfunc(sqrt(M*Lavg*0.5*SNR));
semilogy(EbN0,Pse_DPIM,'r','linewidth',2);
```

4.6   DUAL-HEADER PIM (DH-PIM)

Figure 4.8 shows the symbol structure of OOK, PPM, PIM and DH-PIM waveforms for 4-bit input data. In DH-PIM, the nth symbol $S_n\,(h_n, d_n)$ starts with a header h_n of duration $T_h = (\alpha + 1)T_s$, and concludes with a sequence of d_n empty slots, where T_s is the slot duration and $\alpha > 0$ is an integer. Depending on the most significant bit (MSB) of the input word, two dissimilar headers, H_1 and H_2, which correspond to MSB = 0 and MSB = 1, respectively, are assigned. The header H_1 and H_2 have pulse of durations $0.5\,\alpha T_s$ and αT_s, respectively. Each pulse is followed by a guard band of suitable length $T_g \in \{(\alpha/2 + 1)T_s, T_s\}$ T_g to cater for symbols representing zero. The value of $d_n \in \{0, 1, \ldots, 2^{M-1} - 1\}$ is the decimal value of the input codeword when the symbol starts with H_1, or the decimal value of the 1's complement of the input codeword when the symbol starts with H_2. The header pulse plays a dual role of symbol initiation and time reference for the preceding and succeeding symbols, thus resulting in a built-in symbol synchronization. Therefore, DH-PIM not only removes the redundant time slots that follow the pulse as in the PPM symbol, but it also reduces the average symbol length compared with PIM, thus resulting in an increased data throughput.

The start time of the nth symbol is defined as

$$T_n = T_0 + T_s \left[n(\alpha + 1) \sum_{k=0}^{n-1} d_k \right] \qquad (4.79)$$

The average symbol length $\bar{L}_{\text{DH-PIM}}$ and slot duration $T_{\text{s\_DH-PIM}}$ of DH-PIM$_\alpha$ is given by

$$\bar{L}_{\text{DH-PIM}} = \frac{(2^{M-1} + 2\alpha + 1)}{2} \qquad (4.80)$$

$$T_{\text{s\_DH-PIM}_\alpha} = \frac{2M}{(2^{M-1} + 2\alpha + 1)R_b} \qquad (4.81)$$

DH-PIM average symbol length can be reduced by a proper selection of α, thus offering improved transmission throughput and bandwidth requirements compared to the DPIM, DPPIM and PPM schemes [48].

With reference to Figure 4.24, the DH-PIM pulse train can be expressed mathematically as

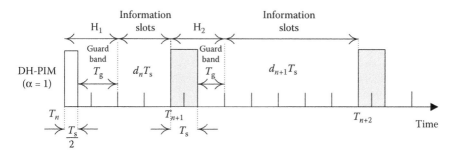

FIGURE 4.24 DH-PIM symbol structure showing the two headers (H_1 and H_2).

$$x(t) = \frac{4\bar{L}_{\text{DH-PIM}}P_{\text{r}}}{3\alpha} \sum_{k=0}^{\infty} \left\{ p\left[\frac{2(t-T_k)}{\alpha T_{\text{s}}} - \frac{1}{2}\right] + h_n p\left[\frac{2(t-T_k)}{\alpha T_{\text{s}}} - \frac{3}{2}\right] \right\} \quad (4.82)$$

where $4\bar{L}_{\text{DH-PIM}}P_{\text{r}}/3\alpha$ is the peak transmitted optical power, $h_n \in \{0, 1\}$ indicating H_1 or H_2, respectively. The program to generate DH-PIM from random binary sequence is given in Program 4.12.

Program 4.12: MATLAB Codes to Generate DH-PIM Symbol Sequence

```
function [DHPIM]=generate_DH_PIM(alpha,M,nsym)
% function to generate DH-PIM symbols

DHPIM = [];
for i=1:nsym
  inpb(i,:) = randint(1,M);
  % random bit generation
end
%%
%*********DH_PIM for even alpha************
DHPIM = [];
if rem(alpha,2)== 0;
  for i=1:nsym
    inpd=bi2de(inpb(i,:),'left-msb');
    if (inpd<=(2^M/2)-1) d(i)=inpd;
      H=[ones(1,alpha/2) zeros(1,((alpha/2)+1))];
      % header
    else d(i) = (2^M-1)-inpd;
      H=[ones(1,alpha) zeros(1,1)];
    end
    DHPIM = [DHPIM H zeros(1,d(i))];
    % information bits
  end

%***************DH_PIM for odd alpha****************
else
  for i=1:nsym
    inpd=bi2de(inpb(i,:),'left-msb');
    if (inpd<=(2^M/2)-1) d(i)= inpd;
      H=[ones(1,2*(alpha/2)) zeros(1,2*((alpha/2)+1))];
    else d(i)= (2^M-1)-inpd;
      H =[ones(1,2*alpha) zeros(1,2)];
    end
      DHPIM = [DHPIM H zeros(1,2*d(i))];
    end
  end
end
```

4.6.1 SPECTRAL CHARACTERISTICS

From (4.82), the Fourier transform of DH-PIM is given by

$$X_N(\omega) = V \sum_{n=0}^{N-1} \int_{-\infty}^{+\infty} \left\{ \text{rect}\left[\frac{2(t-T_n)}{\alpha T_s} - \frac{1}{2} \right] + h_n \text{rect}\left[\frac{2(t-T_n)}{\alpha T_s} - \frac{3}{2} \right] \right\} \cdot e^{-j\omega t} \, dt \qquad (4.83)$$

Therefore,

$$X_N(\omega) = \frac{V}{j\omega} e^{(-j\omega T_0)} (1 - e^{-j\omega T_s \alpha/2}) \sum_{n=0}^{N-1} \left[(1 + h_n e^{-j\omega T_s \alpha/2}) e^{-j\omega T_s n(\alpha+1)} e^{-j\omega T_s \sum_{k=0}^{n-1} d_k} \right] \qquad (4.84)$$

The power spectral density of DH-PIM pulse train can be given by [49]

$$P(\omega) = \begin{cases} \left| 4V^2 \sin^2\left(\frac{\alpha\omega T_s}{4}\right) \left\{ \left[5 - 4\sin^2\left(\frac{\alpha\omega T_s}{4}\right) \right] \right. \right. \\ \left. \left. + \left[9 - 8\sin^2\left(\frac{\alpha\omega T_s}{4}\right) \right] \text{Re}\left(\frac{\psi}{1-\psi}\right) \right\} \middle/ \omega^2 T_s \left(2^{M-1} + 2\alpha + 1 \right); \; \omega \neq \dfrac{2\pi K}{T_s} \\ 0; \qquad\qquad\qquad\qquad\quad \omega = \dfrac{2\pi K}{T_s} \text{ and either } K \text{ even or } \alpha \text{ even} \\ \infty; \qquad\qquad\qquad\qquad\quad \omega = \dfrac{2\pi K}{T_s} \text{ and both } K \text{ odd and } \alpha \text{ odd.} \end{cases}$$

$$(4.85)$$

where K is a positive integer and ψ given by

$$\psi = \frac{1}{2^{M-1}} \left\{ 1 + e^{-j\omega T_s} + e^{-j2\omega T_s} + \cdots + e^{-j\omega(2^{M-1}-1)T_s} \right\} \cdot e^{-j\omega(\alpha+1)T_s} \qquad (4.86)$$

Thus, the spectrum consists of a sinc envelope when $\omega T_s/2\pi$ is not integer, distinct frequency components at the slot frequency and its harmonics when $\alpha\omega T_s/2\pi$ is odd integer and nulls when $\alpha\omega T_s/2\pi$ is even integer; thus the slot component and its harmonics may coincide with the nulls of the *sinc* envelope depending on the values of α (see Figure 4.25 and Program 4.12). Consequently, the existence of the slot components and the locations of nulls are affected by the pulse shape. The presence of the slot frequency component is very useful at the receiver where a phase-locked loop circuit could be used to extract it for synchronization. For comparison, the PSDs of the OOK, PPM, DPIM and DH-PIM are shown in Figure 4.26.

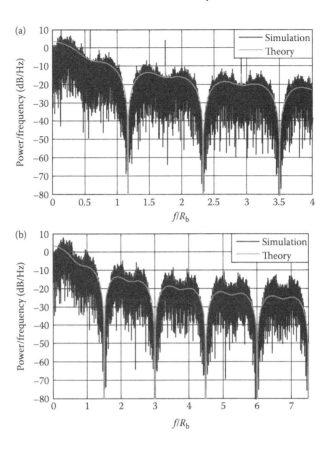

FIGURE 4.25 Predicted and simulated power spectral density for (a) 16-DH-PIM$_1$ and (b) 16-DH-PIM$_2$.

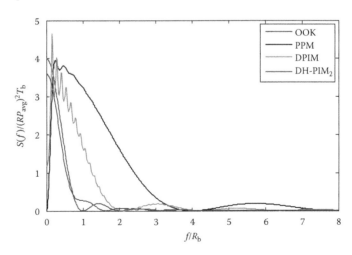

FIGURE 4.26 (**See colour insert.**) Power spectral density of OOK, PPM, DPIM and DH-PIM.

Program 4.13: MATLAB Codes to Calculate PSD of DH-PIM

```
alpha=1;
% value of alpha
M =3;
% bit resolution
Lavg=0.5*(2^(M-1)+2*alpha+1);
% average symbol length
Rb=1;
% normalized bit rate
Ts=M/(Lavg*Rb);
% slot duration
Rs=1/Ts;
% slot rate

p_avg=1;
% average optical power
R=1;
% photodetector responsivity
a=4*R*Lavg*p_avg/(3*alpha);
% electrical pulse amplitude
res=100;
df=Rs/res;
% spectral resolution
f=df:df:8*Rb;
% frequency vector
x=f*Ts;

% psd calculation
w=2*pi*f;
temp=(sin(alpha*w*Ts/4)).^2;
G=((1-exp(-1i*w*Ts*2^(M-1))).*exp(-1i*w*Ts*(alpha+1)))./...
  ((1-exp(-1i*w*Ts))*2^(M-1));
P=4*a^2*temp.*((5-4*temp)+(9-8*temp).*real(G/(1-G)))./...
  (w.^2*Ts*(2^(M-1)+2*alpha+1));
a=max(P);
if rem(alpha,2)==1
  for K=1:2:floor(f(end)/Rs)
    P(K*res)=4*P(K*res);
  end
end
P=P/(((p_avg*R)^2)*Tb);
```

4.6.2   Error Performance on Gaussian Channels

With reference to the DH-PIM system block diagram shown in Figure 4.27, the DH-PIM waveform $x(t)$ is scaled by the average optical power P_r and is transmitted over a channel. White Gaussian noise is added to the signal before being detected at the receiver. The receiver employs a unit-energy filter, which is matched to $x(t)$

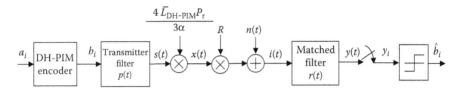

FIGURE 4.27   Block diagram of the matched filter-based receiver for the DH-PIM scheme.

followed by a sampler at a rate of $1/T_s$. Finally, a simple threshold detector is employed to recover the detected symbol.

Like the case of DPIM, the symbol length is variable in DH-PIM. Hence, an error is not necessarily confined to the symbol in which it occurs, and packet error probability would be a more appropriate metric for performance evaluation. However, the packet error probability can be approximated using the slot error probability.

Assuming equal occurrence of H_1 and H_2, in the absence of noise, the photocurrent $I_p = 4\bar{L}_{\text{DH-PIM}}RP_r/3\alpha$. Following the similar approach adopted for DPIM, the slot error probability of DH-PIM is

$$P_{\text{sle-DH-PIM}} = p(0)p_{e0} + p(1)p_{e1} \tag{4.87}$$

where p_{e1} is the probability of erasure error, p_{e0} is the probability of false alarm

$$p(1) = \frac{3\alpha}{4\bar{L}_{\text{DH-PIM}}} \tag{4.88}$$

$$p(0) = 1 - p(1) = \frac{4\bar{L} - 3\alpha}{4\bar{L}_{\text{DH-PIM}}} \tag{4.89}$$

Since the probability of receiving a pulse is less than the probability of receiving no pulse, the optimum threshold level will be above $0.5I_p$, and may be defined in terms of I_p as

$$I_{\text{T-opt}} = \alpha_{\text{T-opt}} = kI_p \tag{4.90}$$

where $0 < k < 1$ represents a threshold factor.

For the case of transmitting an empty slot, the probability of a false alarm is given as

$$p_{e0} = \int_{kI_p}^{\infty} \frac{1}{\sqrt{2\pi}\sigma} e^{-y^2/2\sigma^2} \, dy \tag{4.91}$$

$$p_{e0} = Q\left[k\frac{I_p}{\sigma}\right] \tag{4.92}$$

For the case of transmitting a pulse, the probability of erasure error is given as

$$p_{el} = \int_{-\infty}^{kI_p} \frac{1}{\sqrt{2\pi}\sigma} e^{-(y-I_p)^2/2\sigma^2} \, dy \tag{4.93}$$

$$p_{el} = Q\left[(1-k)\frac{I_p}{\sigma}\right] \tag{4.94}$$

Therefore, incorporating (4.87) in (4.92) and (4.94) gives

$$p_{e0} = Q\left(\sqrt{\frac{2k^2 E_p}{N_0}}\right) \tag{4.95}$$

and

$$p_{el} = Q\left(\sqrt{\frac{2(1-k)^2 E_p}{N_0}}\right) \tag{4.96}$$

Since the channel has no path loss, the average received optical power is equal to the average transmitted optical power P_r. In the absence of noise and for a photodetector responsivity R, the photocurrent is given by

$$I_p = \frac{4\bar{L}_{\text{DH-PIM}} R P_r}{3\alpha} \tag{4.97}$$

Thus, the energy of a pulse is given as

$$E_p = \frac{16(RP_r)^2 M\bar{L}_{\text{DH-PIM}}}{9\alpha^2 R_b} \tag{4.98}$$

Substituting (4.98) in (4.95) and (4.96) gives

$$p_{e0} = Q\left(\frac{\mu k R P_r}{\sqrt{N_0 R_b}}\right) \tag{4.99}$$

$$p_{el} = Q\left(\frac{\mu(1-k)R P_r}{\sqrt{N_0 R_b}}\right) \tag{4.100}$$

where

$$\mu = \sqrt{\frac{32 M \overline{L}_{\text{DH-PIM}}}{9\alpha^2}} \tag{4.101}$$

Substituting (4.91), (4.92), (4.99) and (4.100) in (4.90) gives the probability of slot error for DH-PIM as

$$P_{\text{se}} = \frac{1}{4\overline{L}_{\text{DH-PIM}}}\left[\left(4\overline{L}_{\text{DH-PIM}} - 3\alpha\right)Q\left(\frac{\mu k R P_{\text{r}}}{\sqrt{N_0 R_{\text{b}}}}\right) + 3\alpha Q\left(\frac{\mu(1-k)R P_{\text{r}}}{\sqrt{N_0 R_{\text{b}}}}\right) \right] \tag{4.102}$$

For very small values of error probability, the packet error probability can be approximated using (4.59) and (4.102). The SER of DH-PIM in the AWGN channel can be calculated using Program 4.14.

Program 4.14: MATLAB Codes to Calculate the SER of DH-PIM in the AWGN Channel

```
alpha=2;
% alpha
M=4;
% bit resolution
Lavg=0.5*(2^(M-1)+2*alpha+1);
% Average symbol length
nsym=1e3;
% number of PPM symbols
Rb=1e6;
% Bit rate
Ts=M/(Lavg*Rb);
% slot duration

EbN0=-10:3;
% Energy per slot
EsN0=EbN0+10*log10(M/(1.5*alpha/2));
% Energy per symbol
SNR=10.^(EbN0./10);
% converting db to noraml scale

for ii=1:length(EbN0)
  DH_PIM= generate_DH_PIM(alpha,M,nsym);
  % DHPIM sequence
  Lsig=length(DH_PIM);
  % actual packet length
  DH_PIM_Rx=awgn(DH_PIM,EsN0(ii)+3,'measured');
  % matched filter output in AWGN channel
```

```
% Threshold detections
Rx_DH_PIM_th=zeros(1,Lsig);
Rx_DH_PIM_th(find(DH_PIM_Rx>1/2))=1;
 [No_of_Error(ii) ser(ii)]= biterr(Rx_DH_PIM_th,DH_PIM);
end
figure;
semilogy(EbN0,ser,'ko','linewidth',2);
hold on
% theoritical calculation
DHPIM_factor=(8*M*Lavg)/(9*alpha^2);
Pse_DH_PIM=qfunc(sqrt(DHPIM_factor*SNR));
semilogy(EbN0,Pse_DH_PIM,'r','linewidth',2);
```

4.7 MULTILEVEL DPIM

MDPIM offers greater transmission bit rate and lower bandwidth requirements compared with the PPM and DPIM schemes. It also offers a built-in symbol synchronization and simple slot synchronization compared with PPM and OOK modulation schemes. In MDPIM, each block of M-bit input OOK data $s_i(t)$ with rate $R_b = T_b^{-1}$ is mapped to one of $L = 2^M$ possible symbols of variable length of d time slot T_s. As in DPIM, each symbol starts with a short pulse of one time slot duration and an amplitude of A (if the most significant bit (MSB) of $s_i(t)$ is '0') or $2A$ (if the MSB is '1'), followed by a guard slot $T_g = T_s$ and a succession of empty slots d. The number of empty slots corresponds to the decimal value of $s_i(t)$ or its ones-complement version when the pulse amplitude is $2v$. With reference to Figure 4.8, the MDPIM signal can be expressed as [50]

$$x(t) = A \sum_{n=0}^{\infty} h_n \text{rect} \left[\frac{2(t - T_n) - T_s}{2T_s} \right] \qquad (4.103)$$

where n is the instantaneous symbol number, $h_n A$ represents the signal amplitude and $h_n \in \{1, 2\}$ indicates the pulse amplitude level. The rectangular pulse function is defined as [17]

$$\text{rect}(u) = \begin{cases} 1; & -0.5 < u < 0.5 \\ 0; & \text{otherwise} \end{cases} \qquad (4.104)$$

The start time of the nth symbol is defined as

$$T_n = T_0 + T_s \left[2n + \sum_{k=0}^{n-1} d_k \right] \qquad (4.105)$$

where T_0 is the start time of the first pulse at $n = 0$ and $d_k \in \{0, 1, \ldots, 2^{M-1} - 1\}$ is the number of information time slots in the kth symbol. Throughout this chapter, MDPIM

TABLE 4.2

Mapping of M-bit OOK Data Format into PPM, DPIM, DPPM, DH-PIM$_2$, DAPPM and MDPIM Symbols

OOK ($M = 3$)	8-PPM	8-DPIM (No Guard Slot)	8-DPPM (No Guard Slot)	8-DH-PIM$_2$	DAPPM ($A = 2, L = 4$)	8-MDPIM
000	10000000	1	1	100	1	1 0
001	01000000	10	01	1000	01	1 0 0
010	00100000	100	001	10000	001	1 0 0 0
011	00010000	1000	0001	100000	0001	1 0 0 0 0
100	00001000	10000	00001	110000	2	2 0 0 0 0
101	00000100	100000	000001	11000	02	2 0 0 0
110	00000010	1000000	0000001	1100	002	2 0 0
111	00000001	10000000	00000001	110	0002	2 0

will be referred to as L-MDPIM according to the value of L, in line with other pulse time modulation schemes. Table 4.2 displays the mapping of all possible combinations of a 3-bit OOK code word into 8-PPM, 8-DPIM and 8-DH-PIM$_2$ and 8-MDPIM symbols. Note that for MDPIM, $v = 1$ and 2.

As shown in Table 4.2, MDPIM, similar to DPIM and DH-PIM, not only removes the redundant time slots that follow the pulse as in PPM symbol but it also reduces the average symbol length compared with DPIM and DH-PIM, thus resulting in an increased data throughput. The minimum, maximum and average symbol lengths of MDPIM can be given, respectively, as

$$L_{\min} = 2, \quad L_{\max} = 2^{M-1} + 1$$

$$\bar{L}_{\text{MDPIM}} = \frac{2^{M-1} + 3}{2} \tag{4.106}$$

And the slot duration is defined as

$$T_{s\_\text{MDPIM}} = \frac{2M}{(2^{M-1} + 3)R_b} \tag{4.107}$$

The average symbol length for MDPIM is very similar to that of DH-PIM and much shorter than PPM and DPIM, especially at a bit resolution $M > 5$, where $\bar{L}$ is 1/4 and 1/2 those of PPM and DPIM, respectively, as presented in Figure 4.28.

Of course, for a given bandwidth requirement, MDPIM will offer a much higher transmission rate than PPM, DPIM and DH-PIM$_1$ and slightly higher than DH PIM$_2$. For example, for a bandwidth requirement of 6 MHz, 32-PPM, 32-DPIM, 32-DH-PIM$_1$, 32-DH-PIM$_2$ and 32 MDPIM achieve transmission rates of 114, 208, 192, 350 and 380 packets/s, respectively.

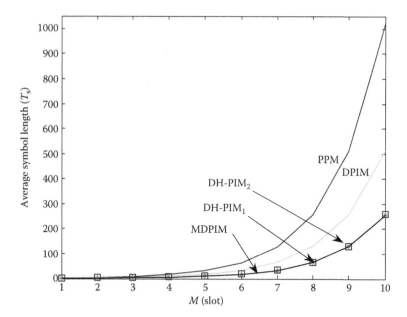

FIGURE 4.28 Average symbol length of PPM, DPIM, DH-PIM$_1$, DH-PIM$_2$ and MDPIM versus M.

The MDPIM transmitter block diagram is shown in Figure 4.29a. The MSB of an M-bit word is first checked. If MSB = 0, then the decimal value of the binary word d is determined, otherwise the data is inverted prior to binary-to-decimal conversion.

Similar to DPIM and DH-PIM, in MDPIM, an error is not necessarily confined to the symbol in which it occurs because of variable symbol length with no fixed symbol boundaries as in PPM. Figure 4.30 shows the probability of slot errors against the SNR for $M = 3$. MDPIM offers similar performance to DH-PIM$_2$ but worth more than DPIM as expected. For example, at P_{se} of 10^{-3}, MDPIM requires about 5 dB additional SNR compared with DPIM.

4.8 COMPARISONS OF BASEBAND MODULATION SCHEMES

4.8.1 POWER EFFICIENCY

The power efficiency and the bandwidth requirement are the benchmark to compare modulation schemes. Since the optical wireless channel is mostly power limited due to the safety issue, the power efficiency is the foremost issue. Power requirement P_{req} is defined as the average optical power required by an ideal system (in the presence of AWGN) to achieve a certain bit rate and error probability. The average power requirement for OOK is

$$P_{avg\_OOK} = \sqrt{\frac{N_0 R_b}{2R^2}} \, Q^{-1}(P_{e\_bit\_OOK}) \tag{4.108}$$

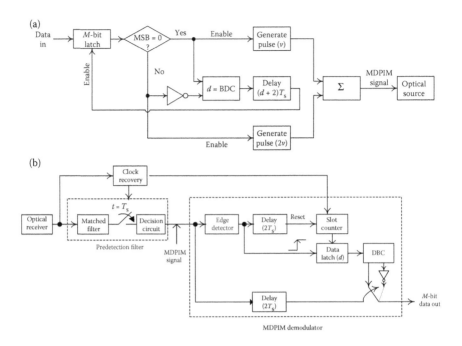

FIGURE 4.29 (a) MDPIM transmitter and (b) MDPIM receiver system block diagrams.

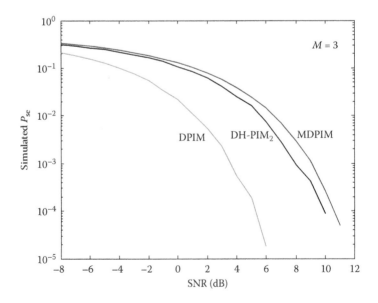

FIGURE 4.30 Slot error rate for MDPIM, DPIM and DH-PIM$_2$ versus SNR.

Equation 4.108 is used as the benchmark against which performances of other modulation schemes would be compared. For the PPM, the average optical power requirements can be approximated from (4.40) as BQ Plates to ASME Specification (SA)

$$P_{\text{avg\_PPM}} = \sqrt{\frac{2N_0 R_b}{R^2 L \log_2 L}} Q^{-1}(P_{\text{sle\_PPM\_H}}) \tag{4.109}$$

Hence, the power efficiency of PPM can be approximated using

$$\eta_{\text{p-PPM}} = \sqrt{\frac{4}{L \log_2 L}} \tag{4.110}$$

The hard decision decoding incurs a 1.5 dB optical power penalty compared to the soft decision decoding, and the average optical power requirement for soft decision decoding is given by

$$n_{\text{p-PPM}} = \sqrt{\frac{2}{L \log_2 L}} \tag{4.111}$$

Similarly, the average optical power requirements and power efficiency for DPIM and DH-PIM can be approximated as

$$P_{\text{avg\_DPIM}} = \sqrt{\frac{2N_0 R_b}{R^2 \overline{L}_{\text{DPIM}} \log_2 L}} \, Q^{-1}(P_{\text{se\_DPIM}}) \tag{4.112}$$

$$\eta_{\text{p-DPIM}} = \sqrt{\frac{8}{(L+1)\log_2 L}} \tag{4.113}$$

$$P_{\text{avg}_{\text{DH-PIM}}} = \sqrt{\frac{9\alpha^2 2N_0 R_b}{16 R^2 M \overline{L}_{\text{DH-PIM}}}} Q^{-1}(P_{\text{se}_{\text{DH-PIM}}}) \tag{4.114}$$

$$\eta_{\text{p-DPIM}} = \sqrt{\frac{9\alpha^2}{2M(2^{M-1} + 2\alpha + 1)}} \tag{4.115}$$

4.8.2 Transmission Bandwidth Requirements

The transmission bandwidth requirements of the baseband modulation schemes can be defined in terms of the minimum bit (slot) duration τ_{min}.

$$B_{\text{req}} = \frac{1}{\tau_{\text{min}}} \tag{4.116}$$

Since the minimum slot duration of PPM, DPIM and DH-PIM are given by (4.27), (4.48) and (4.81), respectively, the bandwidth requirements are given as

$$B_{req\_OOK} = R_b \tag{4.117}$$

$$B_{req\_PPM} = \frac{L}{M} R_b \tag{4.118}$$

$$B_{req\_DPIM} = \frac{L+1}{2M} R_b \tag{4.119}$$

$$B_{req\_DH\text{-}PIM} = \frac{(2^{M-1} + 2\alpha + 1)R_b}{\alpha M} R_b \tag{4.120}$$

Figure 4.31 displays the average optical power requirements and bandwidth requirements of OOK (NRZ and RZ), PPM, DPIM and DH-PIM generated using Program 4.15. The figure clearly shows the trade-off between the bandwidth requirements and average power requirements. For OOK-RZ, the power requirements decrease with duty cycle but the bandwidth increases accordingly. For PPM, DPIM and DH-PIM, average optical power requirement decreases steadily as L increases; however, the bandwidth requirements increase with L. Comparing

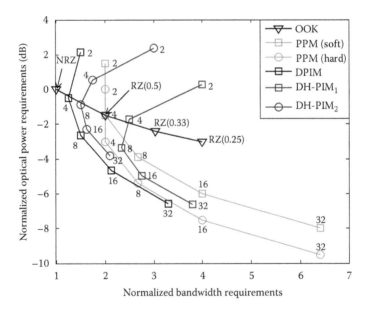

FIGURE 4.31 Optical power requirement normalized to the OOK-NR versus bandwidth requirement normalized to the bit rate for OOK, PPM, PIM, DH-PIM$_1$ and DH-PIM$_2$. The numbers indicate the values of L.

the performance for $L = 2$, it is evident that there is little point in using other modulation techniques compared to OOK-NRZ, since OOK-NRZ outperforms 2-PPM, 2 DPIM and 2-DH-PIM in terms of both power efficiency and bandwidth efficiency.

Program 4.15: MATLAB Codes to Compare Bandwidth Requirements and Power Efficiency of Different Modulation Schemes

```
Rb=1;
P_req_OOK=1;

for M=1:5% bit resolutions

  % *****ppm hard decision decoding***********
  P_req_PPM_hard(M)=10*log10(sqrt(4/(M*2^M)));
  BW_PPM(M)=2^M/M;

  %*****ppm soft decision decoding***********
  P_req_PPM_soft(M)=10*log10(sqrt(2/(M*2^M)));

  % ****************dpim***************
  P_req_DPIM(M)=10*log10(sqrt(8/(M*(2^M+1))));
  BW_DPIM(M)=(2^M+1)/(2*M);

  % ********************DH-PIM1******************
  alpha=1;
  Lavg=(2^(M-1)+2*alpha+1)/2;
  P_req_DHPIM1(M)=10*log10(sqrt(9*alpha^2/(4*M*Lavg)));
  BW_DHPIM1(M)=(2^(M-1)+2*alpha+1)/(alpha*M);

  % ********************DH-PIM2******************
  alpha=2;
  Lavg=(2^(M-1)+2*alpha+1)/2;
  P_req_DHPIM2(M)=10*log10(sqrt(9*alpha^2/(4*M*Lavg)));
  BW_DHPIM2(M)=(2^(M-1)+2*alpha+1)/(alpha*M);

end

duty_cycle=[10.5 0.33 0.25];
BW_OOK=1./duty_cycle;
P_req_OOK=10*log10(sqrt(duty_cycle));

figure;
plot(BW_OOK,P_req_OOK,'-kv','LineWidth',2,
'MarkerSize',10);hold on
plot(BW_PPM,P_req_PPM_hard,'-rs','LineWidth',2,
'MarkerSize',10);
plot(BW_PPM,P_req_PPM_soft,'-ro','LineWidth',2,
'MarkerSize',10);
```

```
plot(BW_DPIM,P_req_DPIM,'-ks','LineWidth',2,'MarkerSize',10);
plot(BW_DHPIM1,P_req_DHPIM1,'-bs','LineWidth',2,
'MarkerSize',10);
plot(BW_DHPIM2,P_req_DHPIM2,'-bo','LineWidth',2,
'MarkerSize',10);
legend('OOK','PPM (soft)', 'PPM(hard)', 'DPIM', 'DH-PIM_1',
'DH-PIM_2');
ylabel('Normalized average optical power requirements');
xlabel('Normalized bandwidth requirements');
```

4.8.3 TRANSMISSION CAPACITY

In the case of PPM, DPIM and DH-PIM assuming an M-bit input word, the number of valid code combinations is given as L. Hence, the transmission capacity can be defined as

$$C_{\text{Tc}} = \frac{L_{\max}}{\bar{L}} R_{\text{b}} \log_2 2^M \qquad (4.121)$$

where $L_{\max}$ is the maximum number slot in a valid symbol.

Since the PPM has a fixed number of slots of length L, (4.121) reduces to

$$C_{\text{Tc-PPM}} = MR_{\text{b}} \qquad (4.122)$$

For the DPIM without GS, $L_{\max} = 2^M$ and $\bar{L} = 0.5(2^M + 1)$; hence, the transmission capacity of DPIM is given by

$$C_{\text{Tc-DPIM}} = \frac{M2^{M+1}}{2^M + 1} R_{\text{b}} \qquad (4.123)$$

Similarly, for DH-PIM, $L_{\max} = 2^{M-1} + \alpha$ and $\bar{L} = 0.5(2^{M-1} + 2\alpha + 1)$; hence, the transmission capacity of DH-PIM is given by

$$C_{\text{Tc-DH-PIM}} = \frac{2M(2^{M-1} + \alpha)}{(2^{M-1} + 2\alpha + 1)} R_{\text{b}} \qquad (4.124)$$

Figure 4.32 shows the transmission capacity of PPM, DPIM, DH-PIM$_1$ and DH-PIM$_2$ normalized to PPM versus M for bandwidth requirements of 1 MHz. The DH-PIM$_2$ achieves the highest capacity, which increases with bit resolution. Although DPIM offers higher capacity at low bit resolution compared to DH-PIM$_1$, the difference becomes negligible at $M > 6$, which is about four times that of PPM.

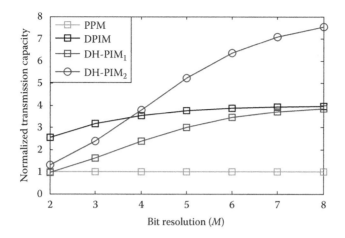

FIGURE 4.32 (**See colour insert.**) Transmission capacity of PPM, DPIM, DH-PIM$_1$ and DH-PIM$_2$ normalized to PPM versus M.

4.8.4 Transmission Rate

In an anisochronous modulation scheme such as DPIM, the symbol length is variable. Hence, an error is not necessarily confined to the symbol in which it occurs. Therefore, it is convenient to base the study on the packet transmission rate rather than the bit rate.

The packet is assumed to have a fixed length of N_{pkt} bits; therefore, the average slot length of the packet is given by

$$L_{pkt} = \frac{\bar{L} N_{pkt}}{M} \tag{4.125}$$

The slot rate $R_s = T_s^{-1}$ and the packet transmission rate $R_{pkt} = R_s L_{pkt}^{-1}$ therefore are

$$R_{pkt} = \frac{R_s M}{\bar{L} N_{pkt}} \tag{4.126}$$

For the PPM and DPIM, the minimum bandwidth requirement is $B_{req} = T_s^{-1}$, and (4.126) can be simplified to

$$R_{pkt-PPM} = \frac{M B_{req}}{N_{pkt} 2^M} \tag{4.127}$$

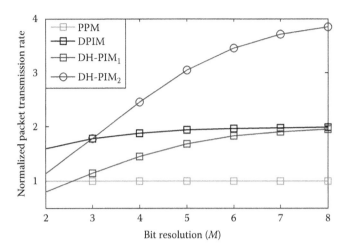

FIGURE 4.33 (**See colour insert.**) Packet transmission rate of PPM, DPIM, DH-PIM$_1$ and DH-PIM$_2$ normalized to PPM versus M.

$$R_{\text{pkt-DPIM}} = \frac{2MB_{\text{req}}}{N_{\text{pkt}}(2^M + 1)} \tag{4.128}$$

However, for DH-PIM, $B_{\text{req}} = 2(\alpha T_s)^{-1}$; hence, packet transmission rate is given by

$$R_{\text{pkt}} = \frac{\alpha M B_{\text{req}}}{N_{\text{pkt}}(2^{M-1} + 2\alpha + 1)} \tag{4.129}$$

Figure 4.33 displays the packet transmission rate of PPM, DPIM, DH-PIM$_1$ and DH-PIM$_2$ normalized to that of PPM versus M. For $M > 6$, DH-PIM$_1$ displays similar transmission rate compared with DPIM and about twice that of PPM; however, DH-PIM$_2$ offers four times the transmission rate compared with PPM.

4.8.5 Peak-to-Average Power Ratio

The significance of the PAPR is twofold. In order to limit the dynamic range of system, and hence to avoid the nonlinearity problem in opto-electronic devices, the PAPR should be as low as possible. On the other hand, to comply with eye-safety limitation, the modulation with higher PAPR is preferable. Hence, care should be taken in selecting the best PAPR ratio.

The peak transmitted optical power of L-PPM is LP_{avg}; hence, PAPR of L-PPM is

$$\text{PAPR}_{\text{PPM}} = 2^M \tag{4.130}$$

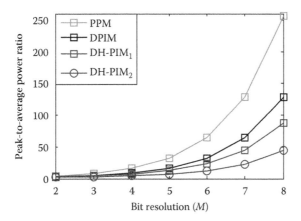

FIGURE 4.34 **(See colour insert.)** PAPR of PPM, DPIM, DH-PIM$_1$ and DH-PIM$_2$ versus M.

Similarly, the PAPR of the DPIM and DH-PIM can be derived as

$$\text{DPIM} = \frac{(2^M + 1)}{2} \tag{4.131}$$

$$\text{PAPR}_{\text{DH-PIM}} = \frac{2(2^{M-1} + 2\alpha + 1)}{3\alpha} \tag{4.132}$$

Figure 4.34 demonstrates the PAPR of PPM, DPIM, DH-PIM$_1$ and DH-PIM$_2$ versus M. The PAPR increases with the bit resolution for all modulation techniques. As can be seen from Equation (4.130), the PAPR of PPM increases exponentially with M. DH-PIM$_2$ offers the least PAPR ratio for all ranges of M.

4.9   SUBCARRIER INTENSITY MODULATION

The subcarrier modulation technique is a maturing, simple and cost-effective approach for exploiting bandwidth in analogue optical communications. In SCM systems, a number of baseband analogue or digital signals are frequency up-converted prior to intensity, frequency, or phase modulation of the optical carrier. For IM-DD (the simplest approach), this results in an optical spectrum consisting of the original optical carrier ω_o, plus two side-bands $\omega_o \pm \omega_{sc}$, where ω_{sc} is the subcarrier frequency. The main drawback of SCM with IM/DD is its poor optical average power efficiency. This is because the SCM electrical signal has both positive and negative values and must take on both values. Therefore, a DC offset must be added in order to satisfy the requirement that $x(t)$ must be nonnegative. As the number of subcarrier signals increase, the minimum value of the SCM signal decreases, becoming more negative, and consequently the required DC bias increases, thus resulting in further deterioration of the optical power efficiency, since the average optical power is proportional to the DC bias [51].

SCM with IM-DD has been adopted for OWC systems because of reduced multipath-induced ISI (due to a number of narrow-band subcarriers) and immunity to the fluorescent-light noise at the DC region of the spectrum [51,52]. In OWC systems, the eye safety also introduces further restriction on the average transmitted optical power; thus, the number of subcarriers adopted is limited. SCM systems have limited optical power budget (lower SNR per channel as a result of limited modulation index due to the light source P-I response, and the eye-safety requirement). In order to improve the power budget as well as the input dynamic range, a second-stage modulation is introduced [53]. As an example, on nondistorting channels with IM/DD and AWGN, binary phase shift keying (BPSK) and quadrature phase shift keying (QPSK) both require 1.5 dB more optical power than OOK [7]. Along with single-subcarrier schemes, multiple-subcarrier modulation techniques also exist, allowing multiple users to communicate simultaneously using frequency division multiplexing [54]. As illustrated in Figure 4.35, each user modulates data onto a different subcarrier frequency, and the frequency division multiplexed sum of all the modulated subcarriers is then used to intensity-modulate an optical source. At the receiver, multiple band-pass demodulators are used to recover the individual data streams.

Through simultaneous transmission of several narrow-band subcarriers, multiple-subcarrier modulation techniques can achieve high aggregate bit rates and improved bandwidth efficiency [1]. However, multiple-subcarrier techniques are less power efficient than single-subcarrier schemes, and the power efficiency worsens as the number of subcarriers is increased. One of the reasons for this reduction in power efficiency is the fact that increasing the number of subcarriers also increases the DC offset required to avoid clipping. By allowing clipping to occur, the power efficiency can be improved slightly, despite the fact that this induces nonlinear distortion [54]. Taking quadrature amplitude modulation (QAM) as an example, for a given number of subcarriers, clipped 4-QAM (with zero DC offset) requires 0.46 dB less average optical power than 4-QAM with a DC offset applied [54].

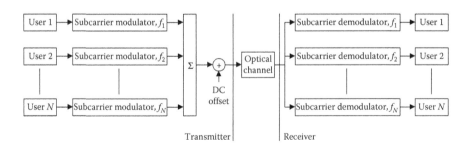

FIGURE 4.35 Basic implementation of multiple-subcarrier modulation.

During one symbol duration

$$m(t) = \sum_{j=1}^{N_s} A_j g(t) \cos(\omega_{cj} t + \theta_j) \tag{4.133}$$

where N_s is the number of subcarriers, $g(t)$ is the rectangular pulse shape function, $\{\omega_{cj}\}_{j=1}^{N_s}$ is the angular frequency and $\{A_j\}_{j=1}^{N_s}$ is the peak amplitude of each subcarrier. At the receiver, a direct detection is employed followed by a standard RF demodulator to extract the data. For a single SIM, the electrical filter used at the receiver must have the same bandwidth as OOK signalling if quadrature-PSK is used. For BPSK-modulated subcarrier, the bandwidth requirement is twice that of OOK. For multiple SIM, the bandwidth requirement apparently increases by a factor of N.

Since each subcarrier contained in $m(t)$ is sinusoidal with both positive and negative values, to modulate the intensity of the LD (or an LED) without clipping, the amplitude of this composite signal must always be greater than the threshold current. A DC offset must therefore be added to $m(t)$ in order to meet this requirement. Consequently, subcarrier modulation schemes are less power efficient than pulse modulation techniques. For instance, in an AWGN-limited direct LOS IR communication link, using a single SIM with either BPSK or QPSK results in 1.5 dB more optical power compared to the OOK [20]. Additionally, the optical modulation depth must be such that the optical source operates within its dynamic range. SIM can also be implemented with the subcarrier signals made orthogonal to one another to achieve orthogonal frequency division multiplexing (OFDM) [55,56]. This is particularly interesting as it can be easily implemented via IFFT/FFT in available signal processing chips [1,51]. By ensuring that each subcarrier transmits at relatively low data rates, the need for an equalizer can be avoided while maintaining the same aggregate data rate. It also offers a greater immunity to near DC noise from the fluorescent lamps [1,51]. However, with a DC offset, a multiple SIM results in a $10\log N$ (dB) increase in optical power requirement compared with a single subcarrier. Also, multiple SIM suffers from both intermodulation and harmonics distortions due to the inherent optical source nonlinearity.

For slower PWM rates, there will be significant spectral aliasing, thus resulting in large subcarrier-induced interference.

Program 4.16: MATLAB Codes to Simulate BER of BPSK Modulation in the AWGN Channel

```
Rb = 1;
% normalized bit rate
fc = Rb*5;
% carrier frequency;
Tb = 1/Rb;
% bit duration
SigLen = 1000;
% number of bits or symbols
```

```
fsamp = fc*10;
% sampling rate
nsamp = fsamp/Rb;
% samples per symbols
Tsamp = Tb/nsamp;
% sampling time

Tx_filter = ones(1,nsamp);
% transmitter filter
bin_data = randint(1,SigLen);
%generating prbs of length SigLen
data_format = 2*bin_data-1;
% BPSK constillation
t = Tsamp:Tsamp:Tb*SigLen;
% time vector
carrier_signal=sqrt(2*Eb/nsamp)*sin(2*pi*fc*t);
% carrier signal

bin_signal = conv(Tx_filter,upsample(data_format,nsamp));
%bin_signal = rectpulse(OOK,nsamp);
% rectangular pulse shaping
bin_signal = bin_signal(1:SigLen*nsamp);
Tx_signal = bin_signal.*carrier_signal;
% transmitted signal

Eb=1;
% energy per bit
Eb_N0_dB = -3:10;
% multiple Eb/N0 values

for ii=1:length(Eb_N0_dB)

  Rx_signal=awgn(Tx_signal,Eb_N0_dB(ii)+3-
    10*log10(nsamp),'measured');
  % additive white gaussian noise
  Rx_output = Rx_signal.*carrier_signal;
  % decoding process

  for jj=1:SigLen
    output(jj) = sum(MF_output((jj-1)*nsamp+1:jj*nsamp));
    % matched filter output
    % alternatively method of matched filter is given in OOK
      simulation
  end

  rx_bin_data=zeros(1,SigLen);
  rx_bin_data(find(output>0))=1;
  [nerr(ii) ber(ii)]=biterr(rx_bin_data,bin_data);
end

figure
semilogy(Eb_N0_dB,ber,'bo','linewidth',2);
```

```
hold on
semilogy(Eb_N0_dB,0.5*erfc(sqrt(10.^(Eb_N0_dB/10)))),
'r-X','linewidth',2);
% theoretical ber, 'mx-');
```

4.10   ORTHOGONAL FREQUENCY DIVISION MULTIPLEXING

Orthogonal frequency division multiplexing (OFDM) is another modulation scheme that can efficiently utilize the available bandwidth [57–59]. The block diagram for an OFDM-based system is as shown in Figure 4.36. It is a special version of subcarrier modulation discussed above in that all the subcarrier frequencies are orthogonal. In an OFDM transmitter, serial data streams are grouped and mapped into N_d constellation symbols $\{X[k]\}_{k=0}^{N_d-1}$ using BPSK, QPSK or M-QAM. N_p pilots are inserted into the data symbols before being transformed into the time domain signal by N-orthogonal subcarriers by means of an IFFT given as

$$m_x[n] = \frac{1}{\sqrt{N}} \sum_{k=0}^{N-1} X[k]e^{\frac{j2\pi nk}{N}}, \quad n = 0, \ldots, N-1 \qquad (4.134)$$

A cyclic prefix of length G (greater than the channel length) is added to the IFFT output to prevent multipath-induced ISI.

In the implementation of the optical OFDM/DMT, the output of the IFFT block feeds straight into a digital-to-analogue converter (DAC) which translates the discrete IFFT sample points into continuous time-varying signal. This continuous time-varying signal is then used to drive the intensity of the optical source, typically LED in this case. The DAC is also usually designed or chosen such that its output is well within the input dynamic range of the driver–LED combination. This is to avoid any signal clipping which might then have a negative impact on the system performance.

At the receiver, after removing the cyclic prefix, $y[n]$ is applied to the FFT. Due to the cyclic prefix, the linear convolution between the transmitted signal and the channel becomes circular convolution; hence, the output of the FFT can be written as multiplication in matrix form given by

$$Y = \text{diag}(\mathbf{X}) \cdot \mathbf{H} + \mathbf{W} \qquad (4.135)$$

where $\mathbf{H} = \mathbf{F} \cdot \mathbf{h}$ is the frequency response of the channel with length L, $[\mathbf{F}]_{n,k} = N^{-\frac{1}{2}}e^{-(j2\pi kn/N)}$ is the FFT matrix, $\mathbf{W}_{N\times1}$ is the white Gaussian noise with $E[\mathbf{W}\mathbf{W}^H] = \sigma_n^2\mathbf{I}_N$, H is the Hermitian transpose and diag is the diagonal matrix.

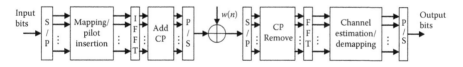

FIGURE 4.36   Block diagram of an optical OFDM.

Although there are reported works which tend to suggest that a certain degree of clipping can be tolerated without causing an intolerable level of performance degradation, clipping will decrease PAPR and increase the average AC power, thus leading to a higher SNR [60,61]. To directly modulate the intensity of an optical source (white LED, for instance), a real, positive signal is required. However, baseband OFDM/DMT signals are generally complex and bipolar. The process of making this signal real and positive is highlighted in the following steps:

1. Hermitian symmetry is imposed on the complex signal $s'(t)$ that feeds into the IFFT block. This ensures that the output of the IFFT block is real.
2. A DC is then added to obtain a unipolar signal, resulting in what is otherwise termed as DC-optical-OFDM (DCO-OFDM). An alternative approach is the use of asymmetrically clipped OFDM (ACO-OFDM). In ACO-OFDM, no DC is added at all. The bipolar real OFDM signal is clipped at the zero level. That is, the entire negative signal is removed. It turns out that if only the odd-frequency OFDM subcarriers are nonzero at the IFFT input, the noise caused by clipping affects only the even subcarrier and the data-carrying odd subcarriers are not impaired at all. The spectral efficiency of ACO-OFDM is however only one-half that of DC-OFDM. This is because only the odd subcarriers are data carrying in ACO-OFDM while all subcarriers are data carrying in DC-OFDM. The real, unipolar signal is then used to directly drive the optical source as shown in Figure 4.37.

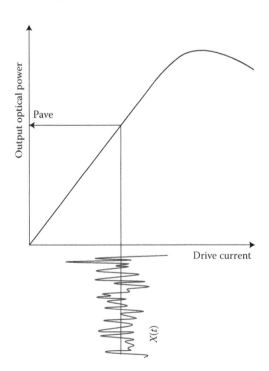

FIGURE 4.37   Real analogue time varying driving an LED.

The LED average driving current (bias point) that corresponds to output optical power P_{ave} in Figure 4.37 is chosen so as to

1. Provide the right level of lighting/illumination in the case of visible light communications
2. Permit the full swing of the input signal
3. Ensure that the signal is kept, as much as possible, within the linear region of the LED characteristics curve

However, the OFDM has a drawback of high PAPR requiring a high dynamic range. In fact, the OFDM suffers most severely due to the nonlinear distortion by the high PAPR. In OFDM system, due to multiamplitude modulation schemes, channel estimation is required for equalization and decoding provided that there is no knowledge of the channel. In one-dimensional channel estimation schemes, pilot insertion could be carried out in the block type and in the comb type. The former uses the least square or the minimum mean-square error (MMSE) to perform estimation in a slow-fading channel [62], whereas the latter uses the LS with interpolation and the maximum likelihood (ML) to perform estimation in a rapidly changing channel [63].

Program 4.17: MATLAB Codes to Simulate OFDM System

```
N=256;
% Number of subcarriers or size of IFFT/FFT
N_data_symbol=128;
% Number of symbol to IFFT
GI = N/4;
% Guard interval 1/4,1/8,1/16,...
M = 4;
% Modulation 2:BPSK, 4:QPSK, (8,16,32,64,128,256):QAM
L = 16;
% Channel length
N_Iteration = 500;
% Number of iteration

SNR= [0:1:15];
% Signal to Noise Ratio in dB

for i=1:length(SNR)
  snr=SNR(i)

  for (k=1:N_Iteration)

    tx_bits = randint(N_data_symbol,1,M);
    % Input Bit Streams to Transmit

    % Modulation
    tx_bits_Mod = qammod(tx_bits,M);
    input_symbol = [zeros((N-N_data_symbol)/2,1); tx_bits_Mod;
      zeros((N-N_data_symbol)/2,1)];
```

```matlab
    % IFFT and Circular Prefix Addition
    ofdm_symbol_ifft=ifft(input_symbol,N);

    % Guard Interval insertion (CP)
    guard_symbol=ofdm_symbol_ifft(N-GI+1:N);

    % Add the cyclic prefix to the ofdm symbol
    ofdm_symbol=[guard_symbol;  ofdm_symbol_ifft];
    %sig_pow=ofdm_symbol'.*conj(ofdm_symbol);

    ofdm_spectrum=ofdm_symbol;
    %ofdm_spectrum=[ofdm_spectrum;  ofdm_symbol];

    % h=randn(L,1)+j*randn(L,1); % Generate random channel
      h(1:L,1)
    %h=h./sum(abs(h)); % Normalization

    h=1; %AWGN

    y1=filter(h,1,ofdm_symbol);
    %y=x*h
    % Adding AWGN Noise
    y=awgn(y1,snr,'measured');
    %y=x*h+n

    % Remove Cyclic prefix
    rx_symbol=y(GI+1:N+GI);

    % The FFT of the time domain signal after the removal of
      cyclic prefix
    rx_symbol_fft=fft(rx_symbol,N);

    % Equalization
    %H_f=fft(h,N);
    %G=1./H_f;

    rx_equalized_zp=rx_symbol_fft;
    rx_equalized=rx_equalized_zp(((N-N_data_symbol)/2)+1:
    (N+N_data_symbol)/2);

    % Demodulate
    rx_bits_zp=qamdemod(rx_equalized,M);
    rx_bits=rx_bits_zp;

    % Comparison
    % Bit Error Rate computation
    [nErr bErr(i,k)]=symerr(tx_bits,rx_bits);
  end
end
```

```
snr_theo=10.^(SNR/10);
Theory_awgn = 0.5*erfc(sqrt(snr_theo));
semilogy(SNR,mean(bErr'),'b',SNR,Theory_awgn,'ro-');
```

4.11   OPTICAL POLARIZATION SHIFT KEYING

Similar to the RF systems, transmission data rate, link range and system reliability of the outdoor FSO systems are severely affected and limited by the atmospheric conditions. Rain, aerosols, gases, fog, temperature and other particles suspended in the atmosphere result in a high optical attenuation. Among all these, fog is the major problem in FSO systems imposing attenuation typically ranging from a few dB/km in a clear atmosphere to over 200 dB/km in a dense fog regime [64]. Exceptionally high attenuation due to the dense fog condition limits the FSO link availability over a long range, reducing the link span to less than 500 m [65]. However, dense (thick) fog is more localized, lasting for a short duration and only occurring in few locations. Another factor that accounts for the FSO performance degradation in a clear atmosphere is the irradiance and phase fluctuation. Under clear air conditions, the link range could be as high as 5–10 km. However, the optical beam propagating through the atmospheric channel may experience fading due to the in-homogeneities in the refractive index of the surrounding air pockets caused by temporal fluctuation in the temperature, pressure and air humidity [66].

The traversing optical beam suffers from both phase and irradiance fluctuation in a turbulent atmosphere. In IM/DD systems, turbulence effects could result in deep irradiance fades that could last up to ~1 − 100 µs [67]. For a link operating at say, 1 Gbps, this could result in a loss of up to 10^5 consecutive bits (a burst error). To avoid this and to reduce the turbulence-induced optical power penalty, the atmospheric turbulence has to be mitigated. A number of modulation schemes in particular based on the IM/DD are widely used in OWC systems. Amplitude shift keying (ASK), frequency shift keying (FSK), phase shift keying (PSK) and differential phase shift keying (DPSK) are the most common band-pass modulation formats adopted for optical and nonoptical communication systems. ASK with the OOK format is the simplest and most widely used but it is highly sensitive to the channel turbulence [68]. To achieve the optimal performance, an adaptive thresholding scheme has to be applied at the receiver, thus increasing the system complexity. Compared to the ASK (OOK), FSK, PSK and DPSK techniques require no adaptive thresholding scheme and offer improved performance in the presence of turbulence [68]. However, angular modulation schemes are highly sensitive to the phase noise, thus requiring a complex synchronization at the receiver [69]. Furthermore, frequency offset in DPSK leads to the additional power penalty owing to delayed and undelayed bits not being in phase [70]. The FSK scheme is bandwidth inefficient and offers inferior BER performance compared to the PSK and DPSK in the additive white Gaussian noise (AWGN) channel [70].

However, the performance of modulation schemes are highly sensitive to the turbulence fluctuation; thus, the need for the adaptive detection technique at the receiver to improve the performance [68]. As an alternative to the standard modulation techniques, there are modulation schemes that exploit the vector characteristics of the propagating optical beam. These schemes rely on using the state of polarization (SOP) of a fully polarized optical beam as the information-bearing parameter, thus exploiting the two orthogonal channels available in a single-mode optical fibre as well as free-space propagation. A variant of PolSK, where two orthogonal input signals maintain their SOP while propagating, is being used in free-space optical communications [71–73]. Binary polarization-modulated DD system offers 2–3 dB lower peak optical power compared to the IM/DD system. However, the coherent light beam may experience a degree of polarization while propagating through the channel [61]. PolSK schemes is considerably insensitive to the laser phase noise at the receiver, provided that the IF filter bandwidth is large enough to avoid phase-to-amplitude noise conversion [62]. In Ref. [72], a polarization-modulated DD system has been proposed to extract the Stokes parameters of the transmitted light for use in binary and multilevel transmissions [71]. A digital coherent optical polarization modulation scheme is outlined in Ref. [74]. For optical beams, polarization states are the most stable properties compared with the amplitude and phase when propagating through a turbulent channel [75]. The experimental results show that the polarization states are maintained over a long propagation link [76]. In this section, both binary PolSK and a coherent multilevel PolSK employing the maximum ratio combining (MRC) technique and the Costas loop are introduced. For the analysis, it is assumed that the link configuration is a direct line of sight with no intersymbol interference, and the system noise is modelled as a complex and additive white Gaussian process.

4.11.1 BINARY POLSK

The block diagram of the PolSK transmitter is shown in Figure 4.38a, where the information is transmitted by switching the polarization of the transmitted optical beam between two linear orthogonal SOPs. The PolSK modulator is based on the lithium niobate ($LiNbO_3$) Mach–Zehnder interferometry (MZI) modulator (see Figure 4.38b) with the operating wavelength of 1550 nm [77]. This MZI consists of two sections: a splitter, followed by a waveguide-based wavelength-dependent phase shift.

The splitter of length d has the propagation matrix $\mathbf{M}_{coupler}$ expressed as

$$\mathbf{M}_{coupler} = \begin{bmatrix} \cos kd & -\sin kd \\ \sin kd & \cos kd \end{bmatrix} \tag{4.136}$$

where k is the coefficient.

The output signals from the two arms will experience a phase difference $\Delta\varnothing$ given by

$$\Delta\varnothing = \frac{2\pi n_1}{\lambda} L_p - \frac{2\pi n_2}{\lambda}(L_p + \Delta L_p) \tag{4.137}$$

Note that this phase difference can arise either through a refractive index difference if $n_1 \neq n_2$ or from a different path length ΔL_p. The propagation matrix $\mathbf{M}_{\Delta\varnothing}$ for the phase shifter with a given $\Delta\varnothing$ is

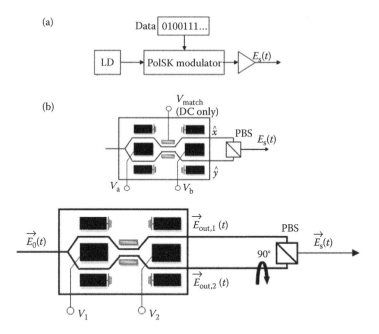

FIGURE 4.38 (a) PolSK transmitter block diagram and (b) LiNbO$_3$ modulator. LD, laser diode; PBS, polarizing beam splitter.

$$\Delta \mathbf{M}_{\Delta\varnothing} = \begin{bmatrix} \exp\left(j\Delta\varnothing_2^-\right) & 0 \\ 0 & \exp\left(j\Delta\varnothing_2^-\right) \end{bmatrix} \quad (4.138)$$

The optical output fields $\vec{E}_{\text{out,1}}(t)$ and $\vec{E}_{\text{out,2}}(t)$ from the two arms can be related to the input field $\vec{E}_0$ through the following relation:

$$\Delta \begin{bmatrix} \vec{E}_{\text{out,1}}(t) \\ \vec{E}_{\text{out,2}}(t) \end{bmatrix} = \mathbf{M} \begin{bmatrix} \vec{E}_0(t) \\ 0 \end{bmatrix} \quad (4.139)$$

where $\mathbf{M}$ is defined as

$$\begin{aligned} \mathbf{M} &= \mathbf{M}_{\Delta\varnothing} \cdot \mathbf{M}_{\text{coupler}} \\ &= \begin{bmatrix} \exp(j\Delta\varnothing/2) & 0 \\ 0 & \exp(-j\Delta\varnothing/2) \end{bmatrix} \begin{bmatrix} \cos kd & -\sin kd \\ \sin kd & \cos kd \end{bmatrix} \\ &= \begin{bmatrix} \exp(j\Delta\varnothing/2)\cos kd & -\exp(j\Delta\varnothing/2)\sin kd \\ \exp(-j\Delta\varnothing/2)\sin kd & \exp(-j\Delta\varnothing/2)\cos kd \end{bmatrix} \end{aligned} \quad (4.140)$$

Expressions (4.139) and (4.140) can be recombined yielding

$$\begin{bmatrix} \vec{E}_{\text{out},1}(t) \\ \vec{E}_{\text{out},2}(t) \end{bmatrix} = M \begin{bmatrix} \vec{E}_0(t) \\ 0 \end{bmatrix} = \begin{bmatrix} \exp(j\Delta\varnothing/2)\cos kd \\ \exp(-j\Delta\varnothing/2)\sin kd \end{bmatrix} \vec{E}_0(t) \tag{4.141}$$

As illustrated in Figure 4.38, V_a and V_b are used to control the amount of light launched in either $\hat{x}$ or $\hat{y}$ polarization and control the length, and thus the relative phase of the two polarizations, respectively. The third electrode V_{match} applied to the 3 dB coupler is used for wavelength matching. Since the splitter considered here only splits the power equally for $2kd = \pi/2$, (4.141) becomes

$$\begin{bmatrix} \vec{E}_{\text{out},1}(t) \\ \vec{E}_{\text{out},2}(t) \end{bmatrix} = M\vec{E}_{\text{in}}(t)$$

$$= \frac{1}{\sqrt{2}} \begin{bmatrix} \exp\left(j\Delta\varnothing_2^-\right) \\ \exp\left(-j\Delta\varnothing_2^-\right) \end{bmatrix} \vec{E}_0(t) \tag{4.142}$$

The control signal V_b is expressed in the form of

$$V_b = V_0 + V_m \sin(\omega t) \tag{4.143}$$

where V_0 is the DC component and $V_m \sin(\omega t)$ is the modulation signal.

$\hat{x}$ and $\hat{y}$ are the axes of polarization used to represent input digital symbols '0' and '1', respectively, thus achieving a constant optical power at the output of the PolSK modulator in order to fully utilize the output power of the laser source. To increase the optical power launched into the FSO channel, an optical amplifier could be used at the output of the PolSK modulator.

Demodulation and detection are accomplished through the analysis of the SOP, which is fully described by the knowledge of the Stokes parameters. This is simply accomplished by looking at the sign of the scalar product of the received SOP vector in the Stokes space with a reference vector representing one of the received SOPs in the absence of noise. Figure 4.39 represents the block diagram of a coherent optical PolSK heterodyne receiver. An optical lens is used to focus the received optical beam onto the photodetector. The received signal $E_r(t)$ can be viewed in both cases as two orthogonal ASK signals, related to orthogonal components of the transmitted optical field. The local oscillator $E_{lo}(t)$ is linearly polarized at $\pi/4$ with respect to the receiver reference axes.

Uncorrelated $E_r(t)$ and $E_{lo}(t)$ signals are given by

$$E_r(t) = \sqrt{P_r}\, e^{i(\omega_r t + \varnothing_r(t))} \{[1 - m(t)] \cdot x + m(t) \cdot y\} \tag{4.144}$$

$$E_{lo}(t) = \sqrt{P_{lo}}\, e^{i(\omega_{lo} t + \varnothing_{lo}(t))} \{x + y\} \tag{4.145}$$

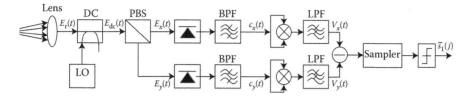

FIGURE 4.39 Block diagram of the coherent optical PolSK heterodyne receiver. LO, local oscillator; DC, directional coupler; BPF, band-pass filter; LPF, low-pass filter.

where P_r and P_{lo} are the received signal and local oscillator signal powers, respectively. ω_r, $\varnothing_r(t)$ and ω_{lo}, $\varnothing_{lo}(t)$ are the angular frequencies and phase noises for the received and local oscillator fields, respectively, and $m(t)$ is the message signal.

$E_r(t)$ and $E_{lo}(t)$ are mixed using an unbalanced directional coupler with a transfer matrix given by [69]

$$[S]_{dc} = \begin{bmatrix} \alpha_{dc} & \sqrt{1-\alpha_{dc}^2} \\ -\sqrt{1-\alpha_{dc}^2} & \alpha_{dc} \end{bmatrix} \tag{4.146}$$

where α_{dc} is the power splitting ratio.

Therefore, the optical field $E_{dc}(t)$ at the coupler output and consequently at the PBS input is given by

$$\begin{aligned} E_{dc}(t) &= \alpha_{dc}E_r(t) + \sqrt{1-\alpha_{dc}^2}\,E_{lo}(t) \\ &= \alpha_{dc}\sqrt{P_r}\,e^{i(\omega_r t + \varnothing_r(t))}\left\{[1 - m(t)]\cdot x + m(t)\cdot y\right\} \\ &\quad + \sqrt{\left(1-\alpha_{dc}^2\right)P_{lo}}\,e^{i(\omega_{lo}t+\varnothing_{lo}(t))}\{x + y\} \end{aligned} \tag{4.147}$$

The outputs of the PBS are defined as

$$E_x(t) = \left\{\alpha_{dc}\sqrt{\frac{P_r}{2}}[1-m(t)]e^{i(\omega_r t+\varnothing_r(t))} + \sqrt{\frac{(1-\alpha_{dc}^2)P_{lo}}{2}}\,e^{i(\omega_{lo}t+\varnothing_{lo}(t))}\right\}x \tag{4.148}$$

$$E_y(t) = \left\{\alpha_{dc}\sqrt{\frac{P_r}{2}}m(t)e^{i(\omega_r t+\varnothing_r(t))} + \sqrt{\frac{(1-\alpha_{dc}^2)P_{lo}}{2}}\,e^{i(\omega_{lo}t+\varnothing_{lo}(t))}\right\}y \tag{4.149}$$

Assuming an electron is generated by each detected photon. The outputs of two identical optical receivers are passed through ideal band-pass filters (BPFs) (of a one-sideband bandwidth $W = 2R_b$, where R_b is the data rate) with the outputs defined as

$$c_x(t) = R\alpha_{dc}\sqrt{(1 - \alpha_{dc}^2)P_r P_{lo}}\,[1 - m(t)]\cos(\omega_{IF}t + \varnothing_t(t)) + n_x(t) \quad (4.150)$$

$$c_y(t) = R\alpha_{dc}\sqrt{(1 - \alpha_{dc}^2)P_r P_{lo}}\,m(t)\cos(\omega_{IF}t + \varnothing_t(t)) + n_y(t) \quad (4.151)$$

where R is the photodiode responsivity, $\omega_{IF} = \omega_r - \omega_{lo}$ and $\varnothing_t(t) = \varnothing_r(t) - \varnothing_{lo}(t)$ are the intermediate angular frequency (IF) and the IF phase noise, respectively. The system noises $\{n_x(t), n_y(t)\}$ are modelled as independent, uncorrelated AWGN noises with zero mean and variance $\sigma_n^2 = WN_0$, where N_0 is the one-sideband noise power spectral density.

An ideal square-law demodulator composed of electrical mixers, a low-pass filter, a sampler and a threshold detector are used to recover the information signal. Note that the phase noise contribution is not included because of using the square-law demodulation scheme [69].

The SOP of a fully polarized optical beam can be described through the Stokes parameters [78].

$$S = \begin{pmatrix} S_0 \\ S_1 \\ S_2 \\ S_3 \end{pmatrix} \quad (4.152)$$

The components have the following physical interpretations: S_0 is the optical intensity; S_1 is the intensity difference between horizontal ($S_1 = +ve$) and vertical ($S_1 = -ve$) polarized components; S_2 indicates the preference for $+45°$ (positive S_2) or $-45°$ SOPs; S_3 is the preference for right-hand circular and left-hand circular polarizations. Since the optical field is linearly polarized and its power is unchanged, the Stokes parameters are expressed as

$$S_0 = |c_x(t)|^2 + |c_y(t)|^2 = \frac{R^2\alpha_{dc}^2(1 - \alpha_{dc}^2)P_r P_{lo}}{2} + n_0(t)$$

$$S_1 = |c_x(t)|^2 - |c_y(t)|^2 = \frac{R^2\alpha_{dc}^2(1 - \alpha_{dc}^2)P_r P_{lo}[1 - 2m(t)]}{2} + n_1(t)$$

$$(4.153)$$

$$S_2 = 2|c_x(t)| \times |c_y(t)|\cos(0) = 0 + n_2(t)$$

$$S_3 = 2|c_x(t)| \times |c_y(t)|\sin(0) = 0 + n_3(t)$$

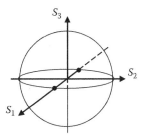

FIGURE 4.40 SOPs at the output of the PolSK receiver.

where S_0, S_1, S_2 and S_3 are the estimation Stokes parameters and $\{n_i(t)\}_{i=0,1,2,3}$ are the noise contribution which are independent of the received SOP and have the same variance. Note that the proposed 2PolSK refers only to the parameter S_1. A digital symbol '0' is assumed to have been received if S_1 is above the threshold level of 0 and '1' otherwise. Two orthogonal SOPs map onto opposite points at S_1 on the equator with respect to the origin in the Poincare sphere shown in Figure 4.40.

The following hypotheses must be presumed in such a way that the quantum limit of the proposed receiver can be determined [69]:

- To neglect the penalty induced by the unbalanced directional coupler, its coefficient α_{dc} is chosen to be close to unity
- The power of the LO power is assumed to be sufficiently high
- The responsivity of the PD is assumed to be equal to unity
- Filters do not cause any signal distortion and only limit the noise power and eliminate the undesired signal components
- PDs and filters on different electronic branches at the receiver are assumed to be identical

4.11.2 Bit Error Rate Analysis

We only consider a direct LOS of FSO link; thus, only the background radiation modelled as an AWGN is considered. Assuming IID data transmission, the total probability of error P_{ec} conditioned on the received irradiance is given by

$$P_{ec} = \frac{1}{2} P(e|0) + \frac{1}{2} P(e|1)$$
$$= P(e|0) \tag{4.154}$$

where $P(e|0)$ is the conditional bit error probability for receiving a '1' provided a '0' was sent.

Noise signals $\{n_x(t), n_y(t)\}$, including the background noise and the quantum noise, can be expressed as [27]

$$n_x(t) = n_{xp}(t)\cos(\omega_{IF}t + \varnothing_t(t)) - n_{xq}(t)\sin(\omega_{IF}t + \varnothing_t(t))$$

$$n_y(t) = n_{yp}(t)\cos(\omega_{IF}t + \varnothing_t(t)) - n_{yq}(t)\sin(\omega_{IF}t + \varnothing_t(t))$$

(4.155)

where $\{n_{xi}(t), n_{xq}(t)\}$ and $\{n_{yi}(t), n_{yq}(t)\}$ are the phase and quadrature components, respectively, having a normal distribution with a zero mean and a variance of σ_n^2.

Given $m(t) = 0$ and $K = R\alpha_{dc}\sqrt{(1 - \alpha_{dc}^2)P_rP_{lo}}$, (4.150) and (4.151) are given by

$$\langle c_x(t)\rangle = \{K + n_{xi}(t)\}\cos(\omega_{IF}t + \varnothing_t(t)) - n_{xq}(t)\sin(\omega_{IF}t + \varnothing_t(t))$$

$$\langle c_y(t)\rangle = n_{yi}(t)\cos(\omega_{IF}t + \varnothing_t(t)) - n_{yq}(t)\sin(\omega_{IF}t + \varnothing_t(t))$$

(4.156)

The baseband outputs $V_x(t)$ and $V_y(t)$ for the upper and the lower arms (Figure 4.39), respectively, are given as

$$V_x(t) = \sqrt{[K + n_{xi}(t)]^2 + n_{xq}^2(t)}$$

$$V_y(t) = \sqrt{n_{yi}^2(t) + n_{yq}^2(t)}$$

(4.157)

$V_x(t)$ and $V_y(t)$ have fixed mean values and the same variance given by

$$E[V_x(t)] = K$$

$$E[V_y(t)] = 0$$

(4.158)

$$\sigma_x^2 = \sigma_y^2 = \sigma_n^2$$

With $\omega_{IF} \ll \omega_r$, the PDF of $V_x(t)$ and $V_y(t)$ can be described by the Rice and the Rayleigh probability functions, respectively [27]

$$p(V_x) = \left\{\frac{V_x}{\sigma_n^2}I_0\left(\frac{KV_x}{\sigma_n^2}\right)\exp\left[-\frac{V_x^2 + K^2}{2\sigma_n^2}\right]\right\}$$

$$p(V_y) = \frac{V_y}{\sigma_n^2}\exp\left(-\frac{V_y^2}{2\sigma_n^2}\right)$$

(4.159)

where I_0 is the zero-order modified Bessel function of the first kind [27].

The conditional BER for $m(t) = 0$ can be derived as

$$P_{ec} = \int_0^\infty p(V_x) \left[\int_{V_x}^\infty p(V_y) dV_y \right] dV_x = \int_0^\infty \left\{ \frac{V_x}{\sigma_n^2} I_0 \left(\frac{KV_x}{\sigma_n^2} \right) \exp \left[-\frac{2V_x^2 + K^2}{2\sigma_n^2} \right] \right\} dV_x \quad (4.160)$$

By invoking changes of variables $m = \sqrt{2}V_x/\sigma_n$ and $n = k / \left(\sqrt{2}\sigma_n \right)$ and substituting into (4.160), P_{ec} now becomes

$$P_{ec} = \frac{1}{2} e^{-n^2/2} \int_0^\infty m I_0(mn) e^{-(m^2+n^2)/2} \, dm \quad (4.161)$$

Defining the Q-function as

$$Q(n,0) = \int_0^\infty m I_0(mn) e^{-(m^2+n^2)/2} \, dm = 1 \quad (4.162)$$

P_{ec} is represented as

$$P_{ec} = \frac{1}{2} \exp \left(-\frac{n^2}{2} \right) = \frac{1}{2} \exp \left(-\frac{K^2}{4\sigma_n^2} \right) \quad (4.163)$$

The electrical SNR at the output of the BPF is defined as

$$\text{SNR}(P_r) = \left(R\alpha_{dc} \sqrt{\left(1 - \alpha_{dc}^2\right) P_r P_{lo}} / \sqrt{2}\sigma_n \right)^2 \quad (4.164)$$

P_{ec} can be expressed in terms of the SNR by substituting (4.164) into (4.163):

$$SP_{ec}(P_r) = \frac{1}{2} \exp \left(-\frac{\text{SNR}(P_r)}{2} \right) \quad (4.165)$$

This result is the same as the BER expression of FSK. With regard to the system sensitivity, PolSK and FSK techniques have complete equivalence [79].

4.11.3 MPolSK

The schematic diagram of the MPolSK optical coherent transmitter is illustrated in Figure 4.41. The laser beam is linearly polarized at an angle of $\pi/4$ with respect to the transmitter reference axes. The linearly polarized beam is then launched into a polarization beam splitter which is considered as two ideal linear polarizers oriented orthogonal

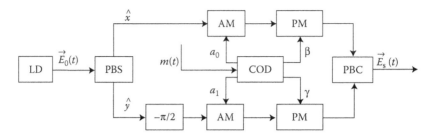

FIGURE 4.41 Block diagram of the MPolSK transmitter for coherent optical communication systems. LD, laser diode; COD, encoder; PBS, polarization beam splitter; PBC, polarization beam combiner; AM, external amplitude modulator; PM, LiNbO$_3$ device-based external phase modulator.

to each other. This yields the horizontal ($\hat{x}$-polarization) and vertical ($\hat{y}$-polarization) SOPs with equal amplitude and zero phase differences. The $\hat{y}$-polarization component is phase shifted by $\pi/2$. Both orthogonal polarized components are amplitude and phase modulated synchronously before being fed into the polarization beam combiner.

The emitted optical field $\vec{E}_s(t)$ is thus given as

$$\vec{E}_s(t) = \sqrt{\frac{P}{2}} e^{j(\omega_s t + \varphi_s)} \{a_0 e^{j\beta} \cdot \hat{x} + j a_1 e^{-jy} \cdot \hat{y}\} \qquad (4.166)$$

where P, ω_s and φ_s are the power, angular frequency and the laser phase noise of the transmitted optical carrier, respectively. The unit vectors $\hat{x}$ and $\hat{y}$ represent the polarization direction. The phase-modulation functions β and γ are equal to π and zero corresponding to the transmission of space and mark, respectively. The amplitudes a_0 and a_1 for $\hat{x}$ and $\hat{y}$ polarization components are expressed as

$$a_0 = a_1 = \left(2m + 1 - \sqrt{M}\right)d, \quad m = 0,1,\ldots,\sqrt{M} - 1 \qquad (4.167)$$

where m is the decibel information, d is half the distance between adjacent symbols in one polarization axis and M is the number of signal-point in the constellation and $\sqrt{M}$ must be an integer. The proposed MPolSK modulation format can be described as a two-dimensional multilevel amplitude modulation in the orthogonal polarization axis. The transmitted field is constant over the symbol interval and each symbol is associated with a value of the optical field. The instantaneous transitions between subsequent symbol intervals are assumed.

The block diagram of the MPolSK diversity coherent receiver is shown in Figure 4.42. The background noise is limited by an ideal optical band-pass filter (BPF) with a narrow bandwidth, typically 1 nm.

The received optical signal is expressed in the following form:

$$\vec{E}_r(t) = \sqrt{\frac{P_r}{2}} e^{i[\omega_s t + \varphi_t(t)]} \{a_0 e^{j\beta} \cdot \hat{x} + j a_1 e^{-jr} \cdot \hat{y}\} \qquad (4.168)$$

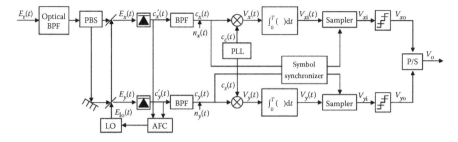

FIGURE 4.42 Block diagram of the MPolSK coherent optical receiver. Optical BPF, optical band-pass filter; BPF, electric band-pass filter; LO, local oscillator; AFC, automatic frequency control; PLL, Costas-loop-based phase locked loop circuit; P/S, parallel-to-serial converter.

where P_r and $\varphi_t(t)$ are the received optical power and phase noise of the laser carrier propagating through the atmospheric turbulence medium, respectively. Both parameters are time-variant statistical quantities due to the turbulence.

The local oscillator (LO) used is the same as the transmitter laser module. The LO output $\vec{E}_{lo}(t)$, linearly polarized at $\pi/4$ with respect to the receiver reference axes, is mixed with the received field $\vec{E}_r(t)$, and is expressed as

$$\vec{E}_{lo}(t) = \frac{\sqrt{P_{lo}}}{2} e^{i(\omega_{lo}t + \varphi_{lo})} \{\hat{x} + \hat{y}\} \tag{4.169}$$

where P_{lo}, ω_{lo} and φ_{lo} represent the power, angular frequency and the phase noise of the local oscillator, respectively.

The slow-frequency fluctuations associated with the LO is compensated by an automatic frequency control (AFC) module with its control signal derived from the IF electric signal [80].

$\vec{E}_r(t)$ is split into $\hat{x}$ and $\hat{y}$ axes and are combined with the corresponding components of LO. Thus, the optical fields $\vec{E}_x(t)$ and $\vec{E}_y(t)$ are given by

$$\vec{E}_x(t) = \left\{ \sqrt{\frac{P_r}{2}} a_0 e^{i(\omega_s t + \beta + \varphi_t(t))} + \frac{\sqrt{P_{lo}}}{2} e^{i(\omega_{lo}t + \varphi_{lo})} \right\} \hat{x}$$

$$\tag{4.170}$$

$$\vec{E}_y(t) = \left\{ \sqrt{\frac{P_r}{2}} a_1 e^{i\left(\omega_s t + r - \frac{\pi}{2} + \varphi_t(t)\right)} + \frac{\sqrt{P_{lo}}}{2} e^{i(\omega_{lo} + \varphi_{lo})} \right\} \hat{y}$$

After passing through two identical photodetectors and electrical BPFs, the constant terms are filtered out and the additive noise is limited. The bandwidth and centre frequency of BPFs are $B_{bp} = 2(R_s + k_F B_L)$ and ω_{IF}, respectively. R_s and B_L are the symbol rate and linewidth of the laser sources, respectively. The parameter k_F is chosen to pass the signal through the filter with a minimum distortion. Electrical current signals appearing on the x and y of BPF channels following photodetection with a unit detector area are expressed as [80]

$$c_x(t) = R\sqrt{\frac{P_r P_{lo}}{2}}\, a_0 \cos(\beta)\cos[\omega_{IF}t + \varphi_{IF}(t)] + n_x(t)$$

$$(4.171)$$

$$c_y(t) = R\sqrt{\frac{P_r P_{lo}}{2}}\, a_1 \cos(\gamma)\sin[\omega_{IF}t + \varphi_{IF}(t)] + n_y(t)$$

where R represents the photodiode responsivity, $\omega_{IF} = \omega_s - \omega_{lo}$ and $\varphi_{IF}(t) = \varphi_t(t) - \varphi_{lo}$ are the intermediate angular frequency and phase noise, respectively. Note that $\sin(\beta)$ and $\sin(\gamma) = 0$ as $\{\beta, \gamma\} = \{0, \pi\}$. $n_x(t)$ and $n_y(t)$ are assumed to be statistically independent, and stationary additive white Gaussian noise (AWGN) with a zero-mean and an equal variance $0.5\sigma_n^2$. Both noise terms are uncorrelated such that $n_x(t) \cdot n_y(t) = 0$. The noise terms in (4.171) can be expressed as [27]

$$n(t) = \frac{n^I(t)\cos[\omega_{IF}t + \varphi_{IF}(t)]}{-n^Q(t)\sin[\omega_{IF}t + \varphi_{IF}(t)]}$$

$$(4.172)$$

where n^I and n^Q are the in-phase and quadrature components of the noise $n(t) \sim (0, \sigma_n^2/2)$.

Assuming using an ideal square-law elements, or multipliers and BPF filter bandwidth being large compared with the IF beat linewidth to avoid phase noise to amplitude noise conversion, the phase noise due to the laser LO can be considered to negligible.

$c_x(t)$ and $c_y(t)$ are then multiplied with the local carrier signal $c_c(t)$ and $c_s(t)$ generated by the Costas-loop-based PLL circuit [27]:

$$c_c(t) = \cos[\omega_{IF}t + \varphi_{PLL}(t)]$$
$$c_s(t) = \sin[\omega_{IF}t + \varphi_{PLL}(t)]$$

$$(4.173)$$

where $\varphi_{PLL}(t)$ represents the estimate of $\varphi_{IF}(t)$.

The outputs of the mixers are passed through identical matched filters (MF) to reject the higher frequency components, with outputs expressed as

$$V_{xi}(t) = [c_x(t) \cdot c_c(t)] * h_m(t) = R\sqrt{\frac{P_r P_{lo}}{2}}\, a_0 \cos(\beta)\cos[\Delta\varphi(t)] + n'_x(t)$$

$$(4.174)$$

$$V_{yi}(t) = [c_y(t) \cdot c_s(t)] * h_m(t) = R\sqrt{\frac{P_r P_{lo}}{2}}\, a_1 \cos(r)\cos[\Delta\varphi(t)] + n'_y(t)$$

where $*$ indicates the convolution operation, and $h_m(t)$ is the impulse response of MF with a bandwidth of $B_m = R_s + k_F B_L$, $\Delta\varphi(t) = \varphi_{IF}(t) - \varphi_{PLL}(t)$ is the phase tracking error. Subsequently $V_{xi}(t)$ and $V_{yi}(t)$ are passed though a sampler at a sampling time $t =$ the symbol period T, with the outputs given by

$$V_{xi} = \int_0^T V_{xi}(t)dt = \pm R\sqrt{\frac{P_r P_{lo}}{2}}a_0\cos(\Delta\varphi) + n_x'$$

$$(4.175)$$

$$V_{yi} = \int_0^T V_{yi}(t)dt = \pm R\sqrt{\frac{P_r P_{lo}}{2}}a_1\cos(\Delta\varphi) + n_y'$$

where the plus sign in $\pm$ denotes the transmission of a space, and the minus sign represents a mark.

4.11.4 DIFFERENTIAL CIRCLE POLARIZATION SHIFT KEYING

PolSK modulation is especially attractive for peak power limited systems as it is a constant envelop modulation scheme and is highly insensitive to the laser phase noise. However, the polarization coordinate of the transmitter must be aligned precisely with the receiver in linear PolSK schemes. Circular PolSK (CPolSK) is based on two rotation states of circular polarization requiring no coordinate alignment, thus offering the much-needed installation flexibility for outdoor OWC systems in particular for moving objects [75]. It has been shown that CPolSK schemes with the direct detection offers ~3 dB power gains compared to OOK [75]. However, the results have also shown that the coherent detection provides better performance than the direct detection. The improvement is about 10^{-3} BER advance for the same received signal power [81]. Coherent DCPolSK FSO systems are adopted to ameliorate the turbulence-induced irradiance fluctuation and to circumvent the phase noise, thus improving the BER performance.

The schematic of DCPolSK optical coherent transmitter is depicted in Figure 4.43. The emitted optical field of the carrier is linearly polarized along $\pi/4$ with respect to the reference axis of the modulator. The differentially encoded information bits $m(t)$ are used to modulate the field phase of the laser in such a way that the field phase is equal to zero and π for transmission of a space and a mark, respectively.

The emitted optical signal is defined as

$$\vec{E}_s(t) = \sqrt{\frac{P}{2}}e^{j[\omega_s t + \varphi_s(t)]}\left\{e^{j\Delta\varnothing/2}\cdot\hat{x} + e^{j-\Delta\varphi/2}\cdot\hat{y}\right\} \qquad (4.176)$$

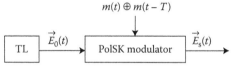

FIGURE 4.43 Block diagram of DCPolSK transmitter for the coherent optical communication system. TL, transmitting laser; PolSK, external polarization modulator.

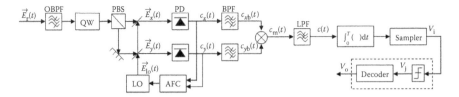

FIGURE 4.44 Block diagram of a heterodyne receiver for a DCPolSK-modulated signal. OBPF, optical band-pass filter; QW, quarter-wave plate; LO, local oscillator; AFC, automatic frequency control circuit; PBS, polarization beam splitter; PD, photodiode; BPF, electric band-pass filter; LPF, electric low-pass filter.

The left and right circular polarizations are produced by changing the field phase of $\Delta\varnothing = -\pi/2$ and $\pi/2$ with respect to the transmission of the differentially encoded space and mark, respectively.

The structure of the heterodyne DCPolSK receiver is illustrated in Figure 4.44. The optical band-pass filter with a narrow linewidth, typically 1 nm, is there to reduce the background noise without introducing any distortion. The quarter wave retarder, made of birefringent materials, changes the light travelling through it by one quarter of a wavelength with respect to the other polarization components. Therefore, a circularly polarized light passing through a quarter wave retarder becomes a linearly polarized light with a polarization axis along $-\pi/4$ and $\pi/4$ depending on which the polarization component $(\hat{x}, \hat{y})$ is retarded. The local oscillator laser source, linearly polarized at $\pi/4$, is identical to the transmitter laser. To compensate for slow-frequency fluctuations in the local oscillator, the automatic frequency control circuit is used to control the laser bias current [80].

The received optical signal $\vec{E}_r(t)$ is given as

$$\vec{E}_r(t) = \sqrt{\frac{P_r}{2}} e^{i[\omega_s t + \varphi_r(t)]} \{ e^{j(\Delta\varnothing/2)} \hat{x} + e^{j(-\Delta\varnothing/2)} \hat{y} \} \qquad (4.177)$$

where P_r and $\varphi_r(t)$ represent optical power and phase noise of the received signal, respectively. Both variables are time-variant statistics resulting from the turbulence fluctuation.

The local oscillator $\vec{E}_{lo}(t)$ is given in (4.169). After passing through the quarter-wave plate, the optical signal $\vec{E}_r(t)$ is split by the polarization beam splitter and is mixed with $\vec{E}_{lo}(t)$ resulting in

$$\vec{E}_x(t) = \left\{ \sqrt{\frac{P_r}{2}} e^{i[\omega_s t + (\Delta\varnothing/2) + (\pi/4) + \varphi_r(t)]} + \sqrt{\frac{P_{lo}}{2}} e^{i[\omega_{lo} t + \varphi_{lo}(t)]} \right\} \hat{x}$$

$$\vec{E}_y(t) = \left\{ \sqrt{\frac{P_r}{2}} e^{i[\omega_s t - (\Delta\varnothing/2) - (\pi/4) + \varphi_r(t)]} + \sqrt{\frac{P_{lo}}{2}} e^{i[\omega_{lo} t + \varphi_{lo}(t)]} \right\} \hat{y} \qquad (4.178)$$

The optical fields are then detected by two identical photodiodes, where the outputs are passed through electrical band-pass filters. The bandwidth and the centre

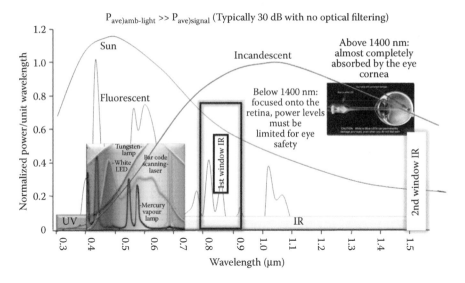

FIGURE 1.2 Normalized power/unit wavelength for optical wireless spectrum and ambient light sources.

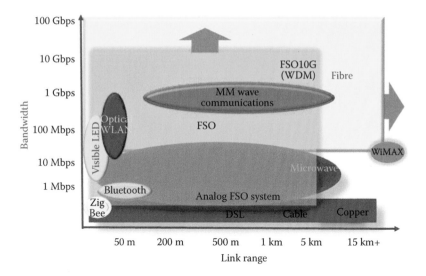

FIGURE 1.3 Bandwidth capabilities for a range of optical and RF technologies.

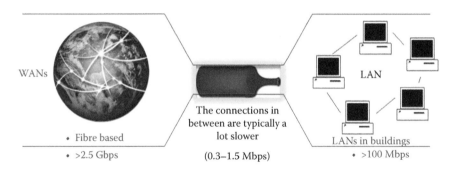

FIGURE 1.5 Access network bottleneck.

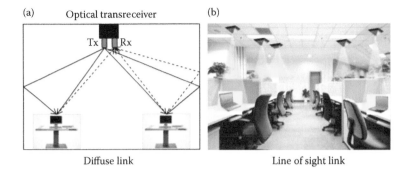

Diffuse link   Line of sight link

FIGURE 1.11 Optical wireless LANs: (a) diffuse and (b) line of sight.

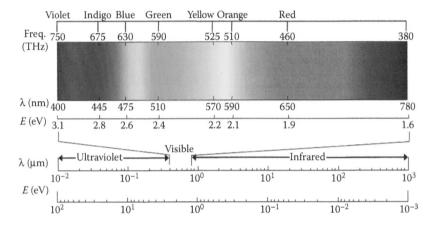

FIGURE 2.1 Wavelength and energy of ultraviolet, visible and infrared portion of the electromagnetic spectrum.

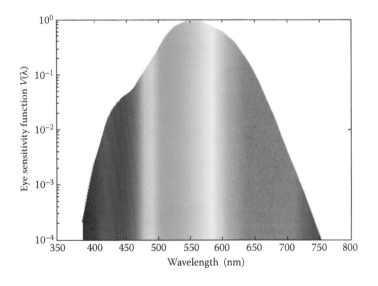

FIGURE 2.10 Eye sensitivity function based on the 1978 CIE data.

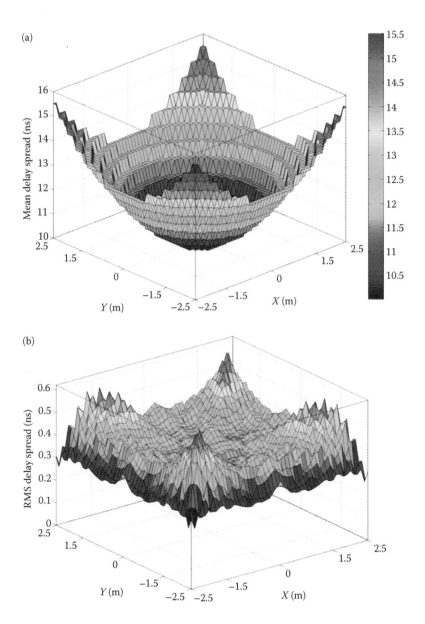

FIGURE 3.7 Channel delay spread: (a) mean delay spread, (b) RMS delay spread with LOS component, (c) RMS delay spread without LOS component and (d) maximum data rate distributions.

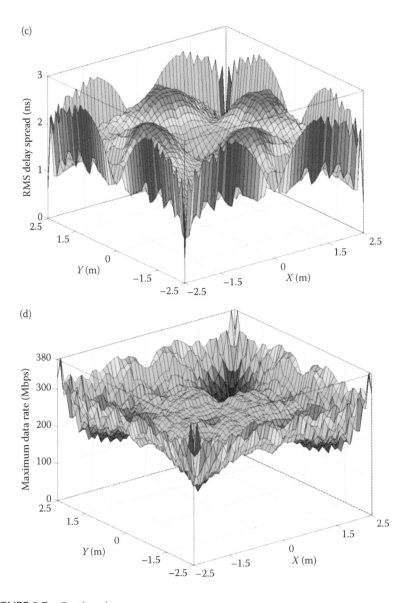

FIGURE 3.7 Continued.

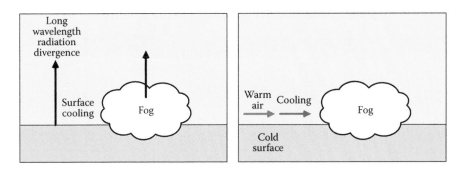

FIGURE 3.16 Schematic depiction of fog formation: (a) radiation and (b) advection.

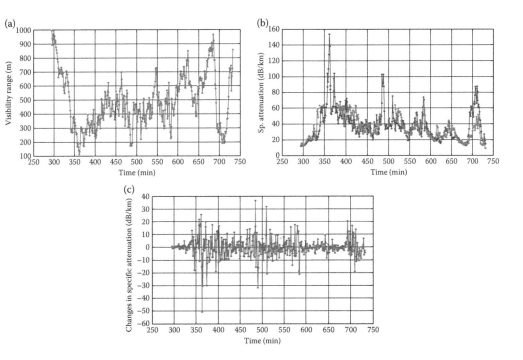

FIGURE 3.20 Time profiles of (a) visual range, (b) specific attenuation and (c) differences in specific attenuation during a fog event occurred in Milan on 11th January 2005. In (b) two profiles are shown: the measured laser attenuation (red curve) and the attenuation as estimated from visual range (blue curve). (Adapted from F. Nadeem et al., *1st International Conference on Wireless Communication, Vehicular Technology, Information Theory and Aerospace & Electronic Systems Technology*, 2009, pp. 565–570.)

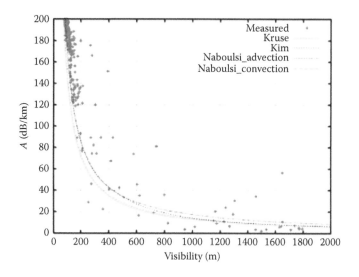

FIGURE 3.23 Measured attenuation coefficient as a function of visibility range at λ = 830 nm in early 2008, Prague, Czech Republic. (Adapted from M. Grabner and V. Kvicera, Experimental study of atmospheric visibility and optical wave attenuation for free-space optics communications, http://ursi-france.institut-telecom.fr/pages/pages_ursi/URSIGA08/papers/F06p5.pdf, last visited 2nd Sept. 2009.)

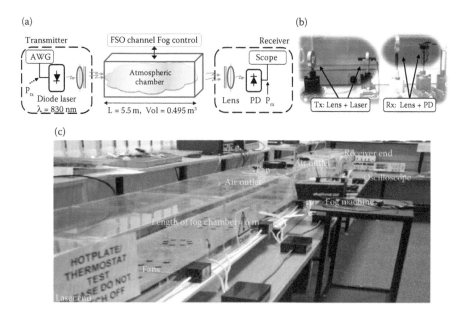

FIGURE 3.34 (a) Block diagram of the FSO experiment set-up, (b) the simulation chamber and (c) the laboratory chamber set up.

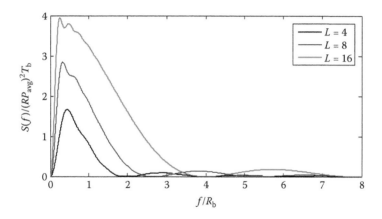

FIGURE 4.9 PSD of PPM for $L = 4$, 8 and 16.

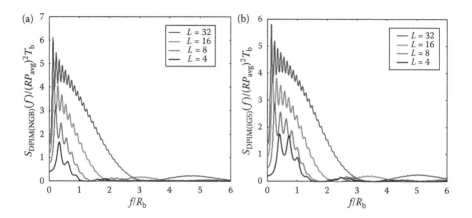

FIGURE 4.14 PSD of (a) DPIM(NGB) and (b) DPIM(1GS) for $L = 4$, 8, 16 and 32.

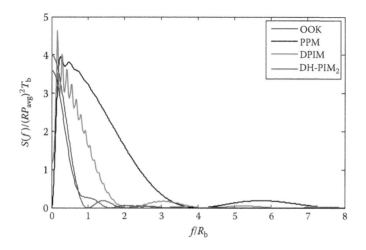

FIGURE 4.26 Power spectral density of OOK, PPM, DPIM and DH-PIM.

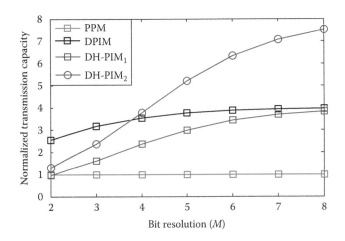

FIGURE 4.32 Transmission capacity of PPM, DPIM, DH-PIM$_1$ and DH-PIM$_2$ normalized to PPM versus M.

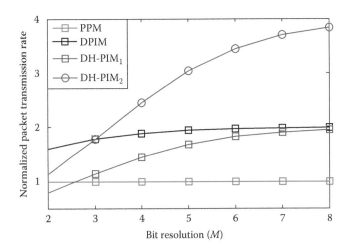

FIGURE 4.33 Packet transmission rate of PPM, DPIM, DH-PIM$_1$ and DH-PIM$_2$ normalized to PPM versus M.

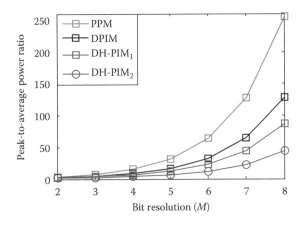

FIGURE 4.34 PAPR of PPM, DPIM, DH-PIM$_1$ and DH-PIM$_2$ versus M.

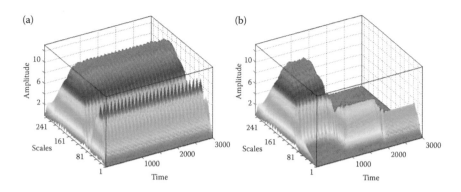

(a)

(b)

FIGURE 5.22 The CWT of the signal of (a) nonstationary signal and (b) stationary signal.

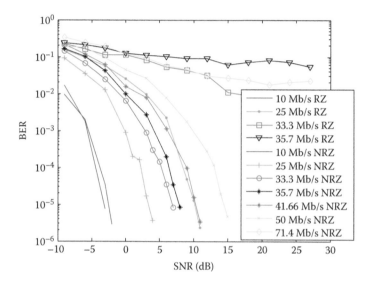

FIGURE 5.47 BER against SNR for OOK–NRZ/RZ with $0.6/T_b$ and $0.35/T_b$ low-pass filtering.

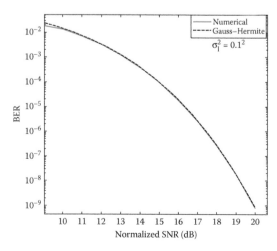

FIGURE 6.13 BER against the normalized SNR using numerical and 20th-order Gauss–Hermite integration methods in weak atmospheric turbulence for $\sigma_I^2 = 0.1^2$.

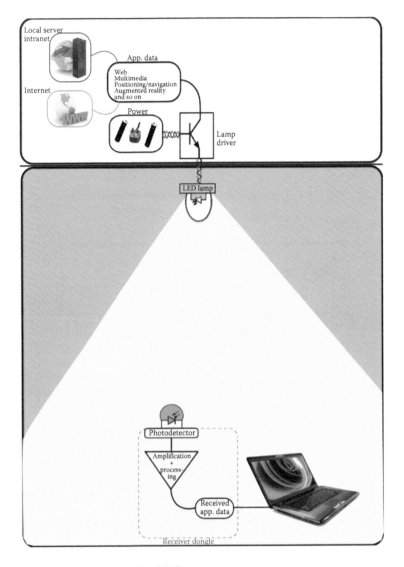

FIGURE 8.1 An illustration of the VLC concept.

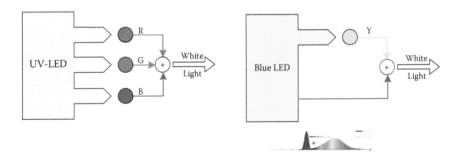

FIGURE 8.3 Two approaches for generating white emission from LEDs.

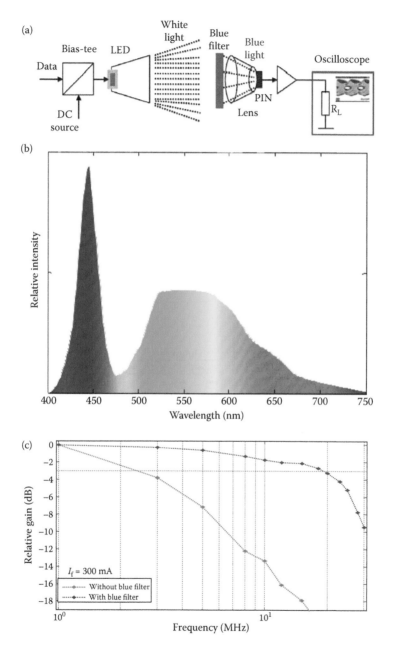

FIGURE 8.4 (a) VLC link, (b) LED optical spectrum of Osram Ostar white-light LED and (c) modulation bandwidth, with and without blue filtering.

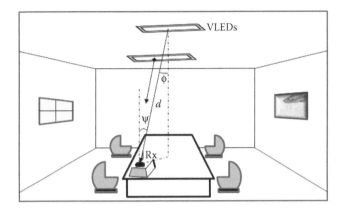

FIGURE 8.6 Illumination of VLEDs.

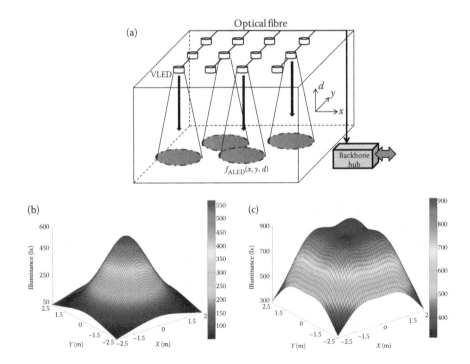

FIGURE 8.7 (a) LED array, and illuminance distribution for (b) one transmitter and (c) four transmitters.

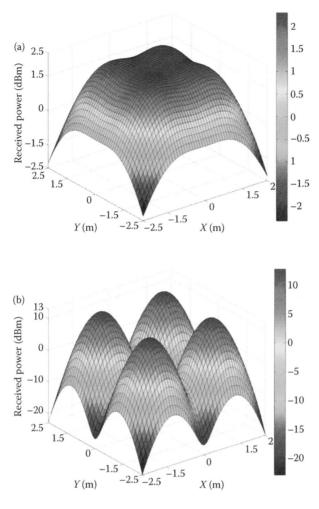

FIGURE 8.8 Optical power distribution in received optical plane for a FWHM of (a) 70° and (b) 12.5°.

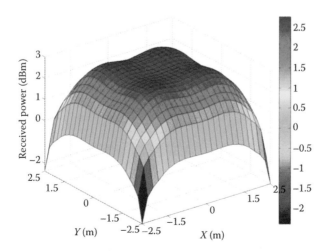

FIGURE 8.12 The distribution of received power with reflection.

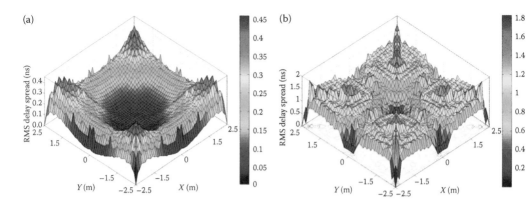

FIGURE 8.13 RMS delay distribution for (a) a single transmitter positioned at (2.5,2.5) and (b) for four transmitters positioned at (1.25,1.25), (1.25,3.75), (3.75,1.25), (3.75,3.75).

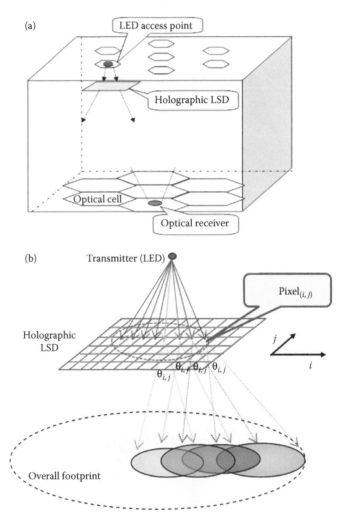

FIGURE 8.25 Indoor cellular VLC system: (a) block diagram and (b) transmitter with a holographic LSD.

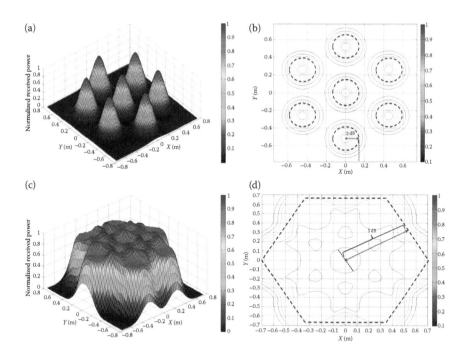

FIGURE 8.26 Predicted normalized power distribution at the receiving plane: (a) without LSD and (c) with a 30° LSD. Predicted power contour plot at the receiving plane: (b) without LSD and (d) with a 30° LSD.

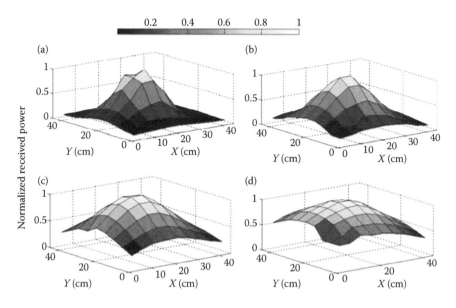

FIGURE 8.27 Spatial distribution of received power: (a) without LSD, (b) with 10° LSD, (c) with 20° LSD and (d) with 30° LSD.

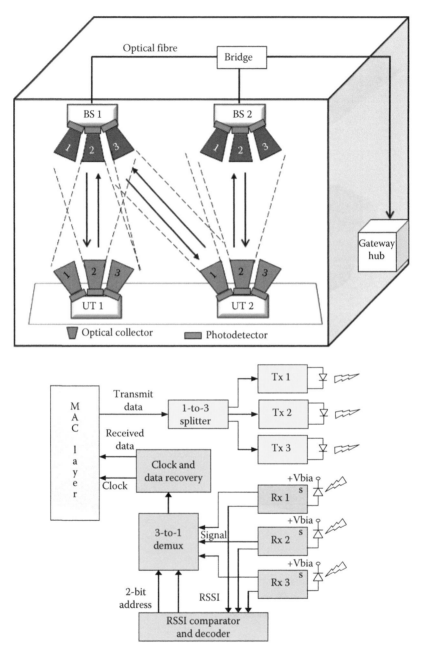

FIGURE 8.28 A one-dimensional optical wireless cellular system: (a) block diagram and (b) functional block diagram of a BS/UT module.

frequency of the BPFs are $2R_s$ and ω_{IF}, respectively, where R_s is the symbol rate. Thus, the electric currents $c_{xb}(t)$ and $c_{yb}(t)$ are given as

$$c_{xb}(t) = c_x(t) * h_{BP}(t) = \sqrt{\frac{R^2 P_r P_{lo}}{2}} \cos\begin{bmatrix} \omega_{IF}t + \varphi_{IF}(t) \\ +\Delta\varnothing/2 + \pi/4 \end{bmatrix} + n_x(t)$$

$$c_{yb}(t) = c_y(t) * h_{BP}(t) = \sqrt{\frac{R^2 P_r P_{lo}}{2}} \cos\begin{bmatrix} \omega_{IF}t + \varphi_{IF}(t) \\ -\Delta\varnothing/2 - \pi/4 \end{bmatrix} + n_y(t)$$

(4.179)

where $\omega_{IF} = \omega_s - \omega_{lo}$ and $\varphi_{IF}(t) = \varphi_t(t) - \varphi_{lo}(t)$ are the frequency and phase noise of the intermediate signal, respectively.

The noise terms $n_x(t)$ and $n_y(t)$ are defined in (4.155). After multiplication and filtering, the baseband electric current $c(t)$ is expressed as

$$\begin{aligned} c(t) &= c_m(t) * h_{LP}(t) \\ &= \frac{R^2 P_r P_{lo}}{4} \cos(\Delta\varnothing + \pi/2) \\ &\quad + \sqrt{\frac{R^2 P_r P_{lo}}{8}} \left\{ \begin{array}{l} \left[n_x^I(t) + n_y^I(t) \right]\cos(\Delta\varnothing/2 + \pi/4) \\ + \left[n_y^Q(t) - n_x^Q(t) \right]\sin(\Delta\varnothing/2 + \pi/4) \end{array} \right\} \end{aligned}$$

(4.180)

where $h_{LP}(t)$ denotes the impulse response of the low-pass filter with a bandwidth of R_s. Note that the intermediate phase noise disappears due to the multiplication and filtering, illustrating DCPolSK insensitivity to the phase noise. It follows that the signal $c(t)$ is integrated over the symbol period T and sampled at time $t = T$ to obtain the decision variable V_i. The signal V_i is then compared with the zero threshold level yielding the signal V_j, thus leading to the detection of the transition between two adjacent symbols, by which information is encoded.

4.11.5 ERROR PROBABILITY ANALYSIS

An error occurs when $V_i < 0$ for transmission of a space with $\Delta\varnothing = -\pi/2$, thus

$$V_i = \sqrt{\frac{R^2 P_r P_{lo}}{8}} \left(\sqrt{\frac{R^2 P_r P_{lo}}{2}} + n_x^I + n_y^I \right)$$

(4.181)

Defining the variable $Z = \sqrt{R^2 P_r P_{lo}/2} + n_x^I + n_y^I$ with the mean of $E[Z] = \sqrt{R^2 P_r P_{lo}/2}$ and a variance $\sigma_z^2 = \sigma_n^2$ the error rate before the code converter is thus expressed as

$$P_{e1} = \frac{1}{\sqrt{2\pi}\sigma_z} \int_{-\infty}^{0} e^{-\frac{\left(Z - \sqrt{\frac{R^2 P_r P_{lo}}{2\sigma_z^2}}\right)^2}{2\sigma_z^2}} dZ = \frac{1}{2}\text{erfc}\left(\sqrt{\frac{r}{2}}\right)$$

(4.182)

where $r = R^2 P_r P_{1o}/2\sigma_n^2$ is the SNR at the demodulator output. The decoder determines the changes between adjacent symbols. An error occurs if only one of two adjacent symbols is erroneous. Thus, the error probability becomes

$$P_{e2} = 2P_{e1}(1 - P_{e1}) = \text{erfc}\left(\sqrt{\frac{r}{2}}\right)\left[1 - \frac{1}{2}\text{erfc}\left(\sqrt{\frac{r}{2}}\right)\right] \qquad (4.183)$$

APPENDIX 4.A

4.A.1   DERIVATION OF SLOT AUTOCORRELATION FUNCTION OF DPIM(1GS)

Let a_n be a DPIM(1GS) slot sequence. Assume that, in any given slot n, a_n may take a value of either 0 or 1. Thus

$$a_n \in \{0, 1\} \quad \text{for all } n \qquad (4.A.1)$$

The autocorrelation function of this slot sequence is given by [25]

$$R_k = \overline{a_n a_{n+k}} = \sum_{i=1}^{I} (a_n a_{n+k})_i P_i \qquad (4.A.2)$$

where P_i is the probability of getting the ith $a_n a_{n+k}$ product, and I is the number of possible values for the product.

When $k = 0$, the possible products are $1 \times 1 = 1$ and $0 \times 0 = 0$, and consequently $I = 2$. The probability of getting a product of 1 is $1/\overline{L}_{\text{DPIM}}$, and the probability of getting a product of 0 is $(\overline{L}_{\text{DPIM}} - 1)/\overline{L}_{\text{DPIM}}$. Thus

$$R_0 = \sum_{i=1}^{2} (a_n a_n)_i P_i = 1 \cdot \frac{1}{\overline{L}_{\text{DPIM}}} + 0 \cdot \frac{(\overline{L}_{\text{DPIM}} - 1)}{\overline{L}_{\text{DPIM}}} = \overline{L}_{\text{DPIM}}^{-1} \qquad (4.A.3)$$

where from (4.3), $\overline{L}_{\text{DPIM}} = (L + 3)/2$.

By including a guard slot in each symbol, it is not possible for pulses to occur in adjacent slots. Thus, when $k = 1$, the probability of getting a product of 1 is 0, and therefore

$$R_1 = 0 \qquad (4.A.4)$$

When $k = 2$, the probability of getting a product of 1 is simply the probability of a 1 multiplied by the probability that that 1 represents the start of a shortest-duration symbol. Thus

$$R_2 = 1 \cdot \overline{L}_{\text{DPIM}}^{-1} \cdot L^{-1} \qquad (4.A.5)$$

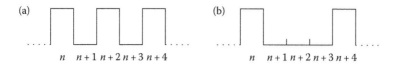

FIGURE 4A.1 Possible DPIM(1GS) slot sequences resulting in a product of 1 when $k = 4$. (a) $L = 2$ and (b) $L > 2$.

Similarly, when $k = 3$, the probability of getting a product of 1 is simply the probability of a 1 multiplied by the probability that that 1 represents the start of a symbol containing a guard slot and one empty slot. Thus

$$R_3 = 1 \cdot \overline{L}_{\text{DPIM}}^{-1} \cdot L^{-1} \tag{4.A.6}$$

When $k = 4$ and $L = 2$, the only way to obtain a product of 1 is to have a 1 in slot n which represents the start of a shortest-duration symbol, which is then followed by another shortest-duration symbol, as illustrated in Figure 4A.1a. For $L > 2$, along with the sequence just described for $L = 2$, a product of 1 may also be obtained by having a 1 in slot n which represents the start of a symbol containing a guard slot and two empty slots, as illustrated in Figure 4A.1b.

Thus, for $k = 4$,

$$R_4 = \begin{cases} 1 \cdot \overline{L}_{\text{DPIM}}^{-1} \cdot L^{-1} \cdot L^{-1} = \overline{L}_{\text{DPIM}}^{-1} \cdot L^{-2} & \text{for } L = 2 \\ 1 \cdot \overline{L}_{\text{DPIM}}^{-1} \cdot L^{-1} \cdot L^{-1} + 1 \cdot \overline{L}_{\text{DPIM}}^{-1} \cdot L^{-1} = \overline{L}_{\text{DPIM}}^{-1} (L^{-2} + L^{-1}) & \text{for } L > 2 \end{cases} \tag{4.A.7}$$

By continuing this process to obtain further values of R_k for various values of L, it may be observed that, in the limit $2 \le k \le L + 1$:

$$R_k = R_{k-1} + L^{-1} R_{k-2} \tag{4.A.8}$$

This is a second-order linear recurrence relationship. In order to derive an expression for R_k in terms of L, first assume a basic solution of $R_k = Am^k$, where m is to be determined. Hence

$$Am^k = Am^{k-1} + L^{-1} Am^{k-2} \tag{4.A.9}$$

Assuming $A \ne 0$ and $m \ne 0$, this may be simplified to

$$m^2 = m + L^{-1} \tag{4.A.10}$$

$$m^2 - m - L^{-1} = 0 \tag{4.A.11}$$

Therefore

$$m = \frac{1 \pm \sqrt{1 + 4L^{-1}}}{2}$$

(4.A.12)

The general solution is therefore given by

$$R_k = Am_1^k + Bm_2^k$$

(4.A.13)

where

$$m_1 = \frac{1 + \sqrt{1 + 4L^{-1}}}{2} \quad \text{and} \quad m_2 = \frac{1 - \sqrt{1 + 4L^{-1}}}{2}$$

(4.A.14)

Now, $R_0 = \bar{L}_{\text{DPIM}}^{-1}$ and $R_1 = 0$. Substituting these into (4.A.13) results in

$$\bar{L}_{\text{DPIM}}^{-1} = A + B$$
$$0 = Am_1 + Bm_2$$

(4.A.15)

Solving for A and B leads to

$$A = \frac{-\bar{L}_{\text{DPIM}}^{-1} m_2}{\sqrt{1 + 4L^{-1}}} \quad \text{and} \quad B = \frac{\bar{L}_{\text{DPIM}}^{-1} m_1}{\sqrt{1 + 4L^{-1}}}$$

(4.A.16)

Therefore,

$$R_k = \frac{\bar{L}_{\text{DPIM}}^{-1}}{\sqrt{1 + 4L^{-1}}} \left(-m_2 m_1^k + m_1 m_2^k \right)$$

(4.A.17)

$$R_k = \frac{\bar{L}_{\text{DPIM}}^{-1} (-m_1 m_2)}{\sqrt{1 + 4L^{-1}}} \left(m_1^{k-1} - m_2^{k-1} \right)$$

(4.A.18)

Since $m_1 m_2 = -L^{-1}$, (4.A.18) is given as

$$R_k = \frac{\bar{L}_{\text{DPIM}}^{-1} L^{-1}}{\sqrt{1 + 4L^{-1}}} \left(m_1^{k-1} - m_2^{k-1} \right)$$

(4.A.19)

Thus, substituting for m_1 and m_2, which are defined in (4.A.14), the full expression for R_k is

$$R_k = \left(\frac{\overline{L}_{\text{DPIM}}^{-1} L^{-1}}{\sqrt{1 + 4L^{-1}}} \right) \left[\left(\frac{1 + \sqrt{1 + 4L^{-1}}}{2} \right)^{k-1} - \left(\frac{1 - \sqrt{1 + 4L^{-1}}}{2} \right)^{k-1} \right] \quad \text{for } 2 \le k \le L+1$$

(4.A.20)

By calculating further values of R_k for $k > L + 1$, for various values of L, it may be observed that R_k is given by the following summation:

$$R_k = \frac{1}{L} \sum_{i=1}^{L} R_{k-1-i} \quad \text{for } k > L + 1$$ (4.A.21)

For reference, the ACF of DPIM(1GS) is plotted in Figure 4A.2 for various values of L. Note that all four plots were generated using $a_n \in \{0, 1\}$.

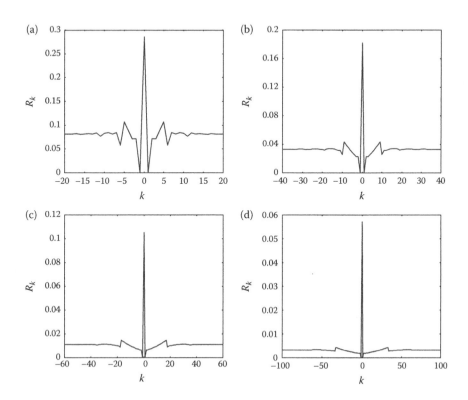

FIGURE 4A.2 ACF of DPIM(1GS) for (a) $L = 4$, (b) $L = 8$, (c) $L = 16$ and (d) $L = 32$.

APPENDIX 4.B

4.B.1   PSD OF DH-PIM

4.B.1.1   Fourier Transform of DH-PIM

The mathematical model of a DH-PIM pulse train has been presented in Section 4.2. The truncated transmitted signal $x_N(t)$ of N symbols is given by (4.79) as

$$x_N(t) = V \sum_{n=0}^{N-1} \left\{ \text{rect}\left[\frac{2(t - T_n)}{\alpha T_s} - \frac{1}{2} \right] + h_n \text{rect}\left[\frac{2(t - T_n)}{\alpha T_s} - \frac{3}{2} \right] \right\} \qquad (4.B.1)$$

where V is the pulse amplitude, $h_n \in \{0, 1\}$ indicating H_1 or H_2 respectively, n is the instantaneous-symbol number and T_n is the start time of the nth symbol defined in (4.76).

The Fourier transform of the truncated signal (4.B.1) can be written as

$$X_N(\omega) = V \sum_{n=0}^{N-1} \int_{-\infty}^{+\infty} \left\{ \text{rect}\left[\frac{2(t - T_n)}{\alpha T_s} - \frac{1}{2} \right] + h_n \text{rect}\left[\frac{2(t - T_n)}{\alpha T_s} - \frac{3}{2} \right] \right\} \cdot e^{-j\omega t} \, dt \qquad (4.B.2)$$

where ω is the angular frequency.

Therefore

$$X_N(\omega) = V \sum_{n=0}^{N-1} \left\{ \int_{T_n}^{T_n + \frac{\alpha T_s}{2}} e^{-j\omega t} \, dt + h_n \int_{T_n + \frac{\alpha T_s}{2}}^{T_n + \alpha T_s} e^{-j\omega t} \, dt \right\} \qquad (4.B.3)$$

Thus, the Fourier transform of the truncated DH-PIM signal can be written as

$$X_N(\omega) = \frac{V}{j\omega} e^{(-j\omega T_0)} (1 - e^{-j\omega T_s \alpha/2}) \sum_{n=0}^{N-1} \left[(1 + h_n e^{-j\omega T_s \alpha/2}) e^{-j\omega T_s n(\alpha+1)} e^{-j\omega T_s \sum_{k=0}^{n-1} d_k} \right] \qquad (4.B.4)$$

4.B.1.2   Power Spectral Density of DH-PIM

The power spectral density of the signal can be obtained by averaging over a large number of symbols N and then performing the limiting operation as given in [27]

$$P(\omega) = \lim_{N \to \infty} \frac{E[X_N(\omega) \cdot X_N^*(\omega)]}{E[T_N - T_0]} \qquad (4.B.5)$$

where $E[x]$ is the expected value of x and $X_N^*(\omega)$ is the complex conjugate of $X_N(\omega)$.
From (4.76), the expected value of $(T_N - T_0)$ is given by

$$E[T_N - T_0] = NT_s \left[1 + \alpha + \frac{2^{M-1} - 1}{2} \right], \qquad (4.B.6)$$

and

$$E[X_N(\omega) \cdot X_N^*(\omega)]$$

$$= \left(\frac{V}{\omega}\right)^2 \left(\frac{e^{(-j\omega T_0)}}{j} \cdot \frac{e^{(j\omega T_0)}}{-j}\right) \left[\left(1 - e^{-j\omega T_s \alpha/2}\right)\left(1 - e^{j\omega T_s \alpha/2}\right)\right] \cdot S_N(\omega),$$

$$E[X_N(\omega) \cdot X_N^*(\omega)] = 4V^2 \frac{\sin^2\left(\dfrac{\alpha\omega T_s}{4}\right)}{\omega^2} \cdot S_N(\omega) \qquad (4.B.7)$$

where

$$S_N(\omega) = E\left[\sum_{n=0}^{N-1}\sum_{q=0}^{N-1}\left((1 + h_n e^{-j\omega T_s \alpha/2})(1 + h_q e^{j\omega T_s \alpha/2})e^{-j\omega T_s (\alpha+1)(n-q)}e^{-j\omega T_s\left[\sum_{k=0}^{n-1} d_k - \sum_{k=0}^{q-1} d_k\right]}\right)\right].$$

$$(4.B.8)$$

Expression (4.B.8) is best evaluated by splitting it into three regions:

1. $S_{N1}(\omega)$, where $q < n$
2. $S_{N2}(\omega)$, where $q = n$
3. $S_{N3}(\omega)$, where $q > n$

and then summing them up as

$$S_N(\omega) = S_{N1}(\omega) + S_{N2}(\omega) + S_{N3}(\omega) \qquad (4.B.9)$$

Here, we set out to find the expressions for $S_{N1}(\omega)$, $S_{N2}(\omega)$ and $S_{N3}(\omega)$ as outlined in (4B.9).

1. Taking $q < n$ in (4.B.8) gives

$$S_{N1}(\omega) = \sum_{n=1}^{N-1}\sum_{q=0}^{n-1} E\left[(1 + h_n e^{-j\omega T_s/2})(1 + h_q e^{j\omega T_s/2})e^{-j\omega T_s(\alpha+1)(n-q)}e^{-j\omega T_s\sum_{k=q}^{n-1} d_k}\right]$$

$$(4.B.10)$$

$$S_{N1}(\omega) = \sum_{n=1}^{N-1}\sum_{q=0}^{n-1} E\left[\left(1 + h_n h_q + h_n e^{-j\omega T_s/2} + h_q e^{j\omega T_s/2}\right) \cdot e^{-j\omega T_s(\alpha+1)(n-q)}e^{-j\omega T_s\sum_{k=q}^{n-1} d_k}\right]$$

$$S_{N1}(\omega) =$$

$$\sum_{n=1}^{N-1}\sum_{q=0}^{n-1}\left\{ E\left(1 + h_n h_q + h_n e^{-j\alpha\omega T_s/2} + h_q e^{j\alpha\omega T_s/2}\right) E\left[e^{-j\omega T_s \sum_{k=q}^{n-1} d_k} \right] e^{-j\omega T_s(\alpha+1)(n-q)} \right\} \quad (4.B.11)$$

$h_n \in \{0, 1\}$ and $h_q \in \{0, 1\}$, so the expected values of h_n and h_q are

$$E[h_n] = E[h_q] = \frac{1}{2} \qquad (4.B.12)$$

Furthermore, $h_n h_q \in \{0, 0, 0, 1\}$, so the expected value of $h_n h_q$ is

$$E[h_n h_q] = \frac{1}{4} \qquad (4.B.13)$$

Consequently, the first factor reduces to

$$E\left[1 + h_n h_q + h_n e^{-j\alpha\omega T_s/2} + h_q e^{j\alpha\omega T_s/2}\right] = 1 + \frac{1}{4} + \frac{e^{j\alpha\omega T_s/2} + e^{-j\alpha\omega T_s/2}}{2}$$

$$E\left[1 + h_n h_q + h_n e^{-j\alpha\omega T_s/2} + h_q e^{j\alpha\omega T_s/2}\right] = \frac{5}{4} + \cos\left(\frac{\alpha\omega T_s}{2}\right) = \frac{9}{4} - 2\sin^2\left(\frac{\alpha\omega T_s}{4}\right)$$

$$(4.B.14)$$

and the second factor to

$$E\left(e^{-j\omega T_s \sum_{k=q}^{n-1} d_k} \right) = E\left[\prod_{k=q}^{n-1}\left(e^{-j\omega T_s d_k}\right) \right] = \prod_{k=q}^{n-1}\left(\frac{1}{2^{M-1}}\sum_{d=0}^{2^{M-1}-1} e^{-j\omega T_s d} \right)$$

$$E\left(e^{-j\omega T_s \sum_{k=q}^{n-1} d_k} \right) = \prod_{k=q}^{n-1}\left[\frac{1}{2^{M-1}}\left(\frac{1 - e^{-j\omega T_s 2^{M-1}}}{1 - e^{-j\omega T_s}} \right) \right]$$

$$E\left(e^{-j\omega T_s \sum_{k=q}^{n-1} d_k} \right) = \left[\frac{1}{2^{M-1}}\left(\frac{1 - e^{-j\omega T_s 2^{M-1}}}{1 - e^{-j\omega T_s}} \right) \right]^{n-q}. \qquad (4.B.15)$$

Substituting (4.B.14) and (4.B.15) into (4.B.11) gives

$$S_{N1}(\omega) = \sum_{n=1}^{N-1}\sum_{q=0}^{n-1}\left\{\left[\frac{9}{4} - 2\sin^2\left(\frac{\alpha\omega T_s}{4}\right)\right]\cdot\left(\frac{1}{2^{M-1}}\cdot\frac{1-e^{-j\omega T_s 2^{M-1}}}{1-e^{-j\omega T_s}}\right)^{n-q}e^{-j\omega T_s(n-q)(\alpha+1)}\right\}$$

$$S_{N1}(\omega) = \left[\frac{9}{4} - 2\sin^2\left(\frac{\alpha\omega T_s}{4}\right)\right]\sum_{n=1}^{N-1}\sum_{q=0}^{n-1}\left(\frac{1-e^{-j\omega T_s 2^{M-1}}}{1-e^{-j\omega T_s}}\cdot\frac{e^{-j\omega T_s(\alpha+1)}}{2^{M-1}}\right)^{n-q} \quad (4.B.16)$$

Now, letting

$$G = \left(\frac{1-e^{-j\omega T_s 2^{M-1}}}{1-e^{-j\omega T_s}}\cdot\frac{e^{-j\omega T_s(\alpha+1)}}{2^{M-1}}\right) \quad (4.B.17)$$

and using the results

$$\sum_{n=1}^{N-1}\sum_{q=0}^{n-1}G^{n-q} = \frac{G}{(1-G)^2}\left[N(1-G)-(1-G^N)\right] \quad (4.B.18)$$

in (4.16) gives

$$S_{N1}(\omega) = \left[\frac{9}{4} - 2\sin^2\left(\frac{\alpha\omega T_s}{4}\right)\right]\frac{G}{(1-G)^2}\left[N(1-G)-(1-G^N)\right] \quad (4.B.19)$$

2. Taking $q = n$ in (4.B.8) gives

$$S_{N2}(\omega) = \sum_{n=0}^{N-1}E\left[\left(1+h_n e^{-j\omega T_s\alpha/2}\right)\left(1+h_n e^{j\omega T_s\alpha/2}\right)\right]$$

$$S_{N2}(\omega) = \sum_{n=0}^{N-1}E\left[\left(1+h_n^2 + h_n e^{-j\omega T_s\alpha/2} + h_n e^{j\omega T_s\alpha/2}\right)\right]$$

$h_n^2 \in \{0,1\}$, thus

$$E[h_n^2] = \frac{1}{2} \quad (4.B.20)$$

Hence, $S_{N2}(\omega)$ can be written as

$$S_{N2}(\omega) = \sum_{n=0}^{N-1}\left[\frac{3}{2} + \cos\left(\frac{\alpha\omega T_s}{2}\right)\right]$$

thus

$$S_{N2}(\omega) = \frac{N}{2}\left[5 - 4\sin^2\left(\frac{\alpha\omega T_s}{4}\right)\right] \tag{4.B.21}$$

3. Taking $q > n$ in (4.A.29) gives

$$S_{N3}(\omega) = \sum_{n=0}^{N-2}\sum_{q=n+1}^{N-1} E\left[(1 + h_n e^{-j\omega T_s\alpha/2})(1 + h_q e^{j\omega T_s\alpha/2})e^{-j\omega T_s(\alpha+1)(n-q)}e^{j\omega T_s\sum_{k=n}^{q-1} d_k}\right]$$

which can be rewritten as

$$S_{N3}(\omega) = \sum_{q=1}^{N-1}\sum_{n=0}^{q-1} E\left[(1 + h_n e^{-j\omega T_s\alpha/2})(1 + h_q e^{j\omega T_s\alpha/2})e^{j\omega T_s(\alpha+1)(q-n)}e^{j\omega T_s\sum_{k=n}^{q-1} d_k}\right]$$

Interchanging the letters q and n gives

$$S_{N3}(\omega) = \sum_{n=1}^{N-1}\sum_{q=0}^{n-1} E\left[(1 + h_q e^{-j\omega T_s\alpha/2})(1 + h_n e^{j\omega T_s\alpha/2})e^{j\omega T_s(\alpha+1)(n-q)}e^{j\omega T_s\sum_{k=q}^{n-1} d_k}\right] \tag{4.B.22}$$

Therefore,

$$S_{N3}(\omega) = S_{N1}^*(\omega) \tag{4.B.23}$$

and

$$S_{N1}(\omega) + S_{N3}(\omega) = 2\,\mathrm{Re}\left[S_{N1}(\omega)\right] \tag{4.B.24}$$

Substituting (4.B.13), (4.B.15) and (4.B.24) into (4.B.9) will result in

$$S_N(\omega) = \frac{N}{2}\left[5 - 4\sin^2\left(\frac{\alpha\omega T_s}{4}\right)\right] + \left[\frac{9}{2} - 4\sin^2\left(\frac{\alpha\omega T_s}{4}\right)\right]$$

$$\times \mathrm{Re}\left\{\frac{G}{(1-G)^2}\left[N(1-G) - (1-G^N)\right]\right\} \tag{4.B.25}$$

with G given in (4.B.17).

The power spectral density of the truncated signal can be obtained by substituting (4.B.6) and (4.B.7) into (4.B.5) to produce

$$P(\omega) = \frac{4V^2 \sin^2\left(\dfrac{\alpha\omega T_s}{4}\right)}{\omega^2 T_s \left[1 + \alpha + \dfrac{2^{M-1}-1}{2}\right]} \cdot \lim_{N\to\infty}\left\{\frac{S_N(\omega)}{N}\right\} \qquad (4.B.26)$$

To simplify (4.B.25), the possible values of $S_N(\omega)$ must be investigated as given in (4.B.26), which largely depend on G. G can be rewritten as

$$G = \frac{1}{2^{M-1}}\left\{1 + e^{-j\omega T_s} + e^{-j2\omega T_s} + \cdots + e^{-j\omega(2^{M-1}-1)T_s}\right\} \cdot e^{-j\omega(\alpha+1)T_s}$$

Therefore, the absolute value of this is given as

$$|G| = \frac{1}{2^{M-1}}\left|1 + e^{-j\omega T_s} + e^{-j2\omega T_s} + \cdots + e^{-j\omega(2^{M-1}-1)T_s}\right| \qquad (4.B.27)$$

hence

$$|G| \le \frac{1}{2^{M-1}}\left\{1 + \left|e^{-j\omega T_s}\right| + \left|e^{-j2\omega T_s}\right| + \cdots + \left|e^{-j\omega(2^{M-1}-1)T_s}\right|\right\} \qquad (4.B.28)$$

thus

$$|G| \le 1 \qquad (4.B.29)$$

Therefore, only two cases needed to be investigated as outlined below:

Case 1: where $|G| < 1$
From (4.B.27) and (4.B.28), $|G| < 1$ when $e^{-j\omega T_s} \neq 1$, that is, $\omega \neq (2\pi K/T_s)$ where K is a positive integer.
Here, $\lim\limits_{N\to\infty} G^N = 0$ and therefore, from (4.B.25)

$$\lim_{N\to\infty}\left[\frac{S_N(\omega)}{N}\right] = \frac{1}{2}\left\{\left[5 - 4\sin^2\left(\frac{\alpha\omega T_s}{4}\right)\right] + \left[9 - 8\sin^2\left(\frac{\alpha\omega T_s}{4}\right)\right]\mathrm{Re}\left[\frac{G}{1-G}\right]\right\}$$

$$(4.B.30)$$

substituting (4.B.30) into (4.B.26) results in

$$P(\omega) = \frac{4V^2 \sin^2\left(\dfrac{\alpha\omega T_s}{4}\right)\left\{\left[5 - 4\sin^2\left(\dfrac{\alpha\omega T_s}{4}\right)\right] + \left[9 - 8\sin^2\left(\dfrac{\alpha\omega T_s}{4}\right)\right]\mathrm{Re}\left(\dfrac{G}{1-G}\right)\right\}}{\omega^2 T_s\left(2^{M-1} + 2\alpha + 1\right)}$$

$$(4.B.31)$$

Expression (4.B.31) gives the PSD profile of the DH-PIM signal when it is finite (here, $|G| \neq 1$ and $\omega \neq 0$).

Case 2: where $G = 1$

From (4.B.26), $G = 1$ when $e^{-j\omega T_s} = 1$, that is, $\omega = (2\pi K/T_s)$ where K is a positive integer. Here, expression (4.B.24) is indeterminate, but applying L'Hôpital's rule with $G \to 1$ gives

$$\lim_{G \to 1} \left\{ \frac{G}{(1-G)^2} \cdot \left[N(1-G) - (1 - G^N) \right] \right\} = \frac{N(N-1)}{2}$$

thus from (4.B.25)

$$S_N(\omega) = \frac{N}{4} \left\{ \left[10 - 8\sin^2\left(\frac{\alpha\omega T_s}{4}\right) \right] + (N-1)\left[9 - 8\sin^2\left(\frac{\alpha\omega T_s}{4}\right) \right] \right\} \quad (4.B.32)$$

Substituting (4.B.32) into (4.B.26) results in

$$P(\omega) = \frac{2V^2 \sin^2\left(\dfrac{\alpha\omega T_s}{4}\right)}{\omega^2 T_s \left(2^{M-1} + 2\alpha + 1\right)}$$
$$\lim_{N \to \infty} \left\{ \left[10 - 8\sin^2\left(\frac{\alpha\omega T_s}{4}\right) \right] + (N-1)\left[9 - 8\sin^2\left(\frac{\alpha\omega T_s}{4}\right) \right] \right\} \quad (4.B.33)$$

Depending on the value of K, (4.B.33) will tend to 0 or ∞, as discussed below:

1. For $K = 0$, $\omega = 0$, and applying L'Hôpital's rule (since at $\omega = 0$, (4.B.33) is indeterminate), (4.B.33) will tend towards infinity with N as

$$P(0) = \lim_{N \to \infty} \left\{ \frac{2V^2\{9N+1\}}{T_s\left(2^{M-1} + 2\alpha + 1\right)} \lim_{\omega \to 0} \left[\frac{\sin^2\left(\dfrac{\alpha\omega T_s}{4}\right)}{\omega^2} \right] \right\} \quad (4.B.34)$$

$$P(0) = \frac{\alpha^2 V^2 T_s}{8\left(2^{M-1} + 2\alpha + 1\right)} \lim_{N \to \infty} \{9N + 1\} \to \infty \quad (4.B.35)$$

2. For $K = (2\upsilon/\alpha)$ where $\upsilon \neq 0$ is an integer, $\omega = 2\upsilon(2\pi/\alpha T_s)$ (i.e., even multiple of the slot frequency), (4.A.54) reduces to zero ($P(\omega) = 0$), thus contributing to all the nulls in the spectrum. The location of the nulls thus depends upon α.

3. For all other frequencies of the form $\omega = (2\pi K/T_s)$,

$$\sin^2\left(\frac{\alpha\omega T_s}{4}\right) = 1$$

therefore, (4.B.33) reduces to

$$P(\omega) = \frac{V^2 T_s}{2\pi^2 K^2 \left(2^{M-1} + 2\alpha + 1\right)} \lim_{N\to\infty}(N+1) \qquad (4.B.36)$$

therefore

$$P(\omega) \to \infty$$

Thus giving the potential distinct slot component and its harmonics at $\omega = 2\pi(K/T_s)$, which correspond to the case when α is an odd integer and K is an odd integer.

Therefore, the spectrum consists of a *sinc* envelope when $\omega T_s/2\pi$ is not an integer, distinct frequency components at the slot frequency and its harmonics when $\alpha\omega T_s/2\pi$ is odd integer and nulls when $\alpha\omega T_s/2\pi$ is even integer as discussed above. This confirms that the presence of the slot components and the locations of nulls are affected by the pulse shape. Depending on values of α, the slot component and its harmonics may coincide with the nulls of the *sinc* envelope, as discussed in the next section.

The above results are best summarized by

$$P(\omega) = \begin{cases} \dfrac{4V^2 \sin^2\left(\dfrac{\alpha\omega T_s}{4}\right)\left\{\left[5 - 4\sin^2\left(\dfrac{\alpha\omega T_s}{4}\right)\right] + \left[9 - 8\sin^2\left(\dfrac{\alpha\omega T_s}{4}\right)\right]\text{Re}\left(\dfrac{G}{1-G}\right)\right\}}{\omega^2 T_s \left(2^{M-1} + 2\alpha + 1\right)}; \\ \qquad\qquad \omega \neq \dfrac{2\pi K}{T_s} \\[2em] 0; \qquad\qquad\qquad\qquad\qquad \omega = \dfrac{2\pi K}{T_s} \text{ and either } K \text{ even or } \alpha \text{ even} \\[2em] \infty; \qquad\qquad\qquad\qquad\qquad \omega = \dfrac{2\pi K}{T_s} \text{ and both } K \text{ odd and } \alpha \text{ odd.} \end{cases}$$

$$(4.B.37)$$

4.B.1.3 Further Discussion on the PSD Expression

4.B.1.3.1 DC Component

The DC component of the power spectral density is given in Equation 4.B.35. From Equation 4.B.35 and assuming that N is a limited and very large number, the DC component is

$$P_{DC} \approx \frac{\alpha^2 V^2 T_s (9N + 1)}{8 (1 + 2\alpha + 2^{M-1})} \tag{4.B.38}$$

For $M = 2$ and $\alpha = 1$, the DC component of the power spectral density of 4-DH-PIM$_1$ is given from Equation 4.B.38 as

$$P_{DC,4\text{-DH-PIM}_1} = \frac{V^2 T_s (9N + 1)}{40} \tag{4.B.39}$$

Therefore, by dividing the expression in Equation 4.B.38 by that in Equation 4.B.39, the DC component of DH-PIM normalized to that of 4-DH-PIM$_1$ can be given by

$$P_{DC\text{-nor}} = \frac{5\alpha^2}{1 + 2\alpha + 2^{M-1}} \tag{4.B.40}$$

Therefore, the normalized DC component depends on the values of α and M.

4.B.1.3.2 Slot Component

The amplitude of the slot component tends to infinity when N tends to infinity, and assuming N is a limited and very large number, Equation 4.B.36 can be written as

$$P_{slot}(\omega) \approx \frac{V^2 T_s (N + 1)}{2\pi^2 K^2 (2^{M-1} + 2\alpha + 1)} \tag{4.B.41}$$

The peak amplitude of the fundamental slot component of 4-DH-PIM$_1$ is given from Equation 4.B.41 as

$$P_{slot,4\text{-DH-PIM}_1}(\omega) = \frac{V^2 T_s (N + 1)}{10\pi^2 K^2} \tag{4.B.42}$$

Therefore, by dividing Equation 4.B.41 on Equation 4.B.42, the peak amplitude of the fundamental slot component of DH-PIM normalized to that of 4-DH-PIM$_1$ can be given by

$$P_{slot\text{-nor}} = \frac{5}{1 + 2\alpha + 2^{M-1}} \tag{4.B.43}$$

Therefore, the amplitude of the normalized fundamental slot component is a function of M and α as shown in Section 4.6.

4.B.1.3.3 Slot Recovery

The presence of the slot component in the spectrum suggests that the slot synchronization can be achieved using a phase-locked loop (PLL) circuit which can be employed at the receiver to extract the slot frequency directly from the incoming DH-PIM data stream similar to that in PPM and DPIM. For all even values of α, the slot components are masked by the nulls and therefore, at the receiver end a simple PLL circuit is not capable of extracting the slot frequency. However, the slot frequency can be extracted by employing a nonlinear device followed by a PLL circuit.

REFERENCES

1. J. M. Kahn and J. R. Barry, Wireless infrared communications, *Proceedings of IEEE*, 85, 265–298, 1997.
2. D. A. Rockwell and G. S. Mecherle, Optical wireless: Low-cost, broadband, optical access [Online]. Available: www.freespaceoptic.com/WhitePapers/optical_wireless.pdf.
3. J. R. Barry, *Wireless Infrared Communications*, Boston: Kluwer Academic Publishers, 1994.
4. T. Lueftner, C. Kroepl, M. Huemer, J. Hausner, R. Hagelauer and R. Weigel, Edge-position modulation for high-speed wireless infrared communications, *IEE Proceedings Optoelectronics*, 150, 427–437, 2003.
5. N. Hayasaka and T. Ito, Channel modeling of nondirected wireless infrared indoor diffuse link, *Electronics and Communications in Japan*, 90, 9–19, 2007.
6. R. J. Green, H. Joshi, M. D. Higgins and M. S. Leeson, Recent developments in indoor optical wireless systems, *IET Communications*, 2, 3–10, 2008.
7. W. Hirt, M. Hassner and N. Heise, IrDA-VFIr(16 Mbits/s): Modulation code and system design, *IEEE Personal Communications*, 8, 58–71, 2001.
8. I. Millar, M. Beale, B. Donoghue, K. W. Lindstrom and S. Williams, The IrDA standards for high-speed infrared communications, *Hewlett-Packard Journal*, 2, 1–8, 1998.
9. U. Gliese, T. N. Nielsen, S. Norskov and K. E. Stubkjaer, Multifunctional fiber-optic microwave links based on remote heterodyne detection, *IEEE Transactions on Microwave Theory and Techniques*, 46, 458–468, 1998.
10. J. J. O'Reilly, P. M. Lane and M. H. Capstick, Optical generation and delivery of modulated mm-waves for mobile communications, *Analogue Optical Fibre Communications*, B. Wilson, Z. Ghassemlooy and I. Darwazeh, Eds., 1st ed. London: The Institute of Electrical Engineers, 1995.
11. K. Sato and K. Asatani, Speckle noise reduction in fiber optic analog video transmission using semiconductor laser diodes, *IEEE Transactions on Communications*, 29, 1017–1024, 1981.
12. D. M. Pozar, *Microwave and RF Wireless Systems*, New Jersey: John Wiley, 2001.
13. L. P. De Jong and E. H. Nordholt, An optical led transmitter for baseband video with JFET non-linearity compensation, *Sensors and Actuators*, 5, 1–11, 1984.
14. J. P. Frankurt, Analogue transmission of TV channels on optical fibres with non-linearties correction by regulated feed forward, *Electrical and Electronic Engineering*, 12, 298–304, 1984.
15. K. K. Wong, T. O'Farrell and M. Kiatweerasakul, The performance of optical wireless OOK, 2-PPM and spread spectrum under the effects of multipath dispersion and artificial light interference, *International Journal of Communication Systems*, 13, 551–557, 2000.

16. B. Wilson and Z. Ghassemlooy, Optical PWM data link for high quality analogue and video signals, *Journal of Physics E: Scientific Instrument*, 18, 841–845, 1987.

17. N. M. Aldibbiat, Z. Ghassemlooy and R. McLaughlin, Dual header pulse interval modulation for dispersive indoor optical wireless communication systems, *IEE Proceedings—Circuits, Devices and Systems*, 149, 187–192, 2002.

18. D. Zwillinger, Differential PPM has a higher throughput than PPM for the band-limited and average-power-limited optical channel, *IEEE Transactions on Information Theory*, 34, 1269–1273, 1988.

19. D. Shiu and J. M. Kahn, Differential pulse position modulation for power-efficient optical communication, *IEEE Transactions on Communication*, 47, 1201–1210, 1999.

20. M. J. N. Sibley, Dicode pulse-position modulation: A novel coding scheme for optical-fibre communications, *IEE Proceedings—Optoelectronics*, 150, 125–131, 2003.

21. G. Lee and G. Schroeder, Optical pulse position modulation with multiple positions per pulsewidth, *IEEE Transactions on Communications*, 25, 360–364, 1977.

22. U. Sethakaset and T. A. Gulliver, Differential amplitude pulse-position modulation for indoor wireless optical communications, *EURASIP Journal on Applied Signal Processing*, 2005, 3–11, 2005.

23. T. O'Farrell and K. K. Wong, Complementary sequence inverse keying for indoor wireless infrared channels, *Electronics Letters*, 40, 257–259, 2004.

24. I. D. Association, *Serial Infrared Physical Layer Specification*, Version 1.4, 2001.

25. L. W. Couch, *Digital and Analog Communication Systems*, 6th ed. New Jersey: Prentice Hall, 2001.

26. A. J. C. Moreira, R. T. Valadas and A. M. d. O. Duarte, Performance of infrared transmission systems under ambient light interference, *IEE Proceedings—Optoelectronics*, 143, 339–346, 1996.

27. J. G. Proakis, *Digital Communications*, New York: McGraw-Hill, 2004.

28. M. D. Audeh and J. M. Kahn, Performance evaluation of L-pulse-position modulation on non-directed indoor infrared channels, *IEEE International Conference on Communications*, Louisiana, 1994, pp. 660–664.

29. T. Luftner, C. Kropl, R. Hagelauer, M. Huemer, R. Weigel and J. Hausner, Wireless infrared communications with edge position modulation for mobile devices, *Wireless Communications, IEEE*, 10, 15–21, 2003.

30. H. Sugiyama and K. Nosu, MPPM: A method for improving the band-utilization efficiency in optical PPM, *Journal of Lightwave Technology*, 7, 465–472, 1989.

31. H. Park and J. R. Barry, Performance of multiple pulse position modulation on multipath channels, *IEE Proceedings—Optoelectronics*, 143, 360–364, 1996.

32. P. Hyuncheol and J. R. Barry, Modulation analysis for wireless infrared communications, *C1995 IEEE International Conference on Communications*, Seattle, Washington, 1995, pp. 1182–1186.

33. H. Park and J. R. Barry, Trellis-coded multiple pulse position modulation for wireless infrared communications, *IEEE Transactions on Communications*, 54, 643–651, 2004.

34. R. McEliece, Practical codes for photon communication, *IEEE Transactions on Information Theory*, 27, 393–398, 1981.

35. D. A. Fares, Performance of convolutional codes with multipulse signaling in optical channels, *Microwave and Optical Technology Letters*, 3, 406–410, 1990.

36. S. S. Muhammad, E. Leitgeb and O. Koudelka, Multilevel modulation and channel codes for terrestrial FSO links, *2nd International Symposium on Wireless Communication Systems*, Siena, Italy, 2005, pp. 795–799.

37. J. Anguita, I. Djordjevic, M. Neifeld and B. Vasic, Shannon capacities and error-correction codes for optical atmospheric turbulent channels, *Journal of Optical Networking*, 4, 586–601, 2005.

38. U. Sethakaset and T. A. Gulliver, MAP detectors for differential pulse-position modulation over indoor optical wireless communications, *IEICE Transactions on Fundamentals of Electronics, Communications and Computer Sciences*, E89-A, 3148–3151, 2006.

39. U. Sethakaset and T. A. Gulliver, Performance of differential pulse-position modulation (DPPM) with concatenated coding over optical wireless communications, *IET Communications*, 2, 45–52, 2008.

40. N. M. Aldibbiat, Z. Ghassemlooy and R. McLaughlin, Error performance of dual header pulse interval modulation (DH-PIM) in optical wireless communications, *IEE Proceedings—Optoelectronics*, 148, 91–96, 2001.

41. Z. Ghassemlooy and S. Rajbhandari, Convolutional coded dual header pulse interval modulation for line of sight photonic wireless links, *IET—Optoelectronics*, 3, 142–148, 2009.

42. A. R. Hayes, Digital pulse interval modulation for indoor optical wireless communication systems, PhD thesis, Sheffield Hallam University, UK, 2002.

43. Institute of Electrical and Electronics Engineers, *IEEE Standard 802.3*, 2000 Edition, Institute of Electrical and Electronics Engineers, 16 October 2000.

44. Institute of Electrical and Electronics Engineers, *IEEE Standard 802.11–1997*, 8 November 1997.

45. M. Schwartz, *Information Transmission, Modulation and Noise*, 4th ed. New York: McGraw-Hill, 1990.

46. S. B. Alexander, *Optical Communication Receiver Design*, Washington: SPIE Optical Engineering Press, 1997.

47. G. James, *Modern Engineering Mathematics*, 2nd ed. Essex, UK: Addison-Wesley, 1996.

48. N. M. Aldibbiat, Z. Ghassemlooy and R. McLaughlin, Indoor optical wireless systems employing dual header pulse interval modulation (DH-PIM), *International Journal of Communication Systems*, 18, 285–305, 2005.

49. N. M. Aldibbiat, Optical wireless communication systems employing dual header pulse interval modulation (DH-PIM), PhD thesis, Sheffield Hallam University, UK, 2001.

50. Z. Ghassemlooy and N. M. Aldibbiat, Multilevel digital pulse interval modulation scheme for optical wireless communications, *ICTON2006*, Nottingham, UK, 2006, pp. 149–153.

51. R. You and J. M. Kahn, Average power reduction techniques for multiple-subcarrierintensity-modulated optical signals, *IEEE Transactions on Communications*, 49, 2164–2171, 2001.

52. R. Narasimhan, M. D. Audeh and J. M. Kahn, Effect of electronic-ballast fluorescent lighting on wireless infrared links, *IEE Proceedings—Optoelectronics*, 143, 347–354, 1996.

53. Z. Ghassemlooy, V. R. Wickramasinghe and L. Chao, Optical fibre transmission of a broadband subcarrier multiplexed signal using PTM techniques, *IEEE Transactions on Consumer Electronics*, 42, 229–238, 1996.

54. J. B. Carruthers and J. M. Kahn, Multiple-subcarrier modulation for nondirected wireless infrared communication, *IEEE Journal on Selected Areas in Communications*, 14, 538–546, 1996.

55. O. González, R. Pérez-Jiménez, S. Rodríguez, J. Rabadán and A. Ayala, OFDM over indoor wireless optical channel, *IEE Proceedings—Optoelectronics*, 152, 199–204, 2005.

56. H. Joshi, R. J. Green and M. S. Leeson, Multiple sub-carrier optical wireless systems, *10th Anniversary International Conference on Transparent Optical Networks*, Warsaw, Poland, 2008, pp. 184–188.

57. H. Elgala, R. Mesleh, H. Haas and B. Pricope, OFDM visible light wireless communication based on white LEDs, *IEEE 65th Vehicular Technology Conference*, Dublin, Ireland, 2007, pp. 2185–2189.

58. S. K. Hashemi, Z. Ghassemlooy, L. Chao and D. Benhaddou, Orthogonal frequency division multiplexing for indoor optical wireless communications using visible light LEDs, *6th International Symposium on Communication Systems, Networks and Digital Signal Processing*, Graz, Austria, 2008, pp. 174–178.

59. R. Sang-Burm, C. Jae-Hoon, B. Junyeong, L. HuiKyu and R. Heung-Gyoon, High power efficiency and low nonlinear distortion for wireless visible light communication, *4th IFIP International Conference on New Technologies, Mobility and Security (NTMS)*, Paris, France, 2011, pp. 1–5.

60. E. F. Schubert and J. K. Kim, Solid-state light sources getting smart, *Science*, 308, 1274–1278, 2005.

61. M. A. Khalighi, N. Aitamer, N. Schwartz and S. Bourennane, Turbulence mitigation by aperture averaging in wireless optical systems, *10th International Conference on Telecommunications (ConTEL)*, Zagreb, Croatia, 2009, pp. 59–66.

62. J. Y. Tsao, Solid-state lighting: Lamps, chips, and materials for tomorrow, *Circuits and Devices Magazine, IEEE*, 20, 28–37, 2004.

63. J. Godlewski and M. Obarowska, Organic light emitting devices, *Opto-Electronics Review*, 15, 179–183, 2007.

64. H. Willebrand and B. S. Ghuman, *Free Space Optics: Enabling Optical Connectivity in Today's Network*, Indianapolis: SAMS Publishing, 2002.

65. I. I. Kim, B. McArthur and E. Korevaar, Comparison of laser beam propagation at 785 nm and 1550 nm in fog and haze for optical wireless communications, *SPIE Proceeding: Optical Wireless Communications III*, 4214, 26–37, 2001.

66. W. O. Popoola and Z. Ghassemlooy, BPSK subcarrier intensity modulated free-space optical communications in atmospheric turbulence, *Journal of Lightwave Technology*, 27, 967–973, 2009.

67. K.-S. Hou and J. Wu, A differential coding method for the symmetrically differential polarization shift-keying system, *IEEE Transactions on Communications*, 50, 2042–2051, 2002.

68. W. O. Popoola and Z. Ghassemlooy, BPSK subcarrier intensity modulated free-space optical communications in atmospheric turbulence, *Journal of Lightwave Technology*, 27, 967–973, 2009.

69. S. Betti, G. D. Marchis and E. Iannone, Coherent systems: Structure and ideal performance, *Coherent Optical Communications Systems*, K. Chang, Ed., New York: John Wiley & Sons, Inc, 1995, pp. 242–313.

70. N. Chi, S. Yu, L. Xu and P. Jeppesen, Generation and transmission performance of 40 Gbit/s polarisation shift keying signal, *Electronics Letters*, 41, 547–549, 2005.

71. E. Hu, Y. Hsueh, K. Wong, M. Marhic, L. Kazovsky, K. Shimizu and N. Kikuchi, 4-level direct-detection polarization shift-keying (DD-PolSK) system with phase modulators, *Proceedings of Optical Fiber Communication Conference and Exposition (OFC)*, Atlanta, USA, 23–28 March 2003.

72. S. Betti, G. D. Marchis and E. Iannone, Polarization modulated direct detection optical transmission systems, *Journal of Lightwave Technology*, 10, 1985–1997, 1992.

73. S. Benedetto, R. Gaudino and P. Poggiolini, Direct detection of optical digital transmission based on polarization shift keying modulation, *IEEE Journal on Selected Areas in Communications*, 13, 531–542, 1995.

74. S. Benedetto and P. Poggiolini, Theory of polarization shift keying modulation, *IEEE Transactions on Communications*, 40, 708–721, 1992.

75. X. Zhao, Y. Yao, Y. Sun and C. Liu, Circle polarization shift keying with direct detection for free-space optical communication, *IEEE/OSA Journal of Optical Communications and Networking*, 1, 307–312, 2009.

76. M. M. Karbassian and H. Ghafouri-Shiraz, Transceiver architecture for incoherent optical CDMA network based on polarization modulation, *Journal of Lightwave Technology*, 26, 3820–3828, 2008.

77. S. Benedetto, A. Djupsjobacka, B. Lagerstrom, R. Paoletti, P. Poggiolini and G. Mijic, Multilevel polarization modulation using a specifically designed LiNbO3 device, *IEEE Photonics Technology Letters*, 6, 949–951, 1994.

78. E. Collett, The Stokes polarization parameters, *Polarized Light: Fundamentals and Applications*, New York: Marcel Dekker, Inc., 1993, pp. 33–66.

79. R. Calvani, R. Caponi, F. Delpiano and G. Marone, An experiment of optical heterodyne transmission with polarization modulation at 140 Mbit/s bitrate and 1550 nm wavelength *GLOBECOM '91*, 3, 1587–1591, 1991.

80. S. Betti, G. D. Marchis and E. Iannone, *Coherent Optical Communications Systems*, New York: John Wiley & Son, Inc., 1995.

81. C. Liu, Y. Sun, Y. Yao and X. Zhao, Analysis of direct detection and coherent detection in wireless optical communication with polarization shift keying, *Conference on Lasers & Electro Optics & The Pacific Rim Conference on Lasers and Electro-Optics*, Shanghai, China, 3 August 2009, pp. 1–2.

82. *Safety of Laser Products—Part 1: Equipment Classification and Requirements IEC*, IEC, 2007.

83. KDDI developed high-speed wireless data transmission technology for USB devices, http://en.gigazine.net/index.php?/news/comments/20090715_usb_wireless/, 2009.

5 System Performance Analysis
Indoor

This chapter explores the performance of different modulation schemes under the constraints of noise and interferences. In addition to the AWGN, periodic and deterministic form of noise due to the artificial light sources also exists in indoor optical wireless channels, and they all combine to degrade the link performance severely. Fluorescent lamps emit light strongly at a spectral band of 780–950 nm overlapping with cheap optical transceiver components. Electronic ballast-driven fluorescent lamps have electrical spectrum contents that range up to megahertz making such lamps to be of potentially serious impairment to IR links [1–3]. The diffuse indoor links suffer from the multipath-induced ISI, thus limiting the maximum achievable data rates, for example, 260 Mbps for a typical indoor OWC system [4]. ISI also results in an additional power penalty that increases exponentially with the data rate [1]. The performance of the OOK, PPM and DPIM in the presence of FLI and ISI is investigated in this chapter. To improve the link performance possible mitigation techniques using high-pass filtering, equalization, wavelet transform and the neural network are also outlined in this chapter.

5.1 EFFECT OF AMBIENT LIGHT SOURCES ON INDOOR OWC LINK PERFORMANCE

Infrared transceivers operating in typical indoor environments are subject to intense ambient light, emanating from both natural and artificial sources, thus causing serious performance degradation. The average power of this background radiation generates shot noise, which is accurately modelled as white, Gaussian and independent of the received signal [5]. In addition to this, artificial sources of ambient light also generate a periodic interference signal, which has the potential to significantly degrade link performance. Of all the artificial sources of ambient light, fluorescent lamps, with a high spectral profile at 780–950 nm range, driven by electronic ballasts are potentially the most degrading, since the resulting interference signal contains harmonics of the switching frequency which can extend into the megahertz range [2,3,6]. The fluorescent lamps driven by conventional ballasts are distorted harmonics and have spectral components at multiples of the 50 Hz extending up to 20 kHz [1–3]. The major differences between electronic ballast-driven fluorescent lamps offered by different manufacturers are the switching frequency used, which are typically in the range 20–40 kHz, and the relative strengths of the high-frequency and low-frequency components [7,8]. Unlike lamps directly driven by the power line, emissions from lamps driven by different electronic ballasts are generally not synchronized. Hence, for a given time-averaged fluorescent-induced photocurrent that corresponds to a given level of illumination, the waveform from one or more tubes driven by a single ballast will generally have the greatest possible amplitude excursion and slope, and will thus represent the worst case. The incandescent lamps produce an almost perfect sinusoidal interference at the harmonics of 100 Hz.

Depending on the modulation technique in use, the presence of fluorescent light interference (FLI) affects the link performance of the OWC link differently [6]. Pulse modulation schemes such as PPM, with a low spectral content at or near the DC region, can offer immunity to FLI. On the other hand, modulation schemes such as OOK with high spectral components near the DC region are more likely to be affected by the FLI. The most widely adopted technique used to mitigate the effect of ambient light interference is electrical high-pass filtering (HPF), which may be achieved in practice by tuning the AC coupling between successive amplifier stages [6]. However, electrical HPF introduces a form of ISI known as the baseline wander, that mainly affect modulation techniques with strong frequency components near the DC region [3,9]. The higher the cut-off frequency of the HPF, the greater the attenuation of the interference signal, but also the more severe is the baseline wander. Thus, there exists a trade-off between the extent of FLI rejection and the severity of baseline wander [10].

In a nut shell, when considering the operating environment, there are numerous factors that affect the performance of an indoor OWC system, these include the number, type and location of artificial light sources within a room; the location, orientation and directionality of the transmitter and receiver; natural ambient light, and if so, the size and location of windows. Due to the existence of such a large number of factors, it is convenient to evaluate the system performance by considering a typical indoor environment. Two cases of ambient light conditions are investigated, these are

Case 1: No interference: Natural (solar) ambient light, generating an average photocurrent I_B of 200 µA.

Case 2: FLI: Natural ambient light as in case 1, plus electronic ballast-driven fluorescent light, generating an average photocurrent of 2 µA, thus giving a total average background photocurrent of 202 µA.

The switching frequency of 37.5 kHz is chosen.

5.2 EFFECT OF FLI WITHOUT ELECTRICAL HIGH-PASS FILTERING

In this section, we consider a number of modulation schemes including OOK, PPM and DPIM at various data rates, and how they are distorted by the inclusion of FLI. No form of filtering or other compensating techniques will be employed to remove the interference. To compare different modulation schemes under different channel conditions, two key performance indicators are used as defined below:

1. *Normalized optical power requirement (NOPR)*: The NOPR of a system is calculated by normalizing the optical power required to achieve the desired

bit/slot error probability ξ in the interfering channel with that of OOK system at 1 Mbps in an ideal AWGN channel without interference,

$$\text{NOPR} = \frac{\text{Optical power required to acheive } \xi}{\text{Optical power required to achieve } \xi \text{ for OOK @ 1 Mbps in ideal channel}}$$

(5.1)

2. *Optical power penalty (OPP)*: The OPP of a system is calculated by normalizing the optical power required to achieve the error probability of ξ in the interfering channel with that of the ideal AWGN channel without interference (other system parameters like the modulation-type bit rate remain the same),

$$\text{OPP} = \frac{\text{Optical power required to acheive } \xi}{\text{Optical power required to achieve } \xi \text{ in an ideal AWGN channel}}$$

(5.2)

To achieve viable simulation run times, a number of assumptions are required, as described below:

1. With a switching frequency of 37.5 kHz, there are 750 cycles of the high-frequency component per cycle of the low-frequency component, and hence, the low-frequency component may be assumed to be an offset which is constant over the duration of one high-frequency component cycle. Rather than evaluating all of the 750 offset values, a single offset is used, which is equal to the RMS value of the low-frequency component of the interference signal, taken over one complete cycle, that is, 20 ms. Thus, two new interference signals are generated, one being a single high-frequency cycle plus the offset and the other being a single high-frequency cycle minus the offset. For each bit interval, the error probability is calculated for both signals and the mean value is then taken. This gives a significant reduction in computation time, since only one high-frequency cycle needs to be considered in order to evaluate the error probability.
2. In all simulations in this chapter the actual simulated low- and high-frequency interference is added to the wanted data signal. However, the duration of the simulated signal can be much shorter than one complete cycle of the low-frequency component. Each of the simulations therefore starts at some random point during the low-frequency cycle.
3. In this study, the error probability of 10^{-6} is taken as standard and used in all calculations of OPP and NOPR hereafter.

5.2.1 MATCHED FILTER RECEIVER

The block diagram of a typical OWC communication system employing a matched filter receiver is shown in Figure 5.1. The encoder and decoder must be incorporated at the transmitter and receiver for modulation schemes other than OOK. The transmitter filter has a unit-amplitude rectangular impulse response $p(t)$, with a duration of one bit, T_b. The output of the transmitter filter is scaled by the peak-detected

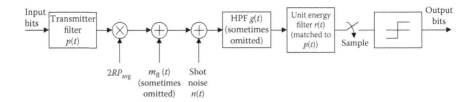

FIGURE 5.1   Block diagram of the OWC system under the influence of FLI.

signal photocurrent $2RP_{avg}$, where R is the photodetector responsivity and P_{avg} is the average received optical signal power. The fluorescent light-induced photocurrent, $m_{fl}(t)$, is then added to the signal, along with the signal-independent shot noise, $n(t)$, which is modelled as white and Gaussian, with a double-sided power spectral density (PSD) $N_0/2 = qI_B$. I_B is the average photocurrent generated by the background light, which is taken as 202 µA. In this section, the HPF is omitted and the detected signal is passed directly to a unit energy filter with an impulse response $r(t)$, which is matched to $p(t)$. The filter output is sampled at the end of each bit period, and a one or zero is assigned depending on whether the signal is above or below the threshold level at the sampling instant. The threshold level is set to its optimum value of $\alpha_{opt} = RP_{avg}\sqrt{T_b}$, which is midway between the expected one and zero levels.

The flowchart for the simulation of the OOK–NRZ is shown in Figure 5.2. Considering a linear system, the error probability in the presence of the FLI can be calculated by separately treating the FLI and the modulating signal at the matched filter input. For the OOK system, the sampled outputs at the matched filter in the absence of any impairments are $2RP_{avg}\sqrt{T_b}$ and 0 for binary '1' and '0', respectively. The output of the matched filter due to the FLI signal, sampled at the end of each bit period, is given as [3]

$$m_k = m_{fl}(t) \otimes r(t)\big|_{t=k\tau} \tag{5.3}$$

where τ is the sampling time which depends on the modulation schemes as described in Chapter 4, and the symbol $\otimes$ denotes convolution.

Since the interfering signal is periodic, the error probability can be estimated by calculating the bit (slot) error probability over the interference duration and averaging over the bit period [6]. By considering every slots over a 20 ms time interval (i.e., one complete cycle of $m_{fl}(t)$) and averaging, the P_{be_OOK} in the presence of the AWGN is given by [3,11]

$$P_{be\_OOK} = \frac{1}{2N_b}\sum_{k=1}^{N_b}Q\left(\frac{RP_{avg}\sqrt{T_b} + m_k}{\sqrt{N_0/2}}\right) + Q\left(\frac{RP_{avg}\sqrt{T_b} - m_k}{\sqrt{N_0/2}}\right) \tag{5.4}$$

where N_b is the total number of bits over a 20 ms interval. For a given packet length D_p, the probability of bit error can be converted into a corresponding packet error rate (PER) as

$$PER = 1 - \left(1 - P_{be\_OOK}\right)^{D_p} \tag{5.5}$$

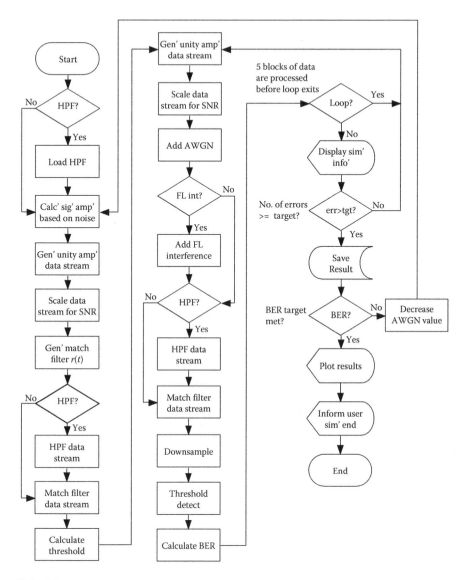

FIGURE 5.2 Flowchart for simulation of OOK shown in Figure 5.1.

The program to simulate the effect of FLI in OOK–NRZ is given in Program 5.1. The NOPR and OPP against the data rates for the OOK–NRZ are given in Figure 5.3. Also shown is the NOPR for OOK in the absence of interference. The NOPR in absence of the inference increases linearly with the logarithm of data rates (note the logarithmic scale in *X*-axis). However, NOPR in the presence of the FLI are almost constant irrespective of data rates (variation of <1 dB) indicating that the FLI is the dominant source of interference. Since the FLI is the dominating factor of the interference, the bit error probability is governed by the photocurrent due to the FLI and

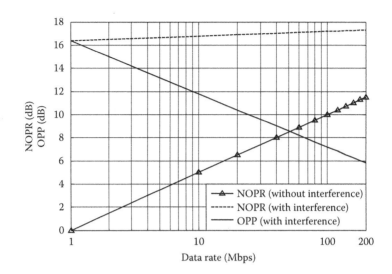

FIGURE 5.3 NOPR and OPP to achieve a BER of 10^{-6} against data rates for the OOK scheme with and without FLI.

hence results in very high OPPs. The OPP for data rates of 1, 10, 100 and 200 Mbps in the presence of FLI are 16.6, 11.6, 7.1 and 6 dB, respectively.

Program 5.1: MATLAB® Codes to Simulate Effect of FLI on OOK

```
q = 1.6e-19;
% Charge of Electron
Ib = 202e-6;
% Background Noise Current + interference
N0 = 2*q*Ib;
% Noise Spectral Density, 2*q*Ib
R = 1;
% Photodetector responsivity
Rb = 1e6;
% Bit rate
Tb = 1/Rb;
% bit duration
sig_length = ceil(20e-3/Tb);
% number of bits
nsamp = 10;
% samples per symbols
Tsamp = Tb/nsamp;
% sampling time

EbN0_db = 30;
% signal-to-noise ratio in dB.
BER = 1;
```

```
% initializing ber
index = 1;
maxerr = 30;
% maximum error per simulation

while (BER > 1E-4)
  terr = 0;
  % total error
  tsym = 0;
  % total bits

  SNR = 10.^(EbN0_db./10);
  % signal-to-noise ratio
  P_avg = sqrt(N0*Rb*SNR/(2*R^2));
  % average transmitted optical power
```

For PIM and DPIM schemes an encoder and decoder are included in Figure 5.1. Following a similar approach adopted for the OOK, the average probability of slot error for PPM with hard decision decoding (HDD) and soft decision decoding (SDD) and DPIM with HDD can be calculated and is given as [11]

$$
P_{se\_PPM\_hard} = \frac{1}{N_{sl}} \sum_{k=1}^{N_{sl}} \left[\frac{1}{L} Q\left(\frac{P_{avg}\sqrt{LT_b \log_2 L/2} + m_k}{\sqrt{N_0/2}} \right) \right.
$$
$$
\left. + \frac{(L-1)}{L} Q\left(\frac{P_{avg}\sqrt{LT_b \log_2 L/2} - m_k}{\sqrt{N_0/2}} \right) \right] \tag{5.6}
$$

$$
P_{se\_DPIM} = \frac{1}{N_{sl}} \sum_{k=1}^{N} \left[\frac{1}{\bar{L}_{DPIM}} Q\left(\frac{P_{avg}\sqrt{\bar{L}_{DPIM}T_b M/2} + m_k}{\sqrt{N_0/2}} \right) \right.
$$
$$
\left. + \frac{(\bar{L}_{DPIM} - 1)}{\bar{L}_{DPIM}} Q\left(\frac{P_{avg}\sqrt{\bar{L}_{DPIM}T_b M/2} - m_k}{\sqrt{N_0/2}} \right) \right] \tag{5.7}
$$

where N_{sl} is the total number of slots over a 20 ms interval and m_k is given by Equation 5.3 with sampling times taken as $T_{s\text{-PPM}}$ and $T_{s\text{-DPIM}}$ for PPM and DPIM, respectively.

For PPM with SDD, rather than considering each slot individually, each symbol consisting of L consecutive slots must be considered as one. Thus, for each symbol, a vector $[m_{iL+1}\ m_{iL+2} \ldots m_{iL+L}]$ is defined, which represents the matched filter outputs due to the interference signal. One is then assigned to each of the L-slot in turn, and the corresponding probability of symbol error is calculated using the union bound. From these L probabilities, the mean probability of symbol error is then calculated. This process is repeated for the next interference signal vector, and so on until all the symbols have been considered. The overall probability of symbol error is then found by averaging all symbols within 20 ms. Thus, the union (upper) bound for error probability for PPM with SDD is given by [3]

$$P_{se\_PPM\_soft} = \frac{1}{N_{sl}} \sum_{k=1}^{N_{sl}/L-1} \sum_{j=1}^{L} \sum_{\substack{k=1 \\ k \neq j}}^{L} Q\left(\frac{P_{avg}\sqrt{LT_b \log_2 L} + m_{iL+j} - m_{iL+k}}{\sqrt{N_0}} \right) \quad (5.8)$$

The NOPR to achieve an SER of 10^{-6} for 4, 8 and 16-PPM(HDD) is given in Figure 5.4. Unlike the ideal cases, the variations in the NOPR for the channel with FLI are small indicating that the FLI is the main source of performance impairment. The NOPR increases with the data rate for all cases; however, the increment is less than 1 dB for 16-PPM and 2 dB for 4-PPM. The OPP is the least for 16-PPM and also decreases with the data rates for all bit resolutions. The OPP at 200 Mbps is ~3 dB for 16-PPM and almost 0.5 dB higher for 8-PPM and a further ~0.5 dB penalty occurs for 4-PPM. For the PPM system with low PSD at or near DC, the OPP is significantly lower when compared to the OOK for all data rates. However, soft decision decoding offers a significant resistance to the FLI (see Figure 5.5). In fact, NOPR for the PPM with and without interference are almost identical for data rates >20 Mbps. The performance improvement at higher data rates is due to the reduced variation of the FLI signal over the duration of one symbol (note that soft decision is carried out based on the relative amplitude of slots within a symbol). In soft detection, it is the values of the FLI samples relative to other samples within the same symbol which are important, rather than the absolute values. This leads to a lower probability of symbol error, thus reducing the power penalty [1].

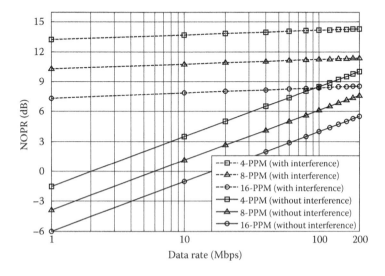

FIGURE 5.4 NOPR to achieve an SER of 10^{-6} against data rates for 4, 8 and 16-PPM with HDD and with/without FLI.

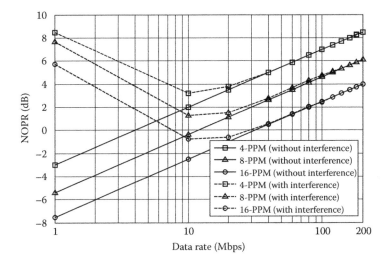

FIGURE 5.5 NOPR to achieve an SER of 10[-6] against data rates of 4, 8 and 16-PPM with SDD and with/without FLI.

As in the previous cases for the OOK and PPM schemes with HDD, a significant power penalty occurs due to FLI for the DPIM modulation scheme as well (see Figure 5.6) and there is little variation in the NOPR for data rates from 1 to 200 Mbps indicating the dominant noise source being the FLI. The NOPRs for 4-DPIM are ~15.3 dB and ~16.4 dB at 1 and 200 Mbps, respectively, which are

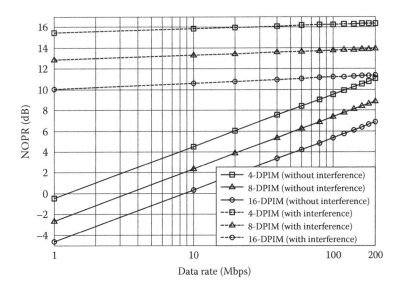

FIGURE 5.6 NOPR against the data rates for 4, 8 and 16-DPIM schemes with and without FLI.

1 dB higher and 2 dB lower for OOK and 4 PPM for respective data rates. As in the case of the PPM(HDD), OPPs are minimum for the 16-DPIM and increase with decreasing bit resolutions. DPIM has power penalties which are slightly higher than those of PPM(HDD) and they range from 14.6 to 16 dB at 1 Mbps to 4.5–5.4 dB at 200 Mbps.

The MATLAB code for determining the NOPR for the DPIM scheme is outlined in Program 5.2.

Program 5.2: MATLAB Codes to Plot NOPR and OPP for DPIM

```
%% optical power penalty for OOK
snr_1Mbps = 10.54;
data_rate = [1 10 20 40 60 80 100 120 140 160 180 200];
%data rates
nf = 5*(log10(data_rate));
% normalization factor
x = log10(data_rate);
% data rate in log scale
NOPR_ook_ideal = 5*(log10(data_rate));
% optical power penalty in ideal channel

%% ********** DPIM AWGN channel*************
NOPR_ideal_4dpim = NOPR_ook_ideal-5*log10(2*2.5/4);
NOPR_ideal_8dpim = NOPR_ook_ideal-5*log10(3*4.5/4);
NOPR_ideal_16dpim = NOPR_ook_ideal-5*log10(4*8.5/4);
semilogx(data_rate,NOPR_ideal_4dpim); hold on
semilogx(data_rate,NOPR_ideal_8dpim,'r');
semilogx(data_rate,NOPR_ideal_16dpim,'k');

%% *************** 4-DPIM in FLI channel ***************
snr_4dpim = [41.7 32.2 29.2 26.45 24.96 23.75 23 22.46 21.81
21.4 20.6 20.6];
% snr required to acheive a ber of 10^-6
snr_diff_4dpim = snr_4dpim-snr_1Mbps;
% difference in SNR compared to the LOS 1 Mbps
NOPR_4dpim = snr_diff_4dpim./2 + nf;
% optical power penalty
p = polyfit(x,NOPR_4dpim,1);
% curve fitting
f4 = polyval(p,x);

%% ***************** 8-DPIM in FLI channel ***************
snr_8dpim = [36.6 27 24.1 21.4 20 18.6 18.1 17.5 16.6 16.4 16.1
15.6];
snr_diff_8dpim = snr_8dpim-snr_1Mbps;
NOPR_8dpim = snr_diff_8dpim./2 + nf;
p = polyfit(x,NOPR_8dpim,1);
f8 = polyval(p,x);
```

```
%% ***************** 16-DPIM in FLI channel ***************
snr_16dpim=[31.1 21.5 18.6 16.1 14.7 13.5 13.1 12.4 11.7 11.2
   11 10.5];
snr_diff_16dpim=snr_16dpim-snr_1Mbps;
NOPR_16dpim=snr_diff_16dpim./2+nf;
p=polyfit(x,NOPR_16dpim,1);
f16=polyval(p,x);

%% plots
semilogx(data_rate,f4,'b--'); hold on
semilogx(data_rate,f8,'r--');
semilogx(data_rate,f16,'k--');
% xlabel('Data rate (Mbps)')
% ylabel('NOPR')%
%% optical power penalty
semilogx(data_rate,f4-NOPR_ideal_4dpim,'b');
semilogx(data_rate,f8-NOPR_ideal_8dpim,'r');
semilogx(data_rate,f16-NOPR_ideal_16dpim,'k');
```

5.3   EFFECT OF BASELINE WANDER WITHOUT FLI

A sequence of pulses which is passed through an HPF experiences a variation in the nominal zero level. This variation in the nominal zero level is a form of ISI known as the baseline wander, which has a detrimental effect on the performance of base-band modulation techniques where a significant power is located at or is close to the DC region. In this section, the OPP required to overcome the effect of baseline wander is investigated. For all the analyses involving high-pass filtering, the HPF is modelled as a first-order RC filter with a 3 dB cut-on frequency and an impulse response $g(t)$. Its response to a single rectangular pulse of amplitude A and duration τ may be expressed as

$$g_{\text{out}}(t) = \begin{cases} Ae^{-t/RC} & 0 \leq t \leq \tau \\ -A(e^{-\tau/RC} - 1)e^{-t/RC} & t > \tau \end{cases} \tag{5.9}$$

where the filter time constant $RC = 1/2\pi f_c$. Due to the principle of superposition of linear systems, if a sequence of such pulses is passed through an HPF, the output is equal to the summation of the individual responses of the pulses within the sequence. Thus, for a bit sequence $A_1 A_2 \ldots A_n$, where $A_{1\ldots n} \in \{0, 1\}$, the output of a first-order RC HPF at the end of the nth bit may be expressed as [12]

$$g_{\text{out}}(t)\big|_{t=n\tau} = \sum_{i=1}^{n} A_i(e^{-2\pi f_c \tau} - 1)(e^{-2\pi f_c \tau})^{i-1} \tag{5.10}$$

To illustrate the effect of baseline wander, consider the OOK–NRZ signal with a rectangular pulse shape as shown in Figure 5.7a. The HPF output $g_{\text{out}}(t)$ with a cut-off frequency of 0.05 times the bit rate and the corresponding baseline wander are

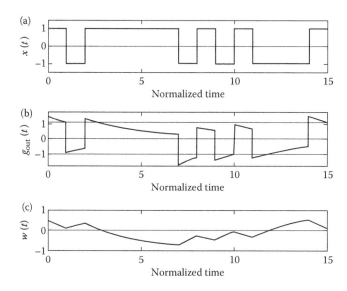

FIGURE 5.7   (a) Transmitted binary signal, (b) high-pass filter output and (c) baseline wander signal.

shown in Figure 5.7b and c, respectively. The MATLAB code used to generate Figure 5.6 is given in Program 5.3. Notice the variation in the amplitude for zero and one level in Figure 5.7b. The average value (i.e., the midpoint between the high and low levels) of the high-pass signal shown changes with time (see Figure 5.7c), and it is this variation that causes the baseline wander.

Program 5.3: MATLAB Codes to Slow the Effect of High-Pass Filter (HPF) and Baseline Wonder in OOK–NRZ signal

```
Rb = 1e6;
% Bit rate
Tb = 1/Rb;
% bit duration
sig_length = 1e2;
% number of bits
nsamp = 10;
% samples per symbols
Tsamp = Tb/nsamp;
% sampling time
Lsym = 1e3;
% number of bits
fc_rb = 5e-2;
% normalized cut-off frequency
A = 1;
% normalized amplitude
```

```
%% ***** Calculate impulse response of filters *****
tx_impulse = ones(1,nsamp)*A;
% transmitter filter
fc = fc_rb*Rb;
% actual cut-on frequency of HPF

t = Tsamp:Tsamp:10*Tb;
% time vector
hpf_impulse(1) = 1*exp(-2*pi*fc*t(1));
% impulse response (see eq (5.9))
for loop = 2:length(t)
   hpf_impulse(loop) = -1*(exp(2*pi*fc*t(1))-1)*exp
(-2*pi*fc*t(loop));
end

%% effect of HPF on OOK
OOK = randint(1,Lsym);
OOK = 2*OOK-1;
% removing dc components
signal = filter(tx_impulse,1, upsample(OOK,nsamp));
% rectangular pulse shaping
hpf_output = filter(hpf_impulse,1,signal);
% hpf output

%% plots
plotstart = (10*nsamp+1);
plotfinish = plotstart+15*nsamp;
t = 0:Tsamp:Tsamp*(plotfinish-plotstart);
subplot(311); plot(t,signal(plotstart:plotfinish),'k');
subplot(312); plot(t,hpf_output(plotstart:plotfinish),'k');
% subplot(413); plot(t,signal(plotstart:plotfinish)-hpf_output
(plotstart:plotfinish),'k')
subplot(313); plot(t,-signal(plotstart:plotfinish)+hpf_output
(plotstart:plotfinish),'k')
```

The effect of baseline wander on the performance of baseband modulation with no FLI can be analysed with reference to Figure 5.1. The histogram plot of the matched filter output with and without HPF for normalized cut-on frequencies of 10^{-3} and 10^{-2} are shown in Figure 5.8 (MATLAB code given in Program 5.4). The solid line in the centre of the plot indicates a value of zero, that is, no difference between expected and actual matched filter outputs, which would be the case in the absence of baseline wander. The increase in the variance of the probability distribution as f_c/R_b increases from 10^{-3} to 10^{-2} is evident from the figure. For relatively low values of f_c/R_b the HPF impulse response spans many bit periods. Accordingly, the ISI introduced by the HPF is comprised of the weighted sum of many independent and identically distributed (i.i.d) binary random variables. Therefore, as a result of the central limit theorem, the distribution can be approximated as Gaussian [3], as confirmed by the overall shape of the histograms. Street et al. [13] used this Gaussian approximation to develop closed-form expressions for the probability of error due to the baseline wander and the

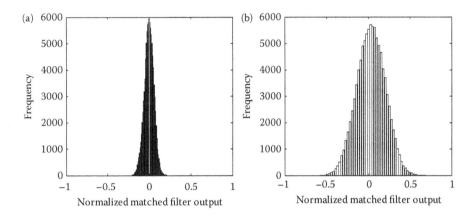

FIGURE 5.8 Histogram of matched filter output for OOK with: (a) $f_c/R_b = 10^{-3}$ and (b) $f_c/R_t = 10^{-2}$.

Gaussian noise for the OOK and Manchester encoding formats. Samaras et al. [12,14] extended this work, using the nonclassical Gauss quadrature rules to determine the probability of error, rather than assuming a Gaussian distribution.

Program 5.4: MATLAB Codes to Plot Distribution
OOK due to the BWL Effect

```
Rb = 1e6;
% Bit rate
Tb = 1/Rb;
% bit duration
nsamp = 10;
% samples per symbols
Tsamp = Tb/nsamp;
% sampling time
Lsym = 1e5;
% number of bits
fc_rb = 1e-3;
%normalized cut-off frequency
fc = fc_rb*Rb;
% cut-on frequency of HPF
p_ave = 1;
p_peak = 2*p_ave;
% peak power of TX'd signal for OOK

% ***** Calculate impulse response of filters *****
tx_impulse = ones (1,nsamp) *p_peak;
mf_impulse = ones (1,nsamp) * (1/sqrt (Tb));
t = Tsamp:Tsamp:200*Tb;
hpf_impulse (1) = 1*exp (-2*pi*fc*t (1));
```

```
for loop = 2:length(t)
   hpf_impulse(loop) = -1*(exp(2*pi*fc*t(1))-1)*exp
(-2*pi*fc*t(loop));
end

% ***** calculate overall impulse respose*****
temp1 = conv(tx_impulse,hpf_impulse);
temp2 = conv(temp1,mf_impulse);
temp2 = temp2*Tsamp;
system_impulse = temp2(nsamp:nsamp:200*nsamp);
%discrete impulse response

% ***** Do analysis on a per sequence basis *****
expected_one = 0.5*p_peak*sqrt(Tb);
expected_zero = -0.5*p_peak*sqrt(Tb);

OOK = randint(1,Lsym);
OOK = 2*OOK-1;
% removing dc components
mf_output = filter(system_impulse,1,OOK)/(2*expected_one);
mf_output_one = mf_output(find(OOK==1)); % output for
transmitted bit of 1;
mf_output_zero = mf_output(find(OOK==-1));

nbin = 51;
[n_zero,xout] = hist(mf_output_zero,nbin);
[n_one,xout] = hist(mf_output_one,nbin);
% combined histogram
% both expected outputs are shifted to zero
% note that removing dc value makes energy for zero and one
identical
expect_one = xout(find(n_one==max(n_one)));
n_total = n_zero+n_one;

Fig.; bar(xout,n_zero);
% histogram for zero bits
Fig.; bar(xout,n_one);
% histogram for zero bits
Fig.;
bar(xout-expect_one,n_total);
set(0,'defaultAxesFontName', 'timesnewroman','defaultAxesFont
Size',12)
xlabel('Normalized matched filter output');
ylabel('Frequency')
```

To analyse the effect of HPF, the discrete-time equivalent impulse response of the cascaded transmitter filter, receiver filter and HPF needs to be calculated. The resulting impulse response c_j decays rapidly to zero with time; hence it can be truncated without a significant loss of accuracy [3]. The BER is then approximated using the truncated length J and is calculated by averaging the error rate over all possible symbol sequences of length J. The discrete-time equivalent impulse response, truncated to have a duration of J-bit, is given as [3]

$$
c_j = \begin{cases} p(t) \otimes r(t) \otimes g(t)\big|_{t=j\tau}, & 1 < j < J \\ 0, & \text{otherwise} \end{cases}
$$
(5.11)

where $g(t)$ is the impulse response of HPF.

Considering K distinct bit sequences of length J, denoted as $\mathbf{a}_1, \mathbf{a}_2, \ldots, \mathbf{a}_K$, let $a_{i,J}$ represent the value of the Jth bit in sequence $\mathbf{a}_i$, where $a_{i,J} \in \{0,1\}$. When $\mathbf{a}_i$ is passed through the system, the matched filter output, sampled at the end of the Jth bit period, is given by

$$
A_{i,J} = 2RP_{\text{avg}} a_i \otimes c_j\big|_{j=J}
$$
(5.12)

The average bit error probability for the OOK system for the Jth bit is given by [3,11]

$$
P_{\text{be\_OOK}} = \frac{1}{K} \sum_{i=1}^{K} Q\left(\frac{|A_{i,J}|}{\sigma_n}\right)
$$
(5.13)

where σ_n is the standard deviation of the Gaussian, nonwhite, zero mean, shot noise samples at the matched filter output, given as [12,15]

$$
\sigma_n = \sqrt{\frac{N_0}{2} \frac{(1 - e^{-2\pi f_c \tau})}{2\pi f_c} \frac{1}{\tau}}
$$
(5.14)

where f_c is the cut-off frequency of HPF.

PPM with HDD may be evaluated in the same way as OOK using (5.11) with the sampling times replaced by $t = jT_b M/L$. Since the PPM slot sequence is not i.i.d. and the HPF cut-on frequencies are not necessarily small compared to the bit rate, the probability distribution cannot be assumed to be Gaussian. Considering K-distinct PPM slot sequences of length J, denoted as $\mathbf{b}_1, \mathbf{b}_2, \ldots, \mathbf{b}_K$, let $b_{i,J}$ represent the value of the Jth bit in sequence $\mathbf{b}_i$, where $b_{i,J} \in \{0,1\}$. When $\mathbf{b}_i$ is passed through the system, the matched filter output, sampled at the end of the Jth slot period, is given by

$$
B_{i,J} = LRP_{\text{avg}} \mathbf{b}_i \otimes c_j\big|_{j=J}
$$
(5.15)

The probability of slot error is then found by averaging overall K-sequence:

$$
P_{\text{se\_PPM\_hard}} = \frac{1}{K} \sum_{i=1}^{K} Q\left(\frac{|B_{i,J} - \alpha|}{\sigma_n}\right)
$$
(5.16)

where α is the threshold level, set midway between one and zero levels in the absence of any baseline wander, given as

$$\alpha = RP_{\text{avg}} \sqrt{LT_{\text{b}} \log_2 L} \left(\frac{1}{2} - \frac{1}{L} \right) \tag{5.17}$$

and σ_n is the standard deviation of the zero-mean nonwhite Gaussian noise, given as

$$\sigma_n = \sqrt{\frac{N_0}{2} \frac{\left(1 - e^{-2\pi f_c T_{\text{s-PPM}}}\right)}{2\pi f_c} \frac{1}{T_{\text{s-PPM}}}} \tag{5.18}$$

For PPM with SDD, the method is similar, but rather than just considering the Jth slot in each sequence, the next whole symbol after the Jth slot is considered. Therefore, slightly longer sequences must be generated. For a sequence k, let the next whole symbol after the Jth slot be denoted as $b_{i,p}\, b_{i,p+1} \cdots b_{i,p+L}$, where $p \geq J$. The corresponding outputs of the system are given by Equation 5.15 with the sampling times replaced by $j = p, p+1, \ldots, p+L$. Assuming that, for the final symbol of each sequence under consideration, the 'one' was transmitted in slot $(p+w)$, where $1 \leq w \leq L$, the probability of symbol error for PPM is given as

$$P_{\text{e\_symb\_PPM\_soft}} = \frac{1}{K} \sum_{i=1}^{K} \sum_{\substack{j=1 \\ j \neq p+w}}^{L} Q\left(\frac{\left| B_{i,p+w} - B_{i,p+j} \right|}{\sigma_n} \right) \tag{5.19}$$

Using similar approach applied to the OOK and PPM schemes, the probability of slot error for DPIM is found by averaging the overall K-sequence:

$$P_{\text{se\_DPIM}} = \frac{1}{K} \sum_{i=1}^{K} Q\left(\frac{\left| B_{i,J} - \alpha \right|}{\sigma_n} \right) \tag{5.20}$$

where α is the threshold level, set midway between expected one and zero levels in the absence of any baseline wander, given as

$$\alpha = RP_{\text{avg}} \sqrt{\overline{L}_{\text{DPIM}} T_{\text{b}} \log_2 L} \left(\frac{1}{2} - \frac{1}{L_{\text{DPIM}}} \right) \tag{5.21}$$

and σ_n is the standard deviation of the zero-mean nonwhite Gaussian noise, given by (5.18).

Figures 5.9 through 5.12 shows a plot of normalized average optical power requirement (NOPR) versus f_c/R_{b} for OOK, PPM(HDD), PPM(SDD) and DPIM, respectively. Figure 5.9 clearly demonstrates the susceptibility of OOK to the

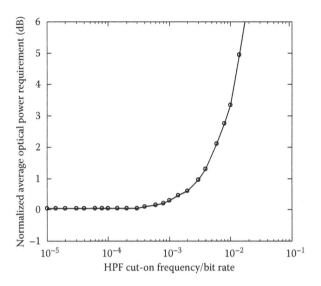

FIGURE 5.9   Normalized optical power requirement versus f_c/R_b for OOK.

baseline wander. Power penalties are incurred for normalized cut-on frequencies above ~10^{-3}, and a 3 dB average OPP is introduced when HPF f_c is ~1% of R_b. However, PPM is much more resistant to the effects of baseline wander compared with OOK. For both methods of detection, power penalties are not incurred until the HPF f_c reaches ~10% of R_b, which is several orders of magnitude higher than OOK. Other than the ~1.5 dB reduction in the OPP which the SDD offers over the threshold

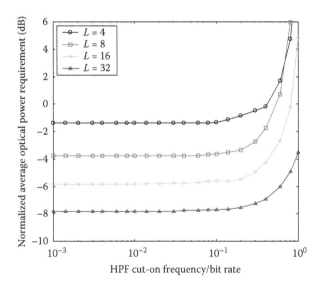

FIGURE 5.10   Normalized average optical power requirement versus f_c/R_b for PPM (HDD).

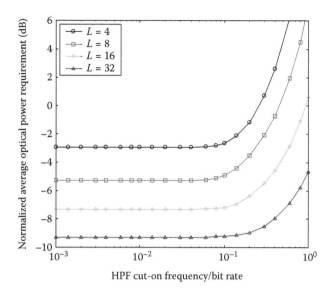

FIGURE 5.11 Normalized average optical power requirement versus f_c/R_b for PPM (SDD).

detection, there is little difference between the two sets of curves. As expected, higher orders are slightly more resistant to the baseline wander since the bandwidth requirement is greater and consequently, there is less power below f_c and this is a small fraction of the total power. DPIM offer more resistant to the baseline wander than OOK, with the higher orders achieving the greatest robustness. This is mainly

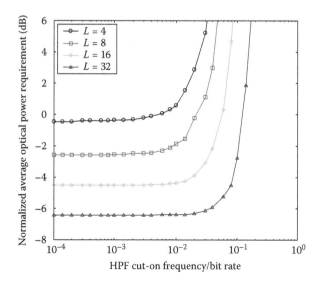

FIGURE 5.12 Normalized average optical power requirement versus f_c/R_b for DPIM.

due to the increase in the bandwidth requirement. For lower orders, power penalties start to become more apparent at the normalized cut-on frequencies of ~0.01, which is an order of magnitude higher than OOK. Compared with the performance of PPM, DPIM is more susceptible to the baseline wander. As discussed in previous chapter, the susceptibility of various modulation schemes to the baseline wander can be explained by observing the PSD profiles. Since OOK contains a large proportion of its power within the DC region, then it is the most susceptible to the baseline wander followed by DPIM and PPM. There is a progressive decrement in the power at low frequencies with increase in the order of DPIM and PPM, and hence higher order PPM and DPIM are less susceptible to the baseline wander compared to the low orders.

5.4 EFFECT OF FLI WITH ELECTRICAL HIGH-PASS FILTERING

As discussed earlier an electrical HPF can diminish the effects of FLI, whether this is implemented in the analogue or digital domain. Also, the choice of cut-on frequency is a trade-off between the extent of FLI rejection and the severity of the baseline introduced by the HPF and the attenuation of the interfering signal; the optimum choice being that which minimizes the overall power penalty. While f_c is an important parameter, other filter characteristics such as the stop band attenuation and the roll-off are also likely to have an effect on the performance that may produce variations in the optimum f_c. Given that the optimum f_c changes with data rate, one solution would be to simulate a system over all envisaged data rates and f_c to determine the correct cut-on frequency to use. This is obviously less than practical and time consuming; a compromise method being to choose a limited number of data rates and a range of f_c for a given filter to determine as close to the optimum as possible. In this section, the optimum f_c for HPF which minimizes the overall power penalty, is estimated using a method that combines the analysis carried out in the previous two sections. Using the optimum cut-on frequencies the optical power requirements are calculated, and the effectiveness of HPF as a means of mitigating the effect of FLI is assessed. To determine the optimum f_c, Figure 5.1 is adopted with the inclusion of both the FLI signal and HPF. The sampled output due to the FLI signal after passing through the matched filter and HPF is given by

$$m_j = m(t) \otimes r(t) \otimes g(t)\big|_{t=jT_b} \tag{5.22}$$

The sampled output due to Jth bit is given by Equation 5.12 and the overall output signal is superposition of the C_j and m_j. Hence, the overall probability of error is found by averaging over M-bit and J-sequence [3,11]:

$$P_{be\_OOK} = \frac{1}{2M \cdot 2^J} \sum_{k=1}^{M} \sum_{a_j \in \{0,1\}^J} \left[Q\left(\frac{A_{i,J} + m_k}{\sigma_n}\right) + Q\left(\frac{-A_{i,J} - m_k}{\sigma_n}\right) \right] \tag{5.23}$$

Similarly, the probability of slot error for PPM can be approximated as

$$P_{se\_PPM\_hard} = \frac{1}{M \cdot 2^J} \sum_{k=1}^{M} \sum_{b_j \in \{0,1\}^J} \left[\frac{1}{L} Q \left(\frac{B_{i,J} + m_k - \alpha}{\sigma_n} \right) \right.$$
$$\left. + \frac{(L-1)}{L} Q \left(\frac{\alpha - B_{i,J} - m_k}{\sigma_n} \right) \right] \qquad (5.24)$$

where α and σ_n are the threshold level and standard deviation of the noise given in Equations 5.17 and 5.18, respectively.

As in previous section, the probability of symbol error can be obtained by averaging:

$$P_{e\_symb\_PPM\_soft} = \frac{1}{N/L} \frac{1}{K} \sum_{n=0}^{(N/L)-1} \sum_{i=1}^{K} \sum_{\substack{j=1 \\ j \neq p+w}}^{L} Q \left(\frac{B_{i,p+w} + m_{nL+w} - B_{i,p+j} - m_{nL+w}}{\sigma_n} \right) \qquad (5.25)$$

Similarly, the slot error probability of DPIM in the presence of FLI with HPF is given by

$$P_{se\_DPIM} = \frac{1}{M 2^J} \sum_{k=1}^{M} \sum_{b_j \in \{0,1\}^J} \left[\frac{1}{\bar{L}_{DPIM}} Q \left(\frac{B_{i,J} + m_k - \alpha}{\sigma_n} \right) \right.$$
$$\left. + \frac{(\bar{L}_{DPIM} - 1)}{\bar{L}_{DPIM}} Q \left(\frac{\alpha - B_{i,J} - m_k}{\sigma_n} \right) \right] \qquad (5.26)$$

in order to determine the optimum f_c of HPF and to achieve an SER of 10^{-6} for a range of f_c/R_b the NOPR is calculated as shown in Figures 5.13 through 5.16 for OOK, PPM(HDD), PPM(SDD) and DPIM, respectively. It is evident that high-pass filtering gives virtually no reduction in the NOPR for bit rates of 1 and 10 Mbit/s. For the purpose of calculating power requirements, optimum normalized cut-on frequencies of 1.4×10^{-4} and 2×10^{-4} were identified for 1 and 10 Mbit/s, respectively. These cut-on frequencies represent the maximum values that can be used without introducing power penalties due to baseline wander. At a bit rate of 100 Mbit/s, by using a normalized f_c of $\sim 7 \times 10^{-3}$, it is clear that high-pass filtering can yield a reduction in the average optical power requirement.

However, the electrical HPF yields reductions in the average optical power requirements of PPM(HDD), even when operating at just 1 Mbit/s. The narrowness of the troughs in the 1 Mbit/s curves of Figure 5.14 suggest that little variation can be tolerated for f_c of HPF if the optimum performance is to be achieved. At 10 Mbit/s the troughs are broader, making the actual choice of f_c/R_b not quite so critical. At 100 Mbit/s there is a floor in the power requirement curves, indicating that there is a region in which f_c is high enough to attenuate the interference signal

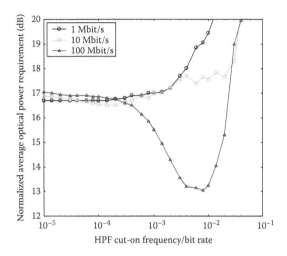

FIGURE 5.13 Normalized average optical power requirement versus normalized HPF cut-on frequency for OOK at various bit rates.

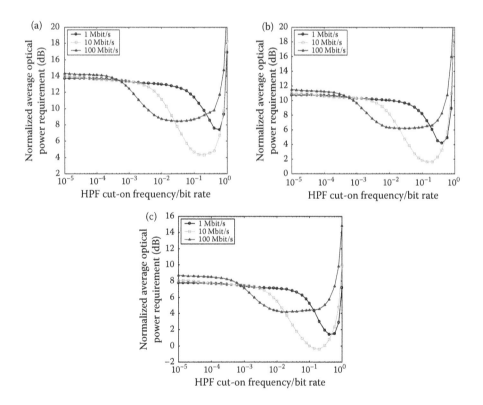

FIGURE 5.14 Normalized average optical power requirement versus normalized HPF cut-on frequency for PPM hard decision decoding at various bit rates with: (a) $L = 4$, (b) $L = 8$ and (c) $L = 16$.

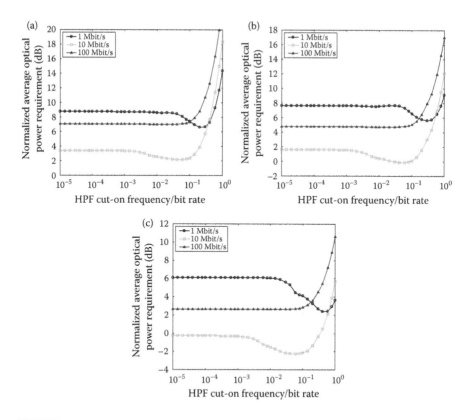

FIGURE 5.15 Normalized average optical power requirement versus normalized HPF cut-on frequency for PPM soft decision decoding at various bit rates with: (a) $L = 4$, (b) $L = 8$ and (c) $L = 16$.

without introducing baseline wander induced performance detritions. Within this region, the fluorescent light-induced power penalty is due to the additional shot noise only, and selecting the f_c of HPF anywhere within this region will give approximately the same level of performance. It is evident from Figure 5.15 that electrical HPF yields reductions in the average optical power requirements of PPM(SDD) operating at 1 Mbit/s. At 10 Mbit/s, from the analysis carried out in Section 5.2, PPM(SDD) is found to suffer only small power penalties without the use of electrical HPF. Consequently, the reduction in the average optical power requirements of Figure 5.15 appears modest, and the broad troughs suggest that the HPF f_c does not have to be very precise in order to minimize the average optical power requirements. At 100 Mbit/s, PPM(SDD) is immune to the FLI without the use of electrical HPF, and consequently there is no reduction in the average optical power requirement.

From Figure 5.16, it is evident that when operating at 1 Mbit/s in the presence of FLI, electrical HPF introduces no reduction in the average optical power requirements for DPIM. At 10 Mbit/s, high-pass filtering is more effective, thus resulting

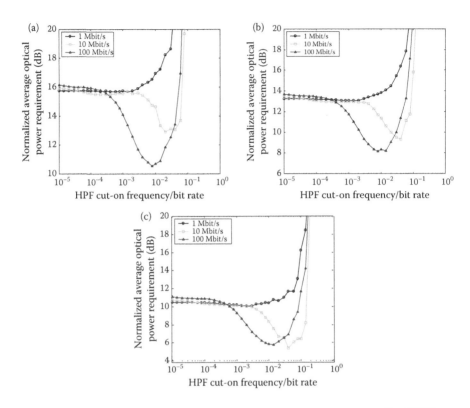

FIGURE 5.16 Normalized average optical power requirement versus normalized HPF cut-on frequency for DPIM at various bit rates with: (a) $L = 4$, (b) $L = 8$ and (c) $L = 16$.

in reduced average optical power requirements. Compared with PPM(HDD), DPIM has lower optimum HPF cut-on frequencies. This is due to DPIM higher suscepti-bility to the baseline wander. At 100 Mbit/s, electrical high-pass filtering once again results in reduced average optical power requirements. The broad troughs suggest that the HPF f_c does not have to be exact in order to achieve the near maxi-mum reduction.

Using these optimum normalized cut-on frequencies, the average optical power requirements for OOK, PPM(HDD), PPM(SDD) and DPIM at bit rates of 1, 10 and 100 Mbit/s are shown in Figures 5.17 through 5.20. Also shown are the power requirements with no FLI or HPF.

By comparing Figure 5.17 with Figure 5.3, it is clear that for bit rates of 1 and 10 Mbit/s, filtering is not effective in reducing the FLI-induced power penalty. This is due to the fact that OOK is very susceptible to the baseline wander and conse-quently, only low normalized cut-on frequencies are possible, which at low-to-medium bit rates are not effective in attenuating the interference signal. At a bit rate of 100 Mbit/s, high-pass filtering is more effective, giving a ~4.4 dB reduction in the average optical power requirement. However, this still leaves a power penalty of ~3 dB compared with the same bit rate with no FLI.

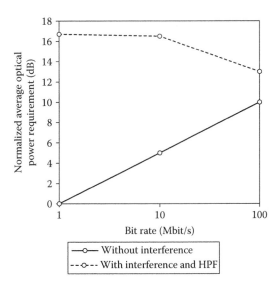

FIGURE 5.17   Normalized average optical power requirement versus bit rate for OOK with FLI and optimized high-pass filtering.

Unlike OOK, the electrical high-pass filtering results in reduced average optical power requirements for PPM(HDD), even when operating at 1 Mbit/s data rate. For 1 Mbit/s the narrowness of the troughs in Figure 5.14 suggest that little variation in f_c of HPF can be tolerated if the optimum performance is to be achieved. At 10 Mbit/s

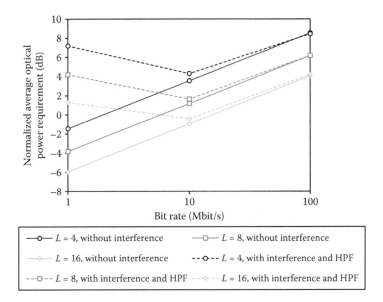

FIGURE 5.18   Normalized average optical power requirement versus bit rate for PPM (hard) with FLI and optimized high-pass filtering.

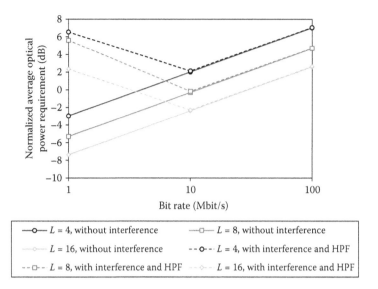

FIGURE 5.19 Normalized average optical power requirement versus bit rate for PPM (soft) with FLI and optimized high-pass filtering.

the troughs are broader, thus making the actual choice of f_c/R_b not quite so critical. At 100 Mbit/s there is a floor in the power requirement curves, indicating that there is a region in which f_c is high enough to attenuate the interference signal sufficiently without introducing the baseline wander affect. Within this region, FLI-induced power penalty is due to the additional shot noise only, and selecting the HPF f_c anywhere within this region will give approximately the same level of performance.

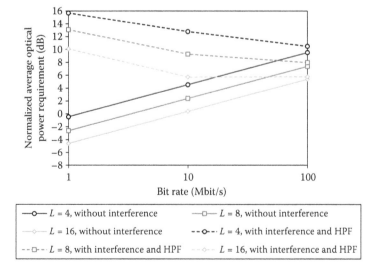

FIGURE 5.20 Normalized average optical power requirement versus bit rate for DPIM with FLI and optimized high-pass filtering.

In PPM(SDD) with no HPF there is a small power penalty at data rate >10 Mbits/s (Figure 5.5). Consequently, there is only a modest reduction in the average optical power requirement with high-pass filtering. At 100 Mbit/s, PPM(SDD) with no filtering shows immunity to FLI, and consequently no reduction in the average optical power requirement (see Figure 5.19). As previously observed for OOK and PPM(HDD) schemes, it is evident from Figure 5.6 that with FLI the average optical power requirements for DPIM are very similar for all three bit rates. Relative to the power requirements with no FLI power penalties decreases with the increase in bit rate. In DPIM power penalties are slightly higher than those of PPM(HDD), ranging from 14.2–16.2 dB to 5.3–7 dB at 1 and 100 Mbit/s, respectively.

5.5 WAVELET ANALYSIS

The topic of wavelets is multifaceted and highly mathematical, and a subject that is arguably dominated by researchers with a pure or applied mathematical background. Difficulties due to the mathematical complexities of wavelet analysis are alluded to by some tutorials [16] and by the authors of some of the many mathematically dominated texts published on the subject. Discussion on wavelet theory is by necessity limited to its application in this work. Evidently, the first recorded mention of what has become to be known as a 'wavelet' seems to have been in 1909 in a thesis by Alfred Haar. The concept of wavelets in its present theoretical form was first proposed by Jean Morlet and the team at the Marseille Theoretical Physics Center working under Alex Grossmann in France. Since then, a number of notable names have worked on the subject such as Yves Meyer with the main algorithm being credited to work undertaken by Stephane Mallat in 1988 [17]. Since that time, work on the subject has become widespread and covers disciplines too numerous to mention. In later years research became particularly active in the United States with fundamental work being undertaken by Ingrid Daubechies, Ronald Coifman and Victor Wickerhauser. A comprehensive history of the subject can be found in Ref. [18]. The feature-rich mapping of signals to wavelet coefficients has resulted in an explosion of papers and potential applications in many fields of science and engineering from stock market applications and human motor behaviour to optimization of the JPEG2000 standard [19]. However, it is possible to generalize these applications into three main areas, these being: data compression, denoising and feature extraction. In some cases it is a subtle combination of these elements that form a particular application. In this work it is arguably the combined abilities of denoising and feature identification that are employed. The identification and denoising properties of wavelets are also raising medical research interests in diagnostic applications as in Refs. [20–24]. The subject of communications engineering is well subscribed to by researchers with the subject of wavelets being applied in almost every facet of the field. In the field of communication engineering, wavelet packet transform (WPT) is proposed for multiple accesses technique [25,26], modulation techniques as an alternative to the subcarrier modulation [27,28], symbol synchronization [29], signal estimation [23], CDMA system [30], channel characterization [31,32], adaptive denoising [33], mitigate ISI [16], channel equalization [30,34–36] and traffic analysis [33].

5.5.1 THE CONTINUOUS WAVELET TRANSFORM

Although the Heisenberg uncertainty principle applies to any transform, multi-resolution analysis (MRA) minimizes its impact as not every spectral component is resolved equally as in the short-term Fourier transform (STFT). A continuous wavelet transform (CWT) decomposes signals over dilated and translated mother wavelet $\psi(t)$. A mother wavelet is well localized in time and frequency and have zero mean:

$$\int_{-\infty}^{\infty} \psi(t)\, dt = 0 \tag{5.27}$$

the family of mother wavelets is normalized using the norm $\|\psi_{s,T}(t)\| = 1$. The family of wavelets is obtained by scaling ψ by s and translating by τ as shown below:

$$\psi_{s,\tau}(t) = \frac{1}{\sqrt{s}} \psi\left(\frac{t - \tau}{s}\right) \tag{5.28}$$

To scale a wavelet means to stretch or dilate. As the wavelet is stretched in the horizontal axis, it is squashed in vertical direction to ensure that the energy contained within the scaled wavelets is the same as the original mother wavelet [37]. The translation moves the wavelet along the x-axis. The scaling and translation operations are demonstrated in Figure 5.21 for a Morlet wavelet [32].

The scale in the wavelet analysis is similar to the scale used in maps [16]. The high scales correspond to global information of a signal whereas low scales correspond to detailed information about the signal. The scale is inversely proportional to

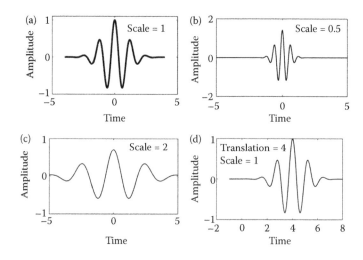

FIGURE 5.21 The scaling and translation to the Morlet wavelet. (a) Mother wavelet, (b) wavelet scaled by 0.5 (compressed), (c) wavelet scaled by 2 (dilated) and (d) wavelet translated by 4.

the conventional frequency, so low scale relates to the high frequency and vice versa. If $s \gg 1$, the window $\left(1/\sqrt{s}\right)\psi(t - \tau/s)$ is large and WT reacts mainly to low frequencies. If $s \ll 1$, the window $\left(1/\sqrt{s}\right)\psi(t - \tau/s)$ is small and WT reacts mainly to high frequencies.

As intimated, there is a relationship between scale and frequency; however, many texts such as Refs. [37,38] suggest it is better to think in terms of pseudofrequency. The relationship between scale and frequency is given by [37,38]

$$s = \frac{F_c \cdot \Delta}{f} \tag{5.29}$$

where Δ is the sample period, f is the pseudofrequency and F_c is the centre frequency of a wavelet in Hertz defined as approximate measure of the oscillatory nature of the basis function at its centre.

The CWT of signal f and reconstruction formula are given by [17,39]

$$F(s,\tau) = \int_{-\infty}^{\infty} f(t) \frac{1}{\sqrt{s}} \psi^* \left(\frac{t - \tau}{s} \right) dt \tag{5.30}$$

$$f(t) = \frac{1}{C_\psi} \int_{-\infty}^{\infty} d\tau \int_{0}^{\infty} \frac{ds}{s^2} F(s,\tau) \frac{1}{\sqrt{s}} \psi \left(\frac{t - \tau}{s} \right) \tag{5.31}$$

where * denotes a complex conjugate and C_ψ depends on the wavelet. The success of the reconstruction depends on this constant called the admissibility constant, and must satisfy the following admissibility condition [40].

$$0 < C_\psi = \int_{-\infty}^{\infty} |\hat{\psi}(\omega)|^2 \frac{d\omega}{\omega} < \infty \tag{5.32}$$

where $\hat{\psi}(\omega)$ is the FT of $\psi(t)$.

Figure 5.22 shows the 3D plot of coefficients of the CWT of signal: (a) stationary and (b) nonstationary; which have identical frequency components as given by

$$x(t) = \cos(20\pi t) + \cos(60\pi t) + \cos(100\pi t) \tag{5.33}$$

$$x(t) = \cos(2\pi f t) \quad \text{where} = \begin{cases} 10 & \text{if } t < 0.5 \\ 30 & \text{if } 0.5 \le t < 1 \\ 50 & \text{otherwise} \end{cases} \tag{5.34}$$

It is intuitive to see how the time-scale map identifies attributes of the original signal, localizing features in time; which is in complete contrast to the FFT spectrums. It is this feature-rich representation of the signal that makes the CWT

(a)   (b)

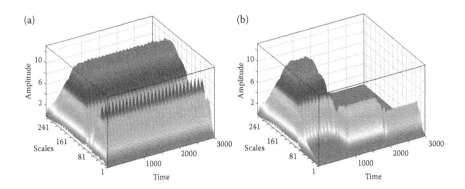

FIGURE 5.22 (**See colour insert.**) The CWT of the signal of (a) nonstationary signal and (b) stationary signal.

and its derivatives to be a popular choice of research and application in signal analysis applications. The multiresolution properties of the CWT are pictorially represented in Figure 5.23. In Figure 5.23, the horizontal axis represents time and the vertical axis represents frequency. Each box represents an equal portion of the time–frequency plane, but giving different proportions to time and frequency. Every box has a nonzero area indicating that an exact point in the time–frequency plane cannot be known. Each rectangular box (both in STFT and WT) has an identical area occupying an identical amount of the time–frequency plane. Each area in the case of CWT has different dimensions for length (frequency) and width (time). For lower frequencies this equates to better frequency resolution but poor time resolution, while higher frequencies have better time but poorer frequency resolution. In contrast the time–frequency map for the STFT would show areas of equal heights and widths; however, different STFT windows (equivalent to mother wavelets) would result in different areas up to a point of some lower bound determined by the uncertainty principle. For the CWT, the resolution in time or frequency is also determined by the choice of wavelet.

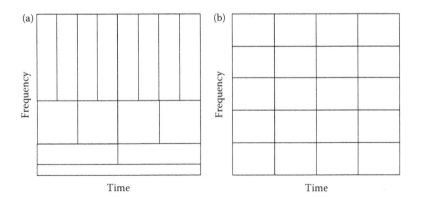

FIGURE 5.23 Time–frequency representation of (a) CWT and (b) STFT.

5.5.2 THE DISCRETE WAVELET TRANSFORM

As with the FFT and STFT, CWT analysis of meaningful signals by hand is almost impossible and computers are used for signal processing. In this case a discretized or time-sampled version of the CWT is used. Fortunately as the mother wavelet is dilated (higher scales and lower frequencies), the sampling rate can be reduced without affecting the results. If synthesis (reconstruction from wavelet coefficients) is required the Nyquist sampling rate must be observed. CWT unmixes parts of the signal which exist at the same instants, but at different scales, giving highly redundant CWT coefficients [32]. The redundancy on the other hand requires a significant amount of computation time without adding any valuable information. CWT has an unmanageable infinite number of wavelets coefficients and for most functions CWT has no analytical solutions and can be calculated only numerically [41]. The redundant coefficients in CWT can be removed by sampling both the scale and time at the powers of two (dyadic sampling), thus leading to the discrete wavelet transform (DWT). DWT provides sufficient information both for analysis and synthesis of the original signal, with a significant reduction in the computation.

The DWT is calculated by applying the concept of the multiresolution analysis (MRA) [42]. The practical implementation of the MRA can be done using filter bank [17]. The process involves using successive, complementary low-pass $g[n]$ and high-pass $h[n]$ filters to split the signal under analysis into its approximation and detail coefficients and down sampling by two. The high-pass part contains generally no significant information about signal and hence they do not decompose further. However, the low-pass parts are further decomposed into two bands, and this process will continue until a satisfactory level of information is obtained. Figure 5.24 represents processing of splitting the spectrum using a filter bank and a corresponding

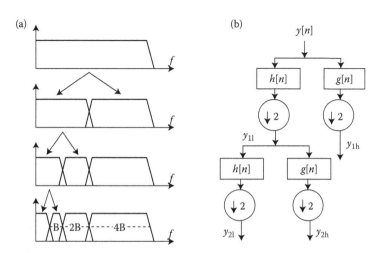

FIGURE 5.24 (a) Decomposition of a signal by DWT and (b) the DWT decomposition process.

wavelet decomposition tree with the approximation and detail coefficients. The first level of approximation y_{11} and detail coefficients y_{1h} are given by [32]

$$y_{1h}(k) = \sum_n y(n)g(2k - n) \tag{5.35}$$

$$y_{11}(k) = \sum_n y(n)h(2k - n) \tag{5.36}$$

The high-pass and low-pass filters satisfy the condition of the quadrature mirror filter, that is, the sum of the magnitude response of filters is equal to one for all frequencies. The approximation coefficients can further be decomposed into different DWT coefficient levels with a maximum level of $\log_2 L$, where L is the signal length. The original signal can be reconstructed by following the inverse of decomposition process or by using [43]:

$$x'(n) = \sum_k (y_{kh}(n) \cdot g(2k - n)) + (y_{kl}(n) \cdot h(2k - n)) \tag{5.37}$$

5.5.3 DWT-Based Denoising

The receiver design based on the DWT for the OOK system is given in Figure 5.25. This is very similar to the receiver structure shown in Figure 5.1 except for the wavelet denoising block. Wavelet denoising involves decomposition of the signal $y(n)$ into different DWT levels, processing of the DWT coefficients and reconstruction.

In the 1st level of decomposition, $y(n)$ is split into low-pass and high-pass signals as given by Equations 5.35 and 5.36, respectively, whereas in the 2nd stage, the low-pass band signal y_{11} is further divided into low-pass y_{21} and band-pass y_{2h} components. Filtering and decimation process is continued until the required level is reached. The task here is to separate the interference and the modulating signal in the DWT domain and remove the interfering signal from the reconstructed signal. Since the FLI signal spectral component is mostly based at lower frequency region, the received signal is decomposed until DWT coefficients of interfering signal are concentrated on the approximation coefficients. For removing the interfering signal from the received signal, the approximation coefficients, which correspond to the

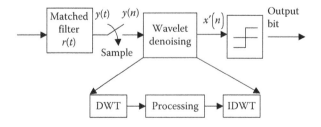

FIGURE 5.25 DWT-based receiver in the presence of artificial light interference.

interfering signal, are then made equal to zero so that the reconstructed signal is interference free, that is,

$$y_{\gamma h}(k) = 0 \qquad\qquad (5.38)$$

where γ is the number of the decomposition levels. The signal is then reconstructed back using the inverse of the decomposing process as given by Equation 5.37. DWT decomposes the signal in the logarithmic scale; therefore, it is difficult to define a precise band of frequencies that correspond to the lowest level of approximations. On the other hand, due to the spectral overlapping of the interfering signal and modulating signal, it is thorny to precisely define the cut-off frequency for the optimal performance. In the previous section, the performance of the system employing an HPF with a given f_c that resulted in the least BLW effect was discussed. However, the definition of such precise cut-off frequency is not possible using the DWT scheme. Therefore, the number of decomposition level γ is calculated in such a way that the lowest level of approximation signal is within 0.5 MHz range in order to make $f_c \sim 0.5$ MHz (except for a data rate <20 Mbps, for which f_c is taken as 0.3 MHz). The cut-off frequency is chosen as 0.5 MHz as studies showed that it provides near-optimal performance in presence of FLI for a digital HPF [43]. The number of decomposition level γ_{dl} is calculated using the following expressions:

$$\gamma_{dl} = -\left\lfloor \log_2\left(\frac{F_s}{f_c}\right) \right\rfloor \qquad\qquad (5.39)$$

where $\lfloor . \rfloor$ is the floor function. It is to be noted here that f_c varies with the data rate and the decomposition level.

DWT offers flexibility in the analysis where different mother wavelets could be adopted as necessary for particular application. The NOPRs linked to different mother wavelets are listed in Table 5.1. Only selected mother wavelets from the Daubechies (db), Symlet (sym), discrete Meyer wavelet (dmey), biorsplines (bior) and Coiflets (coif) families with the best and the worst performance are listed. The Haar wavelet (db1) and the db8 show the worst and best performance, respectively, followed closely by the db10 and sym9. Hence, the db8 is adopted for discussion in this chapter.

The MATLAB code to simulate the wavelet-based denoising is given in Program 5.5. See Program 5.1 for all the parameters and calculation of error probability for a range of SNR. A similar approach can be applied to other modulation schemes including PPM and DPIM.

The NOPR for OOK schemes with the DWT denoising in the presence of FLI is shown in Figure 5.26. Also shown is the NOPR for the ideal channel without interference. Since no improvement can be achieved below data rates of 10 Mbps, such rates are not considered here. Since OOK is very susceptible to the BLW, only low normalized cut-off frequencies (normalized to sampling rate) are possible, which are not effective in reducing the interference. At data rates above 10 Mbps, DWT is very effective in reducing the interference. At a bit rate of 10 Mbps, DWT offers a reduction of ~5.5 dB in NOPR compared to the system without filtering. Higher

TABLE 5.1

NOPR to Achieve a BER of 10^{-6} at a Data Rate of 200 Mbps for a DWT-Based Receiver with Different Mother Wavelets

Mother Wavelets	SNR (dB)	NOPR (dB)
db1	22.01	15.7
db2	20.9	15.2
db5	16.1	12.9
db8	15.2	12.3
db10	16.2	12.8
Coif2	16.6	13.0
Coif5	15.4	12.4
Sym3	17.2	13.3
Sym4	15.7	12.6
Sym7	15.3	12.4
Dmey	15.7	12.6
Bior2.2	16.7	13.1

Source: Adapted from S. Rajbhandari, Z. Ghassemlooy and M. Angelova, *IEEE/OSA Journal of Lightwave Technology*, 27, 4493–4500, 2009.

reduction of ~7.4 dB is observed at 40 Mbps. However, this still leaves an OPP of ~1.6 dB compared to the same bit rate without interference. The OPP is further reduced at higher data rates with a value of ~1.2 dB at 100 Mbps. Note that optimum HPF still leaves a power penalty of 3 dB at 100 Mbps.

Program 5.5: MATLAB Codes to Simulate Wavelet-Based Denoising for OOK–NRZ in the Presence of FLI

```
OOK = randint (1,sig_length);
% random signal generation
OOKm = [zeros(1,Lfilter) OOK zeros(1,Lfilter)];
Tx_signal = rectpulse(OOKm,nsamp)*i_peak;
% Pulse shaping function (rectangular pulse)
Rx_signal = R*Tx_signal + sgma*randn(1,length(Tx_signal));
% received signal (y = x + n)
% %****************Effect of FL****************
start_time = abs(rand(1,1))*10E-3;
end_time = start_time + Tb*nsamp*sig_length;
Ib = 2E-6;
% average current due to the fL
i_elect = fl_model(Ib,Tsamp,start_time,end_time);
```

```
% FLI model
Rx_OOK_fl = Rx_signal + i_elect(1:length(Rx_signal));
% Interference due to FL

MF_out = conv(Rx_OOK_fl,rt)*Tsamp;
% matched filter output
MF_out_downsamp = MF_out(nsamp:nsamp:end);
% sampling at end of bit period
MF_out_downsamp = MF_out_downsamp(1:length(OOKm));
% truncation
%% **************** wavelet denoising of the signal
[C,L] = wavedec(MF_out_downsamp,Lev,wname);
cA = appcoef(C,L,wname,Lev);
C(1:length(cA)) = 0;
Rx_OOK = waverec(C,L,wname); % reconstructed signal
Rx_OOK = Rx_OOK(Lfilter + 1:end-Lfilter);
Rx_OOK = mapminmax(Rx_OOK);

Rx_th = zeros(1,sig_length);
Rx_th(find(Rx_OOK > 0)) = 1;
% thresholding
```

The NOPR to achieve an error probability of 10^{-6} with the DWT denoising for 4, 8 and 16-PPM with the HDD scheme for a data rate range of 1–200 Mbps is demonstrated in Figure 5.27. With reference to Figure 5.4, it is clear that DWT denoising shows marked reduction in the NOPR (~10.7, 12.1 and 12.4 dB for 4, 8 and 16 PPM, respectively) at a data rate of 1 Mbps. Also note the significant reduction in power requirements at a data rate of 1 Mbps compared to HPF (Figure 5.18). Since the PPM

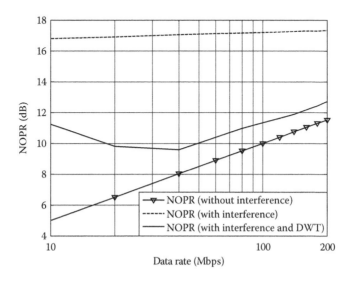

FIGURE 5.26 Normalized optical power requirement against data rates for OOK modulation scheme with and without DWT denoising in the presence of the FLI.

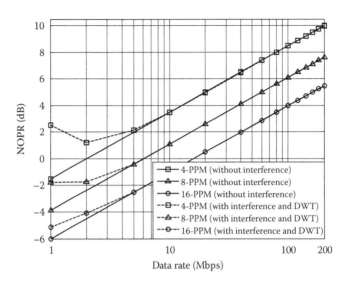

FIGURE 5.27 NOPR versus the data rates for 4, 8 and 16 PPM with the hard decoding scheme with DWT denoising in the presence of the FLI.

frequency spectrum is mostly at the higher frequency region with no DC component at all, a higher f_c for the HPF could be used without experiencing the BLW effect. Hence, for the PPM scheme the DWT denoising technique offers improvement even at low frequencies. Above a data rate of 10 Mbps for 4-PPM and 5 Mbps for 8 and 16-PPM, the DWT-based denoising completely eliminates the FLI-induced power penalties.

The NOPR to achieve an error probability of 10^{-6} with the DWT denoising for 4, 8 and 16-PPM (SDD) scheme for a data rate range of 1–200 Mbps is given in Figure 5.28.

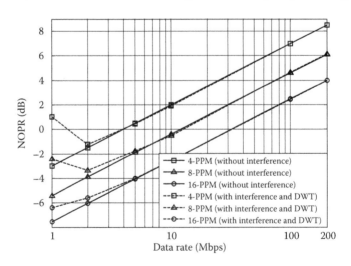

FIGURE 5.28 NOPR versus the data rates for 4, 8 and 16 PPM with the soft decoding scheme with DWT denoising in the presence of the FLI.

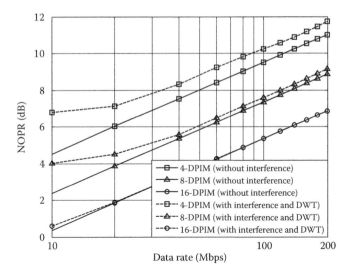

FIGURE 5.29   NOPR versus the data rates for 4, 8 and 16 DPIM schemes with DWT denoising in the presence of the FLI.

Compared with Figure 5.5, DWT offers no improvement above a data rate of 20 Mbps as PPM(SDD) is immune to the FLI. DWT provides a significant improvement even at 1 Mbps with power penalties of ~4, ~2.9 and ~1 dB for 4, 8 and 16-PPM, respectively, compared to the same bit rate with no interference. The power penalties without DWT are 11.45 and 13.1 dB, thus illustrating the effectiveness of the DWT denoising. At 10 Mbps FLI-induced power penalty is 1.1–2.1 dB, which are completely eliminated when using the DWT denoising.

The NOPR to achieve an SER of 10^{-6} with the DWT denoising for 4, 8 and 16-DPIM at a data rate range of 1–200 Mbps is given in Figure 5.29. Comparing the performance of the OOK scheme in Figure 5.26 and PPM(HDD) in Figure 5.27, it can be observed that the performance of DPIM with DWT is intermediate between OOK and PPM. Lower orders of DPIM show performance similar to OOK with constant OPPs compared to the ideal case. However, higher orders of DPIM show almost zero power penalties. The phenomenon is due to progressive reduction of the DC and low-frequency components with increasing bit resolutions. Compared to the performance without FLI, the power penalties with DWT are ~0.7 dB at a data rate >40 Mbps, which reduces to 0.4 dB for 8-DPIM. For 16-DPIM at data rate >20 Mbps the power penalties are completely reduced to zero when adopting the DWT denoising scheme.

5.5.4   COMPARATIVE STUDY OF DWT AND HPF

This section provides the comparative study of adopting the DWT and HPF for reduction of FLI. Because of spectral overlapping of the modulating and interference signals, it is in fact impossible to reduce the effect of FLI without incurring OPPs especially for modulating techniques with a high DC component. Since OOK has the highest DC and the lower frequency components, it suffers most adversely from the application of the filtering. Hence, the task here is to design high-quality denoising techniques with reduced complexity as well as a minimal information loss. A case study for OOK with HPF and DWT denoising is presented to illustrate the effectiveness of the proposed schemes. It is simple to conclude that a filtering technique that provides the least OPP is the best for other modulation techniques as well, because OOK suffers most severely from BLW. Digital HPF shows significantly improved performance compared to its analogue counterpart, and is therefore adopted. For the performance of the OOK and other modulation techniques with the analogue HPF, readers should consult Refs. [3,6,11].

The OPPs against the decomposition level γ for DWT for data rates of 20, 50, 100 and 200 Mbps is given in Figure 5.30. The OPPs for different filter cut-off frequencies and for the same data rates for HPF are given in Figure 5.31. For DWT the lowest

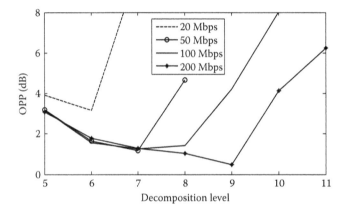

FIGURE 5.30 The OPP against the decomposition level for the DWT-based receiver at data rates of 20, 50, 100 and 200 Mbps.

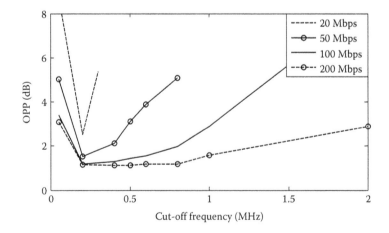

FIGURE 5.31 The OPP against cut-off frequencies for a HPF-based receiver at data rates of 20, 50, 100 and 200 Mbps.

threshold level is at $\gamma = 9$ for 200 Mbps. Since DWT decomposes signal in a logarithmic scale, the value of γ at which the OPP is the minimal increases by one when the data rate doubles. In the case of the HPF, the optimum f_c is ~0.2 MHz for all data rates. Note that it is not possible to define a precise f_c for DWT (as only the integer value of γ is permissible). The results shown in Figures 5.30 and 5.31 indicate that the wavelet-based receiver, with no parameter optimization, displays similar or better performance compared to the best performance achieved with the HPF. This makes the denoising process faster and more convenient.

In addition to the performance improvement, a key advantage of the DWT is the reduced system complexity. For optimum performance, the value of decomposition level γ varies from 6 to 9 (see Figure 5.31). For an input signal length of n, the total number of floating point operations (only multiplication is taken into consideration here) for the first level of decomposition is nJ. For the second stage, the length of input signal is $n/2$ and the total number of operations is $nJ/2$ and so on. Hence, the DWT requires a maximum of $2nJ$ operations for analysis and synthesis meaning the maximum number of floating point operations is $4nJ$. Since $J = 15$ for Daubechies 8 'db8', the maximum number of the operations is $60n$. For an HPF of order L, the total number of floating point operations is $nL/2$. At a date rate of 200 Mbps and with f_c of 0.2 MHz the filter order L is 2148, thus illustrating much reduced complexity of the DWT. Moreover, the realization of the DWT is also simpler compared to the HPF as a repetitive structure is used at each level of analysis and synthesis. Hence, only a 15th-order filter is required to realize for the DWT compared to a filter of 2148 orders for the HPF.

5.5.5 EXPERIMENTAL INVESTIGATIONS

The nondirected LOS-OWC system is deployed in a typical $6 \times 5 \times 3$ m³ laboratory room environment with a receiver located at a height of 1 m above the floor. The schematic diagram of the experimental set-up is given in Figure 5.32 (note that it is not to scale). A laser diode (LD) operating at a wavelength of 830 nm with a maximum optical output power of 10 mW is directly modulated using an input data source. The pseudorandom bit sequence (PRBS) of $2^{10}-1$ bits is generated using an arbitrary waveform generator (AWG) and is converted into a non-return-to-zero

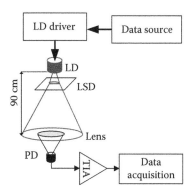

FIGURE 5.32 The schematic diagram of the experimental set-up for an indoor OWC link.

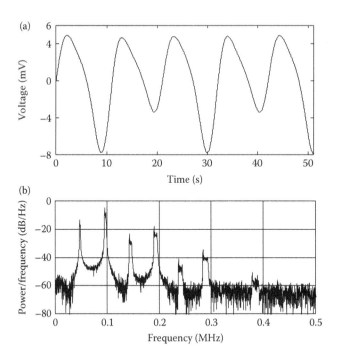

FIGURE 5.33 (a) Time waveform and (b) spectrum of the interference produced by an artificial light source.

(NRZ) OOK format prior to intensity modulation of a laser diode. A holographic diffuser of 10° full-width half-maximum (FWHM) is used to ensure eye safety as well as increase the optical footprint. The receiver is positioned at the centre of the optical footprint where the received optical power is at its maximum value. The receiver consists of an optical concentrator (a focal length of 5 cm and a diameter of 2.5 cm), and a photodiode (PD) with a daylight filter at 800–1100 nm wavelength range. The peak spectral sensitivity of the PD is 0.59 A/W at 950 nm wavelength with an active area of 7 mm². The receiver optics (lens and PD) are adjusted to obtain the maximum optical gain. The photocurrent at the PD is amplified using a commercial transimpedance amplifier (TIA), followed by a data acquisition system in order to acquire, process and analyse the real-time data.

The experiment is carried out in two settings: (a) in complete darkness and (b) in the presence of the ambient light source. The ambient light source is a florescent light lamp located at the ceiling 3 m above the floor and 2 m above the receiver. The receiver is located directly underneath the light source and the illuminance of the light sources at the receiver position is ~350 lux. The waveform and the PSD of the fluorescent light measured using the OSRAM (SFH205F) photodetector are given in Figure 5.33. The waveform of FLI is a distorted sinusoidal signal at a fundamental frequency of 50 kHz with harmonics extending up to 0.5 MHz (see Figure 5.33b). The time waveform of the 10 Mbps OOK–NRZ signal in the presence of FLI is depicted in Figure 5.34. It can clearly be observed that the photocur-

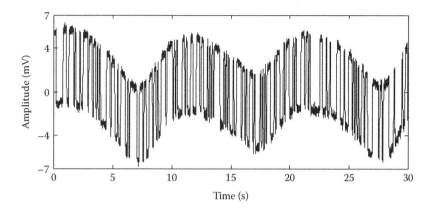

FIGURE 5.34 The time waveform of the received signal at 10 Mbps in the presence of the FLI.

rent due to the inference is higher than the modulating signal. As a result, it is impossible to achieve an OWC link with a low error probability without incorporating a denoising technique.

To quantify the OWC link performance, the Q factor is measured in the presence and absence of FLI. The Q-factor is a measure of the optical signal to noise ratio (OSNR) and hence can be used to access the quality of the link. The Q factor for binary modulation scheme is given by

$$Q = \frac{v_H - v_L}{\sigma_H - \sigma_L} \tag{5.40}$$

where v_H and v_L are the mean received voltages and σ_H and σ_L are the standard deviations for the 'high' and 'low' level signals, respectively.

To verify the experimental results; computer simulation of the OWC system under test is also carried out using the measured parameters including the noise variance, the FLI photocurrent, the average modulating signal photocurrent and the impulse response of system (transmitter, channel and receiver). The AC coupling capacitor at the receiver, which induces a baseline wander (BLW) effect in baseband modulation schemes [6,9], is modelled as a first-order analogue RC–HPF with a cut-off frequency of 10 kHz.

The measured and simulated Q-factors against the data rate in the presence and absence of FLI are given in Figure 5.35. The simulation results and the practical measurements closely follow each other with a difference of <1.5 dB for all measured data rates. Though, the Q-factor should theoretically decrease with the data rate, the highest Q-factor is observed between 10 and 15 Mbps. The Q-factor is lower at lower data rates; this is due to the BLW effects that can be verified by the eye diagrams given in Figure 5.35b and c at data rates of 1 and 10 Mbps, respectively. Notice the higher eye-opening at 10 Mbps compared to 1 Mbps due to the BLW effect. On the other hand, the measurement shows that the 3 dB cut-off frequency of the overall system (combined impulse response of the transmitter, channel and receiver) is ~13 MHz, resulting in a maximum Q-factor in the range of 10–15 Mbps. Note that

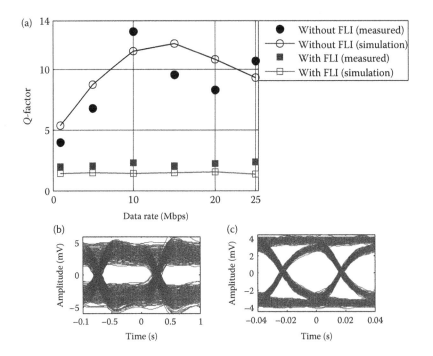

FIGURE 5.35 (a) The measured and simulated Q-factors against the data rate in the presence and absence of FLI and eye diagrams of the received OOK–NRZ in the absence of FLI, (b) at 1 Mbps and (c) 10 Mbps.

the Q-factor in the presence of FLI is very low (<2) irrespective of the data rate, thus indicating the adverse effect of FLI in indoor OWC systems.

To reduce the effect of FLI, a DWT-based denoising technique is proposed in Ref. [44]. In this study, the practical verification of the proposed system is carried out. Three methods of denoising are studied here: (a) analogue HPF, (b) digital HPF and (c) DWT-based denoising. The analogue HPF is modelled as a first-order RC filter with a 3 dB cut-off frequency f_c of 0.3 MHz. The response of HPF to a single rectangular pulse of amplitude A and duration T_b can be expressed as [9]

$$g(t) = \begin{cases} Ae^{-t/RC} & 0 \le t \le T_b \\ -A(e^{-T_b/RC} - 1)e^{-t/RC} & t > T_b \end{cases} \tag{5.41}$$

where the filter time is constant $RC = 1/2\pi f_c$.

The second approach is based on using a second-order digital Butterworth filter with a normalized cut-off frequency (normalized to the sampling frequency) of 0.3 MHz. The third approach employs a DWT-based denoising scheme. The detailed algorithm of the wavelet-based denoising is discussed in Section 5.5.3.

The simulation was carried out using the MATLAB routine with DWT as in Ref. [44]. The simulation results show that the cut-off frequency of 0.3 MHz is the

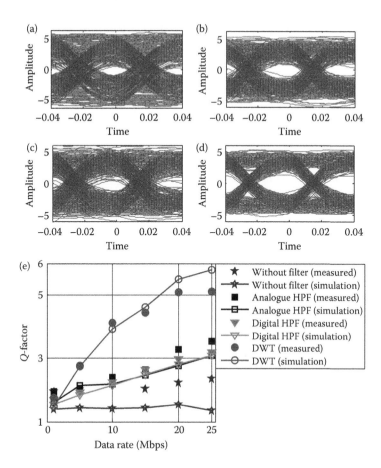

FIGURE 5.36 Eye-diagrams of the received OOK–NRZ at 25 Mbps in the presence of FLI (a) without filter, (b) analogue HPF, (c) digital HPF, (d) DWT and (e) the measured and simulated Q-factors against the data rate for indoor OWC systems in the presence of FLI with HPF and wavelet denoising schemes.

optimal for the fluorescent light source utilized in this experiment irrespective of the data rate. The measured and simulated Q-factors against the data rate in the presence of FLI with and without denoising are illustrated in Figure 5.36. Also shown are the eye diagrams of the received signal with and without denoising at 25 Mbps (Figure 5.36a–d). The wide eye-opening for the DWT-based denoising case compared to the HPF clearly shows the advantage of DWT, which can also be verified from the Q factor versus the data rate plot in Figure 5.36e. The DWT approach outperforms the HPF approach for the entire range of data rates. A difference of ~2 in the Q-factor at data rates higher than 10 Mbps is notable, proving the effectiveness of the DWT. This leads to the improvement of the bit error rate from ~10^{-4} with an HPF to ~10^{-7} with the DWT at 25 Mbps data rate. There is no significant difference between the digital and analogue HPFs and hence any of these filters can be incorporated. It is noteworthy that there is a close match between values of simulated and measured Q-factors.

5.6 LINK PERFORMANCE FOR MULTIPATH PROPAGATION

5.6.1 OOK

The block diagram of an unequalized OOK system is given in Figure 5.37. The input bits, assumed to be i.i.d. and uniform on $\{0, 1\}$, are passed to a transmitter filter, which has a unit amplitude rectangular impulse response $p(t)$, with a duration of one bit T_b. The pulses are then scaled by the peak transmitted optical signal power $2P_{avg}$, where P_{avg} is the average transmitted optical signal power.

The optical signal $x(t)$ is then subjected to the multipath distortion, hence the channel output $\varphi(t)$ in the absence of noise is given by

$$\varphi(t) = x(t) \otimes h(t) \tag{5.42}$$

where the symbol $\otimes$ denotes convolution.

The receiver front consists of the unit energy filter $r(t)$, followed by a sampler and a threshold detector detail of which can be found in Chapter 4. The input signal to the matched filter is given by

$$z(t) = R\varphi(t) + n(t) \tag{5.43}$$

Note that this filter is optimum only when there is no multipath dispersion. The output of the matched filter is sampled at the bit rate, and the samples are passed to a threshold detector, which assigns a one or zero to each bit depending on whether the sampled signal is above or below the threshold level, thus generating an estimate of the transmitted bit sequence.

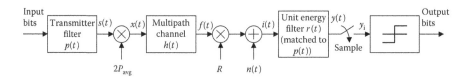

FIGURE 5.37 The block diagram of the unequalized OOK system.

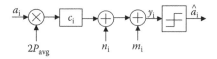

FIGURE 5.38 Discrete-time equivalent system for the communication system.

The system can be modelled using the discrete time model as given in Figure 5.38. The discrete-time impulse response c_i of the cascaded system (the transmitter filter, channel and the receiver filter) is given as

$$c_i = p(t) \otimes h(t) \otimes r(t)\big|_{t=iT_b} \qquad (5.44)$$

with the normalization $\Sigma_i c_i = 1$ and sampling times are shifted to maximize the zero-sample h_0.

The noise samples n_i with zero mean and variance σ^2 are given by

$$n_i = n(t) \otimes r(t)\big|_{t=iT_b} \qquad (5.45)$$

Unless the channel is nondispersive, c_i contains a zero tap, a single precursor tap and possibly multiple postcursor taps. The magnitude of the zero tap is larger than the magnitudes of the other taps. In a nondispersive channel, the optimum sampling point, that is, that which minimizes the probability of error, occurs at the end of each bit period. However, in dispersive channels, the optimum sampling point changes as the severity of ISI changes. In order to isolate the power penalty due to ISI, two assumptions are made. Firstly, perfect timing recovery is assumed. This is achieved by shifting the time origin so as to maximize the zero tap, c_0. Secondly, an optimal decision threshold is assumed. For OOK, basic symmetry arguments can be used to deduce that the optimum threshold level lies midway between expected one and zero levels, regardless of the severity of ISI.

Suppose that c_i contains ζ taps. Let $\mathbf{a}_i$ be an m-bit sequence, and $a_{j,m-1}$ the value of the $(m-1)$th slot (penultimate slot) in the sequence $\mathbf{a}_i$, where $a_{j,m-1} \in \{0,1\}$. Unless the channel is nondispersive, for a $\mathbf{a}_i$ sequence of m slots, c_i will contain m taps (**slots**): a single precursor tap, a zero tap, which has the largest magnitude, and $(m-2)$ postcursor taps (see Figure 5.39). All slots within a given sequence a_i are affected by dispersion of pulses appearing within S_i as well as adjacent sequences $\mathbf{a}_{i-1}$, or $\mathbf{a}_{i+1}$, except for the penultimate slot, which is only affected by pulse dispersion appearing within the sequence $\mathbf{a}_i$. Sequences that fall outside the boundaries of $\mathbf{a}_i$ will not contribute to the dispersion on the penultimate slot of $\mathbf{a}_i$. Therefore, when calculating the optical power requirement, only the penultimate slot will be considered for each sequence.

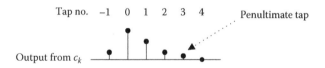

FIGURE 5.39 Discrete-time impulse response c_k for a sequence of six taps.

Let $\mathbf{a}_i$ be an ζ-bit sequence and a_i be the value of the penultimate bit in that sequence, where $a_i \in (0, 1)$. Let y_i denote the receiver filter output corresponding to the penultimate bit, which, in the absence of noise, is given by

$$y_i = 2RP_{\mathrm{avg}} a_i \otimes h_k \big|_{k=\zeta} \tag{5.46}$$

where ζ is the number of channel taps and $a_i = \{a_1, a_2 \ldots a_\zeta\}$ is the ζ bit sequence.

Thus, the probability of error for the penultimate bit in $\mathbf{a}_i$ is given by

$$
\varepsilon_i =
\begin{cases}
Q\left(\dfrac{\alpha_{\mathrm{opt}} - y_i}{\sqrt{N_0/2}} \right) & \text{if } a_i = 0 \\[4mm]
Q\left(\dfrac{y_i - \alpha_{\mathrm{opt}}}{\sqrt{N_0/2}} \right) & \text{if } a_i = 1
\end{cases}
\tag{5.47}
$$

where α_{opt} is the optimum threshold level, set to the midway value of $RP_{\mathrm{avg}} \sqrt{T_b}$. The average BER is approximated by averaging overall possible bits sequence of length ζ, and is given by [45]

$$P_e = \frac{1}{2^\zeta} \sum_{i \in s_0} Q\left(\frac{0.5RP_{\mathrm{avg}} \sqrt{T_b} - y_i}{\sqrt{N_0/2}} \right) + \frac{1}{2^\zeta} \sum_{i \in s_1} Q\left(\frac{y_i - 0.5RP_{\mathrm{avg}} \sqrt{T_b}}{\sqrt{N_0/2}} \right) \tag{5.48}$$

where s_0 and s_1 correspond to bit '0' and '1', respectively.

Program 5.6: MATLAB Codes to Simulate Ceiling Bounce Model and Received Signal Eye Diagram

```
Rb=200e6;
% Bit rate
Tb=1/Rb;
% bit duration
sig_length = 1e3;
% number of bits
nsamp = 10;
% samples per symbols
Tsamp = Tb/nsamp;
% sampling time

%% *****Channel Impulse response (using Ceiling bounce
model)*************
Dt = 0.5;
% Normalized delay spread
Drms = Dt*Tb;
```

```
%RMS delay spread
a = 12*sqrt(11/13)*Drms;
K = 30*nsamp;
% number of channel taps
k = 0:K;
h = ((6*a^6)./(((k*Tsamp)+a).^7));
% channel impulse response
h = h./sum(h);
% normalizing for conservation of energy

% *************** filter definitions*******
pt = ones(1,nsamp)
% tranmitter filter
c = conv(pt,h);

OOK = randint(1,sig_length);
% random signal generation
Tx_signal = rectpulse(OOK,nsamp);
Pulse shaping function (rectangular pulse);
channel_output = conv(Tx_signal,h);
% channel output
eyediagram(channel_output, 3*nsamp);
```

Figure 5.40 shows the effect of multipath propagation on OOK–NRZ signalling with a normalized delay spread D_T of 0.4, where variations in signal amplitude due to preceding pulses can easily be seen. It is intuitive to see that for a given severity of channel-induced distortion higher data rates will cause further variation in signal amplitudes. It can also be noted from the figure that the best sampling point would be at the end of the bit period.

Based on the assumption of the perfect timing recovery and channel delay, the program to calculate the error probability of OOK–NRZ in a multipath channel using the ceiling bounce model (see Program 5.6 for the code to simulate this channel

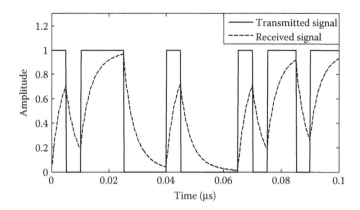

FIGURE 5.40 Example of multipath distortion on OOK–NRZ at normalized delay spread of 0.4.

model) is given in Program 5.7. An alternative approach to the simulation based on the discrete-time equivalent system (Figure 5.38) is given in Program 5.8. Note that the simulation time for Program 5.8 is significantly lower than that of Program 5.7.

Program 5.7: MATLAB Codes to Simulate the Error Probability of OOK–NRZ in a Multipath Channel. The Simulation Is Based on the System Block Diagram Given in Figure 5.37

```
%% *****Channel Impulse response (using ceiling bounce
model*************
Dt=0.1;
% Normalized delay spread
Drms=Dt*Tb;
%RMS delay spread
a=12*sqrt(11/13)*Drms;
K=30*nsamp;
% number of channel taps
k=0:K;
h=((6*a^6)./(((k*Tsamp)+a).^7));
% channel impulse response
h=h./sum(h);
% normalizing for conservation of energy

%% system impulse response
pt=ones(1,nsamp)*i_peak;
% transmitter filter
rt=pt;
% receiver filter matched to pt
c=conv(pt,h);
c=conv(c,rt);
% overall impulse response of system
delay=find(c==max(c));
% channel delay

%% multipath simulation
OOK=randint(1,sig_length);
% random signal generation
Tx_signal=rectpulse(OOK,nsamp)*i_peak;
% Pulse shaping function (rectangular pulse)
channel_output=conv(Tx_signal,h);
% channel output, without noise
Rx_signal=channel_output+sgma*randn(1,length(channel_output));
% received signal with noise
%Rx_signal=awgn(channel_output,EbN0+3-10*log10(nsamp),
  'measured');

%% Matched filter simulation
MF_out=conv(Rx_signal,rt)*Tsamp;
% matched filter output
```

```
MF_out_downsamp=MF_out(delay:nsamp:end);
% sampling at end of bit period
MF_out_downsamp=MF_out_downsamp(1:sig_length);
% truncation
Rx_th=zeros(1,sig_length);
Rx_th(find(MF_out_downsamp>Ep/2))=1;
```

**Program 5.8: MATLAB Codes to Simulate the Error Probability
of OOK–NRZ in a Multipath Channel Based on a
Discrete-Time Equivalent System (Figure 5.38)**

```
%% system impulse response
Tx_filter=ones(1,nsamp);
Rx_filter=fliplr(Tx_filter);
c=conv(Tx_filter,h);
c=conv(c,Rx_filter);
delay=find(c==max(c));
if delay>nsamp;
  hi(1)=c(delay-nsamp);
  % taking precursor tap
else hi(1)=0;
end
hi=[hi(1) c(delay:nsamp:end)];
% channel impulse response
hi=hi/sum(hi);
% normalization

OOK=randint(1,sig_length);
% random signal generation
Rx_signal=conv(OOK, hi);
% Received signal, after matched filter (without noise)
Rx_signal=Rx_signal(2:sig_length+1);
% truncation to avoid precursors
MF_out=awgn(Rx_signal,EbN0_db+3,'measured');
% MF output with noise
Rx_th=zeros(1,sig_length);
Rx_th(find(MF_out>0.5))=1;
```

The unequalized NOPR for OOK to achieve a BER of 10^{-6} in a diffuse channel with D_T of [0,1] is given in Figure 5.41. Note that the data rate (bit duration) used is irrelevant as the RMS delay spread is normalized to bit duration. It is clearly apparent from Figure 5.41 that the unequalized NOPR of OOK increases exponentially with D_T. A NOPR of 3 dB is incurred when D_T is ~0.23. When D_T is 0.51, the NOPR is ~18 dB beyond which the target BER of 10^{-6} is practically impossible to achieve with a reasonable optical power.

At lower error probability, the degradation due to the multipath propagation is dominated by the worst-case bit sequence, which consists of a single bit preceded and followed by a long string of zeros and the unequalized OPP can approximated as [46]

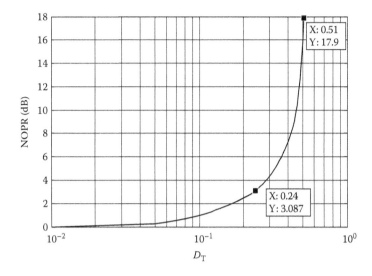

FIGURE 5.41 NOPR against the normalized delay spread D_T for unequalized OOK in a diffuse indoor OWC channel.

$$\text{unequalized OPP (dB)} = 10 \log_{10} \left[\frac{Q^{-1}(2^\zeta \xi)}{(2h_0 - 1)Q^{-1}(\xi)} \right] \tag{5.49}$$

It can be noted that at low error probability, the OPP depends only on the h_0 and number of channel taps ζ. The simulation and approximated (using (5.49)) OPPs for the OOK against the channel tap with highest amplitude h_0 are given in Figure 5.42. There is a close match between the theoretical and approximated BER results for a

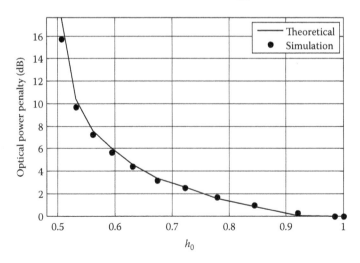

FIGURE 5.42 Theoretical and simulated OPPs for the unequalized OOK system against the h_0.

range of h_0 indicating that Equation 5.49 gives a good approximation. The OPP increases with the decreasing value of h_0. This is expected for a channel with severe ISI where h_0 reduces. When the $h_0 \leq 1/2$, the expression (5.49) yields infinite power penalty that can be seen from asymptotic increment in the OPP with decreasing h_0.

5.6.2 PPM

To analyse the effect of multipath-induced ISI, the discretization approach discussed in the section above can be adopted. The discrete-time equivalent system for PPM scheme is given in Figure 5.43. Note that both HDD and SDD can be adopted as necessary. The detailed descriptions of the matched filter-based receiver for the PPM system including SDD and HDD schemes are discussed in Chapter 4.

To analyse the slot error probability for the PPM(HDD) system, one can adopt the same analysis approach outlined for OOK. However, due to the difference in the sampling rate, the discrete-time impulse response c_i of the cascaded system (the transmitter filter, channel and the receiver filter) is given as

$$c_i = p(t) \otimes h(t) \otimes r(t)\Big|_{t=i\frac{T_b M}{L}} \tag{5.50}$$

Hence, the received signal y_i in the absence of noise, is given by

$$y_i = LRP_{\text{avg}} b_i \otimes h_k\Big|_{k=\zeta} \tag{5.51}$$

Comparing (5.44) and (5.50), it can be observed that for the same data rate, the PPM will have a higher channel span due to a shorter pulse duration, and hence results in higher ISI for the same normalized delay spread (normalized to bit duration). As in the case of OOK, to estimate the error probability of PPM in a multipath channel, one needs to consider all possible combination of the PPM symbols within the channel span ζ, calculate the slot error probability of individual slot and average it over the entire channel span. Hence, the slot error probability of PPM is given by

$$P_{\text{se\_PPM\_hard}} = \frac{1}{2^\zeta} \sum_i^{2^\zeta} \left[\frac{1}{L} Q\left(\frac{y_i - \alpha_{\text{opt}}}{\sqrt{N_0/2}} \right) + \frac{(L-1)}{L} Q\left(\frac{\alpha_{\text{opt}} - y_i}{\sqrt{N_0/2}} \right) \right] \tag{5.52}$$

where α_{opt} is the optimum threshold level.

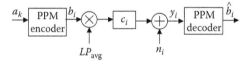

FIGURE 5.43   Discrete-time equivalent system for the PPM scheme.

Unlike OOK, α_{opt} does not lie in the middle of the zero and one level due to unequal probability of occurrence of the empty slots and pulses. In fact, α_{opt} is a complicated function of the bit resolution and the discrete-time impulse response h_i. An iterative approach to calculate the optimum threshold level α_{opt} and approximate $P_{se\_PPM\_hard}$ is taken into consideration in Ref. [11]. The iterative approach requires a significant computational time, thus not practical for $\zeta > 20$ [11]. In this work a constant threshold level α_{th} is used which is given by

$$\alpha_{th} = 0.5 - 0.25\sum_{i=1}^{\infty}h_i \qquad (5.53)$$

The summation $\sum_{i=1}^{\infty}h_i$ provides the estimation of the energy of pulse that is spread to its adjacent pulses due to the dispersion. For an LOS link, the summation is zero and hence $\alpha_{th} = 0.5$. The dispersive channel will have a nonzero summation value. The value of α_{th} in a dispersive channel is less than 0.5.

Using SDD, a pulse is detected depending upon its relative amplitude within a symbol. Assuming that one occurs in slot q of each symbol, the average symbol error probability for unequalized PPM(SDD) is given by [11,47]

$$P_{symbol\_PPM\_soft} = \frac{1}{2^{\zeta}}\sum_{i=1}^{\zeta}\sum_{\substack{p=1 \\ p\neq q}}^{L}Q\left(\frac{y_{(i-1)L+q} - y_{(i-1)L+p}}{\sqrt{N_0}}\right) \qquad (5.54)$$

The NOPR to achieve an SER of 10^{-6} for unequalized 4, 8 and 16-PPM with SDD and HDD schemes are given in Figure 5.44. The threshold level given in Equation 5.53

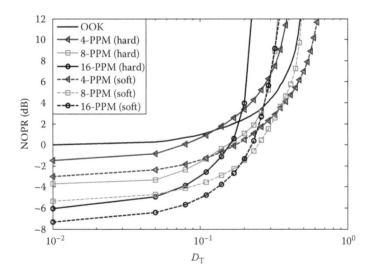

FIGURE 5.44 NOPR against the normalized delay spread for 4, 8 and 16-PPM with hard and soft decision decoding schemes in a dispersive channel.

is used for the HDD scheme. Also shown is the NOPR for OOK for comparisons. For any given order and delay spread, the SDD scheme has a lower power requirement than the HDD. For an ideal channel this difference is ~1.5 dB, but for multipath channels the difference increases with D_T. The higher-order PPM shows a sharp increase in NOPR with increasing D_T for both SDD and HDD schemes. This is due to decrease in the slot duration for higher-order PPM, effectively increasing the ISI. For example, for $D_T > 0.19$, 4-PPM(HDD) offers lower NOPR to achieve an SER of 10^{-6} compared to the 16-PPM(HDD). For the SDD scheme, the intersection point for 4-PPM and 16-PPM is ~0.24. It can also be observed that OOK offers the least OPP compared to all orders of the PPM(HDD) for a highly dispersive channel ($D_T > 0.27$) and 4-PPM (SDD) offers the least power requirement for higher D_T values. The SDD scheme offers improved resistance to the ISI and the difference in OPPs for SDD and HDD increases with D_T. The difference in NOPR for the SDD and HDD schemes increases from 1.5 dB for $D_T = 0$ to ~8.5 dB for $D_T = 0.4$ for 4-PPM with the SDD always requiring lower NOPR. The difference is even higher for 8 and 16-PPM with a noticeable difference of ~12.5 dB at $D_T = 0.23$ for 16 PPM. This clearly demonstrates the advantage of SDD over HDD in a diffuse channel.

Despite the improved power efficiency of SDD, it is evident that OOK outperforms 16-PPM for D_T beyond ~0.3, and offers a similar average OPR to 8 PPM at D_T ~0.3. Furthermore, compared with PPM(HDD), OOK offers a lower average OPR than any orders of PPM considered for normalized delay spreads above ~0.27. This clearly demonstrates the severity of multipath-induced ISI in pulse modulation schemes.

5.6.3 DPIM

Following a similar approach adopted for PPM, the effect of multipath-induced ISI on the DPIM scheme can be approximated. Similar to OOK and PPM schemes, for calculation of the error probability all possible DPIM sequences of at least ζ slots are generated. The SER is approximated by averaging over all possible slot sequences of length ζ and is as follows:

$$P_{se\_DPIM} = \frac{1}{2^\zeta} \sum_{i}^{2^\zeta} \left[\frac{1}{\bar{L}_{DPIM}} Q\left(\frac{y_i + \alpha_{opt}}{\sqrt{N_0/2}} \right) + \frac{\left(\bar{L}_{DPIM} - 1\right)}{\bar{L}_{DPIM}} Q\left(\frac{\alpha_{opt} - y_i}{\sqrt{N_0/2}} \right) \right] \quad (5.55)$$

For each value of D_T in the range 10^{-3}–0.4, the average optical power requirements for various orders of DPIM(NGB) are shown in Figure 5.45. For comparison, the average OPR for OOK is also shown in the figure. In the absence of multipath dispersion or when the normalized delay spread is small, PPM with a hard decision decoding has a lower average OPR than DPIM for any given order. This is due to the lower average duty cycle of PPM, which results in increased power efficiency. However, due to its lower bandwidth requirement, DPIM has a lower ISI-induced power penalty than PPM. Consequently, as D_T increases, the average OPR curves for the two schemes intersect, and beyond the point of intersection, DPIM offers lower

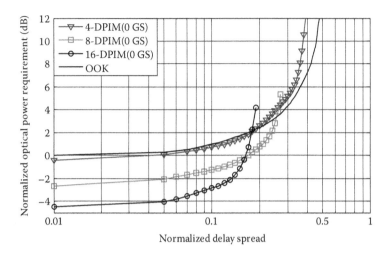

FIGURE 5.45 NOPR against the normalized delay spread for 4, 8 and 16-DPIM with and without guard slots in a dispersive channel.

OPR of the two schemes. As an example, at $D_T = 10^{-2}$, PPM has average OPRs that are lower than DPIM by 1 and 1.1 dB for $L = 4$ and $L = 16$, respectively. However, when $D_T = 0.2$, it is the DPIM that outperforms PPM by ~0.8 dB and ~0.6 dB for $L = 4$ and $L = 8$, respectively. In contrast, for a given order, PPM(SDD) always achieves a lower OPR than DPIM. Compared with OOK, all orders of DPIM considered offer a lower average OPR for D_T below ~0.18. However, beyond ~0.26 it is the OOK which yields the lowest OPR.

For a DPIM slot sequence propagating through a multipath channel, the postcursor ISI is the most severe in slots immediately following a pulse. Hence, placing guard slot(s) immediately after the pulse increases the immunity of DPIM to the multipath-induced ISI [5]. At the receiver, on detection of a pulse, the time slot(s) within the guard band are automatically assigned as zeros, regardless of whether or not the sampled output of the receiver filter is above or below the threshold level. Thus, the postcursor ISI present in these slot(s) has no effect on system performance, provided the pulse initiating the DPIM symbol is correctly detected.

5.7 MITIGATION TECHNIQUES

One of the major difficulties with the diffuse indoor IR environment is intersymbol interference caused by the many reflections a transmitted signal undergoes before arriving at the receiver. A signal reflected once would have the minimum delay, except in the special LOS case, and maximum power provided that all surfaces attenuate the signal equally. Unfortunately, other paths within the room will reflect the signal many more times resulting in a delayed and attenuated additional contribution to the once reflected signal seen at the receiver; these contributions result in a 'smearing' of the pulse shape.

Modulation schemes with a low duty cycle are most suitable for indoor diffuse OWC systems due to the limited power available to portable devices and for eye safety requirements. In an ideal LOS channel reducing the modulation duty cycle can result in an improved SNR performance for a given BER. However, in non-LOS multipath channels low duty cycle signals (i.e., shorter pulse duration) are affected more severely due to increased channel dispersion and the subsequent ISI caused by preceding pulses.

The matched filter receiver with a fixed threshold-level detection discussed in the previous section clearly showed that optical power penalty increase exponentially with channel delay spread and the power penalty is higher for modulation schemes with shorter pulse duration. Given the performance constraints of a simple threshold-based receiver in a noisy multipath channel, one needs to find alternative mitigation scheme in order to reduce the degrading effects of ISI and noise. Most techniques employed in diffuse indoor OWC systems are implemented in the electrical domain and are borrowed from the more established RF technologies. In this section a selection of these methods is reviewed and described.

5.7.1 FILTERING

One of the most simple and frequently implemented methods used to improve performance of communication systems is filtering. In IR systems an optical filter placed in front of the photodiode are used to reduce all out-of-band natural and artificial light signal, thus improving the SNR. Additionally electrical analogue or digital filters are to improve the SNR by rejecting high-frequency components that are not associated with the received signal. One of such filters is the Matched filter, which is the optimum filter for digital communications in the AWGN channel. The perfect way of implementing this scheme is by making the normalized gain response of the receiver filter $H_r(f)$ identical to the amplitude spectrum of the pulse $G_r(f)$ as in Equation 5.56 [48]. This well-known technique is abundantly documented in the literature [48,49] and can easily be approximated in systems and simulations by the simple integrate and dump circuitry; as the average power of the noise is zero, such circuitry maximizes the SNR in the same way.

$$|H_r(f)| = |G(f)| \qquad (5.56)$$

The presence of multipath-induced ISI will severely affect the performance of a matched filtered-based receiver if it is matched to the transmitter filter and the channel distortion is not taken into consideration; therefore, matching will no longer be accurate. Often, standard low-pass, high-pass or band-pass filters are used to lessen the effects of noise in communication systems. For a simple indoor diffuse IR system, a low-pass filter would be used to reject high frequency noise components. Carruthers and Kahn [50], suggest a normalized f_c of $0.6/T_b$ for the unequalized OOK–NRZ data format. However, simulations of such filters in a basic threshold-based receiver and a multipath channel show the optimum normalized f_c is nearer to $0.7/T_b$. Figure 5.46 illustrates the BER against the SNR for 50 Mb/s OOK–NRZ and

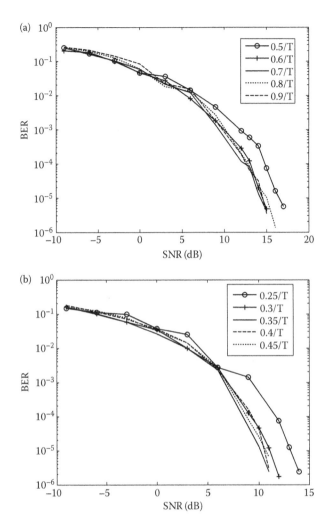

FIGURE 5.46 BER against the SNR for (a) 50 Mb/s OOK–NRZ with various cut-off frequencies and (b) 25 Mb/s OOK–RZ with various cut-off frequencies.

25 Mbit/s OOK–RZ data formats for a range of f_c showing almost the same profile except for f_c of $0.5/T_b$.

Figure 5.47 depicts the BER performance of a threshold-based receiver against the SNR for OOK–NRZ/RZ with f_c of $0.6/T_b$ and $0.35/T_b$, respectively. It can be seen that the lower data rates NRZ show improved BER performance compared with higher data rates. Although the filtered RZ case is an improvement over the unfiltered case, its BER performance advantage over NRZ has disappeared completely at the lower data rates. Given these results, it is clear that simple low-pass filtering alone cannot compensate for the multipath-induced distortion in short duration pulse modulation schemes such as OOK–RZ and PPM.

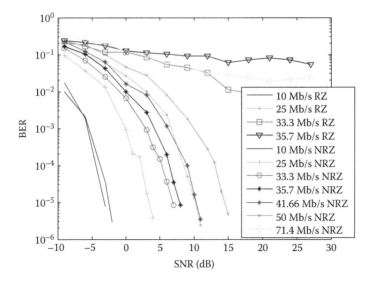

FIGURE 5.47 (See colour insert.) BER against SNR for OOK–NRZ/RZ with $0.6/T_b$ and $0.35/T_b$ low-pass filtering.

5.7.2 EQUALIZATION

Equalization is concerned with compensating for imperfections in the channel. In the simplest case the equalizer filter has the inverse characteristics of the channel, essentially restoring the transmitted pulse shape at the receiver. There are two main methods as well as other related variations of equalizers that are frequently used in communication systems. These are the zero forcing equalizer (ZFE) and the decision feedback equalizer (DFE) both implemented as digital filters.

5.7.2.1 The Zero Forcing Equalizer

The ZFE implemented as a transversal filter or the FIR filter consists of a taped delay line with an input X_k, usually tapped at intervals of 'T' seconds, where 'T' is the symbol interval. The output of each tap is weighted by a variable gain factor a_N and each weighted output is summed to form the final output of the filter y_k at a particular time kT (see Figure 5.48).

There are $(2N + 1)$ taps where N is chosen large enough to span the ISI with corresponding weights $a_{-N} \ldots a_0 \ldots a_N$. The minimization of ISI is achieved by considering only those inputs that appear at the correct sample times. For convenience, $x(kT) = x_k$ and $y(kT) = y_k$. For zero ISI we require [48]:

$$y_k = \begin{cases} 1 & \text{for } k = 0 \\ 0 & \text{elsewhere} \end{cases} \tag{5.57}$$

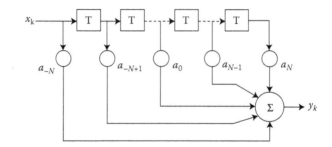

FIGURE 5.48   Structure of the ZFE.

The output y_k can be expressed in terms of the inputs and tap weights:

$$y_k = \sum_{n=-N}^{N} a_n x_{k-n} \tag{5.58}$$

Therefore,

$$y_k = \begin{cases} 1 & \text{when } k = 0 \\ 0 & \text{when } k = \pm 1, \pm 2, \ldots \pm N \end{cases} \tag{5.59}$$

A transversal equalizer can force the output to go to zero at N-point either side of the peak output and we can solve the $(2N+1)$ equations in its matrix form as follows:

$$\begin{bmatrix} x_0 & x_{-N+1} & \cdots & x_{-N} & \cdots & x_{-2N+1} & x_{-2N} \\ x_{N-1} & x_0 & \cdots & x_{-N+1} & \cdots & x_{-N} & x_{-2N+1} \\ \vdots & \vdots & & & & & \\ x_N & x_{N-1} & \cdots & x_0 & \cdots & x_{-N+1} & x_{-N} \\ \vdots & \vdots & & & & & \\ x_{2N-1} & x_N & \cdots & x_{N-1} & \cdots & x_0 & x_{-N+1} \\ x_{2N} & x_{2N-1} & \cdots & x_N & \cdots & x_{N-1} & x_0 \end{bmatrix} \begin{bmatrix} a_{-N} \\ a_{-N+1} \\ \vdots \\ a_0 \\ \vdots \\ a_{N-1} \\ a_N \end{bmatrix} = \begin{bmatrix} 0 \\ 0 \\ \vdots \\ 1 \\ \vdots \\ 0 \\ 0 \end{bmatrix} \tag{5.60}$$

$$a = X^T q \tag{5.61}$$

where a is the filter coefficient array, X is the sample point matrix and q is the output array.

A major drawback of the ZFE is that it ignores the presence of additive noise and its use may result in a significant noise enhancement. This can be understood by noting that the equalizer filter response is the inverse of the channel characteristics $G_\theta(f) = 1/C(f)$; so the equalizer introduces large gains where $C(f)$ is small, that is, boosting noise in the process. Relaxing the zero ISI condition and selecting a channel equalizer characteristic such that the combined power of the residual ISI and the additive noise at the equalizer output is minimized can improve the performance of a ZFE. This is achieved by using an equalizer that is optimized based on the minimum mean square error (MMSE) criterion [51]. Although the characteristics of an indoor OWC channel are quasi-static they are not uniform for any given receiver transmitter pair. For example, moving a receiver from one part of a room to another will alter the impulse response $h(t)$ of the channel as seen by the receiver. We cannot therefore embody a common set of filter coefficients at the receiver. It is common that a training pulse is transmitted to the receiver in order to adjust the weights of the filter taps. Such a filter is usually referred to as a preset equalizer since the coefficients are calculated prior to transmission. A further variation on the ZFE is the adaptive equalizer where the tap weights are adjusted with time to compensate for fluctuations in the channel response.

5.7.2.2 Minimum Mean Square Error Equalizer

Based on the classical equalization theory, the most common cost function is the mean-squared error between the desired signal and the output of the equalizer [52]. An equalizer that minimizes this cost function is therefore known as the MMSE as shown in Figure 5.49. This type of equalizer also relies on periodically transmitting a known training sequence in order for the algorithm to minimize the mean square error.

The output from the equalizer is given as

$$y_k = \sum_{n=0}^{N-1} w_n x_{k-n} \tag{5.62}$$

Taking the expected 'E' or mean values of the squared error between the output and training sequence, dropping the 'k' for notational clarity and writing in terms of the correlation matrix $\mathbf{R}$ and the cross-correlation vector $\mathbf{p}$ gives

$$E[e^2(k)] = E[d^2(k)] + \mathbf{w}^T\mathbf{R}\mathbf{w} - 2\mathbf{w}^T\mathbf{p} \tag{5.63}$$

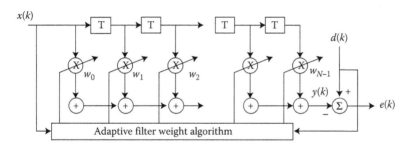

FIGURE 5.49   Structure of the minimum mean square error equalizer.

where

$$\mathbf{R} = E[\mathbf{x}\mathbf{x}^T] = E \begin{bmatrix} x_k \\ x_{k-1} \\ x_{k-2} \end{bmatrix} [x_k \ x_{k-1} \ x_{k-2}] = E \begin{bmatrix} (x_k^2) & (x_k x_{k-1}) & (x_k x_{k-2}) \\ (x_{k-1} x_k) & (x_{k-1}^2) & (x_{k-1} x_{k-2}) \\ (x_{k-2} x_k) & (x_{k-2} x_{k-1}) & (x_{k-2}^2) \end{bmatrix} \quad (5.64)$$

And

$$\mathbf{p} = E[d_k \mathbf{x}_k] = E \begin{bmatrix} d_k x_k \\ d_k x_{k-1} \\ d_k x_{k-2} \end{bmatrix} = \begin{bmatrix} p_0 \\ p_1 \\ p_2 \end{bmatrix} \quad (5.65)$$

In general, for an 'N' weight filter

$$\mathbf{p} = \begin{bmatrix} p_0 \\ p_1 \\ \vdots \\ p_{N-1} \end{bmatrix} \quad (5.66)$$

The optimum weight vector is then found by setting the partial derivative gradient vector to zero (i.e., $\nabla = 2\mathbf{R}\mathbf{w} - 2\mathbf{p} = 0$); therefore,

$$\mathbf{w}_{opt} = \mathbf{R}^{-1}\mathbf{p} \quad (5.67)$$

5.7.2.3 Decision Feedback Equalizer

This is a possible improvement on the ZFE and MMSE. It is a nonlinear device that uses previous decisions to counter the ISI caused by the previously detected symbol on the current symbol to be detected. The DFE consists of two filters, a feedforward filter (prefilter) and a feedback filter (ISI estimator) (see Figure 5.50 [52]). The feed-forward filter is generally a fractionally spaced FIR filter with adjustable tap coefficients and has a form identical to the linear ZFE [51]. The feedback filter is an FIR filter with symbol spaced taps having adjustable coefficients, its input being the set

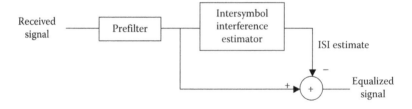

FIGURE 5.50 Schematic DFE structure.

of previously detected symbols. The output of the feedback filter is subtracted from the output of the feedforward filter to form the input to the detector. The DFE does suffer from performance degradation due to the feedback errors; however, the errors do not cause catastrophic failure and occur at relatively higher BER, out of the usual operating area.

5.8   EQUALIZATION AS A CLASSIFICATION PROBLEM

Consider a baseband communication system as shown in Figure 5.37. The discrete received signal is given by

$$y_i = \sum_{n=0}^{\zeta} h_n a_{n-i} + n_i = b_i + n_i \tag{5.68}$$

where b_i is the noise-free channel output, n_i is the AWGN and h_n is the channel taps given by the ceiling bounce model.

The Lth-order equalizer has L-taps with an equally spaced delay of τ for the symbol space equalizer. The channel output can be written in a vector form as

$$Y_i = \begin{bmatrix} y_i \ y_{i-1} \dots y_{i-L+1} \end{bmatrix}^T \tag{5.69}$$

where T means the transpose operation.

Hence, the channel has L-dimensional observation space. Depending on the output vector Y_i, the equalizer attempts to classify the receiver vector into one of the two classes: binary zero and binary one. The equalization problem is hence forming a decision boundary that corresponds to the transmitted symbol. Therefore, determining the values of the transmitted symbol with the knowledge of the observation vectors is basically a classification problem [53].

The linear decision boundary may be utilized when the patterns are linearly separable. However, practical channels are not linearly separable and hence a linear boundary region on such channel is not the optimal. In general, the optimal decision boundary is nonlinear and the realization of the nonlinear decision boundary can be achieved by using ANN with a nonlinear transfer function.

5.9   INTRODUCTION TO ARTIFICIAL NEURAL NETWORK

Artificial neural network (ANN) is a mathematical and computer model which is loosely based on the biological neural networks. ANN, with a nonlinear statistical modelling capacity, is extensively used for modelling complex relationships between inputs and outputs and for pattern classification. Although introduced in 1948 [54], the extensive studies of ANN started only in the early 1980s after important theoretical results related to ANN were attained. Now ANN finds its application in diverse areas such as the computing [55], medicine [56–58], finance [59], control system [60], statistical modelling [61] and engineering [62].

ANN consists of simple processing units, neurons (or cells) interconnected in predefined manner. The neurons communicate by sending signals to each other. A neuron is limited to functionality of classifying only linearly separable classes [63]. However, ANN with many neurons can perform a complicated task such as pattern classification, nonlinear mapping. In fact, ANNs with sufficient number of neurons are universal approximators [64].

5.9.1 NEURON

Each neuron in the ANN does a simple task of modifying the input(s) by some pre-defined mathematical rule. The neuron has N inputs $\{x_i; i = 1, \ldots, N\}$, a weight w_i associating with each input and an output y. There may be an additional parameter w_0 known as the bias which can be thought as weight associated with a constant input x_0 always set to 1. The functional block diagram of a neuron is shown in Figure 5.51.

The intermediate output a can be calculated mathematically as

$$a = \sum_i w_i x_i \tag{5.70}$$

where $i = 0, \ldots, N$ if there is a bias and $i = 1, \ldots, N$ otherwise.

The output y is a function of the activation a and is given by

$$y = f(a) \tag{5.71}$$

The activation or transfer function $f(.)$ depends on the application but is generally differentiable. Some popular activation functions are described below.

1. *Linear function:* defined as $y = ma$. For $m = 1$, the function is known as an identity function as the output is the exact copy of input.
2. *Binary threshold function:* limits the activation to 1 or 0 depending on the net input relative to some threshold function. Considering a threshold level of θ, the output is given by

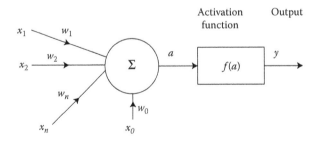

FIGURE 5.51 The schematic diagram of neuron showing inputs, a bias, weights and an output.

$$y = \begin{cases} 1 & \text{if } a \geq \theta \\ 0 & \text{if } a < \theta \end{cases} \tag{5.72}$$

3. *Sigmoid function (logistic and tanh):* are commonly used activation functions for nonlinear processing and pattern classification. The output of the log-sigmoid function is a continuous function in the range of 0–1 defined as

$$y = \frac{1}{1 + e^{-a}} \tag{5.73}$$

The tan-sigmoid function is a variation of the log-sigmoid function as output ranges from −1 to +1, given by

$$y = \tanh(a) \tag{5.74}$$

5.9.2 ANN ARCHITECTURES

The neurons can be interconnected in different and complex manners. The simplest way is to arrange neurons in a single layer. However, there are other topologies, of which the most common are briefly described below.

1. *Single-layer feedforward network:* It contains only an output layer connected to an input layer (Figure 5.52). The number of neurons in each layer varies depending on the number of inputs and outputs requirement for a particular application, but the network is strictly feedforward, that is, signal flows from the input layer to the output layer only. Single-layer networks are limited in capacity because of classification capability in linearly separable classes [63].
2. *Multilayer feedforward network:* In order to extract the higher-order statistics of data, a higher number of neurons and layers are necessary [65]. Hence, multilayer networks with capability of forming complex decision

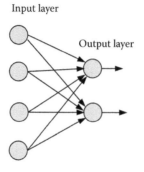

FIGURE 5.52 A single-layer feedforward network with two output neurons.

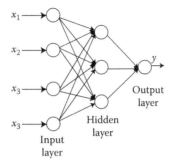

FIGURE 5.53   Fully connected feedforward multilayer network.

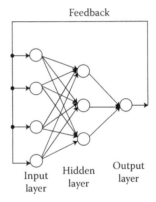

FIGURE 5.54   Neural network with feedback connection.

regions is utilized in different applications. Provided there is a sufficient number of neurons a 2-layer ANN can be used as a universal approximator mapping any input–output data set [64]. Figure 5.53 shows a fully connected 2-layer network. Every multilayer network consists of (i) an input layer with no processing taking place, thus not counted as part of a network layer, (ii) a hidden layer and (iii) an output layer.

3. *Recurrent neural network (RNN):* Unlike a multilayer network, here the output is feedback to the input layer as shown in Figure 5.54. The network is very useful in time-series prediction. The training is more difficult in the recurrent network because of its chaotic behaviour [66]. Elman and the Hopfield networks are common recurrent network architecture [66].

5.10   TRAINING NETWORK

The training of ANN is accomplished by providing ANN with the input and desired output data sets. The ANN can adjust its free parameters (weights and bias) based on the input and the desired output to minimize the cost function $E(n)$ (difference between the desired response $d(n)$ and actual response $y(n)$ of the network). The training process depends on two factors: a training data set and a training algorithm. The training data set should be representative of the task to be learnt. A poorly selected training set may increase the learning time. The training algorithm requires a minimization of the cost function. The convergence of cost function to the local minima instead of the global minimum has been an issue and algorithms based on the adaptive learning rate can improve the learning rate as well as a convergence to the global maximum [67,68].

The learning can be classified into supervised and unsupervised learning. In supervised learning, the network is provided with an input–output pair that represents the task to be learnt. The network adjusts its free parameters so that the cost function is minimized. In unsupervised learning or self-organization, the desired response is not present and the network is trained to respond to clusters of pattern within the input. In this paradigm the ANN is supposed to discover statistically salient features of the input population [69].

In this study, multilayer feedforward ANN is used for adaptive channel equalization for the indoor OWC. The supervised learning is more suitable of the equalization as the training sequence can be used to approximate channel more accurately. A popular supervised training algorithm known as the backpropagation (BP) is described below.

5.10.1 BACKPROPAGATION LEARNING

The BP supervised learning algorithm is the most popular training algorithm for multilayer networks. It adjusts the weight of ANN to minimize the cost function $E(n)$:

$$E(n) = \|d(n) - y(n)\|^2 \tag{5.75}$$

BP algorithm performs a gradient descent on $E(n)$ in order to reach a minimum. The weights are updated as

$$w_{ij}(n+1) = w_{ij}(n) - \eta \frac{\partial E(n)}{\partial w_{ij}(n)} \tag{5.76}$$

where w_{ij} is the weight from the hidden node i to the node j and η is the learning rate parameter. The performance of the algorithm is sensitive to η. If η is too small, the algorithm takes a long time to converge and if η is too large, the system may oscillate causing instability [65]. Hence, adaptive leaning rates are adopted for faster convergence [67,68].

The BP can be summarized as [70]:

1. Step 1: Initialize the weights and thresholds to small random numbers.
2. Step 2: Present the input vectors, $x(n)$ and desired output $d(n)$.
3. Step 3: Calculate the actual output $y(n)$ from the input vector sets and calculate $E(n)$.

4. Step 4: Adapt weight based on (5.76).
5. Step 5: Go to step 2.

5.11   THE ANN-BASED ADAPTIVE EQUALIZER

The architecture of the ANN-based channel equalizer is depicted in Figure 5.55. The sample output y_i is passed through tap delay lines (TDLs) prior to being presented to the ANN for channel equalization. The output of the ANN is sliced using $\alpha_{th} = 0.5$ to generate a binary data sequence. In the DF structure, the decision output is fed back to the ANN. The length of both the forward and feedback TDLs depend on the channel span, which also depends on the delay spread. Since the ceiling bounce model [50] shows an exponential decay in the channel components, few TDLs are adequate for the optimum performance. The ANN needs to be trained in a supervised manner to adjust its free parameters.

These are open questions: What should be the size of the network? What architecture should be chosen? How many layers should be in the network? There are neither satisfactory answers to nor a predefined rule for these questions. The number of neurons in the input and output layers depends on the input vector and the desired outputs. However, the number of hidden layers and neurons can be varied. Increasing the hidden layers results in increased system complexity and the training time; however, too few neurons may result in an unsatisfactory performance. The goal of optimizing ANN is to find an architecture which takes less time to train, less storage requirement and less complexity without performance degradation. There are two major approaches for designing an optimal network structure: (i) build a larger network and prune the number of neurons with reduced connections, and (ii) start with a small network and neurons and continue to grow until the desired level of performance is achieved [71].

Theoretically there is no reason for using more than two hidden layers; practically the ANN structure with a single hidden layer is sufficient. Using two hidden layers rarely improves the model, and it may introduce a greater risk of converging to a local minima of cost function [72]. If sufficient neurons in a hidden layer are used, then it is not necessary to use more than one hidden layer [73]. Due to the risk of converging to the local minima, a single hidden layer is utilized with sufficient number of neurons. Theoretically, it is difficult to determine the exact number of neurons required in the hidden layer. Hence, in this study, six neurons in the hidden layer are used as simulation results have shown that it could provide near-optimum results for all range of delay spread for all modulation techniques, though the number of neurons can be reduced for a less dispersive channel.

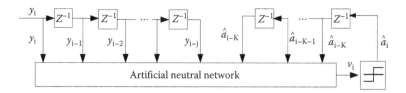

FIGURE 5.55   ANN-based DFE structure.

TABLE 5.2
List of Training Algorithms for ANN

Training Algorithm	Average Training Time (s)
Levenberg–Marquardt	1.88
BFGS algorithm	7.00
Variable learning rate	4.2
Resilient backpropagation	2.64
Scaled conjugate gradient	2.83
Conjugate gradient with Powell/Beale restarts	3.07
Fletcher–Powell conjugate gradient	3.90
Polak–Ribiére conjugate gradient	2.46

Source: Adapted from S. Rajbhandari, Application of wavelets and artificial neural network for indoor optical wireless communication systems, PhD, Northumbria University, Newcastle upon Tyne, 2010.

Another key issue in applying the ANN in supervised learning method is the learning algorithm. Although the philosophy of all training algorithms is the same, to minimize the cost function, they differ in the algorithm to obtain the minimization of cost function. Some training algorithms converge faster, but require larger memory space while some guarantee convergence to the global minimum of cost function [67]. The training algorithms available in MATLAB 7.4.0 are listed in Table 5.2. For the comparative study, a 2-layer ANN with six and one neurons in the hidden and output layers, respectively, is selected. The SNR is chosen as 24 dB for all simulation with a different combination of transfer functions. For each training algorithm, a total of nine simulations are carried out and the average time required for successful training is calculated and listed in the second column of Table 5.3. The Levenberg–Marquardt algorithm requires the least time to train whereas the variable learning rate (traingdx) requires the longest time. However, it is to be noted that the memory requirements for the Levenberg–Marquardt algorithm is larger than for any other algorithms [74]. Another key factor in selecting the training algorithm is the performance evaluation under different conditions. The BER performance for a

TABLE 5.3
ANN Parameters for Equalizations

Parameters	Values
ANN type	Feedforward BP MLP
Number of hidden layer	1
Transfer functions (hidden layer)	Log-sigmoid
Transfer functions (output layer)	Linear
Training length	1000
Training algorithm	Scaled conjugate gradient

range of SNRs and the channel delay spreads are simulated for training algorithms requiring a training time of <3 s. It is found that the scaled conjugate gradient algorithm provides the most consistence performance with improved performance compared to the Levenberg–Marquardt in a highly dispersive channel. Based on the optimization discussed above, a feedforward back-propagation ANN with the parameters given Table 5.3 is chosen for all the simulations hereafter.

The program to simulate the BER probability using ANN-based linear and ANN-based DFE is given in Programs 5.9 and 5.10, respectively. Note that only the training and equalization part of the simulation is given. For complete system simulation, follow Program 5.1, and programs given in previous chapters. Note that simulation for PPM and DPIM can be carried out using a similar procedure.

Program 5.9: MATLAB Codes for the ANN-Based Linear Equalizer for OOK–NRZ

```
%% system impulse response
Tx_filter=ones(1,nsamp);
Rx_filter=fliplr(Tx_filter);
c=conv(Tx_filter,h);
delay=find(c==max(c));
hi=c(delay:nsamp:end);
% channel impulse response
hi=hi/sum(hi);
% normalization

%% ANN parameters
ff_TDL=3;
% feed forward length
ff_zeros=zeros(1,ff_TDL);
nneu=6;
% number of neurons in hidden layer

%% Training Neural Network
tlen=1500;
%training length
OOK=randint(1, tlen);
% random signal generation
Rx_signal=awgn(filter(hi,1,OOK),EbN0_db+3,'measured');
% Received signal, notice matched filter is not optimum
training_input=input_ANN([ff_zeros Rx_signal],ff_TDL,tlen);
% training input
net=newff(minmax(training_input),[nneu 1],{'logsig','purelin'},
'trainscg');
net=init(net);
net.trainParam.epochs=1000;
net.trainParam.goal=1e-30;
net.trainParam.min_grad=1e-30;
net.trainParam.show=NaN;
```

```
[Net] =train(net,training_input,OOK);
%% Simulation of ANN output and BER
OOK = randint(1,sig_length);
% random signal generation
Rx_signal = awgn(filter(hi,1,OOK),EbN0_db + 3,'measured');
% Received signal, notice matched filter is not optimum
ann_input = input_ANN([ff_zeros Rx_signal],ff_TDL,sig_length);
ANN_output = sim(Net,ann_input);
Rx_th = zeros(1,sig_length);
Rx_th(find(ANN_output > 0.5)) = 1;

function [training_set] =input_ANN_linear(training_input,
  ff_TDL,sig_length)
% function to generate the ANN input
 start =1;
 finish = ff_TDL + 1;
 training_set = zeros(ff_TDL + 1,sig_length);
  for i = 1:sig_length
    training_set( :,i) =training_input(start:finish)';
   start = start + 1;
   finish = finish + 1;
  end
end
```

Program 5.10: MATLAB Codes for the ANN-Based DFE for OOK–NRZ

```
%% ANN parameters
ff_TDL = 3;
% feed forward length
fb_TDL = 2;
% feedback length
ff_zeros = zeros(1,ff_TDL);
fb_zeros = zeros(1,fb_TDL);
nneu = 6;
% number of neurons in hidden layer

%% Training ANNN
tlen = 1500;
%training length
OOK = randint(1, tlen);
% random signal generation
Rx_signal = awgn(filter(hi,1,OOK),EbN0_db + 3,'measured');
% Received signal, notice matched filter is not optimum
training_input = ANN_input_dfe([ff_zeros Rx_signal],ff_TDL,fb_
TDL,tlen,[fb_zeros OOK]);
net = newff(minmax(training_input),[6 1],{'logsig','purelin'},
'traincgb');
net = init(net);
net.trainParam.epochs = 1000;
```

```
net.trainParam.goal = 1e-30;
net.trainParam.min_grad = 1e-30;
net.trainParam.show = NaN;
[Net] = train(net,training_input,OOK);

%% ANN based DFE
OOK = randint(1,sig_length);
 % random signal generation
 Rx_signal = awgn(filter(hi,1,OOK),EbN0_db + 3,'measured');
 % Received signal, notice matched filter is not optimum
 Rx_signal = [ff_zeros Rx_signal];
 Rx_th = fb_zeros;
 for j = 1:sig_length
 ann_input = [Rx_signal(j:j + ff_TDL) Rx_th(j:j + fb_TDL-1)]';
 ann_output(j) = sim(Net,ann_input);
 if ann_output(j) >0.5, Rx_th = [Rx_th 1];
 else Rx_th = [Rx_th 0];
 end
 end
 % bit error calculation
 nerr = biterr(OOK,Rx_th(fb_TDL + 1:end));

function [training_set] = ANN_input_dfe(Rx,ff_TDL,fb_TDL,sig_len,
Rx_th)
% function to generate input for DFE
training_set = [];
for i = 1:sig_len
 training_set( :,i) = [Rx(i:i + ff_TDL) Rx_th(i:i + fb_TDL-1)]';
end
end
```

The NOPR of OOK–NRZ for the ANN-based linear and DF equalizers for a range of D_T are shown in Figure 5.56. Also shown is the NOPR for the unequalized case for a matched filter-based receiver. The DF scheme offers the least NOPR for a range of D_T closely followed by the linear case. Unlike the unequalized system where the irreducible NOPR can be observed at $D_T > 0.5$, the equalized cases do not show such case even for $D_T = 2$ for both linear and DF equalizers. The linear and DF schemes show identical performance for $D_T < 1$ varying only at $D_T > 1$. At lower D_T values, typically <0.01, the equalizers show no improvement in the performance. However, the effectiveness of the equalizer in reducing the NOPR becomes apparent with increasing D_T. The optical power gain achieved with the equalizer increases as the channel became more dispersive. For example, DFE yields NOPR reduction of ~0.8 and ~6 dB for D_T of 0.1 and 0.4, respectively. Compared to the LOS ($D_T=0$) case, the OPP for D_T of 2 is ~6 dB for DFE, OPP increases with D_T. Two factors have contributed to the power penalty: (i) ever presence of the residual ISI particularly when channel is highly dispersive and (ii) incorrect decisions being feedback to ANN [76].

The structure of the linear equalizer for the PPM scheme is essentially similar to that of the OOK equalizer. The hard or soft decision decoding schemes can be

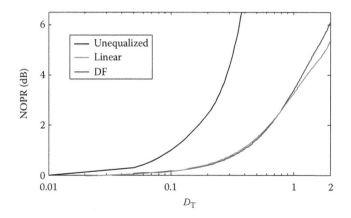

FIGURE 5.56 NOPR versus the normalized delay spread for unequalized and ANN equalized (linear and decision feedback) OOK schemes.

carried out to the equalized output for regeneration of PPM symbols. Similar to the OOK, a threshold value of 0.5 is used for the HDD scheme. However, in the SDD, the entire slots within the symbol are compared and a slot with the maximum amplitude is assigned one and all other slots are automatically set to zero. The DF structure for HDD is simple to implement as the hard decision output can be directly feedback to the ANN. However, the DF is difficult to implement for SDD as the receiver has to buffer a PPM symbol before making a decision. The delay in the decoding means a soft decision output cannot be feedback. One simple approach is to feedback the hard decision output. The key issue for the SDD is the feedback of the hard decision. As a result, the error in the hard decision tends to propagate. Thus, a performance improvement is difficult to achieve using SDD compared to the HDD though complexity increased significantly as both hard and soft decision decoding is necessary at the receiver.

The NOPR to achieve an SER of 10^{-6} for the PPM(HDD) scheme with an ANN-based adaptive equalizer for a range of D_T is shown in Figure 5.57. Also shown is the NOPR for the unequalized PPM. Similar to the unequalized case, the NOPR for the equalized cases also shows an exponential growth. However, the gradients of the curves are significantly reduced and do not show the infinite power penalties even at a D_T of 2. Unlike the unequalized cases where higher-order PPM shows a sharp rise in the NOPR with the increasing value of D_T, equalized systems show almost similar profile for all bit resolutions with reduced increment in the power penalty. As expected, the reduction in NOPR for the equalizer system compared to the unequalized case increases with increasing value of D_T. For example, the reduction in NOPR for the equalized system compared to the unequalized case is 0 dB and ~7.3 dB at D_T values of 0.01 and 0.3, respectively, for 4-PPM. However, the gain is larger for 8-PPM with ~9.6 dB difference at D_T of 0.3. Higher gain is achieved at higher values of D_T. However, comparison is not possible for the higher D_T values as the unequalized cases require very high NOPR. The NOPR of DFE for PPM with hard decision

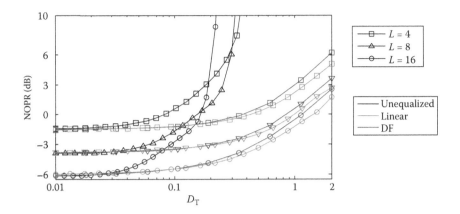

FIGURE 5.57 NOPR against the normalized delay spread D_T for unequalized and ANN-based equalizers for PPM(HDD) scheme.

shows a similar trend to OOK. The DFE offers a reduction in OPP compared to the linear equalizer at $D_T > 0.2$.

The NOPR against D_T for the 4, 8 and 16-PPM(SDD) with an ANN-based adaptive linear equalizer is depicted in Figure 5.58. Also shown is the NOPR for the unequalized cases. As in other equalized cases, the equalized system illustrates a significantly improved performance compared to unequalized cases and showing no infinite NOPR for the D_T of 2. Since the soft decision decoding provides natural immunity to the ISI even without an equalizer, the reduction in NOPR using the equalizer is higher for the hard decision decoding. For example, the equalized system offers a reduction of ~7.3 and ~4 dB in NOPR for 4-PPM system at D_T of 0.3 for the hard and soft decision decoding schemes, respectively. It should be noted that higher gain in the equalized hard decision should not be interpreted as ineffectiveness of the equalizer for the soft decoding. Rather the soft decoding without an equalizer also provides some resistance

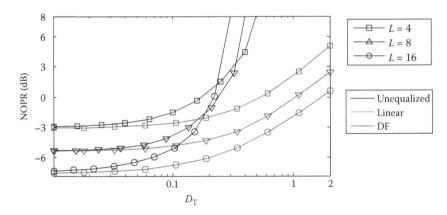

FIGURE 5.58 NOPR against the normalized delay spread D_T for unequalized and ANN-based linear equalizers for PPM(SDD) scheme.

to the ISI, meaning reduced gain with an equalizer. For example, the NOPRs for the unequalized 4-PPM at a D_T of 0.3 are ~6.9 and ~2.5 dB for the hard and soft decision decoding schemes, respectively. For the equalized systems, the NOPRs are ~−0.42 and ~−1.51 dB for the hard and soft decision decoding, respectively. This clearly indicates that the soft decoding offer reduced NOPR for both the equalized and unequalized systems. The simulation confirmed that DFE with soft decoding does not provide any improvement in performance compared to the hard decoding due to the hard decision in the feedback loop and is hence not reported here.

Note that unlike the unequalized cases, where the higher order PPM system shows a sharp rise in the OPP, the equalized system shows almost a similar profile for all M. The OPP is the highest for 16-PPM with both soft and hard decoding cases due to shortest slot durations. However, equalized 4-PPM and 8-PPM illustrate identical OPPs. Two factors are involved here: (a) the slot duration and (b) the probability of two consecutive pulses. Since 8-PPM has shorter pulse duration, the OPP due to this factor should be higher than that for 4-PPM. On the other hand, due to longer symbol length, the probability of two consecutive pulses is significantly lower for 8-PPM compared to 4-PPM, hence the OPPs related to two consecutive pulses is higher for 4-PPM (avoiding two consecutive pulses can provide significant performance improvement as in the case of DPIM [11]).

The NOPR to achieve an SER of 10^{-6} against the D_T for equalized 4, 8 and 16 DPIM with an ANN-based equalizers is outlined in Figure 5.59. Also shown is the NOPR of the unequalized system. Similar to other equalized systems, ANN-based equalizer shows a significant reduction in NOPR compared to the unequalized cases for $D_T > 0.1$. Linear equalizer offers ~6 dB reduction in NOPR for 4 DPIM compared to the unequalized case at $D_T = 0.3$. Note that it is not possible to completely remove the ISI and hence a nominal OPP occurs for the equalized case if $D_T < 0.1$ in comparison with the LOS channel. However, higher OPP is visible at $D_T > 0.1$ and the OPP is ~6 dB at $D_T = 2$ for all cases. As in the previous cases, the DF structure provides improved performance compared to the linear structure for a highly dispersive channel and a difference of ~0.6 dB is observed at D_T of 1.5.

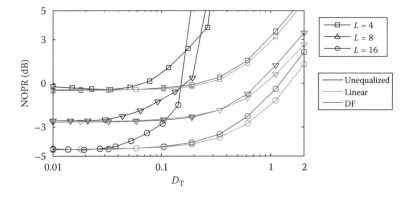

FIGURE 5.59 NOPR versus the normalized RMS delay spread for unequalized and ANN equalizers for the DPIM scheme.

5.11.1 COMPARATIVE STUDY OF THE ANN- AND FIR-BASED EQUALIZERS

The comparative studies of the linear and DF-based equalizers based on the ANN and the traditional FIR filters are carried out in this section. To evaluate the performance of the traditional and the ANN-based equalizers, the MSE is calculated between the equalizer outputs and the desired outputs. MSEs are calculated in identical channel conditions and hence can be used to measure for effectiveness of the equalizer. To calculate the MSE, 1000 random bits are transmitted through a diffuse channel with D_T of 2 at a data rate of 200 Mbps using the OOK modulation scheme. For simplicity as well as for more comprehensive comparisons, the channel is assumed to be noise free and hence error in the equalizer outputs is solely due to dispersion in the channel. The noise-free received sequence is used for training of the ANN and the traditional equalizer. The outputs of the equalizers are compared with the desired output to calculate MSEs. The resulting MSEs for both equalizers are given in Figure 5.60. The MSE for the traditional linear equalizer is in the range of 10^{-4}–10^{-1} while for the ANN equalizer is in the range 10^{-8}–10^{-4}. This indicates the effectiveness of the ANN as an equalizer compared to the traditional equalizer.

The bit error probability of the ANN and the traditional linear equalizers for the OOK modulation scheme in a dispersive channel with D_T of 2 at a data rate of 200 Mbps is given in Figure 5.61. The figure reveals that the traditional linear equalizer can match the ANN-based equalizer even in a highly dispersive channel but the advantage of the ANN is in terms of reduced training length (TL). The simulation results show that the number of training symbols required for the traditional equalizer is significantly higher than that of the ANN equalizer especially in a highly dispersive channel. BER performance indicates that the ANN trained using 200 bits offers almost identical performance to that of the traditional equalizer trained using

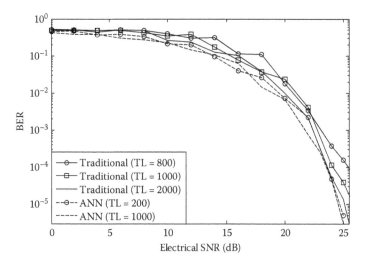

FIGURE 5.60 BER against the electrical SNR for the traditional and ANN linear equalizers for the OOK scheme at a data rate of 20 Mbps for a channel with D_T of 2 with different training lengths (TLs).

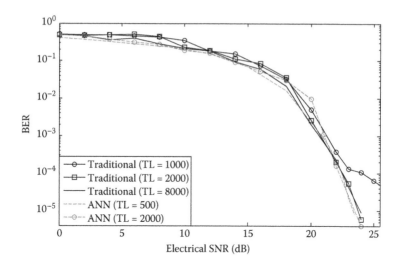

FIGURE 5.61   BER against the electrical SNR for the traditional and ANN DF equalizers for the OOK modulation scheme at a data rate of 200 Mbps for a channel with D_T of 2 with different training lengths.

1000 bits for higher SNR values. Lower training length means reduced training time, less complexity and improved throughput.

The bit error probability of the ANN and the traditional DFE for the OOK modulation scheme in a dispersive channel with D_T of 2 at a data rate of 200 Mbps is given in Figure 5.62. As with the linear equalizer, the DF ANN equalizer requires significantly lower training time compared to the traditional equalizer. It is found that the traditional DFE is significantly difficult to train in a highly dispersive channel and the LMS does not converge for almost all ranges of step sizes. Hence, a normalized LMS [77] is used to train the traditional equalizer. On the other hand, all the training algorithms provide almost similar performance for the ANN-based receiver with a training length of 500 matching the performance of the traditional equalizer with a training length of 2000. This also simplifies the ANN structure and parameter optimization compared to the traditional equalizers.

5.11.2 Diversity Techniques

Owing to the problems of high ambient noise and multipath dispersion inherent in diffuse IR systems and the limitation of traditional techniques, alternative methods of mitigating these effects are being investigated including angle diversity [78,79], multispot systems [80–84], sectored receivers [85,86] and code combining [87,88] to name a few. Most of these schemes seek to eliminate noise or ISI by limiting the field of view of the optical receiver, cutting out the noise and limiting the available paths to the receiver thus reducing the ISI. Such techniques could potentially lead to severe shadowing if one transmitter—receiver pair was used in such systems. To overcome this, it is usual to employ several transmitting beams or receiver elements. These schemes undoubtedly have the potential to offer performance improvements, but at the expense

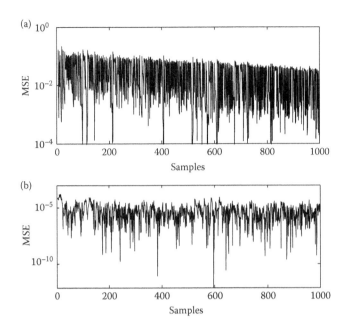

FIGURE 5.62 The MSE between the actual and target outputs from the linear equalizers for: (a) FIR filter equalizer and (b) ANN equalizer.

of system complexity, more difficult deployment and additional hardware cost. However, there are alternative mitigation techniques in the electrical domain that are attractive for OWC diffuse links. It is envisaged that these alternatives scheme could be embodied in many diversity schemes to improve the system performance further.

REFERENCES

1. J. M. Kahn and J. R. Barry, Wireless infrared communications, *Proceedings of IEEE*, 85, 265–298, 1997.
2. A. J. C. Moreira, R. T. Valadas and A. M. d. O. Duarte, Optical interference produced by artificial light, *Wireless Networks*, 3, 131–140, 1997.
3. R. Narasimhan, M. D. Audeh and J. M. Kahn, Effect of electronic-ballast fluorescent lighting on wireless infrared links, *IEE Proceedings—Optoelectronics*, 143, 347–354, 1996.
4. F. R. Gfeller and U. Bapst, Wireless in-house data communication via diffuse infrared radiation, *Proceedings of the IEEE*, 67, 1474–1486, 1979.
5. Z. Ghassemlooy, A. R. Hayes and B. Wilson, Reducing the effects of intersymbol interference in diffuse DPIM optical wireless communications, *IEE Proceedings—Optoelectronics*, 150, 445–452, 2003.
6. A. J. C. Moreira, R. T. Valadas and A. M. d. O. Duarte, Performance of infrared transmission systems under ambient light interference, *IEE Proceedings—Optoelectronics*, 143, 339–346, 1996.
7. S. Lee, Reducing the effects of ambient noise light in an indoor optical wireless system using polarizers, *Microwave and Optical Technology Letters*, 40, 228–230, 2004.

8. A. R. Hayes, Z. Ghassemlooy, N. L. Seed and R. McLaughlin, Baseline-wander effects on systems employing digital pulse-interval modulation, *IEE Proceedings—Optoelectronics*, 147, 295–300, 2000.

9. Z. Ghassemlooy, Investigation of the baseline wander effect on indoor optical wireless system employing digital pulse interval modulation, *IET Communications*, 2, 53–60, 2008.

10. U. Sethakaset and T. A. Gulliver, Differential amplitude pulse-position modulation for indoor wireless optical communications, *EURASIP Journal on Applied Signal Processing*, 2005, 3–11, 2005.

11. A. R. Hayes, Digital pulse interval modulation for indoor optical wireless communication systems, PhD Sheffield Hallam University, UK, 2002.

12. K. Samaras, A. M. Street, D. C. O'Brien and D. J. Edwards, Error rate evaluation of wireless infrared links, *Proceedings of IEEE International Conference on Communications*, Atlanta, USA, 1998, pp. 826–831.

13. A. M. Street, K. Samaras, D. C. Obrien and D. J. Edwards, Closed form expressions for baseline wander effects in wireless IR applications, *Electronics Letters*, 33, 1060–1062, 1997.

14. S. Khazraei, M. R. Pakravan and A. Aminzadeh-Gohari, Analysis of power control for indoor optical wireless code-division multiple access networks using on-off keying and binary pulse position modulation, *IET Communications*, 4, 1919–1933, 2010.

15. K. Samaras, D. C. O'Brien, A. M. Street and D. J. Edwards, BER performance of NRZ–OOK and Manchester modulation in indoor wireless infrared links, *International Journal of Wireless Information Networks*, 5, 219–233, 1998.

16. R. Polikar, *The Engineer's Ultimate Guide to Wavelet Analysis: The Wavelet Tutorial*, 2005, 28/02/2008. Available: http://users.rowan.edu/~polikar/WAVELETS/WTtutorial.html

17. S. G. Mallat, A theory for multiresolution signal decomposition: The wavelet representation, *IEEE Transactions on Pattern Analysis and Machine Intelligence*, 11, 674–693, 1989.

18. B. B. Hubbard, *The World According to Wavelets: The Story of a Mathematical Technique in the Making*. Wellesley, MA: A. K. Peters, 1996.

19. A. Grinsted, J. C. Moore and S. Jevrejeva, Application of the cross wavelet transform and wavelet coherence to geophysical time series, *Nonlinear Processes in Geophysics*, 11, 561–566, 2004.

20. A. Jamin and P. Mahonen, Wavelet packet modulation for wireless communications, *Wireless Communications and Mobile Computing*, 5, 1–18, 2005.

21. A. R. Lindsey, Wavelet packet modulation for orthogonally multiplexed communication, *IEEE Transactions on Signal Processing*, 45, 1336–1339, 1997.

22. B. G. Negash and H. Nikookar, Wavelet-based multicarrier transmission over multipath wireless channels, *IEE Electronics Letters*, 36, 1787–1788, 2000.

23. M. Sablatash, J. H. Lodge and C. J. Zarowski, Theory and design of communication systems based on scaling functions, wavelets, wavelet packets and filter banks, *The 8th Intl. Conf. on Wireless Communication*, Alberta, Canada, 1996, pp. 640–659.

24. F. Dovis, M. Mondin and F. Daneshgaran, The modified Gaussian: A novel wavelet with low sidelobes with applications to digital communications, *IEEE Communications Letters*, 2, 208, 1998.

25. M. M. Akho-Zahieh and O. C. Ugweje, Wavelet packet based MC/MCD-CDMA communication system, *Electronics Letters*, 42, 644–645, 2006.

26. H. Zhang, H. H. Fan and A. Lindsey, Wavelet packet waveforms for multicarrier CDMA communications, *IEEE International Conference on Acoustics, Speech, and Signal Processing*, OH, USA, 2002, pp. III-2557–III-2560.

27. H. Zhang, H. H. Fan and A. Lindsey, A wavelet packet based model for time-varying wireless communication channels, *IEEE Third Workshop on SPAWC*, Taiwan, 2001, pp. 50–53.

28. M. Martone, Wavelet-based separating kernels for array processing of cellular ds/cdma signals in fast fading, *IEEE Transaction on Communications*, 48, 979–995, 2000.

29. X. Fernando, S. Krishnan, H. Sun and K. Kazemi-Moud, Adaptive denoising at infrared wireless receivers, *The International Society for Optical Engineering*, 5074, 199–207, 2003.

30. S. Gracias and V. U. Reddy, Wavelet packet based channel equalization, *Sadhana*, 21, 75–89, 1996.

31. S. Mallat, *A Wavelet Tour of Signal Processing*, 2nd ed., San Diego: Academic Press, 1999.

32. C. S. Burrus, R. A. Gopinath and H. Guo, *Introduction to Wavelets and Wavelet Transforms: A Primer*, New Jersey: Prentice-Hall, 1998.

33. J. P. Gazeau, *Wavelet Analysis in Signal and Image Processing*, Newcastle upon Tyne, UK: Northumbria University, 2007.

34. F. Liu, X. Cheng, J. Xu and X. Wang, Wavelet based adaptive equalization algorithm, *Global Telecommunications Conference 1997*, Phoenix, AZ, USA, 1997, pp. 1230–1234.

35. A. K. Pradhan, S. K. Meher and A. Routray, Communication channel equalization using wavelet network, *Digital Signal Processing*, 16, 445–452, 2006.

36. N. G. Prelcic, F. P. Gonzalez and M. E. D. Jimenez, Wavelet packet-based subband adaptive equalization, *Signal Processing*, 81, 1641–1662, 2001.

37. C. M. Leavey, M. N. James, J. Summerscales and R. Sutton, An introduction to wavelet transforms: A tutorial approach, *Insight*, 45, 344–353, 2003.

38. M. Misiti, Y. Misiti, G. Oppenheim and J.-M. Poggi, *Wavelet Toolbox: User's Guide*. MA, USA: The MathWorks Inc., 2008.

39. G. Strang and T. Nguyen, *Wavelets and Filter Banks*, 2nd ed., Wellesley, MA, USA: Wellesley-Cambridge Press, 1996.

40. R. Polikar, The story of wavelets, *Physics and Modern Topics in Mechanical and Electrical Engineering*, Wisconsin, USA: World Scientific and Eng. Society Press, 1999, pp. 192–197.

41. C. Valens, *A Really Friendly Guide to Wavelets*, 1991. Available: http://pagesperso-orange.fr/polyvalens/clemens/wavelets/wavelets.html

42. R. J. Dickenson and Z. Ghassemlooy, BER performance of 166 Mbit/s OOK diffuse indoor IR link employing wavelets and neural networks, *Electronics Letters*, 40, 753–755, 2004.

43. R. J. Dickenson, Wavelet analysis and artificial intelligence for diffuse indoor optical wireless communication, PhD, School of Computing, Engineering and Information Sciences, Northumbria University, Newcastle upon Tyne, 2007.

44. S. Rajbhandari, Z. Ghassemlooy and M. Angelova, Effective denoising and adaptive equalization of indoor optical wireless channel with artificial light using the discrete wavelet transform and artificial neural network, *IEEE/OSA Journal of Lightwave Technology*, 27, 4493–4500, 2009.

45. J. R. Barry, J. M. Kahn, W. J. Krause, E. A. Lee and D. G. Messerschmitt, Simulation of multipath impulse response for indoor wireless optical channels, *IEEE Journal on Selected Areas in Communications*, 11, 367–379, 1993.

46. J. M. Kahn, W. J. Krause and J. B. Carruthers, Experimental characterization of non-directed indoor infrared channels, *IEEE Transactions on Communications*, 43, 1613–1623, 1995.

47. M. D. Audeh and J. M. Kahn, Performance evaluation of L-pulse-position modulation on non-directed indoor infrared channels, *IEEE International Conference on Communications*, New Orleans, Louisiana, USA, 1994, pp. 660–664.

48. I. Otung, *Communication Engineering Principles*, Basingstoke, England: Palgrave Macmillan, 2001.

49. F. G. Stremler, *Introduction to Communication Systems*, 3rd ed., New York: Addison-Wesley, 1990.
50. J. B. Carruthers and J. M. Kahn, Modeling of nondirected wireless infrared channels, *IEEE Transaction on Communication*, 45, 1260–1268, 1997.
51. J. G. Proakis, M. Salehi and G. Bauch, *Contemporary Communication Systems Using MATLAB*, 2nd ed., Boston: PWS Publishing Company, 2003.
52. A. Burr, *Modulation and Coding for Wireless Communications*. UK: Prentice-Hall, 2001.
53. L. Hanzo, C. H. Wong and M. S. Yee, Neural networked based equalization, in *Adaptive Wireless Transceivers*, Chichester, West Sussex: Wiley-IEEE Press, 2002, pp. 299–383.
54. A. K. Jain, M. Jianchang and K. M. Mohiuddin, Artificial neural networks: A tutorial, *Computer*, 29, 31–44, 1996.
55. A. Nogueira, M. Rosario de Oliveira, P. Salvador, R. Valadas and A. Pacheco, Using neural networks to classify Internet users, *Advanced Industrial Conference on Telecommunications/Service Assurance with Partial and Intermittent Resources*, Washington, DC, USA, 2005, pp. 183–188.
56. A. N. Ramesh, C. Kambhampati, J. R. T. Monson and P. J. Drew, Artificial intelligence in medicine, *Annals of The Royal College of Surgeons of England*, 86, 334–338(5), 2004.
57. H. B. Burke, Evaluating artificial neural networks for medical applications, *International Conference on Neural Networks*, Houston, TX, USA, 1997, pp. 2494–2495.
58. S.-C. B. Lo, J.-S. J. Lin, M. T. Freedman and S. K. Mun, Application of artificial neural networks to medical image pattern recognition: Detection of clustered microcalcifications on mammograms and lung cancer on chest radiographs, *The Journal of VLSI Signal Processing*, 18, 263–274, 1998.
59. S. Walczak, An empirical analysis of data requirements for financial forecasting with neural networks, *Journal of Management Information Systems*, 17, 203–222, 2001.
60. Y. Fu and T. Chai, Neural-network-based nonlinear adaptive dynamical decoupling control, *IEEE Transactions on Neural Networks*, 18, 921–925, 2007.
61. G. Dorffner, Neural networks for time series processing, *Neural Network World*, 6, 447–468, 1996.
62. A. Patnaik, D. E. Anagnostou, R. K. Mishra, C. G. Christodoulou and J. C. Lyke, Applications of neural networks in wireless communications, *IEEE Antennas and Propagation Magazine*, 46, 130–137, 2004.
63. D. MacKay, *Information Theory, Inference, and Learning Algorithms*, 1st ed., Cambridge, UK: Cambridge University Press, 2003.
64. K. Hornik, M. Stinchcombe and H. White, Multilayer feedforward networks are universal approximators, *Neural Networks*, 2, 359–366 1989.
65. S. Haykin, *Neural Networks: A Comprehensive Foundation*, 2nd ed., New Jersey, USA: Prentice-Hall, 1998.
66. D. P. Mandic and J. A. Chambers, *Recurrent Neural Networks for Prediction,* Chichester, West Sussex: John Wiley and Sons Ltd., 2001.
67. L. Behera, S. Kumar and A. Patnaik, On adaptive learning rate that guarantees convergence in feedforward networks, *IEEE Transactions on Neural Networks*, 17, 1116–1125, 2006.
68. A. Toledo, M. Pinzolas, J. J. Ibarrola and G. Lera, Improvement of the neighborhood based Levenberg–Marquardt algorithm by local adaptation of the learning coefficient, *IEEE Transactions on Neural Networks*, 16, 988–992, 2005.
69. U. Windhorst and H. Johansson, *Modern Techniques in Neuroscience Research*, Berlin: Springer-Verlag, 1999.
70. R. P. Lippmann, An introduction to computing with neural nets, *IEEE ASSP Magazine*, 4, 4–22, 1987.

71. S. Sitharama Iyengar, E. C. Cho and V. V. Phoha, *Foundations of Wavelet Networks and Applications*, 1st ed., Boca Raton, Florida: Chapman & Hall/CRC, 2002.

72. E. A. Martínez–Rams and V. Garcerán–Hernánde, Assessment of a speaker recognition system based on an auditory model and neural nets, *IWINAC 2009*, Santiago de Compostela, Spain, 2009, 488–498.

73. S. Trenn, Multilayer perceptrons: Approximation order and necessary number of hidden units, *IEEE Transactions on Neural Networks*, 19, 836–844, 2008.

74. H. Demuth, M. Beale and M. Hagan, *Neural Network Toolbox 5 User's Guide*, The Mathworks, Natick, Massachusetts, 2007, pp. 5-34-51.

75. S. Rajbhandari, Application of wavelets and artificial neural network for indoor optical wireless communication systems, PhD, Northumbria University, Newcastle upon Tyne, 2010.

76. J. G. Proakis, *Digital Communications*, 4th ed., New York: McGraw-Hill, Inc., 2001.

77. S. Haykin, *Adaptive Filter Theory*. New Jersey, USA: Prentice-Hall International, 2001.

78. R. T. Valadas, A. M. R. Tavares and A. M. Duarte, Angle diversity to combat the ambient noise in indoor optical wireless communication systems, *International Journal of Wireless Information Networks*, 4, 275–288, 1997.

79. C. Y. F. Ho, B. W. K. Ling, T. P. L. Wong, A. Y. P. Chan and P. K. S. Tam, Fuzzy multi-wavelet denoising on ECG signal, *Electronics Letters*, 39, 1163–1164, 2003.

80. H. Elgala, R. Mesleh, H. Haas and B. Pricope, OFDM visible light wireless communication based on white LEDs, *Vehicular Technology Conference, 2007. VTC2007-Spring. IEEE 65th*, Dublin, Ireland, 2007, 2185–2189.

81. A. G. Al-Ghamdi and J. M. H. Elmirghani, Analysis of diffuse optical wireless channels employing spot-diffusing techniques, diversity receivers, and combining schemes, *IEEE Transactions on Communications*, 52, 1622–1631, 2004.

82. J. K. Kim and E. F. Schubert, Transcending the replacement paradigm of solid-state lighting, *Opt. Express*, 16, 21835–21842, 2008.

83. M. Hoa Le, D. O'Brien, G. Faulkner, Z. Lubin, L. Kyungwoo, J. Daekwang, O. YunJe, and W. Eun Tae, 100 Mb/s NRZ visible light communications using a postequalized white LED, *IEEE Photonics Technology Letters*, 21, 1063–1065, 2009.

84. P. Djahani and J. M. Kahn, Analysis of infrared wireless links employing multibeam transmitters and imaging diversity receivers, *IEEE Transactions on Communications*, 48, 2077–2088, 2000.

85. C. R. A. T. Lomba, R. T. Valadas and A. M. de Oliveira Duarte, Sectored receivers to combat the multipath dispersion of the indoor optical channel, *Sixth IEEE International Symposium on Personal, Indoor and Mobile Radio Communication*, Toronto, Canada, 1995, 321–325.

86. D. O'Brien, L. Zeng, H. Le-Minh, G. Faulkner, O. Bouchet, S. Randel, J. Walewski et al., Visible light communication, in *Short-Range Wireless Communications: Emerging Technologies and Applications*, R. Kraemer and M. Katz, Eds., Chichester, West Sussex: Wiley Publishing, 2009.

87. K. Akhavan, M. Kavehrad and S. Jivkova, High-speed power-efficient indoor wireless infrared communication using code combining—Part II, *IEEE Transaction on Communications*, 50, 1495–1502, 2002.

88. L. Kwonhyung, P. Hyuncheol and J. R. Barry, Indoor channel characteristics for visible light communications, *IEEE Communications Letters*, 15, 217–219.

6 FSO Link Performance under the Effect of Atmospheric Turbulence

This chapter analyses the effect of atmospheric turbulence on the different modulation techniques used in FSO. Atmospheric turbulence is known to cause signal fading in the channel, and the existing mathematical models for describing the fading have already been introduced in Chapter 3. There are many different types of modulation schemes that are suitable for optical wireless communication systems as discussed in Chapter 4; the emphasis in this chapter will, however, be on the effect of atmospheric turbulence-induced fading on the following techniques: on–off keying (OOK), pulse position modulation (PPM) and phase shift keying premodulated subcarrier intensity modulation. Since the average emitted optical power is always limited, the performance of modulation techniques is often compared in terms of the average received optical power required to achieve a desired bit error rate at a given data rate. It is very desirable for the modulation scheme to be power efficient, but this is, however, not the only deciding factor in the choice of a modulation technique. The design complexity of its transmitter and receiver and the bandwidth requirement of the modulation scheme are all equally important. Although it can be argued that the optical carrier has abundant bandwidth, the modulation bandwidth available for communication is ultimately limited by the optoelectronic devices.

The classical modulation technique used for FSO is OOK [1,2]. This is primarily because of the simplicity of its design and implementation. It is not surprising therefore that the majority of the work reported in the literature [3,4] is based on this signalling technique. However, the performance of a fixed threshold-level OOK in atmospheric turbulence is not optimal, as will be shown in the following section. In atmospheric turbulence, an optimal-performing OOK requires the threshold level to vary in sympathy with the prevailing irradiance fluctuation and noise, that is, to be adaptive. The PPM requires no adaptive threshold and is predominantly used for deep space free-space optical communication links because of its enhanced power efficiency compared to the OOK signalling [5–10]. The PPM modulation technique, however, requires a complex transceiver design due to tight synchronization requirements and a higher bandwidth than the OOK. The performance analysis of these modulation schemes and that of the SIM scheme in a number of atmospheric turbulence channels (log-normal, gamma–gamma and negative exponential) will be further investigated in this chapter. The SIM also requires no adaptive threshold and does not require as much bandwidth as PPM, but suffers from a high peak-to-average-power ratio, which translates into poor

power efficiency. Choosing a modulation scheme for a particular application therefore entails trade-offs among these listed factors.

6.1 ON–OFF KEYING

OOK is the dominant modulation scheme employed in commercial terrestrial wireless optical communication systems. This is primarily due to its simplicity and resilience to the innate nonlinearities of the laser and the external modulator. OOK can use
either non-return-to-zero (NRZ) or return-to-zero (RZ) pulse formats. In NRZ-OOK,
an optical pulse of peak power $\alpha_e P_T$ represents a digital symbol '0' while the transmission of an optical pulse of peak power P_T represents a digital symbol '1'. The
optical source extinction ratio, α_e, has the range $0 \le \alpha_e < 1$. The finite duration of
the optical pulse is the same as the symbol or bit duration T_b. With OOK-RZ, the
pulse duration is lower than the bit duration, giving an improvement in power efficiency over NRZ-OOK at the expense of an increased bandwidth requirement. In all
the analyses that follow, the extinction ratio, α_e, is equal to zero and NRZ-OOK,
which is the scheme deployed in present commercial FSO systems, is assumed unless
otherwise stated.

6.1.1 OOK in a Poisson Atmospheric Optical Channel

If the received average power is given by $P_r = P_t \exp(-\gamma_T L)$, where γ_T represents the
overall channel attenuation, then the average received photoelectron count is given
by [11]

$$\langle n \rangle = \frac{\eta \lambda T_b P_r}{hc} \tag{6.1}$$

where h and c are the Planck's constant and the speed of light in vacuum, respectively, and η is the quantum efficiency of the photodetector. However, the instantaneous count n, unlike the average count, is not constant. As mentioned in Chapter 2,
it varies with time due to the following reasons:

1. The quantum nature of the light/photodetection process, which suggests
   that the instantaneous number of counts n follows the discrete Poisson distribution with an associated quantum noise of variance $\langle n \rangle$ (the mean and
   variance of a Poisson distribution are the same).
2. The received signal field varies randomly due to the effect of scintillation.

This implies that the number of counts is now doubly stochastic and based on
the log-normal turbulence model of Chapter 3, the probability of n counts is derived
as [12]

$$p_1(n) = \int_0^\infty \frac{\left(\eta \lambda T_b P_r / hc\right)^n \exp(-\eta \lambda T_b P_r / hc)}{n! \sqrt{2\pi \sigma_i^2} P_r} \exp\left[-\frac{1}{2\sigma_i^2}\left(\ln \frac{P_r}{P_0} + \frac{\sigma_i^2}{2}\right)^2\right] dP_r \tag{6.2}$$

where P_0 is the received average power in the absence of atmospheric turbulence and σ_I^2 is the strength of the power fluctuation indicator. When an optical pulse is transmitted (that is bit '1' sent), a decision error occurs when the number of counts, n, is less than a predetermined threshold count, n_{th}. Thus, the probability of detecting bit '0' when bit '1' is transmitted is given by [12]

$$p_1(n < n_{th}) = \sum_{n=0}^{n_{th}-1} \left[\int_0^\infty \frac{(\eta\lambda T_b(P_r + P_{Bg}))^n \exp(-\eta\lambda T_b(P_r + P_{Bg})/hc)}{(hc)^n n! \sqrt{2\pi\sigma_I^2} P_r} \right.$$

$$\left. \times \exp\left(-\frac{1}{2\sigma_I^2}\left(\ln\frac{P_r}{P_0} + \frac{\sigma_I^2}{2} \right)^2 \right) dP_r \right] \tag{6.3}$$

where P_{Bg} is the power of the background radiation that falls within the receiver's field of view and $n_b = \eta\lambda TP_{Bg}/hc$.

Similarly, the probability of detecting bit '1' when bit '0' is transmitted is derived as [12]

$$p_0(n > n_{th}) = \sum_{n=n_{th}}^\infty \frac{(\eta\lambda T_b P_{Bg}/hc)^n \exp(-\eta\lambda TP_{Bg}/hc)}{n!}$$

$$= 1 - \sum_{n=0}^{n_{th}-1} \frac{(\eta\lambda T_b P_{Bg}/hc)^n \exp(-\eta\lambda T_b P_{Bg}/hc)}{n!} \tag{6.4}$$

It should be noted from Equation 6.4 that atmospheric turbulence has no impact when no optical power is transmitted. If bits '1' and '0' are assumed to be equally likely to be transmitted, then the system theoretical bit error rate P_e becomes

$$P_e = 0.5[p_0(n > n_{th}) + p_1(n < n_{th})] \tag{6.5}$$

For an optimal performance, n_{th} is the value of n that satisfies expression (6.6) obtained by invoking the maximum likelihood symbol-by-symbol detection condition.

$$(P_{Bg})^n \exp(-\eta\lambda TP_{Bg}/hc) = \left\{ \int_0^\infty \frac{(P_r + P_{Bg})^n \exp(-\eta\lambda T_b(P_r + P_{Bg})/hc)}{\sqrt{2\pi\sigma_I^2} P_R} \right.$$

$$\left. \times \exp\left[-\frac{1}{2\sigma_I^2}\left(\ln\frac{P_r}{P_0} + \frac{\sigma_I^2}{2} \right)^2 \right] dP_r \right\} \tag{6.6}$$

The impact of scintillation on the achievable bit error rate (BER) of the system is shown in Figure 6.1, which is obtained by combining Equations 6.3 through 6.5, while the values of n_{th} used is obtained from the solution of Equation 6.6.

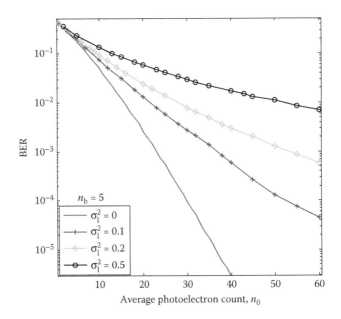

FIGURE 6.1 BER against the average photoelectron count per bit for OOK-FSO in a Poisson atmospheric turbulence channel for $\sigma_I^2 = [0, 0.1, 0.2, 0.5]$.

In the figure, the BER is plotted against the average count $n_0 = \eta \lambda T P_0 / hc$. The penalty incurred due to scintillation is quite evident from the plot. For example, with respect to no scintillation condition, over 20 additional photoelectron counts per bit are needed to maintain the same BER of 10^{-4} in a channel characterized by $\sigma_I^2 > 0.1$. Consequently, when designing a terrestrial laser communication link, an adequate margin, based on the results shown in Figure 6.1, should be provided to cater for the scintillation effect.

Figure 6.1 also shows that increasing the average count can help ameliorate the effect of turbulence and result in improved BER in very weak turbulence, but as the strength of turbulence increases, increased average photoelectron count has a less significant impact. For example, at $\sigma_I^2 = 0.1$ and $n_0 = 10$, the BER is about 10^{-1} but this decreases to less than 10^{-3} when n_0 is increased to 40. In comparison, when $\sigma_I^2 = 0.5$, the BER is higher than 10^{-1} for $n_0 = 10$ and remains higher than 10^{-2} when n_0 is increased to 40. Note that in the absence of turbulence, $n_0 = \langle n \rangle$ and the result presented in Figure 6.1 should be viewed as the theoretical performance lower bound since the photomultiplication process has been assumed ideal.

6.1.2   OOK IN A GAUSSIAN ATMOSPHERIC OPTICAL CHANNEL

With large signal photoelectron counts, and by taking the detection thermal noise into account, the generated signal current probability distribution can be approximated as the tractable Gaussian distribution. Without any loss of generality, the receiver area may be normalized to unity such that the optical power may be represented by the optical intensity, I. If R represents the responsivity of the PIN photodetector, the received signal in an OOK system is therefore given by

$$i(t) = RI\left[1 + \sum_{j=-\infty}^{\infty} d_j g(t - jT_b)\right] + n(t) \tag{6.7}$$

where $n(t) \sim N(0, \sigma^2)$ is the additive white Gaussian noise, $g(t - jT_b)$ is the pulse shaping function and $d_j = [-1, 0]$. At the receiver, the received signal is fed into a threshold detector which compares the received signal with a predetermined threshold level. A digital symbol '1' is assumed to have been received if the received signal is above the threshold level and '0' otherwise.

The probability of error, as derived in Chapter 4, is given as

$$P_{ec} = Q\left(\frac{i_{th}}{\sigma}\right) \tag{6.8}$$

where $Q(x) = 0.5\mathrm{erfc}\,(x/\sqrt{2})$. In the presence of atmospheric turbulence, the threshold level is no longer fixed midway between the signal levels representing symbols '1' and '0'. The marginal probability $p(i/1)$ is then modified by averaging the conditional pdf of $i(t)$ over the scintillation statistics. Note that scintillation does not occur when no pulse is transmitted.

$$p(i/1) = \int_0^\infty p(i/1, I)p(I)\,dI \tag{6.9}$$

Assuming equiprobable symbol transmission and invoking the maximum *a posteriori* symbol-by-symbol detection, the likelihood function L, becomes [13]

$$L = \int_0^\infty \exp\left[\frac{-(i - RI)^2 - i^2}{2\sigma^2}\right]p(I)\,dI \tag{6.10}$$

Figure 6.2 shows the plot of log (L) against the average photocurrent i, at various levels of scintillation and noise variance of 10^{-2}. The threshold level, as would be expected, is at the point where $L = 1$ (i.e., when $\log(L) = 0$).

Based on the log-normal turbulence model, the plot of i_{th} against the log intensity standard deviation for different noise levels is depicted in Figure 6.3 for different noise levels. The threshold is observed to approach the expected value of 0.5 as the scintillation level approaches zero. As an illustration, at a turbulence level $\sigma_l^2 = 0.2$, the probability of bit error P_e obtained is plotted against the normalized SNR $= (RE[I])^2/\sigma^2$ in Figure 6.4; the value of i_{th} used for the adaptive threshold-level graph is obtained from the solution of Equation 6.10. From this figure, the effect of using a fixed threshold level in fading channels results in a BER floor, the values of which depend on the

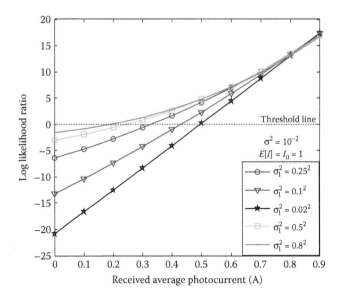

FIGURE 6.2 The likelihood ratio against the received signal for different turbulence levels and noise variance of 10^{-2}.

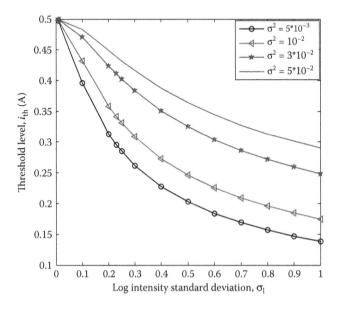

FIGURE 6.3 OOK threshold level against the log intensity standard deviation for various noise levels.

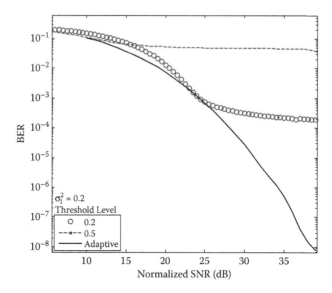

FIGURE 6.4   BER of OOK-based FSO in atmospheric turbulence with $\sigma_I^2 = 0.2$ considering fixed and adaptive threshold levels.

fixed threshold level and turbulence-induced fading strength. With an adaptive threshold, there is no such BER floor and any desired level of BER can thus be realized.

In Figure 6.5, the BER is again plotted against the normalized SNR at various levels of scintillation, including when the threshold is fixed at 0.5. This is intended to show the effect of turbulence strength on the amount of SNR required to maintain a

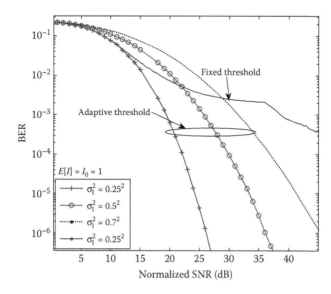

FIGURE 6.5   BER of OOK-FSO with fixed and adaptive threshold at various levels of scintillation, $\sigma_I = [0.2, 0.5, 0.7]$ and $I_0 = 1$.

given error performance level. With a fixed threshold, the BER reaches a floor at a BER that is greater than 10^{-4}, meaning that a lower BER is not achievable at the specified low scintillation level. From this graph, it can be inferred that atmospheric turbulence (i) causes SNR penalty, for example, ~26 dB of SNR is needed to achieve a BER of 10^{-6} due to a very weak scintillation of strength $\sigma_l^2 = 0.25^2$; this however increases by over 20 dB as the scintillation strength increases to $\sigma_l^2 = 0.7^2$, and (ii) implies that adaptive threshold will be required to avoid a BER floor in the system performance.

The results above illustrate that for the OOK-modulated FSO system to perform at its best, the receiver will require knowledge of both the fading strength and the noise level. This can be resolved by integrating into the system an intensity estimation network which can predict the scintillation level based on past events. The implementation of this is not trivial, and as such, commercial FSO designers tend to adopt the fixed threshold approach and include a sufficiently large link margin in the link budget to cater for turbulence-induced fading [1].

6.2   PULSE POSITION MODULATION

This is an orthogonal modulation technique and a member of the pulse modulation family. The PPM modulation technique improves on the power efficiency of OOK but at the expense of an increased bandwidth requirement and greater complexity. In PPM, each block of $\log_2 M$ data bits is mapped to one of M possible symbols. Generally, the notation M-PPM is used to indicate the order. Each symbol consists of a pulse of constant power P_t, occupying one slot, along with $M - 1$ empty slots. The position of the pulse corresponds to the decimal value of the $\log_2 M$ data bits. Hence, the information is encoded by the position of the pulse within the symbol. The slot duration, T_{s_ppm}, is related to the bit duration by the following expression:

$$T_{s\_ppm} = \frac{T_b \log_2 M}{M} \qquad (6.11)$$

The transmitted waveforms for 16-PPM and OOK are shown in Figure 6.6.

A PPM receiver will require both slot and symbol synchronization in order to demodulate the information encoded on the pulse position. Nevertheless, because of its superior power efficiency, PPM is an attractive modulation technique for optical wireless communication systems, particularly in deep space laser communication applications [14]. Assuming that complete synchronization is maintained between the transmitter and receiver at all times, the optical receiver detects the transmitted signal by attempting to determine the energy in each possible time slot. It then selects the signal which corresponds to the maximal energy. In direct photodetection, this is equivalent to 'counting' the number of released electrons in each T_s interval. The photo count per PPM slot can be obtained from

$$K_s = \frac{\eta \lambda P_r T_{s\_ppm}}{hc} \qquad (6.12)$$

where P_r is the received optical power during a slot duration. An APD could be used to give an increase in the number of photon counts per PPM slot but, unfortunately,

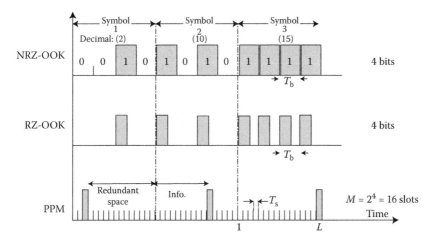

FIGURE 6.6   Time waveforms for 4-bit OOK and 16-PPM.

the photomultiplication process that governs the generation of the secondary electrons is a random process. This implies that a large photomultiplication gain will
eventually lead to a large noise factor and an error-prone performance. For a moder-
ately high received signal, as is the case in commercial and short-range FSO systems,
the BER conditioned on K_s is given by [15]

$$P_{ec} = Q\left(\sqrt{\frac{(\bar{G}q)^2 K_s^2}{(\bar{G}q)^2 F(K_s + 2K_{Bg}) + 2\sigma_{th}^2}}\right) \qquad (6.13)$$

where the parameters are defined as

$K_{Bg} = \eta\lambda P_{Bg}T_s/hc$	Average photon count per PPM slot due to the background radiation of power P_{Bg}
$\bar{G}$	Average APD gain
Q	Electronic charge
$F \approx 2 + \zeta\bar{G}$	Noise factor of the APD
ζ	APD ionization factor
$\sigma_{Th}^2 = (2\kappa T_e q/R_L)(T_{s\_ppm})$	Equivalent thermal noise count within a PPM slot duration [15]
$R_b = 1/T_b$	Bit rate
κ	Boltzmann's constant
R_L	Equivalent load resistance

In the presence of log-normal atmospheric turbulence, the unconditional BER for
a binary PPM-modulated FSO obtained by averaging (6.13) over the scintillation
statistics can be approximated as [8]

$$P_e \approx \frac{1}{\sqrt{\pi}}\sum_{i=1}^{n}w_i Q\left(\frac{\exp(2(\sqrt{2}\sigma_k x_i + m_k))}{F\exp(\sqrt{2}\sigma_k x_i + m_k) + K_n}\right) \qquad (6.14)$$

where $[w_i]_{i=1}^n$ and $[x_i]_{i=1}^n$ are the weight factors and the zeros of an nth-order Hermite polynomial. These values are contained in Appendix 6.A for a 20th-order Hermite polynomial. m_k represents the mean of $\ln(K_s)$, $K_n = (2\sigma_{Th}^2/(\overline{g}q)^2) + 2FK_{Bg}$ and $\sigma_k^2 = \ln(\sigma_N^2 + 1)$. It is noteworthy that the fluctuation of the mean count, K_s, is brought about by the atmospheric turbulence and its ensemble average is given by the following [8]:

$$E[K_s] = \exp\left(\frac{\sigma_k^2}{2} + m_k\right) \tag{6.15}$$

For an M-PPM system, the BER denoted by P_e^M has an upper bound given by [8]

$$P_e^M \leq \frac{M}{2\sqrt{\pi}} \sum_{i=1}^n w_i Q\left(\frac{\exp\left(2\left(\sqrt{2}\sigma_k x_i + m_k\right)\right)}{F\exp\left(\sqrt{2}\sigma_k x_i + m_k\right) + K_n}\right) \tag{6.16}$$

The BER performance of a binary PPM-modulated FSO, using the MATLAB® code in Program 6.1, is shown in Figure 6.7 at different levels of scintillation. The extension of the result is straightforward from Equation 6.16 and hence is not presented here.

As expected, an increase in the atmospheric scintillation results in an increase in the required signal level to achieve a given BER. Increasing the signal strength can be used to minimize the scintillation effect at a low scintillation index, but as turbulence strength increases, it is observed that the BERs all tend towards a high BER asymptotic value.

Program 6.1: MATLAB Codes to Calculate the BER of the Binary PPM Scheme

```
%Evaluation of BER of BPPM FSO (using Equation 6.16) under
weak turbulence using the Gauss-Hermite Quadrature integration
approach.

clear
clc

%****Hermite polynomial weights and roots************
w20=[2.22939364554e-13,4.39934099226e-10,1.08606937077e-7,
7.8025564785e-6,0.000228338636017,0.00324377334224,0.0248105208875,
0.10901720602,0.286675505363,0.462243669601,...

0.462243669601,0.286675505363,0.10901720602,0.0248105208875,
0.00324377334224,0.000228338636017,7.8025564785e-6,
1.08606937077e-7,4.39934099226e-10,2.22939364554e-13];
x20=[-5.38748089001,-4.60368244955,-3.94476404012,
-3.34785456738,-2.78880605843,-2.25497400209,-1.73853771212,
-1.2340762154,-0.737473728545,-0.245340708301,...
```

```
0.245340708301,0.737473728545,1.2340762154,1.73853771212,
2.25497400209,2.78880605843,3.34785456738,3.94476404012,4.60368244955,
5.38748089001];
%********************************************************
Ks1=[140,180,220,260,300];
 for i1=1:length(Ks1)
   Ks=Ks1(i1);

%************Simulation Parameters************************
Rb=155e6; %Bit rate
RL=50; %Load resistance
Temp=300; %Ambient temperature

E_c=1.602e-19; %Electronic charge
B_c=1.38e-23; %Boltzmann constant
%********************************************************

NoTh=(2*B_c*Temp/(2*Rb*RL));

Ioni=0.028;
Kb=10;
gain=150;
F=2+(gain*Ioni);
Kn=((2*NoTh)/(gain*E_c)^2)+(2*F*Kb);
S_I=0.1:0.15:0.9; %Scintillation Index
for i=1:length(S_I)
   SI=S_I(i);
   Sk=log(S_I(i)+1);
   Mk=log(Ks)-(Sk/2);

   Temp=0;
   for j=i:length(x20)

       ANum(j)=(2*x20(j)*sqrt(2*Sk))+(2*Mk);
       Num(j)=exp(ANum(j));
       BDen(j)=(sqrt(2*Sk)*x20(j)+Mk);
       Den(j)=(F*exp(BDen(j)))+Kn;
       Prod(j)=w20(j)*Q(Num(j)/Den(j));
       Temp=Temp+Prod(j);
   end

   BER(i1,i)=Temp/sqrt(pi);
end
end

%*********Plot function*************************
figure
semilogy(S_I,BER)
```

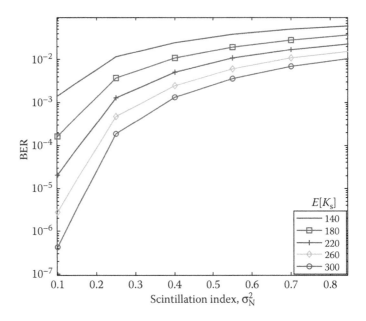

FIGURE 6.7 Binary PPM BER as a function of scintillation index for $K_{Bg} = 10$; $T_e = 300$ K, $\zeta = 0.028$, $R_b = 155$ Mbps and $g = 150$.

6.3 SUBCARRIER INTENSITY MODULATION

Subcarrier intensity modulation is a technique borrowed from the very successful multiple carrier RF communications already deployed in applications such as digital television, LANs, asymmetric digital subscriber line (ADSL), 4G communication systems and optical fibre communications [16,17]. In optical fibre communication networks, for example, the subcarrier modulation techniques have been commercially adopted in transmitting cable television signals and have also been used in conjunction with wavelength division multiplexing [18]. For the seamless integration of FSO systems into present and future networks, which already harbour subcarrier-modulated (or multiple carrier) signals, the study of subcarrier-modulated FSO is thus imperative. Other reasons for studying the subcarrier intensity-modulated FSO systems include

1. It benefits from already developed and evolved RF communication components such as stable oscillators and narrow filters [19].
2. It avoids the need for an adaptive threshold required by optimum-performing OOK-modulated FSO [20].
3. It can be used to increase capacity by accommodating data from different users on different subcarriers.
4. It has comparatively lower bandwidth requirement than the PPM.

There are however some challenges in the implementation of SIM:

1. Relatively high average transmitted power due to
    a. The optical source being ON during the transmission of both binary digits '1' and '0', unlike in OOK where the source is ON during the transmission of bit '1' only.
    b. The multiple subcarrier composite electrical signal, being the sum of the modulated sinusoids (i.e., dealing with both negative and positive values), requires a DC bias. This is to ensure that this composite electrical signal, which will eventually modulate the laser irradiance, is never negative. Increasing the number of subcarriers leads to increased average transmitted power because the minimum value of the composite electrical signal decreases (becomes more negative) and the required DC bias therefore increases [21]. This factor results in poor power efficiency and places a bound on the number of subcarriers that can be accommodated when using multiple SIM.
2. The possibility of signal distortions due to inherent laser nonlinearity and signal clipping due to overmodulation.
3. Stringent synchronization requirements at the receiver side.

It is therefore worthwhile to mention that multiple SIM is only recommended when the quest for higher capacity/users outweighs the highlighted challenges or where FSO is to be integrated into existing networks that already contain multiple RF carriers. Several methods have been researched and documented [21–23] to improve the poor power efficiency of SIM but have not been considered in this chapter.

6.3.1 SIM GENERATION AND DETECTION

In optical SIM links, an RF subcarrier signal $m(t)$, premodulated with the source data $d(t)$, is used to modulate the intensity P_T of the optical source—a continuous wave laser diode. Figure 6.8 illustrates the system block diagram of a SIM-FSO with N subcarriers. The serial-to-parallel converter distributes the incoming data across the N-subcarriers. Each subcarrier carries a reduced symbol rate but the aggregate must be equal to the symbol rate of $d(t)$.

Another obvious possibility, not shown in the figure, is to have different users occupying the N different subcarriers. Prior to modulating the laser irradiance, the source data $d(t)$ are modulated onto the RF subcarriers. For the M-PSK subcarrier modulation shown in Figure 6.8, the encoder maps each subcarrier symbol onto the symbol amplitude $\{a_{ic}, a_{is}\}_{i=1}^{N}$ that corresponds to the constellation in use. Since the subcarrier signal, $m(t)$, is sinusoidal, having both positive and negative values, a DC level b_0 is added to $m(t)$ before it is used to directly drive the laser diode—to avoid any signal clipping.

The following gives the general expression for $m(t)$ in the N-SIM-FSO system:

$$m(t) = \sum_{i=1}^{N} m_i(t) \qquad (6.17)$$

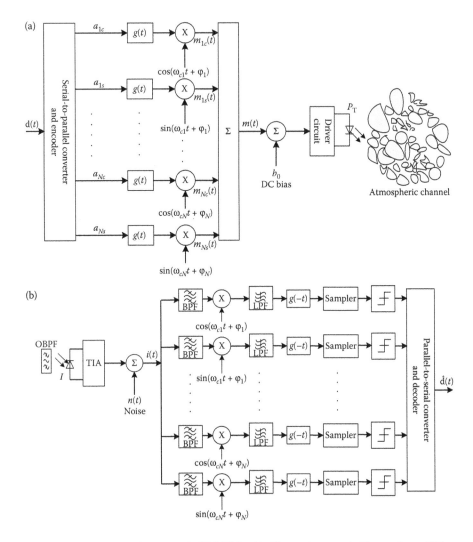

FIGURE 6.8 Block diagram of SIM-FSO. (a) Transmitter and (b) receiver. TIA—transimpedance, OBPF—optical band-pass filter.

During one symbol duration, each RF subcarrier signal is generally represented by

$$m_i(t) = g(t)a_{ic}\cos(\omega_{ci}t + \varphi_i) - g(t)a_{is}\sin(\omega_{ci}t + \varphi_i) \qquad (6.18)$$

where $g(t)$ is the pulse shaping function, and the subcarrier angular frequency and phase are represented by $[\omega_{ci}, \varphi_i]_{i=1}^N$. It follows that each subcarrier can be modulated by any standard RF digital/analogue modulation technique, such as QAM, M-PSK, M-FSK and M-ASK. Using direct detection at the receiver, the incoming optical radiation, P_R, is converted into an electrical signal, $i(t)$. This is followed by a standard

RF demodulator to recover the transmitted symbol as shown in Figure 6.8b. By normalizing the receiver area to unity and representing the received power by irradiance, I, the received signal can be modelled as

$$i(t) = RI[1 + \xi m(t)] + n(t) \tag{6.19}$$

where the optical modulation index $\xi = |m(t)/i_B - i_{Th}|$, as shown in Figure 6.9.

The electrical band-pass filter (BPF) with a minimum bandwidth of $2R_b$ performs the following functions: selection of the individual subcarrier for demodulation, reduction of the noise power and suppression of any slow-varying RI component present in the received signal. For a subcarrier at ω_{ci}, the received signal is

$$i(t) = I_{comp} + Q_{comp} \tag{6.20}$$

where

$$I_{comp} = RI\xi g(t)a_{ic} \cos(\omega_{ci}t + \varphi_i) + n_1(t) \tag{6.21a}$$

$$Q_{comp} = -RI\xi g(t)a_{is} \sin(\omega_{ci}t + \varphi_i) + n_Q(t) \tag{6.21b}$$

$n_1(t)$ and $n_Q(t)$ are the independent additive white Gaussian noise (AWGN) with a zero mean and a variance σ^2. The quadrature components I_{comp} and Q_{comp} are down converted by the reference signals $\cos \omega_c t$ and $\sin \omega_c t$, respectively, and applied to the

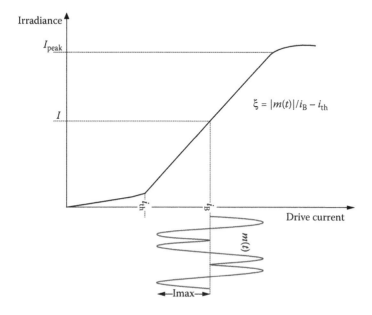

FIGURE 6.9 Output characteristic of an optical source driven by a subcarrier signal showing optical modulation index.

standard receiver architecture. The electrical low-pass filters, which are part of the standard RF receiver, remove any out-of-band (unwanted) signals from the down-converted signal and then pass it onto the decision circuit. In the case of a phase-shift-keying-modulated subcarrier, the decision circuit estimates the phase of the received signal and decides which symbol has been received. By adopting the approach in Ref. [24], the conditional BER expressions can be deduced.

6.3.2 SIM-FSO Performance in Log-Normal Atmospheric Channel

Prior to the analytical estimation of the effect of scintillation on the system performance, the effect of turbulence and noise on a subcarrier signal constellation will first be presented. The simulation will be based on the block diagram of Figure 6.8, with a QPSK-modulated single subcarrier. The constellation at the input of the transmitter is shown in Figure 6.10. The signal is then transmitted through the atmospheric channel with turbulence-induced fading and noise. For this illustration, the electrical SNR = $(\xi RE[I])^2/\sigma^2$ is fixed, for instance, at 2 dB. The received constellation under a very low fading of $\sigma_l^2 = 0.001$ is shown in Figure 6.11, while the effect of turbulence-induced fading is made evident by displaying in Figure 6.12 the received constellation with a much higher fading with $\sigma_l^2 = 0.5$.

With a very low turbulence-induced fading strength $\sigma_l^2 = 0.001$, the constellations form clusters that are clearly confined within their respective quadrants and the chance of erroneous demodulation is very low but as the turbulence level increases to $\sigma_l^2 = 0.5$, the confinement is lost and the constellation points become more

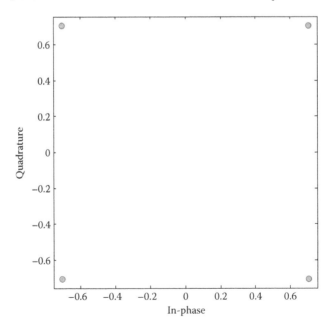

FIGURE 6.10 QPSK constellation of the input subcarrier signal without the noise and channel fading.

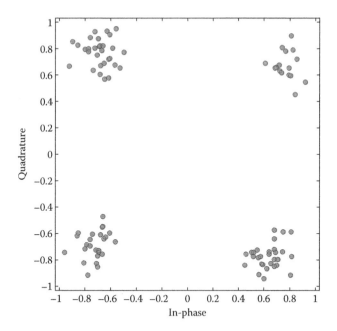

FIGURE 6.11   Received constellation of QPSK premodulated SIM-FSO with the noise and channel fading for SNR = 2 dB and $\sigma_I^2 = 0.001$.

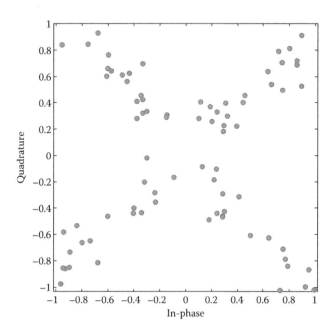

FIGURE 6.12   Received constellation of QPSK premodulated SIM-FSO with the noise and channel fading for SNR = 2 dB and $\sigma_I^2 = 0.5$.

staggered and move towards the centre of the plot. This apparently increases the chance of demodulation error during the symbol detection process. To quantify this observation, the bit error and the outage probabilities of SIM-FSO in atmospheric turbulence will now be presented.

The MATLAB codes for Figures 6.10 through 6.12 are given in Program 6.2. Additional MATLAB codes for this section are given in Appendix 6.B.

Program 6.2: MATLAB Codes Used for Generating Figures 6.10 through 6.12

```
%*****Subcarrier modulation simulation for M-PSK intensity
modulation******%
% %
%**********************************************************%


clear all
clc


%******************SIMULATION PARAMETERS********************
N_sub=1;                   %No of subcarriers
symb=1e2;                  %No of symbols
Rb=155e6;                  %Symbol rate
M=4;                       %M-PSK
Responsivity=1;            %Photodetector responsivity
T=1/Rb;
fmin=1e9;                  %Starting carrier frequency
Fs=50*fmin;                %Sampling frequency
t=0:1/Fs:T;
samples_symb=length(t);    %Number of samples per symbol

fmax=2*fmin; %Single octave operation
if N_sub > 1
  delta=(fmax-fmin)/(N_sub-1); %Frequency spacing of any two
  consecutive subcarriers
else
  delta=0;
end
K=2*pi;


%***********GENERATION OF THE SUBCARRIER SIGNAL*************

[Inphase,Quadrature,Datain]=Basebandmodulation(M,symb,N_sub);
Input1=Inphase+(j*Quadrature); %The input symbols constellation

Ac=1; %Subcarrier signals amplitude
M0=[];M1=[];
for i0=1:N_sub
 j0=i0-1;
 XX=Ac.*cos(K.*t.*(fmin+(j0*delta)));
 XX1=Ac.*sin(K.*t.*(fmin+(j0*delta)));
 M0=[M0;XX];
```

```
 M1=[M1;XX1];
end

for i1=1:N_sub
 I_phase =[];Q_phase=[];
 for j1=1:symb
  I_sig=Inphase(j1,i1).*M0(i1,:);
  I_phase=[I_phase,I_sig];

  Q_sig=Quadrature(j1,i1).*M1(i1,:);
  Q_phase=[Q_phase,Q_sig];
 end
 I_Mod_Data(i1,:)=I_phase;
 Q_Mod_Data(i1,:)=Q_phase;

end
PSK_sig=I_Mod_Data-Q_Mod_Data;
SIM=sum(PSK_sig,1);           %subcarrier multiplexed signal
Mod_index=1/(Ac*N_sub);       %Modulation index
SCM_Tx=(1+Mod_index*SIM);     %The transmitted signal

%************************THE CHANNEL PARAMETERS**************
%************Turbulence parameters*****************
Io=1; %Average received irradiance
I_var=1e-3; %Log irradiance variance
I=Turbulence(I_var,symb,Io,t); %Irradiance using the Log
                                normal turbulence model

SNR_dB=2; %SNR value in dB
SNR=10.^(SNR_dB./10);
Noise_var=(Mod_index*Responsivity*Io)^2./(SNR);
Noise_SD=sqrt(Noise_var);

%**********************RECEIVER DESIGN**********************

%************Filtering to separate the subcarriers************
SCM_Rx=(Responsivity.*I.*SCM_Tx);
%SCM_Rx=(Responsivity.*SCM_Tx);

for i0=1:length(SNR)
 Noise =[];
 Noise=Noise_SD(i0).*randn(size(SCM_Tx));

 if Rb <= 0.5*delta || N_sub == 1
  for i1=1:N_sub
   j1=i1-1; A1=[]; B1=[];
   fc=fmin+j1*delta;
   f_Rb=Rb*2/Fs; %Normalized
   data rate (bandwidth of data)
   fcl=2*(fc-Rb)/Fs; fch=(fc+Rb)*2/Fs; %Normalized cutoff
                                       frequencies
```

```
  w_bpf=[fcl,fch];
  [B1,A1]=butter(1,w_bpf);

%*******Individual subcarrier signal plus noise************
filter_out(i1,:)=filter(B1,A1,SCM_Rx)+Noise;

%******Symbol-by-symbol Coherent Demodulation *************
 for i2=1:symb
  j2=i2-1;
  a2=1+j2*samples_symb;
  b2=i2*samples_symb;

  %*******Coherent demodulation******************
  I_Dem_out=(2*Ac.*cos(2.*pi.*t.*fc)).*filter_out(i1,a2:b2);
  Q_Dem_out=(2*Ac.*sin(2.*pi.*t.*fc)).*filter_out(i1,a2:b2);
  [B2,A2]=butter(2,f_Rb);
  %I_Dem_out2=filter(B2,A2,I_Dem_out);
  %Q_Dem_out2=filter(B2,A2,Q_Dem_out);

  I_Demod_out2(i1,a2:b2)=filter(B2,A2,I_Dem_out);
  Q_Demod_out2(i1,a2:b2)=filter(B2,A2,Q_Dem_out);
  I_symb_end(i1,i2)=I_Demod_out2(i1,end);
  Q_symb_end(i1,i2)=-Q_Demod_out2(i1,end);
 end

 Demod_out2(i1,:)=I_symb_end(i1,:)+(j.*Q_symb_end(i1,:));

 end
else
'ALIASING; subcarriers too close and or too low fmin for the Rb'
 end
end

%Scatter plot-giving the constellation plot
scatterplot(Input1(:,1))
title('The input symbols constellation')
scatterplot(Demod_out2(1,:))
title('The received symbols constellation')

%Note: The codes for the functions: Basebandmodulation and
Turbulence are given at the end of the chapter.
```

6.3.3 BIT ERROR PROBABILITY ANALYSIS OF SIM-FSO

In this section, the BER analysis is presented in a clear but turbulent atmospheric channel; the turbulence-induced irradiance fluctuation considered in this section is based on the log-normal model of Chapter 3. A single BPSK premodulated subcarrier will first be considered and this will be followed by the M-PSK and then the DPSK-modulated subcarrier.

6.3.3.1   BPSK-Modulated Subcarrier

The result of this section will be for a single BPSK premodulated subcarrier $m(t) = g(t)$ $a_c \cos(\omega_c t + \varphi)$, over a symbol duration, where $a_c = [-1,1]$ represents the data symbols '0' and '1'. By employing a coherent demodulation at the subcarrier level, the symbol-by-symbol detection is carried out by multiplying the received signal, given by Equation 6.21a, by a locally generated RF signal of the same frequency ω_c, and phase as follows. Without any loss of generality, the subcarrier phase is equated to zero, that is, $\varphi = 0$. The coherent demodulator output $i_D(t)$ is then given by

$$i_D(t) = I_{comp} \times \cos(\omega_c t)$$
$$= \frac{R\xi I a_c g(t)}{2}[1 + \cos(2\omega_c t)] + n(t)\cos(\omega_c t) \tag{6.22}$$

Passing $i_D(t)$ through a low-pass filter with a bandwidth of $1/T$ suppresses the $\cos(2\omega_c t)$ term without distorting the information-bearing signal. This also reduces the noise variance at the output of the coherent demodulator filter to half its value at the input of the demodulator. Equation 6.22 therefore reduces to

$$i_D(t) = \frac{R\xi I a_c g(t)}{2} + n_D(t) \tag{6.23}$$

where the additive noise $n_D(t) \sim N(0, \sigma^2/2)$. Assuming an equiprobable data transmission such that $p(0) = p(1) = 1/2$, the probability of error conditioned on the received irradiance becomes

$$P_{ec} = p(1)p(e/1) + p(0)p(e/0)$$
$$= 0.5[p(e/1) + p(e/0)] \tag{6.24}$$

The marginal probabilities are given by

$$p(e/1) = \int_{-\infty}^{0} \frac{1}{\sqrt{\pi\sigma^2}} \exp\left\{-\frac{(i_D(t) - \mathcal{K})^2}{\sigma^2}\right\} di_D(t) \tag{6.25a}$$

$$p(e/0) = \int_{0}^{\infty} \frac{1}{\sqrt{\pi\sigma^2}} \exp\left\{-\frac{(i_D(t) + \mathcal{K})^2}{\sigma^2}\right\} di_D(t) \tag{6.25b}$$

where $\mathcal{K} = IR\xi/2$ and $g(t) = 1$ for $0 \le t \le T$ and zero elsewhere. Here, both binary symbols '1' and '0' are affected by the irradiance fluctuation since the optical source is on during the transmission of both data symbols '1' and '0'. This is in contrast to the OOK signalling technique where irradiance fluctuation only affects the data symbol '1'. Based on the antipodal nature of Equation 6.23, the decision threshold level can hence be fixed at the zero mark. This zero-level threshold is irradiance independent and hence not affected by the irradiance fluctuation caused by the atmospheric turbulence. From the foregoing and the apparent symmetry of Equation 6.25a and b, the BER conditioned on the received irradiance can now be written as

$$P_{ec} = \int_0^\infty \frac{1}{\sqrt{\pi\sigma^2}} \exp\left\{-\frac{(i_D(t) + \mathcal{K})^2}{\sigma^2}\right\} di_D(t)$$

$$= 0.5\,\mathrm{erfc}(\mathcal{K}/\sigma) = Q\left(\frac{\mathcal{K}\sqrt{2}}{\sigma}\right) \tag{6.26}$$

At the input of the subcarrier coherent demodulator, the electrical SNR per bit is given by

$$\gamma(I) = \frac{(\xi RI)^2 P_m}{\sigma^2} \tag{6.27}$$

where $P_m = A^2/2T \int_0^T g^2(t)dt$, from which $\sqrt{\gamma(I)} = \sqrt{2}\mathcal{K}/\sigma$. Equation 6.27 can now be expressed in terms of the SNR at the demodulator input as

$$P_{ec} = Q\left(\sqrt{\gamma(I)}\right) \tag{6.28}$$

The unconditional probability P_e is obtained by averaging (6.28) over the lognormal irradiance fluctuation statistics to obtain the following:

$$P_e = \int_0^\infty P_{ec}\,p(I)dI \tag{6.29a}$$

$$P_e = \int_0^\infty Q(\gamma(I))\frac{1}{I\sqrt{2\pi\sigma_l^2}} \exp\left\{-\frac{\left[\ln I/I_0 + \sigma_l^2/2\right]^2}{2\sigma_l^2}\right\}dI \tag{6.29b}$$

A closed-form solution of Equation 6.29b does not exist and using the numerical integration could result in truncating its upper limit. Also, the presence of the argument of the Q function at the lower limit of the Q function integral always poses analytical problems [25]. By combining an alternative representation of the Q function given by Equation 6.30 with the Gauss–Hermite quadrature integration approximation of Equation 6.31, the analytical difficulty involved solving Equation 6.29b can be circumvented.

$$Q(y) = \frac{1}{\pi}\int_0^{\pi/2} \exp\left(-\frac{y^2}{2\sin^2(\theta)}\right)d\theta \quad \text{for } y > 0 \tag{6.30}$$

$$\int_{-\infty}^{\infty} f(x)\exp(-x^2)dx \cong \sum_{i=1}^{n} w_i f(x_i) \tag{6.31}$$

where $[w_i]_{i=1}^{n}$ and $[x_i]_{i=1}^{n}$, whose values are given in Appendix 6.A, are the weight factors and the zeros of an nth-order Hermite polynomial, $He_n(x)$ [26]. The degree of accuracy of (6.10) depends on the order n of the Hermite polynomial. By invoking a change of variable, $y = \ln(I/I_0) + \sigma_1^2/2 / \sqrt{2}\sigma_1$ in Equation 6.29b and combining this with Equations 6.30 and 6.31, the unconditional BER given by Equation 6.29b can be reduced to the following form:

$$P_e \cong \frac{1}{\pi} \int_0^{\pi/2} \frac{1}{\sqrt{\pi}} \sum_{i=1}^{n} w_i \exp\left(-\frac{K_0 \exp\left(2K_1\left[\sqrt{2}\sigma_1 x_i - \sigma_1^2/2\right]\right)}{2\sin^2(\theta)}\right) d\theta \tag{6.32}$$

$$P_e \cong \frac{1}{\sqrt{\pi}} \sum_{i=1}^{n} w_i Q\left(\sqrt{K_0}\exp\left(K_1\left[\sqrt{2}\sigma_1 x_i - \sigma_1^2/2\right]\right)\right) \tag{6.33}$$

The values of K_1 and K_0 are as given in Table 6.1 for different noise-limiting conditions.

In Table 6.1, R_b represents the symbol rate. The BER plots, using both the approximation given by Equation 6.33 and the numerical simulation of the exact solution given by Equation 6.29b against the normalized SNR = $(\xi RE[I])^2/\sigma^2$, are illustrated in Figure 6.13. These results hint that using the 20th-order Gauss–Hermite approximation gives an accurate representation of the exact BER. The Gauss–Hermite integration solution of the BER is however preferred for its simplicity and compactness.

In order to keep the optical source (laser) within its linear dynamic range and avoid signal clipping distortion, the condition $|\xi m(t)| \leq 1$ must always hold. For a given value of ξ, this places an upper bound on the amplitude of each subcarrier. The BER given by Equation 6.33 is plotted against the normalized SNR for different noise-limiting conditions in Figure 6.14, based on the simulation parameters given in Table 6.2. The figure illustrates clearly that for an FSO link with a suitable optical BPF and a narrow FOV detector, the system performance is limited by thermal noise.

TABLE 6.1

Values of K_1 and K_0 for Different Noise-Limiting Conditions

	Noise-Limiting Conditions			
	Quantum Limit	Thermal Noise	Background Noise	Thermal and Background Noise
K_0	$\dfrac{\xi^2 R I_0 P_m}{2qR_b}$	$\dfrac{(\xi R I_0)^2 P_m R_L}{4kT_e R_b}$	$\dfrac{(\xi I_0)^2 R P_m}{2qR_b(I_{sky} + I_{sun})}$	$\dfrac{(\xi R I_0)^2 P_m}{(\sigma_{Bg}^2 + \sigma_{Th}^2)}$
K_1	0.5	1	1	1

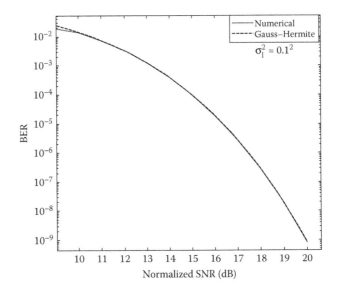

FIGURE 6.13 **(See colour insert.)** BER against the normalized SNR using numerical and 20th-order Gauss–Hermite integration methods in weak atmospheric turbulence for $\sigma_I^2 = 0.1^2$.

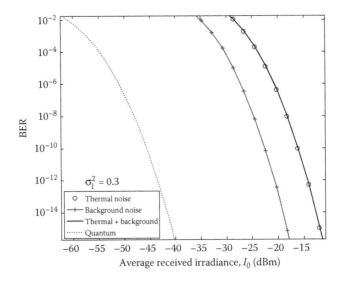

FIGURE 6.14 BER against the average received irradiance in weak turbulence under different noise-limiting conditions for $R_b = 155$ Mpbs and $\sigma_I^2 = 0.3$.

TABLE 6.2
Simulation Parameters

Parameter	Value
Symbol rate R_b	155 Mbps
Spectral radiance of the sky $N(\lambda)$	10^{-3} W/cm²µmSr
Spectral radiant emittance of the sun $W(\lambda)$	0.055 W/cm²µm
Optical band-pass filter bandwidth $\Delta\lambda$ at $\lambda = 850$ nm	1 nm
PIN photodetector field of view (FOV)	0.6 rad
Radiation wavelength λ	850 nm
Number of subcarriers N	1
Link range L	1 km
Index of refraction structure parameter C_n^2	0.75×10^{-14} m$^{-2/3}$
Load resistance R_L	50 Ω
PIN photodetector responsivity $\Re$	1
Operating temperature T_e	300 K
Optical modulation index ξ	1

Moreover, under this thermal noise-limited condition, the SIM-FSO still requires about additional 30 dB of SNR compared with the theoretical quantum limit.

MATLAB codes for generating the BER performance, as shown in Figure 6.14 under different noise conditions are given in Program 6.3. Additional MATLAB codes for this section are given in Appendix 6.B.

Program 6.3: MATLAB Codes for Generating the BER Performance under Different Noise Conditions

```
%Evaluation of BER of SISO BPSK SIM FSO under weak turbulence
using the %Gauss-Hermite Quadrature integration approach.
%Different noise sources considered: Background, thermal
and dark

clear
clc

%*************Simulation Parameters************************
Rb=155e6;         %symbol rate
R=1;              %Responsivity
M_ind=1;          %Modulation index
A=1;              %Subcarrier signal amplitude
RL=50;            %Load resistance
Temp=300;         %Ambient temperature
wavl=850e-9;      %Optical source wavelength

%*******************Background Noise*******************
%Considering 1 cm^2 receiving aperture
sky_irra=1e-3;    % at 850nm wavelength, in W/cm^2-um-sr
```

```
sun_irra=550e-4;        %at 850nm wavelength, in W/cm^2-um
FOV=0.6;                % in radian
OBP=1e-3;               %Optical filter bandwidth in micrometre
Isky=sky_irra*OBP*(4/pi)*FOV^2;   %Sky irradiance
Isun=sun_irra*OBP;      %Sun irradiance

%*************Rytov Variance********************************
Range=1e3;      %Link range in meters
Cn=0.75e-14;    %Refractive index structure
          %parameter
Rhol=1.23*(Range^(11/6))*Cn*(2*pi/wavl)^(7/6); %Log irradiance
                %variance (Must be less than 1
Varl=Rhol;      %Log intensity variance
r=sqrt(Varl);   %log intensity standard deviation

%**********************Physical constants******************
E_c=1.602e-19; %Electronic charge
B_c=1.38e-23; %Boltzmann constant
%*********************************************************

Pd=A^2/2;
K1=((M_ind^2)*R*Pd)/(2*E_c*Rb);   %Quantum limit
K2=((R*M_ind)^2)*Pd*RL/(4*B_c*Temp*Rb); %Thermal noise limit
K3=(Pd*R*M_ind^2)/(2*E_c*Rb*(Isun+Isky)); %Background radiation
                %limit
Ktemp=(4*B_c*Temp*Rb/RL)+(2*E_c*R*Rb*(Isun+Isky));
K4=((R*M_ind)^2)*Pd/Ktemp; %background and thermal
                %noise combined

%***************Hermite polynomial weights and roots*********
w20=[2.22939364554e-13,4.39934099226e-10,1.08606937077e-7,
7.8025564785e-6,0.000228338636017,0.00324377334224, 0.0248105208875,
0.10901720602,0.286675505363,0.462243669601,...

0.462243669601,0.286675505363,0.10901720602,0.0248105208875,
0.00324377334224,0.000228338636017,7.8025564785e-6,
1.08606937077e-7,4.39934099226e-10,2.22939364554e-13];
x20=[-5.38748089001,-4.60368244955,-3.94476404012,
-3.34785456738,-2.78880605843,-2.25497400209,-1.73853771212,
-1.2340762154,-0.737473728545,-0.245340708301,...

0.245340708301,0.737473728545,1.2340762154,1.73853771212,
2.25497400209,2.78880605843,3.34785456738,3.94476404012,
4.60368244955,5.38748089001];
%*****************************************************

%*********************BER evaluation******************

Io=logspace(-10,-4,30); %Average received irradiance
IodBm=10*log10(Io*1e3); %Average received irradiance in
                %dBm
```

```
SNR2=RL*((R.*Io).^2)./(4*B_c*Temp*Rb);
SNR2dB=10*log10(SNR2);
for i1=1:length(Io)

  GH1=0;GH2=0;GH3=0;GH4=0;
  for i2=1:length(x20)
   arg1=sqrt(K1*Io(i1))*exp(0.5*x20(i2)*sqrt(2)*r-Var1/4);
   temp1=w20(i2)*Q(arg1);
   GH1=GH1+temp1;
   arg2=sqrt(K2)*Io(i1)*exp(x20(i2)*sqrt(2)*r-Var1/2);
   temp2=w20(i2)*Q(arg2);
   GH2=GH2+temp2;
   arg3=sqrt(K3)*Io(i1)*exp(x20(i2)*sqrt(2)*r-Var1/2);
   temp3=w20(i2)*Q(arg3);
   GH3=GH3+temp3;
   arg4=sqrt(K4)*Io(i1)*exp(x20(i2)*sqrt(2)*r-Var1/2);
   temp4=w20(i2)*Q(arg4);
   GH4=GH4+temp4;
  end

BER1(i1)=GH1/sqrt(pi);
BER2(i1)=GH2/sqrt(pi);
BER3(i1)=GH3/sqrt(pi);
BER4(i1)=GH4/sqrt(pi);
end

%*****************Plot function************************
figure
subplot(4,1,1)            %Quantum limit case
semilogy(IodBm,BER1)

subplot(4,1,2)            %Thermal noise case
semilogy(IodBm,BER2)

subplot(4,1,3)            %Background radiation case
semilogy(IodBm,BER3)

subplot(4,1,4)            %background and thermal noise case
semilogy(IodBm,BER4)
```

6.3.3.2 M-Ary PSK-Modulated Subcarrier

Here, the data symbols which comprise $\log_2 M$ binary digits are mapped onto one of the M available phases on each subcarrier signal, $m(t)$. Based on the subcarrier coherent demodulation and by following the analytical approach given in Ref. [24], the following conditional BER expressions are obtained:

$$P_{ec} \approx \frac{2}{\log_2 M} Q\left(\sqrt{(\log_2 M)\gamma(I)}\sin(\pi/M)\right) \quad \text{for } M\text{-PSK}, M \geq 4 \quad (6.34a)$$

$$P_{ec} = \frac{2\left(1 - 1/\sqrt{M}\right)}{\log_2 M} Q\left(\sqrt{\frac{3\log_2 M\gamma(I)}{2(M-1)}}\right) \quad \text{for } M\text{-QAM}, \log_2 M \text{ must be even} \quad (6.34b)$$

The unconditional BER, P_e, is thus obtained in a similar fashion, by averaging the conditional bit error rate over the atmospheric turbulence statistics. The resulting BER expression (6.35) for M-PSK has no closed-form solution and can only be evaluated numerically.

$$P_e \cong \frac{2}{\log_2 M} \int_0^\infty Q\left(\sqrt{\gamma(I)\log_2 M} \sin(\pi/M)\right) p(I)\mathrm{d}I \quad (6.35)$$

Whenever a subcarrier coherent detection is used, there is always an ambiguity associated with the estimation of the absolute phase of the subcarrier signal [24]. This poses an implementation challenge for the subcarrier coherent demodulation-based systems; this can however be solved by considering a differential phase shift keying (DPSK)-based SIM-FSO system as discussed below.

6.3.3.3 DPSK-Modulated Subcarrier

This modulation scheme is the most suitable when the absolute phase estimation needed for the subcarrier coherent demodulation is not feasible or too complex to realize. The DPSK premodulated SIM-FSO is demodulated by comparing the phase of the received signal in any signalling interval with the phase of the signal received in the preceding signalling interval [24], as shown in Figure 6.15 for a single-subcarrier FSO system. Accurate demodulation of the present data symbol thus depends on whether the preceding symbol has been correctly demodulated or not. The demodulation of DPSK-based SIM-FSO is feasible during atmospheric turbulence because the turbulence coherence time, which is of the order of milliseconds, is far greater than the typical duration of two consecutive data symbols. This implies that the channel properties are fixed during a minimum of two symbol durations—a prerequisite for noncoherent demodulation of the DPSK subcarrier signal.

The conditional BER of the DPSK premodulated subcarrier is given by [24,27]

$$P_{ec} = 0.5\exp\left(-0.5\gamma(I)\right) \quad (6.36)$$

In the presence of scintillation, the following unconditional BER, P_e, is derived using the Gauss–Hermite quadrature integration approximation of Section 6.3.3.1 as

$$P_e = \int_0^\infty 0.5\exp(-\gamma(I))p(I)\mathrm{d}I \quad (6.37)$$

$$P_e \cong \frac{1}{2\sqrt{\pi}} \sum_{i=1}^n w_i \exp\left(-K^2 \exp\left[x_i 2\sqrt{2}\sigma_1 - \sigma_1^2\right]\right) \quad (6.38)$$

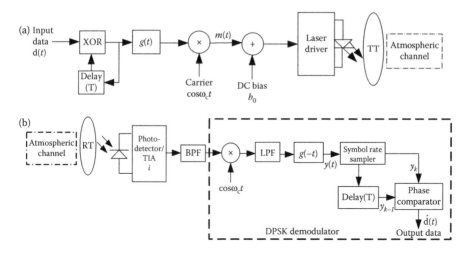

FIGURE 6.15   Block diagram of an FSO link employing DPSK-modulated SIM. (a) Transmitter and (b) receiver. TIP—transimpedance amplifier; TT—transmitter telescope; RT—receiver telescope.

where $= R\xi I_0/2\sqrt{\sigma^2}$. In Figure 6.16, the BER of SIM-FSO based on different modulation techniques on the subcarrier are compared at a scintillation level of $\sigma_l^2 = 0.5^2$.

The performance in a turbulence-free channel is included in the figure for the estimation of turbulence-induced fading penalty. The figure shows clearly the performance superiority of BPSK-modulated SIM in terms of the amount of SNR

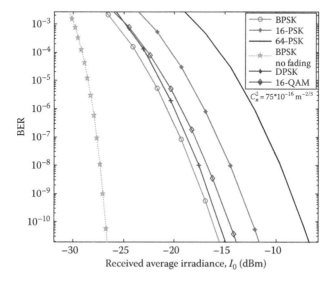

FIGURE 6.16   BER against the received irradiance for SIM-FSO with different subcarrier modulation techniques in weak atmospheric turbulence for $\sigma_l^2 = 0.3$, $\lambda = 850$ nm and link range = 1 km.

required to achieve a given BER. Due to atmospheric turbulence-induced channel fading, a BPSK premodulated SIM-FSO system will incur a power penalty of ~5 dB at a BER of 10^{-6}; the penalty rises to ~10 dB when the error performance level is raised to a BER of 10^{-9}. This penalty is higher for other modulation techniques as shown in the figure. The M-PSK (with $M \geq 4$) modulated subcarrier is known to be more bandwidth efficient [24], while the DPSK is advantageous in that it does not require absolute phase estimation but both are not as power efficient as the BPSK-SIM. The choice of modulation technique at the subcarrier level therefore depends on the application at hand and requires a compromise between simplicity, power and bandwidth efficiencies.

MATLAB codes for simulating the performance for different modulation techniques, including those shown in Figure 6.16, are given in Appendix 6.B.

6.3.3.4 Multiple SIM Performance Analysis

The use of multiple subcarriers is a viable way of increasing system throughput/capacity. To archive this, different data/users are premodulated on different subcarrier frequencies. These are then aggregated and the resulting composite signal used to modulate the intensity of the optical source. For a subcarrier system with N subcarrier signals operating at different frequencies, the inherent nonlinearity of the optical source will result in the transfer of energy among the subcarrier frequencies (i.e., the intermodulation distortion [IMD]). The IMD potentially results in a reduced SNR as it contributes to the amount of unwanted signals within the frequency band of interest. The multiple subcarrier system will be examined by assuming an ideal optical source with no reference to the IMD and the clipping distortion.

By definition, the modulation index/depth is given by

$$\xi \triangleq \frac{|m(t)|}{i'_B} = \sum_{j=1}^{N} \frac{|A_j g(t)\cos(\omega_{cj} + \theta_j)|}{i'_B} \tag{6.39}$$

where $i'_B = i_B - i_{Th}$ as already shown in Figure 6.9. The peak value of the composite signal $m(t)$ occurs when all the individual subcarrier amplitudes add up coherently, that is,

$$\xi = \sum_{j=1}^{N} \left|\frac{A_j}{i'_B}\right| = \sum_{j=1}^{N} \xi_j = N\xi_{sc} \tag{6.40}$$

In Equation 6.19, all the subcarrier signals have been assumed to have the same individual modulation depth $\xi_{sc} = \xi/N$. Since the SNR on each subcarrier is proportional to the square of the modulation depth, there exists at least a loss of $20\log N$ (dB) in electrical SNR (equivalent to $10\log N$ (dB) optical power reduction) on each subcarrier due to the presence of N subcarrier signals with nonoverlapping frequencies.

The BER plots of a BPSK premodulated subcarrier with different numbers of subcarriers are shown in Figure 6.17. The graph is obtained by replacing ξ with ξ_{sc} in

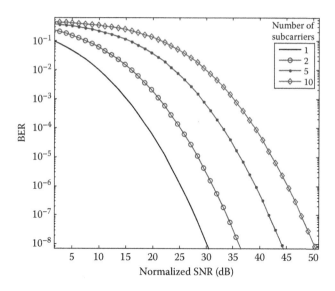

FIGURE 6.17 BER against the normalized SNR for multiple subcarriers FSO system in weak atmospheric turbulence for $N = [1, 2, 5, 10]$ and $\sigma_l^2 = 0.3$.

Equation 6.33. For example, the SNR required to achieve a BER of 10^{-6} with five subcarriers is ~40 dB in an atmospheric channel with a fading strength of $\sigma_l^2 = 0.3$. This increases to ~46 dB with 10 subcarriers under the same channel conditions, depicting a $20 \log N$ (dB) increment over the SNR required by a single-subcarrier FSO system.

In Figure 6.18, these SNR values required to attain a BER of 10^{-6} are plotted against the number of subcarriers for turbulent atmospheric channels of different fading strengths. This shows explicitly that for a given BER value, the SNR values increase with an increase in both number of subcarriers and the fading strength. Based on this, therefore, the multiple SIM-FSO is only recommended whenever the quest for increased throughput/capacity outweighs its accompanying power penalty.

6.3.3.5 Outage Probability in Log-Normal Atmospheric Channels

The outage probability is another metric for quantifying the performance of communication systems in fading channels. A system with an adequate average BER can temporarily suffer from increases in error rate due to deep fades and this 'short outages' is not adequately modelled by the average BER [28]. An alternative performance metric therefore is the probability of outage due to the presence of atmospheric turbulence, which is defined as $P_e > P_e^*$, where P_e^* is a predetermined threshold BER. This is akin to finding the probability that the SNR, $\gamma(I)$, that corresponds to P_e is lower than the threshold SNR (denoted by γ^*) that corresponds to P_e^*. That is

$$P_{out} = P(P_e > P_e^*) \equiv P(\gamma(I) < \gamma^*) \tag{6.41}$$

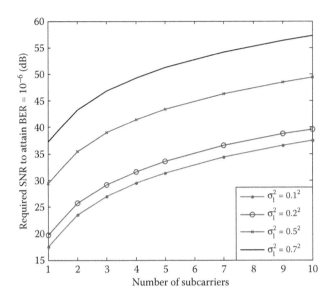

FIGURE 6.18 SNR required to attain a BER of 10^{-6} against the number of subcarriers for BPSK-modulated SIM-FSO system with $\sigma_1 = [0.1, 0.2, 0.5, 0.7]$.

where $\gamma^* = (R\xi I_0)^2/2\sigma^2$ is the average SNR in the absence of atmospheric turbulence for a given noise level. If a parameter m, here called 'power margin', is introduced to account for the extra power needed to cater for turbulence-induced signal fading, then the outage probability is derived as follows:

$$P_{Out} = P\left(m\gamma(I) < \gamma^*\right) = \int_0^{I_0/m} \frac{1}{I\sqrt{2\pi\sigma_1^2}} \exp\left\{-\frac{\left(\ln I/I_0 + \sigma_1^2/2\right)^2}{2\sigma_1^2}\right\} dI$$

$$= Q\left(\frac{1}{\sigma_1} \ln m - \frac{\sigma_1}{2}\right) \tag{6.42}$$

Invoking the Chernoff upper bound, $Q(x) \leq 0.5 \exp(-x^2/2)$, on Equation 6.42 gives an upper-bound value for the outage probability; from this an approximate power margin, m, needed to obtain P_{out} can be obtained as

$$m \approx \exp\left(\sqrt{-2\ln 2P_{out}\sigma_1^2} + \sigma_1^2/2\right) \tag{6.43}$$

This extra power needed to obtain a given outage probability is depicted in Figure 6.19 at various levels of irradiance fluctuation. For example, to achieve an outage probability of 10^{-6}, about 35 dBm of extra power is needed at $\sigma_1^2 = 0.2^2$. This will rise to ~43 and 48 dBm when the irradiance strength is $\sigma_1^2 = 0.5^2$ and $\sigma_1^2 = 0.7^2$,

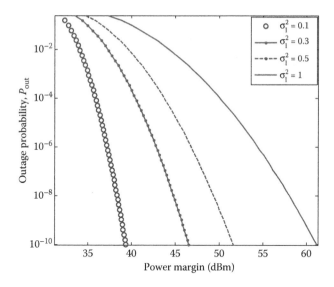

FIGURE 6.19 Outage probability against the power margin for a log-normal turbulent atmospheric channel for $\sigma_I^2 = [0.1, 0.3, 0.5, 1]$.

respectively. The extra margin can also be viewed as the penalty introduced by the atmospheric turbulence, and to reduce it, diversity techniques will be considered in Chapter 7.

The MATLAB codes for generating the outage probability for a log-normal turbulence atmospheric channel are given in Program 6.4.

Program 6.4: MATLAB Code for Figure 6.19

```
%Evaluating the additional power (margin, m) needed to achieve
a given %outage probability, using the Chernoff upper bound.
lognormal %scintillation model used.
clear
clc
Rhol=[0.5]
ro=sqrt(Rhol); %Log intensity standard deviation
for j=1: length(ro)
 r=ro(j);
Pout=logspace(0,-10,50); %Outage probability
for i=1: length(Pout)
 Po=Pout(i);
 arg=sqrt(-2*r^2*log(2*Po))+((r^2)/2);
 m(i)=exp(arg);
 margin(j,i)=10*log10(m(i)*1e3); %Power margin to achieve
 outage prob.
                %in dBm.
end
```

```
end
semilogy(margin,Pout)
xlabel('Power Margin (dBm)')
ylabel('Outage Probability')
  title('SISO')
```

6.3.4 SIM-FSO Performance in Gamma–Gamma and Negative Exponential Atmospheric Channels

To obtain the unconditional BER in a gamma–gamma turbulent atmospheric chan-nel, the irradiance fluctuation statistics in the previous section is replaced appro-priately by the gamma–gamma pdf. For a BPSK premodulated subcarrier, the unconditional BER now becomes

$$P_e = \int_0^\infty Q\left(\sqrt{\gamma(I)}\right) \frac{2(\alpha\beta)^{(\alpha+\beta)/2}}{\Gamma(\alpha)\Gamma(\beta)} I^{\left(\frac{\alpha+\beta}{2}\right)-1} K_{\alpha-\beta}\left(2\sqrt{\alpha\beta I}\right) dI \qquad (6.44)$$

This BER expression can only be evaluated numerically as it does not have a closed-form solution. The values of the parameters α and β, which are used to describe the turbulence strength, are as given in Table 6.3.

In the limit of strong turbulence, that is, in the saturation regime and beyond, the BER is obtained by replacing the pdf in the conditional BER by the fully developed speckle (negative exponential) pdf discussed in Chapter 3. By applying the alterna-tive representation of $Q(.)$ given by Equation 6.30, the unconditional BER in the fully developed speckle regime is derived as

$$P_e = \frac{1}{\pi I_0} \int_0^{\pi/2} \int_0^\infty \exp\left(-\frac{(\xi RI)^2}{4\sigma^2 \sin^2\vartheta} - \frac{I}{I_0}\right) dI\, d\vartheta \qquad (6.45)$$

The multiple integration involved in Equation 6.45 can be conveniently circum-vented, and doing so reduces the BER expression P_e to the following:

TABLE 6.3
Fading Strength Parameters for Gamma–Gamma Turbulence Model

Parameter	Turbulence Regime		
	Weak	Moderate	Strong
σ_1^2	0.2	1.6	3.5
α	11.6	4.0	4.2
β	10.1	1.9	1.4

$$P_e = \frac{1}{\pi} \int_0^{\pi/2} \sqrt{\pi\mathcal{K}_0(\vartheta)} \exp\big(\mathcal{K}_0(\vartheta)\big) \text{erfc}\left(\sqrt{\mathcal{K}_0(\vartheta)}\right) d\vartheta \qquad (6.46)$$

where $\mathcal{K}_0(\vartheta) = (\sigma \sin(\vartheta)/\xi R)^2$ and erfc (.) is the complementary error function. The following upper bound, given by Equation 6.47, is then obtained by maximizing the integrand with the substitution of $\vartheta = \pi/2$ in Equation 6.46.

$$P_e \leq \sqrt{\pi\mathcal{K}_0} \exp(\mathcal{K}_0) Q\left(\sqrt{2\mathcal{K}_0}\right) \qquad (6.47)$$

where $\mathcal{K}_0 = (\sigma/\xi R)^2$. From these BER expressions, the error performance of the system can be predicted for any given value of SNR and turbulence strength (or link range). The numerical simulations of the BER expressions (6.44), (6.46) and the upper bound (6.47) are shown in Figure 6.20, where P_e is plotted against the normalized SNR under different turbulence regimes.

For instance, to achieve a BER of 10^{-6} in a weak atmospheric turbulence, the required SNR is ~29 dB and this rises to ~65 and ~67 dB, respectively, for moderate and intermediate regimes. While in the saturation regime, a staggering ~115 dB (the upper bound value is 4 dB higher) is required to achieve the same level of error performance (i.e., BER of 10^{-6}). Achieving a BER lower than 10^{-6} in the saturation regime requires a phenomenal increase in SNR as seen in Figure 6.20. It is noteworthy that the normalized SNR is in the electrical domain and it is based on the average received irradiance, $E[I]$. Also, the 'kinks' observed in the

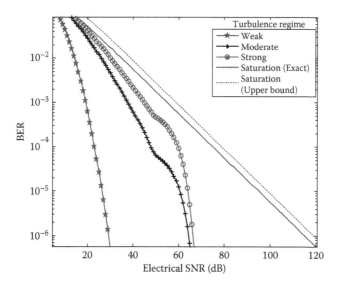

FIGURE 6.20 BER performance against the normalized electrical SNR across all turbulence regimes based on gamma–gamma and negative exponential modes.

curves for strong and moderate turbulence are due to the numerical integration process.

In order to compare the error performance BPSK-SIM with an OOK-modulated FSO system of the same average transmitted optical power, the unconditional BER of the OOK-FSO is modified to become

$$P_e = 0.5 \left[\int_{i_{th}}^{\infty} \frac{1}{\sqrt{\pi\sigma^2}} \exp\left(\frac{-i^2}{\sigma^2}\right) di + \int_0^{\infty} \int_0^{i_{th}} \frac{1}{\sqrt{\pi\sigma^2}} \exp\left[-\frac{(i - 2RI)^2}{\sigma^2}\right] p(I) di \, dI \right] \quad (6.48)$$

In Figure 6.21, the BER performances of the OOK system employing adaptive (optimum) and fixed threshold values of 0.05 and 0.8 are shown alongside that of BPSK-SIM in a weak turbulent atmospheric fading. Although the optimum OOK is marginally superior to BPSK-SIM under the stated conditions, as it requires 1.6 dB electrical SNR less at a BER of 10^{-6}, it does require an accurate knowledge of both the additive noise and fading levels to achieve this performance. With the threshold fixed at say 0.05, the OOK requires about 7 dB electrical SNR more than BPSK-SIM at a BER of 10^{-6}. Also, the BER performance of OOK with a fixed threshold level exhibits a BER floor as shown in Figure 6.21 for $i_{th} = 0.8$ and in the previous work in Ref. [8]. The SIM is therefore recommended in atmospheric turbulence channels as against the fixed threshold OOK currently used in commercial terrestrial FSO systems.

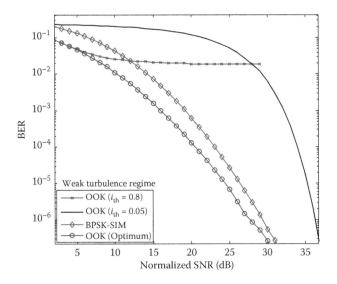

FIGURE 6.21 Error performance of BPSK-SIM and OOK with fixed and adaptive threshold-based FSO in weak turbulence regime modelled using gamma–gamma distribution.

6.3.5   OUTAGE PROBABILITY IN NEGATIVE EXPONENTIAL MODEL ATMOSPHERIC CHANNELS

By following the approach of Section 6.3.3.5, the outage probability in the fully developed speckle is obtained as follows:

$$P_{out} = P\left(m\gamma(I) < \gamma^*\right) = \int_0^{I_0/m} \frac{1}{I_0} \exp\left(-\frac{I}{I_0}\right) dI \qquad (6.49)$$

From Equation 6.49, the power margin m, needed to achieve a given P_{out} saturation regime, is as given by Equation 6.50. This is plotted in Figure 6.22, alongside Equation 6.43 with $\sigma_I^2 = 0.5$, which represents the outage probability in weak atmospheric turbulence. A comparison of the results in this figure reveals that the power margin required to achieve a P_{out} of 10^{-6} in the fully developed speckle regime is about 40 dB higher than that required in the weak turbulence regime with $\sigma_I^2 = 0.5$, and this will increase as the required outage probability level is reduced to below 10^{-6}.

$$m = \left[-\ln\left(1 - P_{out}\right)\right]^{-1} \qquad (6.50)$$

The prohibitive power required in the saturation regime suggests that the establishment of a reliable communication link in this regime is impossible unless the fading effect due to turbulence is mitigated or compensated for.

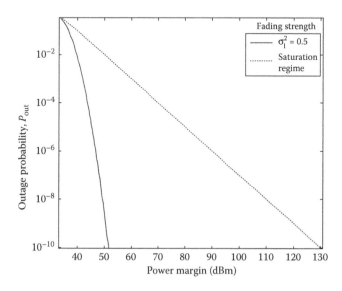

FIGURE 6.22   The outage probability against the power margin in saturation and weak turbulence regimes for $\sigma_I^2 = 0.5$.

6.4   ATMOSPHERIC TURBULENCE-INDUCED PENALTY

In this section, the additional power required due to the presence of turbulence-induced channel fading to achieve a given level of performance will be examined. Without any loss of generality, the result here will be based on the log-normal turbulence model and the BPSK-modulated subcarrier. This can however be extended to other turbulence models and modulation schemes in a straightforward manner. In Figure 6.23, the BER performance is plotted as a function of the normalized SNR for different levels of atmospheric turbulence. As an example, to achieve a BER of 10^{-6} in a channel characterized by $\sigma_I^2 = 0.1$, an SNR of about 24 dB will be required. This increases to ~37 dB as the fading strength increases to $\sigma_I^2 = 0.5$.

Also from this plot, the SNR penalty at any given fading strength and BER can be obtained. The SNR penalty is defined as the difference between the SNR required to achieve a specified BER in the presence and absence of atmospheric turbulence. The SNR penalty is shown in Figure 6.24 for the following BER levels: 10^{-3}, 10^{-6} and 10^{-9}. Expectedly, the SNR penalty increases as the turbulence strength increases and as the benchmark BER value decreases. Beyond $\sigma_I^2 = 0.1$, the SNR penalty increases very sharply, for instance, at a BER of 10^{-6}, when the fading strength rises by 10-folds from $\sigma_I^2 = 0.01$ to $\sigma_I^2 = 0.1$, the penalty only rises from 1 to 6.5 dB. But when the fading strength increases to $\sigma_I^2 = 0.3$, the SNR penalty rises to 15 dB.

The MATLAB script for Figure 6.23 is given in Program 6.5.

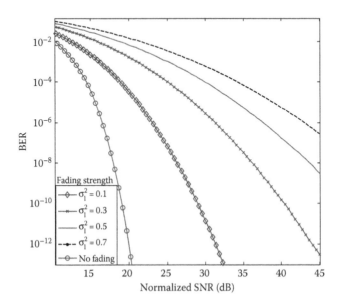

FIGURE 6.23 Error rate performance against normalized SNR for BPSK-SIM-based FSO in weak atmospheric turbulence channel for $\sigma_I^2 = [0, 0.1, 0.3, 0.5, 0.7]$.

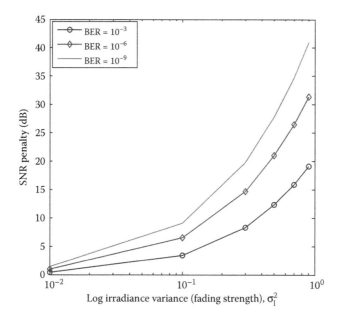

FIGURE 6.24   Turbulence-induced SNR penalty as a function of log irradiance variance for BPSK-SIM-based FSO for BER = [10^{-3}, 10^{-6}, 10^{-9}].

Program 6.5: MATLAB Script for Figure 6.23

```
%Evaluation of BER of SISO BPSK MSM FSO under weak turbulence
using the Gauss-Hermite Quadrature integration approach.

Clear
clc

%*************Parameters************************
R=1;                      %Responsivity
Io=1;                     %Intensity without turbulence
N=1;                      %no of subcarrier
r=sqrt(0.1);              %log intensity standard deviation
Noise=logspace(0,-5,100); %Gaussian noise variance
%*******************************************************
for j=1:length(Noise)
  No=Noise(j);
  SNR(j)=10*log10(((R*Io)^2)/(No));
  K=(R*Io)/(sqrt(2*No)*N);

%****Hermite polynomial weights and roots***********
w20=[2.22939364554e-13,4.39934099226e-10,1.08606937077e-7,
7.8025564785e-6,0.000228338636017,0.00324377334224, 0.0248105208875,
0.10901720602,0.286675505363,0.462243669601,...

0.462243669601,0.286675505363,0.10901720602,0.0248105208875,
0.00324377334224,0.000228338636017,7.8025564785e-6,
1.08606937077e-7,4.39934099226e-10,2.22939364554e-13];
```

```
x20=[-5.38748089001,-4.60368244955,-3.94476404012,
-3.34785456738,-2.78880605843,-2.25497400209,-1.73853771212,
-1.2340762154,-0.737473728545,-0.245340708301,...

0.245340708301,0.737473728545,1.2340762154,1.73853771212,
2.25497400209,2.78880605843,3.34785456738,3.94476404012,
4.60368244955,5.38748089001];
%****************************************************
GH=0;
for i=1:length(x20)
 arg=K*exp(x20(i)*sqrt(2)*r-r^2/2);
 temp=w20(i)*Q(arg);
 GH=GH+temp;
end
BER(j)=GH/sqrt(pi);
end

%*********Plot function*************************
semilogy(SNR,BER)
xlabel('SNR (R*E[I])^2/No (dB)')
ylabel('BER')
%title('BPSK E[I]=Io=1 R=1')
```

Moreover, in Figures 6.25 and 6.26, respectively, the BER and P_{out} metrics of the BPSK-SIM scheme are plotted against the turbulence strength for different values of SNR. These plots show that increasing the signal power can help mitigate the fading caused by the atmospheric turbulence but only in the very weak regime, where $\sigma_l^2 < 0.1$. As the turbulence strength increases beyond this value, the BER and indeed the P_{out} both tend to high asymptotic values that are too high to guarantee a reliable exchange of information via the link. This implies that techniques other than a mere increase in transmitted power will be required to mitigate atmospheric turbulence beyond the very weak regime. To do just that, diversity techniques will be examined in the next chapter. Given in Table 6.4 is a brief summary of the three modulation techniques considered in this chapter.

TABLE 6.4
Comparison of Modulation Techniques

On–Off Keying	Pulse Position Modulation	Subcarrier Intensity Modulation
Simple to implement	Power efficient	Power inefficient
Synchronization not required	Synchronization required	Higher throughput
Adaptive threshold required in fading channels	Adaptive threshold not required in fading channels	Adaptive threshold not required in fading channels
Component nonlinearity not an issue	Component nonlinearity not an issue	Component nonlinearity an issue with multiple subcarriers
Suboptimal with fixed threshold	High bandwidth requirement	Benefits from advances in digital signal processing and matured RF components

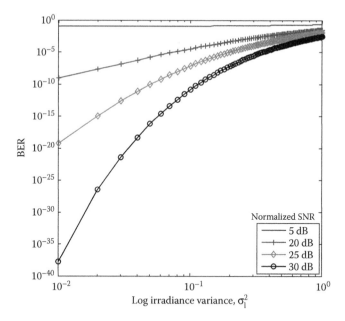

FIGURE 6.25 BER of BPSK-SIM against the turbulence strength in weak atmospheric turbulence for normalized SNR (dB) = [5, 20, 25, 30].

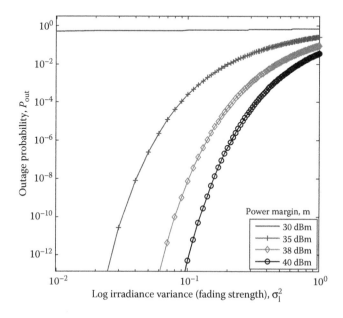

FIGURE 6.26 P_{out} of BPSK-SIM against the turbulence strength in weak atmospheric turbulence for m (dBm) = [30, 35, 38, 40].

APPENDIX 6.A

Zeros and Weights of Gauss–Hermite Integration with $n = 20$

$$\int_{-\infty}^{\infty} f(x)\,dx = \int_{-\infty}^{\infty} e^{-x^2}\left[e^{x^2} f(x)\right]dx \approx \sum_{i=1}^{n} w(x_i)e^{-x_i^2} f(x_i)$$

i	Zeros, x_i	Weight, $w(x_i)$	Total Weight, $w(x_i)e^{x_i^2}$
1	−5.38748089001	2.22939364554E-013	0.898591961453
2	−4.60368244955	4.39934099226E-010	0.704332961176
3	−3.94476404012	1.08606937077E-007	0.62227869619
4	−3.34785456738	7.8025564785E-006	0.575262442852
5	−2.78880605843	0.000228338636017	0.544851742366
6	−2.25497400209	0.00324377334224	0.524080350949
7	−1.73853771212	0.0248105208875	0.509679027117
8	−1.2340762154	0.10901720602	0.499920871336
9	−0.737473728545	0.286675505363	0.493843385272
10	−0.245340708301	0.462243669601	0.490921500667
11	0.245340708301	0.462243669601	0.490921500667
12	0.737473728545	0.286675505363	0.493843385272
13	1.2340762154	0.10901720602	0.499920871336
14	1.73853771212	0.0248105208875	0.509679027117
15	2.25497400209	0.00324377334224	0.524080350949
16	2.78880605843	0.000228338636017	0.544851742366
17	3.34785456738	7.8025564785E-006	0.575262442852
18	3.94476404012	1.08606937077E-007	0.62227869619
19	4.60368244955	4.39934099226E-010	0.704332961176
20	5.38748089001	2.22939364554E-013	0.898591961453

APPENDIX 6.B

6.B.1   MATLAB Scripts for Sections 6.3.2, 6.3.3.2 and 6.3.3.3

6.B.1.1   Section 6.3.2

```
******************%Function Basebandmodulation%**************
function [Inphase,Quadrature,X]=Basebandmodulation(M,symb,N_sub)

X=randint(symb,N_sub,M);
H=modem.pskmod('M',M,'phaseoffset',pi/4,'SymbolOrder',
'gray');
Y=modulate(H,X);
Inphase=real(Y);
Quadrature=imag(Y)


******************%Function Turbulence %*****************
function [I]=Turbulence(Log_Int_var,No_symb,Io,t)
```

```
%Log intensity Variance
Var_l=Log_Int_var;
%Number of symbols
%No_symb=1e4;


%Log normal atmospheric turbulence

l=(sqrt(Var_l).*randn(1,No_symb))-(Var_l/2);
I1=Io.*exp(l);
for i=1:No_symb
 a=1+(i-1)*length(t);
 b=i*length(t);
 I(1,a:b)=I1(i);
end
```

6.B.1.2   Sections 6.3.3.2 and 6.3.3.3

```
%Evaluation of BER of SISO M-QAM SIM FSO under weak turbulence
using the %Gauss-Hermite Quadrature integration approach.
%Considering Background and thermal noise.


clear
clc


%*************Simulation Parameters*************************
M=16;    %Number of levels; M-ary QAM; M should be even
Rb=155e6;        %symbol rate
R=1;             %Responsivity
M_ind=1;         %Modulation index
A=1;             %Subcarrier signal amplitude
RL=50;           %Load resistance
Temp=300;        %Ambient temperature
wavl=850e-9;     %Optical source wavelength


%********************Background Noise*********************
%Considering 1 cm^2 receiving aperture
sky_irra=1e-3;   %at 850nm wavelength, in W/cm^2-um-sr
sun_irra=550e-4; %at 850nm wavelength, in W/cm^2-um
FOV=0.6;  %in radian
OBP=1e-3; %Optical filter bandwidth in micrometre
Isky=sky_irra*OBP*(4/pi)*FOV^2;  %Sky irradiance
Isun=sun_irra*OBP;  %Sun irradiance


%*************Rytov Variance***********************
Range=1e3; %Link range in meters
Cn=0.75e-14; %Refractive index structure parameter
Rhol=1.23*(Range^(11/6))*Cn*(2*pi/wavl)^(7/6);  %Log
irradiance variance (Must be less than 1)
Varl=Rhol; %Log intensity variance
r=sqrt(Varl); %log intensity standard deviation
```

```
%****************Physical constants************
E_c=1.602e-19;  %Electronic charge
B_c=1.38e-23;   %Boltzmann constant
%*****************************************
Pd=A^2/2;
Ktemp=(4*B_c*Temp*Rb/RL)+(2*E_c*R*Rb*(Isun+Isky));
K1=3*log2(M)*((R*M_ind)^2)*Pd/(2*(M-1)*Ktemp);

%*********background and thermal noise combined*************

%****Hermite polynomial weights and roots************
w20=[2.22939364554e-13,4.39934099226e-10,1.08606937077e-7,
7.8025564785e-6,0.000228338636017,0.00324377334224, 0.0248105208875,
0.10901720602,0.286675505363,0.462243669601,...

0.462243669601,0.286675505363,0.10901720602, 0.0248105208875,
0.00324377334224,0.000228338636017, 7.8025564785e-6,
1.08606937077e-7,4.39934099226e-10,2.22939364554e-13];
x20=[-5.38748089001,-4.60368244955,-3.94476404012,
-3.34785456738,-2.78880605843,-2.25497400209,-1.73853771212,
-1.2340762154,-0.737473728545,-0.245340708301,...

0.245340708301,0.737473728545,1.2340762154,1.73853771212,
2.25497400209,2.78880605843,3.34785456738,3.94476404012,4.60368244955,
5.38748089001];
%*************************************************

%******************BER evaluation******************
Io=logspace(-10,-4,30);  %Average received irradiance
IodBm =10*log10(Io*1e3);  %Average received irradiance in dBm
SNR2=((R.*Io).^2)./(Ktemp);
SNR2dB=10*log10(SNR2);
for i1=1:length(Io)

  GH1=0;
  for i2=1:length(x20)

    arg1=sqrt(K1)*Io(i1)*exp(x20(i2)*sqrt(2)*r-Varl/2);
    temp1=w20(i2)*qfunc(arg1);
    GH1=GH1+temp1;
  end
BER1(i1)=2*(1-(1/sqrt(M)))*GH1/(log2(M)*sqrt(pi));
end

%*********Plot function*************************
%figure
semilogy(IodBm,BER1)

xlabel('Received average irradiance,E[I] (dBm)')
ylabel('BER')
```

```
%Evaluation of BER of SISO M-QAM SIM FSO under weak turbulence
using the %Gauss-Hermite Quadrature integration approach.
%Considering Background and thermal noise.

clear
clc

%*************Simulation Parameters*************************
M=16; %Number of levels; M-ary QAM; M should be even
Rb=155e6; %symbol rate
R=1; %Responsivity
M_ind=1; %Modulation index
A=1; %Subcarrier signal amplitude
RL=50; %Load resistance
Temp=300; %Ambient temperature
wavl=850e-9; %Optical source wavelength

%*******************Background Noise***********************
%Considering 1 cm^2 receiving aperture
sky_irra=1e-3; %at 850nm wavelength, in W/cm^2-um-sr
sun_irra=550e-4; %at 850nm wavelength, in W/cm^2-um
FOV=0.6; %in radian
OBP=1e-3; %Optical filter bandwidth in micrometre
Isky=sky_irra*OBP*(4/pi)*FOV^2; %Sky irradiance
Isun=sun_irra*OBP; %Sun irradiance

%*************Rytov Variance******************
Range=1e3; %Link range in meters
Cn=0.75e-14; %Refractive index structure parameter
Rhol=1.23*(Range^(11/6))*Cn*(2*pi/wavl)^(7/6); %Log
irradiance variance (Must be less than 1
Varl=Rhol; %Log intensity variance
r=sqrt(Varl); %log intensity standard deviation

%****************Physical constants***************
E_c=1.602e-19; %Electronic charge
B_c=1.38e-23; %Boltzmann constant
%******************************************
Pd=A^2/2;
Ktemp=(4*B_c*Temp*Rb/RL)+(2*E_c*R*Rb*(Isun+Isky));
K1=3*log2(M)*((R*M_ind)^2)*Pd/(2*(M-1)*Ktemp);

%*********background and thermal noise combined*************
%****Hermite polynomial weights and roots***********
w20=[2.22939364554e-13,4.39934099226e-10,1.08606937077e-7,
7.8025564785e-6,0.000228338636017, 0.00324377334224,0.02481052
08875,0.10901720602,0.286675505363,0.462243669601,...

0.462243669601,0. 286675505363,0.10901720602,0.0248105208875,
0.00324377334224,0.000228338636017,7.8025564785e-6,
1.08606937077e-7,4.39934099226e-10,2.22939364554e-13];
```

```
x20=[-5.38748089001,-4.60368244955,-3.94476404012,
-3.34785456738,-2.78880605843,-2.25497400209,-1.73853771212,
-1.2340762154,-0.737473728545,-0.245340708301,...

0.245340708301,0.737473728545,1.2340762154,1.73853771212,
2.25497400209,2.78880605843,3.34785456738,3.94476404012,4.60368244955,
5.38748089001];

%****************************************************
%*******************BER evaluation*******************

Io=logspace(-10,-4,30);  %Average received irradiance
IodBm =10*log10(Io*1e3);  %Average received irradiance in dBm
SNR2=((R.*Io).^2)./(Ktemp);
SNR2dB=10*log10(SNR2);
for i1=1:length(Io)

  GH1=0;
  for i2=1:length(x20)

    arg1=sqrt(K1)*Io(i1)*exp(x20(i2)*sqrt(2)*r-Var1/2);
    temp1=w20(i2)*qfunc(arg1);
    GH1=GH1+temp1;

  end

BER1(i1)=2*(1-(1/sqrt(M)))*GH1/(log2(M)*sqrt(pi));
end

%*********Plot function**************************
%figure
semilogy(IodBm,BER1)

xlabel('Received average irradiance,E[I] (dBm)')
ylabel('BER')

%Evaluation of BER of SISO DPSK SIM FSO under weak turbulence
using the %Gauss-Hermite Quadrature integration approach.
%Considering Background and thermal noise sources.

clear
clc

%*************Simulation Parameters************************
Rb=155e6; %symbol rate
R=1; %Responsivity
M_ind=1; %Modulation index
A=1; %Subcarrier signal amplitude
RL=50; %Load resistance
Temp=300; %Ambient temperature
wavl=850e-9; %Optical source wavelength
```

```
%*******************Background Noise*********************
%Considering 1 cm^2 receiving aperture
sky_irra=1e-3; %at 850nm wavelength, in W/cm^2-um-sr
sun_irra=550e-4; %at 850nm wavelength, in W/cm^2-um
FOV=0.6; %in radian
OBP=1e-3; %Optical filter bandwidth in micrometre
Isky=sky_irra*OBP*(4/pi)*FOV^2; %Sky irradiance
Isun=sun_irra*OBP; %Sun irradiance

%**************Rytov Variance********************
Range=1e3; %Link range in meters
Cn=0.75e-14; %Refractive index structure parameter
Rhol=1.23*(Range^(11/6))*Cn*(2*pi/wavl)^(7/6); %Log
irradiance variance (Must be less than 1)
Varl=Rhol; %Log intensity variance
r=sqrt(Varl); %log intensity standard deviation

%****************Physical constants***************
E_c=1.602e-19; %Electronic charge
B_c=1.38e-23; %Boltzmann constant
%*********************************************
Pd=A^2/2;
Ktemp=(4*B_c*Temp*Rb/RL)+(2*E_c*R*Rb*(Isun+Isky));
K1=0.5*((R*M_ind)^2)*Pd/Ktemp; %background and thermal noise
combined
%***********************************

%****Hermite polynomial weights and roots***********
w20=[2.22939364554e-13,4.39934099226e-10,1.08606937077e-7,
7.8025564785e-6,0.000228338636017,0.00324377334224,
0.0248105208875,0.10901720602,0.286675505363,0.462243669601,...

0.462243669601,0.286675505363,0.10901720602,0.0248105208875,
0.00324377334224,0.000228338636017, 7.8025564785e-6,
1.08606937077e-7,4.39934099226e-10,2.22939364554e-13];
x20=[-5.38748089001,-4.60368244955,-3.94476404012,
-3.34785456738,-2.78880605843,-2.25497400209,-1.73853771212,
-1.2340762154,-0.737473728545,-0.245340708301,...

0.245340708301,0.737473728545,1.2340762154,1.73853771212,
2.25497400209,2.78880605843,3.34785456738,3.94476404012,4.60368244955,
5.38748089001];
%*********************************************
%*****************BER evaluation*****************

Io=logspace(-10,-2,30); %Average received irradiance
IodBm =10*log10(Io*1e3); %Average received irradiance in dBm
SNR2=RL*((R.*Io).^2)./(4*B_c*Temp*Rb);
SNR2dB=10*log10(SNR2);
for i1=1:length(Io)
 GH1=0;
 for i2=1:length(x20)
```

```
arg1=K1*(Io(i1)^2)*exp(2*x20(i2)*sqrt(2)*r-Var1);
temp1=w20(i2)*exp(-arg1);
GH1=GH1+temp1;

end

BER1(i1)=GH1/(2*sqrt(pi));
end

%*********Plot function***************************
%figure
semilogy(IodBm,BER1)

xlabel('Received average irradiance,E[I] (dBm)')
ylabel('BER')
```

REFERENCES

1. H. Willebrand and B. S. Ghuman, *Free Space Optics: Enabling Optical Connectivity in Today's Network*, Indianapolis: SAMS Publishing, 2002.
2. S. Bloom, E. Korevaar, J. Schuster and H. Willebrand, Understanding the performance of free-space optics, *Journal of Optical Networking*, 2, 178–200, 2003.
3. X. Zhu and J. M. Kahn, Free-space optical communication through atmospheric turbulence channels, *IEEE Transactions on Communications*, 50, 1293–1300, 2002.
4. S. M. Navidpour, M. Uysal and L. Jing, BER performance of MIMO free-space optical links, *60th IEEE Vehicular Technology Conference*, 5, 3378–3382, 2004.
5. I. B. Djordjevic, B. Vasic and M. A. Neifeld, Multilevel coding in free-space optical MIMO transmission with Q-ary PPM over the atmospheric turbulence channel, *IEEE Photonics Technology Letters*, 18, 1491–1493, 2006.
6. M. Razavi and J. H. Shapiro, Wireless optical communications via diversity reception and optical preamplification, *IEEE Transactions on Communications*, 4, 975–983, 2005.
7. S. G. Wilson, M. Brandt-Pearce, Q. Cao and J. H. Leveque, Free-space optical MIMO transmission with Q-ary PPM, *IEEE Transactions on Communications*, 53, 1402–1412, 2005.
8. K. Kiasaleh, Performance of APD-based, PPM free-space optical communication systems in atmospheric turbulence, *IEEE Transactions on Communications*, 53, 1455–1461, 2005.
9. S. Sheikh Muhammad, W. Gappmair and E. Leitgeb, PPM channel capacity evaluation for terrestrial FSO links, *International Workshop on Satellite and Space Communications*, Universidad Carlos III de Madrid, Leganés, Spain, September 2006, pp. 222–226.
10. H. Hemmati, Ed., *Deep Space Optical Communications* (Deep Space Communications and Navigation Series), California: Wiley-Interscience, 2006.
11. G. Keiser, *Optical Fiber Communications*, 3rd ed. New York: McGraw-Hill, 2000.
12. W. O. Popoola, Z. Ghassemlooy, M. S. Awan and E. Leitgeb, Atmospheric channel effects on terrestrial free space optical communication links, *3rd International Conference on Computers and Artificial Intelligence (ECAI 2009)*, Piteşti, România, July 2009, pp. 17–23.
13. W. O. Popoola, Z. Ghassemlooy, J. I. H. Allen, E. Leitgeb and S. Gao, Free-space optical communication employing subcarrier modulation and spatial diversity in atmospheric turbulence channel, *IET Optoelectronics*, 2, 16–23, 2008.
14. H. Hemmati, Interplanetary laser communications, *Optics and Photonics News*, 18, 22–27, 2007.

15. R. M. Gagliardi and S. Karp, *Optical Communications*, 2nd ed. New York: John Wiley, 1995.
16. T. Ohtsuki, Multiple-subcarrier modulation in optical wireless communications, *IEEE Communications Magazine*, pp. 74–79, March 2003.
17. I. B. Djordjevic and B. Vasic, 100 Gb/s transmission using orthogonal frequency-division multiplexing, *IEEE Photonics Technology Letters*, 18, 1576–1578, 2006.
18. G. P. Agrawal, *Fiber-Optic Communication Systems*, 3rd ed. New York: Wiley-Interscience, 2002.
19. H. Rongqing, Z. Benyuan, H. Renxiang, T. A. Christopher, R. D. Kenneth and R. Douglas, Subcarrier multiplexing for high-speed optical transmission, *Journal of Lightwave Technology*, 20, 417–424, 2002.
20. W. O. Popoola, Z. Ghassemlooy and E. Leitgeb, Free-space optical communication using subcarrier modulation in gamma–gamma atmospheric turbulence, *9th International Conference on Transparent Optical Networks (ICTON '07)*, Rome, Italy, July 2007, Vol. 3, pp. 156–160.
21. R. You and J. M. Kahn, Average power reduction techniques for multiple-subcarrier intensity-modulated optical signals, *IEEE Transactions on Communications*, 49, 2164–2171, 2001.
22. S. Teramoto and T. Ohtsuki, Multiple-subcarrier optical communication systems with peak reduction carriers, presented at the *IEEE Global Telecommunications Conference (GLOBECOM)*, San Francisco, USA, 2003.
23. S. Teramoto and T. Ohtsuki, Multiple-subcarrier optical communication systems with subcarrier signal-point sequence, *IEEE Transactions on Communications*, 53, 1738–1743, 2005.
24. J. G. Proakis, *Digital Communications*, New York: McGraw-Hill, 2004.
25. M. K. Simon and M.-S. Alouini, *Digital Communication over Fading Channels*, 2nd ed. New York: John Wiley & Sons Inc., 2004.
26. M. Abramowitz and I. S. Stegun, *Handbook of Mathematical Functions with Formulas, Graphs and Mathematical Tables*, New York, USA: Dover, 1977.
27. W. O. Popoola, Z. Ghassemlooy and E. Leitgeb, BER performance of DPSK subcarrier modulated free space optics in fully developed speckle, *6th International Symposium on Communication Systems, Networks and Digital Signal Processing, (CNSDSP)*, Graz, Austria, 2008, pp. 273–277.
28. V. W. S. Chan, Free-space optical communications, *IEEE Journal of Lightwave Technology*, 24, 4750–4762, 2006.

7 Outdoor OWC Links with Diversity Techniques

Two primary challenges attributed to outdoor OWC (i.e., FSO) communications are (i) building sway and (ii) scattering/absorption-induced attenuation and scintillation-induced link fading. For the former one would need accurate pointing and tracking mechanisms, photodetector arrays and wide beam profiles; though widening the optical beamwidth is at the cost of increased transmit power (see Section 3.3.3). Even in clear atmospheric conditions, links may experience fading due to the turbulence-induced irradiance and phase fluctuations. The signal fluctuation results from random index of refraction variations along the propagation path as discussed in Chapter 3. The phase and irradiance fluctuation suffered by the traversing beam makes optical coherent detection less attractive, simply because it is sensitive to both signal amplitude and phase fluctuations. This then is one of the reasons behind the popularity of direct detection in terrestrial FSO links and of course the direct detection receiver is also much simpler. The scintillation effects could result in deep irradiance fades of 20–30 dB that lasts up to ~1–100 ms [1,2]. For a link operating at say 1 Gbps, this could result in a loss of up to 10^8 consecutive bits (burst error). To avoid this and reduce the power penalty associated with it, the signal fading has to be mitigated. There are a number of options available to mitigate the effect of channel fading including but not limited to: increased power, frequency diversity, spatial and temporal diversity schemes. The first option (i.e., higher power) could be used to provide up to 30 dB of margin, but it is costly and may not be allowed due to the eye safety regulations. The frequency diversity can offer some additional sturdiness but at relatively higher costs due to broadband components. In this chapter, we will extensively discuss the time and spatial diversities as well as the subcarrier intensity modulation (SIM) to mitigate the channel fading in FSO links. The link performance using equal gain combining (EGC), optimal combining or equivalently maximum ratio combing and selection combining (SelC) diversity schemes in log-normal atmospheric channels employing a range of diversity techniques is also outlined in this chapter.

7.1 ATMOSPHERIC TURBULENCE MITIGATION TECHNIQUES

In strong turbulence channels the single-input single-output (SISO) FSO link performance is extremely poor (i.e., not satisfying the typical BER targets) [3–6]. Possible ways of reducing the fading duration and strength include the use of aperture averaging, adaptive optics, error control coding in conjunction with interleaving

[5,7], spatial and time diversities and maximum likelihood sequence detection (MLSD) [8]. The latter suffers from high computational complexity. In aperture averaging, the receiver aperture needs to be far greater than the spatial coherence distance ρ_0 of the atmospheric turbulence in order to receive several uncorrelated signals. This condition is not always achievable in FSO as the spatial coherence distance is of the order of centimetres [9]. For coding to be effective in FSO, it must be robust enough to detect and correct not only random errors but also burst errors because the temporal coherence time τ_0 of atmospheric turbulence is much greater than the typical symbol duration. To mitigate atmospheric turbulence effects, the use of convolution coding with hard-decision Viterbi decoding has been investigated in Ref. [10], while a more efficient Turbo code employing the maximum-likelihood decoding has been reported in Ref. [11]. In addition, the work reported in Ref. [12] explored the use of a much more powerful low-density parity codes (LDPC) to detect and correct the burst error that results from scintillation. In their work, the LDPC-coded SIM in atmospheric turbulence was reported to achieve a coding gain of more than 20 dB compared with similarly coded OOK. It has also been reported in Ref. [13] that the use of space–time block code (STBC) with coherent and differential detection improves the error performance of SIM in atmospheric turbulence. The use of the turbo product code (TPC), as the channel coding scheme, with interleaving have shown to offer good resistance to burst errors and no error floor due to the channel fading in FSO links [14]. However, invoking error control coding could introduce huge processing delays and efficiency degradation in view of the number of redundant bits that will be required [1]. The adaptive optics which is based on the phase conjugation principle is another viable option. It has been used in FSO, as reported in Refs. [14,15], to reverse the wavefront deformation effect of atmospheric turbulence. However, the complexity and cost of its implementation are prohibitive. This perhaps is the reason why its use has been restricted only to deep space FSO where the cost-to-benefit ratio is within acceptable limits.

Spatial diversity, also known as diversity techniques in the RF technology, is another alternative to combat turbulence-induced fading. In such schemes multiple apertures at the transmitter and/or the receiver are used to significantly enhance the performance [16–18]. The use of space diversity in FSO systems has been first proposed in Ref. [19]. Multiple-input multiple-output (MIMO) diversity schemes [20–23] (see Figure 7.1a) are an attractive alternative approach to combat fading and reducing the potential for temporary blockages due to birds, localized fog and so on with their inherent redundancy to improve the performance of wireless communication systems [24,25]. MIMO FSO transmissions using PPM [26] and Q-ary PPM [27] have been used both in Rayleigh and lognormal fading regimes, whereas detailed studies on the BER performance of MIMO FSO links for both independent and correlated lognormal atmospheric turbulence channels are outlined in Ref. [28]. In MIMO systems multiple sources and detectors are physically situated so that the channel fading experienced between source–detector pairs is statistically independent, and thus the diversity benefits. Obviously, depending upon the spacing between the source–detector pairs and on the nature of the channel fading the assumption of the statistically independence may not be valid. For example, with moderate and heavy fog and cloud the link will obviously experience large fades. Alternative means of operation in such

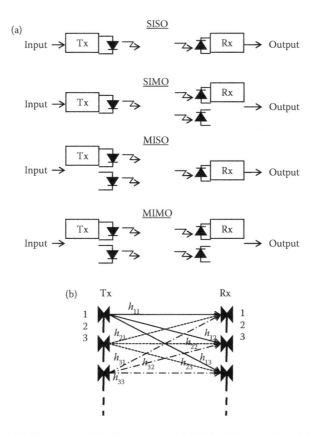

FIGURE 7.1 MIMO system: (a) block diagram and (b) MIMO channel model.

environments must be considered. A more reliable link can be established by transmitting the same information carrying optical radiation through different paths (see Figure 7.1b) and by collecting at the receiver multiple independently faded replicas of the message symbol. Additionally limitation in the transmitted optical power over the free space channel is another good reason for employing MIMO schemes. In a perfect scenario, transmitting high optical power would be the preferred option to support high data rates over a longer link range. However, the eye safety regulation will impose a limit on the transmitted optical power. In Refs. [29,30], it is shown that a very good error rate performance (BER of 10^{-10}) can be achieved for multiple-input single-output (MISO) and single-input multiple-output (SIMO) FSO systems for K-distributed atmospheric turbulence channels with spatial diversity deployed at the transmitter and/or receiver are discussed. In MIMO systems the data rate can be increased by transmitting independent information streams in parallel through the spatial channels [31], and they exhibit a range of benefits known as the diversity gain, the multiplexing gain, and the array gain. Additionally, the MIMO technology also helps at least indirectly with the pointing issue. In the presence of substantial building sway the horizontal rooftop transmit/receive array will experience cyclostationary fading instead of total pointing loss [27].

In MIMO systems, the channel matrix consists of all $\mathcal{MN}$ paths between the $\mathcal{M}$ transmitters and $\mathcal{N}$ receivers. It is modelled as

$$y = Hx + n \tag{7.1}$$

where y and x are the received and transmit $\mathcal{M}$-dimensional signal vectors, respectively, and H and n are the channel matrix (of $\mathcal{M}$ columns and $\mathcal{N}$ rows) and the $\mathcal{N}$-dimensional Gaussian noise vector, respectively.

The temporal diversity is another alternative which can minimize the turbulence influence and thus improve the performance of the FSO transmission link. Depending on the tolerable delay latency, the link can benefit from some degree of time diversity and by employing channel coding and interleaving [32,33]. The combination of encoding abilities of punctured digital video broadcast satellite standard (DVB-S2) and LDPC together with channel interleavers can be used to exploit time diversity in FSO links; this idea has been studied under weak and moderate turbulence conditions [34].

Combining schemes are also used to improve the FSO link performance [29]. The bit error rate of spatial diversity systems employing EGC, optimal combining (OC), or equivalently maximum ratio combining (MRC) and SelC for different modulation techniques other than the SIM have been investigated for the atmospheric turbulence modelled as the K distribution in Ref. [29]. In Ref. [35] the SelC diversity scheme has been applied for independent and identically distributed fading paths in a SIMO FSO link over a gamma–gamma distribution model.

In the conventional SelC diversity schemes the received signal strength is monitored and the link with the highest SNR value is selected. The processing load due to repetitive branch monitoring and switching is reasonably high, leading to high implementation complexities. To overcome the high processing overhead and system complexity switched combining diversity, switch-and-stay combining (SSC) and switch-and-examine combining (SEC) schemes are proposed [36,37]. The EGC diversity scheme has been employed for both independent and mutually correlated lognormal atmospheric turbulence channels and the BER performance is theoretically investigated in Ref. [28].

In this chapter, the performance of FSO systems with spatial and time diversities will be discussed. To illustrate the diversity techniques, only the SIM will be considered. The techniques can however be extended to other modulation techniques. The spatial diversity analysis will be based on the following linear combining techniques: EGC, MRC and SelC. Intersymbol interference will not be considered since terrestrial FSO is basically a line-of-sight technology with negligible delay spread.

7.2 RECEIVER DIVERSITY IN LOG-NORMAL ATMOSPHERIC CHANNELS

The idea of spatial diversity is premised on the fact that for a given $\mathcal{N}$ separated photodetectors, the chance that all the photodetectors will experience deep fade (due to scintillation) simultaneously at any given instant is remote. An important consideration in spatial diversity is the amount of correlation that exists between the signals received by the different photodetectors. Apart from mitigating scintillation, the spatial diversity in an FSO communication link is also advantageous in combating temporary link blockage/outage due to birds or other small object flying cross the link path. It is also a good means of combating misalignment when combined with wide divergence optical sources, thereby circumventing the need for an active tracking. Moreover, it is much easier to provide independent aperture averaging with a multiple aperture system, than in a single aperture where the aperture size has to be far greater than the irradiance spatial coherence distance [1]. In dense fog regime, however, an FSO link with spatial diversity offers limited advantage and an alternative configuration such as the hybrid FSO/RF should be considered. In the following analysis, both instances of correlated and uncorrelated received signals will be considered. Since the spatial coherence length of the atmospheric channel only measures a few centimetres, it follows therefore that the photodetectors only need to be separated by a few centimetres to achieve uncorrelated reception. At the receiver, the beam footprint covers the entire field of view (FOV) of all the $\mathcal{N}$ detectors. The photocurrents $\{i_i(t)\}_{i=1}^{\mathcal{N}}$, as shown in Figure 7.2, are then linearly combined before being sent to the coherent demodulator that separates the composite signal into its constituent subcarriers and then demodulates each subcarrier. The linear combining techniques considered are: MRC, EGC and SelC.

Scintillation is a random phenomenon that changes with time, thereby making the received irradiance time variant, and the coherence time τ_{co}, of the irradiance fluctuation in atmospheric turbulence is known to be of the order of milliseconds [1]. This implies that within a time duration $t < \tau_{co}$, the received signal is constant and time invariant. A typical data symbol duration $T \ll \tau_{co}$ ($T = 1.6$ ns when transmitting at a moderate 625 Mbps symbol rate); it follows therefore, that though the channel is time varying, the received irradiance $\{I_i\}_{i=1}^{\mathcal{N}}$ is time invariant over one symbol duration.

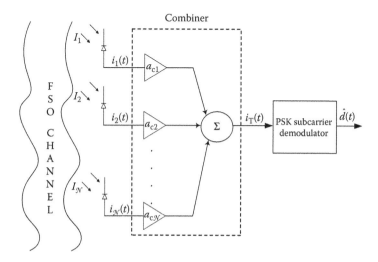

FIGURE 7.2   Block diagram of a spatial diversity receiver with $\mathcal{N}$ detectors.

To facilitate a fair comparison between an FSO link with and without spatial diversity, each detector aperture in the $\mathcal{N}$-detector system is assumed to have an area of $A_D/\mathcal{N}$, where A_D is the detector aperture area with no diversity. It follows therefore, that the background radiation noise on each branch with detector diversity is reduced by a factor of $\mathcal{N}$. Hence, on each branch the additive white Gaussian background noise variance will be $\sigma^2/\mathcal{N}$. The thermal noise on each photodetector is however not reduced by a factor of $\mathcal{N}$. The output of the individual detector during symbol duration is given as

$$i_i(t) = \frac{R}{\mathcal{N}} I_i \left(1 + \sum_{j=1}^{\mathcal{N}} A_j g(t) \cos(\omega_{cj} + \theta_j) \right) + n(t) \quad i = 1, 2 \dots \mathcal{N} \qquad (7.2)$$

The combiner shown in Figure 7.2 scales the signal from each photodetector by a gain factor, $\{a_{ci}\}_{i=1}^{\mathcal{N}}$, before adding them together. Since the photodetectors are required to be a few centimetres apart, and the link range a few kilometres apart, the difference in the propagation delay across the receiver array becomes negligible. The combiner's output thus becomes

$$i_T(t) = \sum_{i=1}^{\mathcal{N}} a_{ci} i_i(t) \qquad (7.3)$$

7.2.1 Maximum Ratio Combining

In the MRC linear combiner, the gain factor $\{a_{ci}\}_{i=1}^{\mathcal{N}}$ is proportional to the received irradiance. The weighted signals are then cophased and summed coherently to obtain the combiner's output current given by Equation 7.3. Without any interference, the MRC is optimal regardless of the fading statistics. This, according to Ref. [36], is because it results in a maximum-likelihood receiver structure. On the other hand, MRC requires the knowledge of the individual received irradiance and phase on each branch, making it unsuitable for noncoherent demodulated subcarriers, such as the DPSK-SIM.

By substituting Equation 7.2 into Equation 7.3 and suppressing the DC components via filtering, the MRC combiner output becomes

$$i_{MRC}(t) = \sum_{i=1}^{\mathcal{N}} \left[\frac{R a_{ci}}{\mathcal{N}} I_i \sum_{j=1}^{\mathcal{N}} A_j g(t) \cos(\omega_{cj} t + \theta_j) \right] + a_{ci} n_i(t) \qquad (7.4)$$

The signal demodulation and the consequent extraction of the transmitted data are done separately for each subcarrier; as such, the photocurrent at a particular subcarrier frequency ω_c is given by

$$i_{MRC}(t) = \sum_{i=1}^{\mathcal{N}} \left[\frac{R a_{ci}}{\mathcal{N}} I_i A g(t) \cos(\omega_c t + \theta) \right] + a_{ci} n_i(t) \qquad (7.5)$$

Since the coherence time of the turbulent atmospheric $\tau_0 \gg T$, the branch irradiance I_i is therefore time invariant over a period T. This leads to the following expression for the signal power S_p at the subcarrier frequency of interest:

$$S_p = \left(\sum_{i=1}^{\mathcal{N}} a_{ci} I_i \right)^2 \frac{1}{T} \int_0^T \left[\frac{RAg(t)\cos(\omega_c t + \theta)}{\mathcal{N}} \right]^2 dt = \left(\frac{RA}{\mathcal{N}\sqrt{2}} \right)^2 \left(\sum_{i=1}^{\mathcal{N}} a_{ci} I_i \right)^2 \quad (7.6)$$

The overall noise is Gaussian, with a zero mean and a variance $\sigma_{MRC}^2 = \sum_{i=1}^{\mathcal{N}} a_{ci}^2 \sigma_i^2 = \sigma_{sc}^2 \sum_{i=1}^{\mathcal{N}} a_{ci}^2$. The electrical SNR at the demodulator input conditioned on the received irradiance is thus derived as

$$\gamma_{MRC}(\overline{I}) = \left(\frac{RA}{\mathcal{N}\sqrt{2}} \sum_{i=1}^{\mathcal{N}} a_{ci} I_i \right)^2 \bigg/ \sigma_{MRC}^2 \quad (7.7)$$

Applying Cauchy inequality [38], $\left(\sum_{i=1}^{\mathcal{N}} a_{ci} I_i \right)^2 \leq \left(\sum_{i=1}^{\mathcal{N}} a_{ci}^2 \right)\left(\sum_{i=1}^{\mathcal{N}} I_i^2 \right)$, to Equation 7.7 results in the following expression for the combiner's output SNR

$$\gamma_{MRC}(\overline{I}) \leq \frac{\left(\dfrac{RA}{\mathcal{N}\sqrt{2}} \right)^2 \left(\sum_{i=1}^{\mathcal{N}} a_{ci}^2 \sigma_{sc}^2 \right)\left(\sum_{i=1}^{\mathcal{N}} I_i^2 / \sigma_{sc}^2 \right)}{\left(\sum_{i=1}^{\mathcal{N}} a_{ci}^2 \sigma_{sc}^2 \right)} \quad (7.8)$$

The left-hand side of the Cauchy inequality is apparently equal to the right-hand side when $a_{ci} \equiv I_i$. For a background noise-limited FSO link, the noise variance on each branch is proportional to the pupil receiver aperture area, $A_D/\mathcal{N}$ and $\sigma_{sc}^2 = \sigma^2/\mathcal{N}$. It should be noted that in arriving at Equation 7.8, the intermodulation distortion due to the inherent nonlinearity of the optical source has not been considered. For an ideal optical source with a modulation index of one, the subcarrier amplitude is constrained by the condition $A < 1/\mathcal{N}$. The optimum electrical SNR for each subcarrier frequency ω_c now becomes

$$\gamma_{MRC}(\overline{I}) = \left(\frac{RA}{\sqrt{2\mathcal{N}}} \right)^2 \left(\sum_{i=1}^{\mathcal{N}} I_i^2 / \sigma^2 \right) = \sum_{i=1}^{\mathcal{N}} \gamma_i(I_i) \quad (7.9)$$

where $\gamma_i(I_i) = (RAI_i)^2 / 2\mathcal{N}\sigma^2$ is the conditional SNR on each diversity branch. The average SNR, $\tilde{\gamma}_{MRC}$, obtained by averaging Equation 7.9 over the scintillation statistics is given as

$$\tilde{\gamma}_{MRC} = \int \gamma_{MRC}(\overline{I})p(\overline{I})\, d\overline{I} \quad (7.10)$$

where $p(\bar{I})$ is the joint pdf of scintillation, given by $p(\bar{I}) = \prod_{i=1}^{\mathcal{N}} p(I_i)$ for $\mathcal{N}$ photo-detectors receiving uncorrelated signals. Similarly, for a BPSK premodulated sub-carrier, the unconditional BER for the subcarrier at frequency ω_c obtained by averaging the conditional error rate over the statistics of the intensity fluctuation across all branches is given by

$$P_{e(MRC)} = \int_0^\infty Q\left(\sqrt{\gamma_{MRC}(\bar{I})}\right) p(\bar{I}) \, d\bar{I} \qquad (7.11)$$

Solving Equation 7.11 involves $(\mathcal{N}+1)$-fold integrations if the classical definition of the Q-function is used, but by using the alternative form of the Q-function and the Gauss–Hermite quadrature integration described in Section 6.3.3.1, this can be simplified to the form below

$$P_{e(MRC)} = \frac{1}{\pi} \int_0^{\pi/2} [S(\theta)]^{\mathcal{N}} \, d\theta \qquad (7.12)$$

where $S(\theta) \approx 1/\sqrt{\pi} \sum_{j=1}^{n} w_j \exp\left((-K_0^2/2\sin^2\theta) \exp[2(x_j\sqrt{2}\sigma_1 - \sigma_1^2/2)]\right)$ and $K_0 = RI_0 A/\sqrt{2\mathcal{N}}\sigma$. With $\mathcal{N} = 1$, expression 7.12 expectedly reduces to the BER with no diversity.

7.2.2 EQUAL GAIN COMBINING

In implementing the EGC spatial diversity technique, the irradiance estimate on each branch is not required but an estimate of the phase of all subcarrier signals on each branch is still very much needed. The EGC combiner samples the photocurrents $\{i_i(t)\}_{i=1}^{\mathcal{N}}$ and sums them coherently with equal weights $\{a_{ci}\}_{i=1}^{\mathcal{N}} = 1$, to produce the decision statistics [24]. With the DC component suppressed via filtering, the photocurrent at the output of the EGC combiner is given by

$$i_{EGC}(t) = \sum_{i=1}^{\mathcal{N}} \left[\frac{R}{\mathcal{N}} I_i \sum_{j=1}^{\mathcal{N}} A_j g(t) \cos(\omega_{cj} t + \theta_j) \right] + n_i(t) \qquad (7.13)$$

The conditional SNR at the output of the EGC combiner obtained is thus obtained as

$$\gamma_{EGC}(\bar{I}) = \left(\frac{RA}{\sqrt{2}\mathcal{N}\sigma}\right)^2 \left(\sum_{i=1}^{\mathcal{N}} I_i\right)^2 < \left(\frac{RA}{\sqrt{2}\mathcal{N}}\right)^2 \left(\sum_{i=1}^{\mathcal{N}} \frac{I_i^2}{\sigma^2}\right) \qquad (7.14)$$

From the foregoing, it is clear that $\gamma_{EGC}(\bar{I}) < \gamma_{MRC}(\bar{I})$. Since the noise variance is proportional to the individual receiver aperture area $\sigma_{EGC}^2 = \Sigma_{i=1}^{\mathcal{N}} \sigma_i^2 = \Sigma_{i=1}^{\mathcal{N}} \sigma^2 / \mathcal{N} = \sigma^2$.

For a log-normal-distributed scintillation, the sum of moderate number of irradiances is known to be another log-normal variable [1,39]. That is, the sum of $\mathcal{N}$ independent irradiance $Z = \Sigma_{i=1}^{\mathcal{N}} I_i = \exp(\mathcal{U})$, where $\mathcal{U}$ is normally distributed with mean $\mu_{\mathcal{U}}$ and variance $\sigma_{\mathcal{U}}^2$. Equation 7.15 gives the pdf of Z while its first and second moments derived in Appendix 7.A are given by Equation 7.16. The application of the central limit is not appropriate here because the number of photodetectors $\mathcal{N}$ is too small to justify its use.

$$P(Z) = \frac{1}{\sqrt{2\pi}\sigma_{\mathcal{U}}} \frac{1}{Z} \exp\left(-\frac{(\ln Z - \mu_{\mathcal{U}})^2}{2\sigma_{\mathcal{U}}^2}\right) \tag{7.15}$$

$$\mu_{\mathcal{U}} = \ln(\mathcal{N}) - \frac{1}{2}\ln\left(1 + \frac{\exp(\sigma_1^2) - 1}{\mathcal{N}}\right) \tag{7.16a}$$

$$\sigma_{\mathcal{U}}^2 = \ln\left(1 + \frac{\exp(\sigma_1^2) - 1}{\mathcal{N}}\right) \tag{7.16b}$$

The average SNR and the unconditional BER for a BPSK premodulated subcarrier are then given by Equations 7.17 and 7.18, respectively

$$\tilde{\gamma}_{EGC} = \int_0^\infty \gamma_{EGC}(Z)p(Z)\,dZ \tag{7.17}$$

$$
\begin{aligned}
P_{e(EGC)} &= \int_0^\infty Q\left(\sqrt{\gamma_{EGC}(Z)}\right) p(Z)\,dZ \\
&= \int_0^\infty \frac{1}{\pi} \int_0^{\pi/2} \exp\left(-\frac{K_1^2}{2\sin^2(\theta)} Z^2\right) p(Z)\,d\theta\,dZ \\
&= \frac{1}{\sqrt{\pi}} \sum_{i=1}^n w_i\, Q(K_1 \exp[x_i \sqrt{2}\sigma_{\mathcal{U}} + \mu_{\mathcal{U}}])
\end{aligned}
\tag{7.18}
$$

where $K_1 = RI_0 A/\sqrt{2}\mathcal{N}\sigma$, where $[w_i]_{i=1}^n$ and $[x_i]_{i=1}^n$, whose values are given in Appendix 6.A of Chapter 6, are the weight factors and the zeros of an nth-order Hermite polynomial, $He_n(x)$ as previously defined in Section 6.3.3.1. With one photodetector, Equation 7.18 gives the same result as for the case with no diversity.

7.2.3 SELECTION COMBINING

Both MRC and EGC spatial diversity techniques discussed so far require the irradiance level and/or the subcarrier signals phase estimates; also a separate receiver chain is needed for each diversity branch thereby adding to the overall complexity of the receiver. The SelC linear combiner on the other hand samples the entire received signal through the multiple branches and selects the branch with the highest SNR or irradiance level, provided the photodetectors receive the same dose of background radiation. The output is equal to the signal on only one of the branches and not the coherent sum of the individual photocurrents as is the case in MRC and EGC. This makes SelC suitable for differentially modulated, noncoherent demodulated subcarrier signals. In addition, the SelC is of reduced complexity compared to the MRC and EGC and its conditional SNR is given by

$$\gamma_{\text{SelC}}(I) = \frac{R^2 A^2 I_{\text{max}}^2}{2\mathcal{N}\sigma^2} \tag{7.19}$$

where $I_{\text{max}} = \max(I_1, I_2, \dots I_{\mathcal{N}})$. The pdf of the received irradiance, $p(I_{\text{max}})$, given by Equation 7.20, is obtained by first determining its cumulative density function (cdf) and then differentiating as detailed in Appendix 7.B.

$$p(I_{\text{max}}) = \frac{2^{1-\mathcal{N}} \mathcal{N} \exp(-y^2)}{I\sigma_1 \sqrt{2\pi}} [1 + \text{erf}(y)]^{\mathcal{N}-1} \tag{7.20}$$

where $y = (\ln(I/I_0) + \sigma_1^2/2)/\sqrt{2}\,\sigma_1$. The average SNR and the unconditional BER, for a BPSK premodulated subcarrier in a turbulent atmospheric channel, are given by

$$\tilde{\gamma}_{\text{SelC}} = \int_0^\infty \frac{R^2 A^2 I_{\text{max}}^2}{2\mathcal{N}\sigma^2} p(I_{\text{max}})\, dI_{\text{max}} \tag{7.21}$$

$$P_{\text{e(SelC)}} = \frac{2^{1-\mathcal{N}} \mathcal{N}}{\sqrt{\pi}} \sum_{i=1}^n w_i [1 + \text{erf}(x_i)]^{\mathcal{N}-1} Q\left(K_0 \exp\left[x_i \sigma_1 \sqrt{2} - \sigma_1^2/2\right]\right) \tag{7.22}$$

where $K_0 = RI_0 A/\sqrt{2\mathcal{N}}\,\sigma$. For a binary DPSK premodulated SIM, the unconditional bit error obtained using the Gauss–Hermite quadrature integration approach is derived as

$$P_{\text{e(SelC)}} = \int_0^\infty \frac{1}{2} \exp\left(-\frac{\gamma_{\text{SelC}}(I)}{2}\right) p(I_{\text{max}})\, dI$$

$$\cong \frac{\mathcal{N}}{2^{\mathcal{N}}\sqrt{\pi}} \sum_{i=1}^n w_i [1 + \text{erf}(x_i)]^{\mathcal{N}-1} \exp\left(-K_2^2 \exp\left(2x_i \sigma_1 \sqrt{2} - \sigma_1^2\right)\right) \tag{7.23}$$

where $K_2 = RI_0 A/2\sigma\sqrt{\mathcal{N}}$.

The SelC diversity schemes may not be best for high-speed FSO links. This is because SelC is characterized by the high processing load owing to repetitive branch monitoring and switching occurrences, thus resulting in augmented implementation complexities. To reduce the high processing load (thus the complexity) *switched combining* diversity, switch-and-stay combining (SSC) and switch-and-examine combining (SEC) diversity schemes are introduced. In SSC, once the existing received SNR drops below a certain threshold level the combiner switches to the next branch, regardless of SNR for the new branch even if it is less than the original branch [24]. In SSC and SEC diversity schemes there is no need for continual monitoring of all receiving signals, thus leading to a much simplified receiver design compared to SelC, but at the cost of inferior performance [40].

7.2.4 EFFECT OF RECEIVED SIGNAL CORRELATION ON ERROR PERFORMANCE

For a receiver array system in which the photodetectors are separated by a spatial distance s which is less than the spatial coherence distance of the channel at the receiver plane (i.e., $s < \rho_0$), the photodetectors will experience correlated irradiance fluctuations. This means that the multiple receivers can all experience similar irradiance fading at the same time. The effect of this on the system error performance will be investigated by considering the best-case scenario with an optimum MRC combiner and only one subcarrier signal. The BER conditioned on the received irradiance is

$$P_{ec} = Q\left(\sqrt{\gamma_{MRC}(\overline{I})}\right) = \frac{1}{\pi} \int_0^{\pi/2} \exp\left(-\frac{K_0^2}{2I_0^2 \sin^2\theta} \sum_{i=1}^{\mathcal{N}} I_i^2\right) d\theta \tag{7.24}$$

However, in the presence of scintillation, the unconditional BER becomes $P_e = E[P_{ec}]$, which is the average of Equation 7.24 over the joint pdf of the irradiance fluctuation. On the basis of Tatarski's [41] infinite power-series expression for the correlation coefficient of an optical wave travelling through a turbulent atmosphere given by Equation 7.25, and shown graphically in Figure 7.3, it can be inferred that

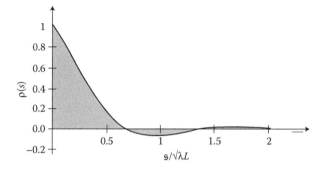

FIGURE 7.3 Correlation coefficient for a weak turbulent field as a function of transverse separation. (G. R. Osche: *Optical Detection Theory for Laser Applications*, New Jersey: Wiley, 2002. Copyright Wiley-VCH Verlag GmbH & Co. KGaA. Reproduced with permission.)

the correlation coefficient is inversely proportional to the spatial separation whenever $s/\sqrt{\lambda L} < 1$, where $\rho_0 \approx \sqrt{\lambda L}$.

$$\rho(s) = 1 - 2.36(2\pi s/\lambda L)^{5/6} + 1.71(2\pi s/\lambda L) - 0.024(2\pi s/\lambda L)^2$$
$$+ 0.00043(2\pi s/\lambda L)^4 + \cdots \tag{7.25}$$

From the foregoing, the following expression is obtained for the covariance matrix:

$$C_X = \begin{bmatrix} \sigma_x^2 & \cdots & \rho \dfrac{s_{12}}{s_{1\mathcal{N}}} \sigma_x^2 \\ \vdots & \ddots & \vdots \\ \rho \dfrac{s_{12}}{s_{\mathcal{N}1}} \sigma_x^2 & \cdots & \sigma_x^2 \end{bmatrix} \tag{7.26}$$

where s_{ij} is the spatial separation between photodetectors i and j and ρ is the correlation coefficient between two photodetectors with a spatial separation s_{12}. The joint pdf of the received irradiance is then given by

$$p(I_1, I_2 \ldots I_{\mathcal{N}}) = \frac{\exp(-\mathbb{X} C_X^{-1} \mathbb{X}^T/8)}{2^{\mathcal{N}} \prod_{i=1}^{\mathcal{N}} I_i (2\pi)^{\mathcal{N}/2} |C_X|^{1/2}} \tag{7.27}$$

where $\mathbb{X} = [\ln I_1/I_0, \ln I_2/I_0 \ldots \ln I_{\mathcal{N}}/I_0]$ is a row matrix. The expression for the unconditional BER is then obtained as

$$P_e = \int Q\left(\sqrt{\gamma_{\mathrm{MRC}}(\bar{I})}\right) p(I_1, I_2 \ldots I_{\mathcal{N}}) \, d\bar{I} \tag{7.28}$$

7.2.5 OUTAGE PROBABILITY WITH RECEIVER DIVERSITY IN A LOG-NORMAL ATMOSPHERIC CHANNEL

Here, the outage probability of a SIM-FSO link with an array of photodetectors is presented. The emphasis in this section will be on the EGC linear combining because of its simplicity. The result can however be easily extended to other linear combining schemes in a similar manner. With an array of photodetectors, the outage probability becomes: $P_{\mathrm{Out}} = P(P_{e(\mathrm{EGC})} > P_e^*) \equiv P(\gamma_{\mathrm{EGC}}(\bar{I}) < \gamma^*)$. If m_{EGC} represents the extra power margin required to achieve a given P_{Out} and γ^* is the SNR in the absence of atmospheric turbulence, then the outage probability is derived as

$$P_{\mathrm{Out}} = Q\left(\frac{\ln(m_{\mathrm{EGC}}/\mathcal{N}) + \mu_u}{\sigma_u}\right) \tag{7.29}$$

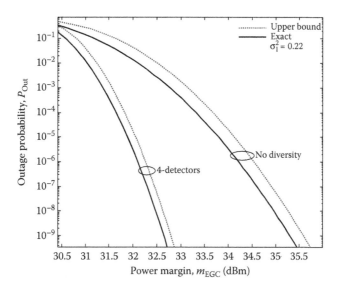

FIGURE 7.4   The exact outage probability and its upper bound against the power margin with EGC spatial diversity in weak turbulent atmospheric channel for $\sigma_I^2 = 0.2^2$ and $\mathcal{N} = [1, 4]$.

By applying the Chernoff bound $Q(x) \leq 0.5 \exp(-x^2/2)$, to Equation 7.29, an approximate value of m_{EGC} to achieve a given P_{Out} is obtained as

$$m_{EGC} \approx \mathcal{N} \exp\left(\sqrt{-2\sigma_u^2 \ln(2P_{Out})} - \mu_u\right) \tag{7.30}$$

Figure 7.4 shows how the upper bound of the outage probability given by Equation 7.30 compares with the exact solution obtained from Equation 7.29. The upper bound requires less than 0.5 dB of power more than the exact solution, and it appears to become tighter as the number of independent photodetector increase.

Expression 7.31, which is the ratio of Equations 6.23 to 7.30, gives the diversity gain based on the EGC linear combining for a given outage probability. The equation is the ratio of the link margin without spatial diversity to that with spatial diversity.

$$\text{Gain} = \frac{1}{\mathcal{N}} \exp\left(\sigma_I^2/2 + \mu_u + \sqrt{-2\sigma_I^2 \ln(2P_{Out})} - \sqrt{-2\sigma_u^2 \ln(2P_{Out})}\right) \tag{7.31}$$

7.3   TRANSMITTER DIVERSITY IN A LOG-NORMAL ATMOSPHERIC CHANNEL

In this section, the error performance of a SIM-FSO with a multiple optical transmitter and a single photodetector is discussed. The sources are assumed sufficiently spaced so that the photodetector receives uncorrelated optical radiations. To ensure a fair comparison and to maintain a constant power requirement, it is assumed that the power available for a single-transmitter system is equally shared among $\mathcal{M}$-laser transmitters. This requirement is similar to that specified in the preceding section. As such, the irradiance from each optical source is reduced by a factor of $\mathcal{M}$ compared to a single-transmitter system. An alternative approach will be for each source in the array to transmit the same power as in a single-transmitter system, in this instance, the power requirement is increased by a factor of $\mathcal{M}$. Based on the former, the received signal is obtained as

$$i(t) = \sum_{i=1}^{\mathcal{M}} \left[\frac{R}{\mathcal{M}} I_i \sum_{j=1}^{\mathcal{N}} A_j g(t) \cos(\omega_{cj} t + \theta_j) \right] + n(t) \tag{7.32}$$

Since the optical sources in the array are only separated by few centimetres, the phase shift experienced by the received irradiance due to the path difference is therefore negligible. The SNR on each subcarrier, conditioned on the received irradiance is derived as

$$\gamma_{\text{MISO}}(\bar{I}) = \left(\frac{RA}{\sqrt{2}\mathcal{M}\sigma} \right)^2 \left(\sum_{i=1}^{\mathcal{M}} I_i \right)^2 \tag{7.33}$$

From the obvious similarity between Equations 7.14 and 7.33, it can then be concluded that the unconditional BER for a SIMO system is the same as that of single source with the EGC combined multiple photodetectors.

7.4 TRANSMITTER–RECEIVER DIVERSITY IN A LOG-NORMAL ATMOSPHERIC CHANNEL

In consistency with the earlier assumptions, the total transmitted power is equal to the transmitted power when a single optical source is used with the same bit rate. In addition, the combined aperture area of the $\mathcal{N}$ photodetectors is the same as the no-spatial-diversity case. Moreover, the $\mathcal{M}$ optical sources and the $\mathcal{N}$ photodetectors are assumed to be well spaced to avoid any correlation in the received signals. First, the received signals are combined using the EGC, and from the preceding section, a MISO system with $\mathcal{M}$ laser sources is said to be identical to an EGC-combined SIMO with $\mathcal{M}$ photodetectors; these combined lead to the following as the conditional SNR of the SIM-FSO in MIMO configuration

$$\gamma_{\text{MIMO}}(\bar{I}) = \left(\frac{RA}{\sqrt{2}\mathcal{M}\mathcal{N}\sigma} \right)^2 \left(\sum_{i=1}^{\mathcal{N}} \sum_{j=1}^{\mathcal{M}} I_{ij} \right)^2 = \left(\frac{RA}{\sqrt{2}\mathcal{M}\mathcal{N}\sigma} \right)^2 \left(\sum_{i=1}^{\mathcal{M}\mathcal{N}} I_i \right)^2 \tag{7.34}$$

This expression is the same as that of an EGC combiner with a total of $\mathcal{MN}$ photodetectors. Hence, the unconditional BER is obtained by replacing $\mathcal{N}$ in Equation 7.18 by $\mathcal{MN}$.

If, however, the received signals are combined using the MRC linear combining scheme, the conditional SNR on each receiver branch will be

$$\gamma_i(I_i) = \left(\frac{RA}{\sqrt{2\mathcal{N}\,\mathcal{M}\sigma}} \sum_{j=1}^{\mathcal{M}} I_{ij} \right)^2 \tag{7.35}$$

Considering the fact that the sum of independent log-normal random variables is another log-normal distribution [39], the unconditional BER becomes

$$P_e = \frac{1}{\pi} \int_0^{\pi/2} [S(\theta)]^{\mathcal{N}} \, d\theta \tag{7.36}$$

where $S(\theta) \approx \left(1/\sqrt{\pi}\right)\Sigma_{j=1}^{n} w_j \exp\left(-(K_2^2/2\sin^2\theta)\exp[2(x_j\sqrt{2}\sigma_u + \mu_u)]\right)$ and $K_2 = RI_0A/\sqrt{2\mathcal{N}\,\mathcal{M}\sigma}$, while σ_u^2 and μ_u are as previously defined by Equation 7.16 with $\mathcal{N}$ replaced by $\mathcal{M}$.

7.5 RESULTS AND DISCUSSIONS OF SIM-FSO WITH SPATIAL DIVERSITY IN A LOG-NORMAL ATMOSPHERIC CHANNEL

To analyse the results obtained so far, a 20th-order Hermite polynomial is assumed while the log irradiance variance is assumed to vary between 0.2^2 and unity, (i.e., $0.2^2 \le \sigma_l^2 \le 1$). By plotting the BER expressions against the normalized SNR $= (R\xi E[I])^2/\sigma^2$, the link margin $m_{\mathcal{N},\sigma_l}$, defined as SNR with no diversity/ SNR with $\mathcal{N}$ photodetectors at a BER of 10^{-6} and turbulence strength σ_l^2, is obtained. In Figure 7.5, $m_{\mathcal{N},\sigma_l}$ for BPSK-SIM is plotted against $\mathcal{N}$ for the EGC and SelC techniques for various values of σ_l^2, while $m_{\mathcal{N},\sigma_l}$ for a DPSK-modulated SIM is plotted against the fading strength in Figure 7.6.

In both situations, the use of SelC in weak turbulence results in negative link margins. For instance, Figure 7.6 shows that for $\sigma_l < 0.4$, SelC spatial diversity results in between -2 and -7 dB link margin for $2 \le \mathcal{N} \le 10$, similarly in Figure 7.5, it results in a margin of up to -5 dB at $\sigma_l = 0.2$. The negative link margin experienced with SelC can be attributed to the fact that at a very low turbulence level, the effect of reducing the received intensity by a factor of $\mathcal{N}$ on each branch is dominant over the turbulence-induced intensity fluctuation.

However, as the turbulence level increases beyond $\sigma_l = 0.4$, SelC spatial diversity starts to yield a positive link margin; producing up to 9 dB link margin with two photodetectors in a DPSK premodulated SIM link. And for $\sigma_l > 0.2$ in Figure 7.5, SelC proves worthwhile with positive link margins but the gains are still

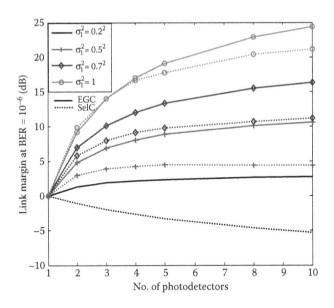

FIGURE 7.5 BPSK-SIM link margin with EGC and SelC against number of photodetectors for various turbulence levels and a BER of 10^{-6}.

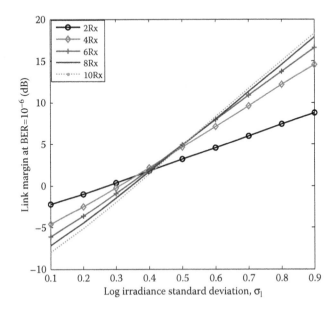

FIGURE 7.6 DPSK-SIM with SelC spatial diversity link margin against turbulence strength for $\mathcal{N} = [2, 4, 6, 8, 10]$.

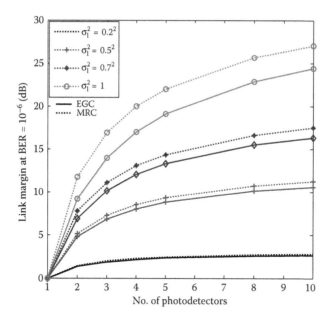

FIGURE 7.7 BPSK-SIM diversity link margin with EGC and MRC against number of photodetectors for various turbulence levels and a BER of 10^{-6}.

lower than that obtainable from EGC by about ~1 to ~6 dB depending on the number of photodetectors used. Based on the foregoing, SelC spatial diversity will not be recommended for use on short link FSO experiencing weak irradiance fluctuation.

The performance of EGC and MRC linear combiners is compared in Figure 7.7, this figure shows very clearly, that the link margin obtainable using the EGC is between 0 and ~2 dB (depending on the turbulence severity) lower than using the complex MRC.

Using two photodetectors with optimal MRC in atmospheric turbulence with $0.2^2 < \sigma_l^2 < 1$ has the potential to reduce the SNR required to achieve a BER of 10^{-6} between ~2 and ~12 dB. With up to four independent photodetectors, however, the theoretical link margin for the MRC combiner increases to ~20 dB as shown in Figure 7.7. Another inference from this figure is that the spatial diversity gain (link margin) becomes more pronounced as scintillation increases; using two detectors with MRC at the turbulence level, $\sigma_l = 0.2$ results in a link margin which is ~10 dB lower than at $\sigma_l = 1$.

Also for $\mathcal{N} \geq 4$, the marginal link margin per unit detector ($m_{\mathcal{N},\sigma_l} - m_{\mathcal{N}-1,\sigma_l}$) reduces drastically as the graphs begin to flatten out. For instance, increasing $\mathcal{N}$ from 4 to 10 with MRC across the turbulence levels $0.2^2 \leq \sigma_l^2 \leq 1$ only results in a small increase of 0 and ~6 dB link margins, while increasing $\mathcal{N}$ from 1 to 4 over the same turbulence range results between ~3 and ~22 dB diversity gains.

The plot of the diversity gain (7.31) at a P_{Out} of 10^{-6} is shown in Figure 7.8 for different number of photodetectors and log intensity variance. This plot illustrates the

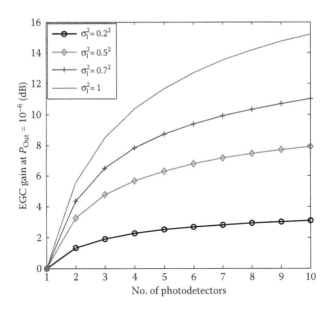

FIGURE 7.8 EGC diversity gain in log-normal atmospheric channel against the number of photodetectors at P_{Out} of 10^{-6} and $\sigma_I^2 = [0.2^2, 0.5^2, 0.7^2, 1]$.

gain based on the outage probability metric. From the plot, using two photodetectors result in ~4 dB gain at $\sigma_I^2 = 0.7^2$ and this rises to ~8 dB with $\mathcal{N} = 4$. Beyond four photodetectors, the graphs start to plateau, implying a reduction in the amount of gain recorded for each additional photodetector.

To illustrate the impact of signal correlation on the error performance, the combination of Equations 7.26 through 7.28 is plotted in Figure 7.9 for $\mathcal{N} = [2, 3]$, $\rho = [0, 0.1, 0.3, 0.6]$ and $\sigma_I^2 = 0.5^2$. For instance, at a BER of 10^{-6}, the use of two photodetectors with correlation coefficients $\rho = 0.3$ and 0.6 requires additional ~1.6 and ~3 dB of SNR, respectively compared to when $\rho = 0$. With three photodetectors, the additional SNR required to achieve the same BER of 10^{-6} is ~2.4 and ~4.6 dB for $\rho = 0.3$ and 0.6, respectively. This shows that the correlation effect results in a power penalty and this result, by extension, buttresses the need for the photodetector separation to be greater than the spatial coherence length ρ_0 in order to get the most of the spatial diversity technique.

Furthermore, in Figure 7.10, the plot of Equation 7.36 against the normalized SNR is shown at a turbulence level $\sigma_I^2 = 0.3$ for different values of $\mathcal{N}$ and $\mathcal{M}$. It can be inferred from the plot that at a BER of 10^{-6}, using a 2×2 MIMO requires ~0.4 dB of SNR more than employing a 1×4-MIMO configuration. However, spacing four photodetectors to ensure that the received signals are uncorrelated is far more demanding and cumbersome than spacing two photodetectors. Also, to achieve a BER of 10^{-6}, the use of 4×4-MIMO system requires about ~4 and 1 dB less SNR compared with using a lone source with 4 and 8 photodetectors, respectively.

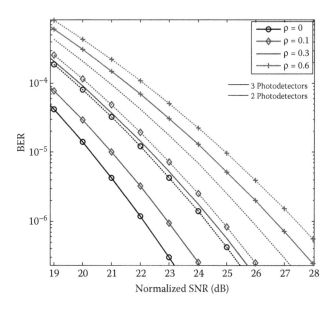

FIGURE 7.9 Error performance of BPSK-SIM at different values of correlation coefficient for $\mathcal{N} = [2, 3]$ and $\sigma_l^2 = 0.5^2$.

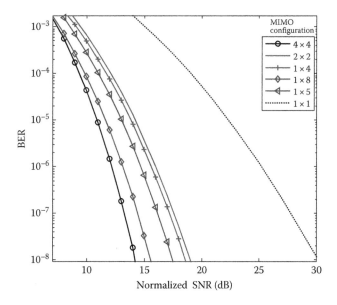

FIGURE 7.10 Error performance of BPSK-SIM with MIMO configuration in turbulent atmospheric channel for $\sigma_l^2 = 0.3$.

7.6 SIM-FSO WITH RECEIVER DIVERSITY IN GAMMA–GAMMA AND NEGATIVE EXPONENTIAL ATMOSPHERIC CHANNELS

In this section, the performance of SIM-FSO will be analysed in terms of BER and the outage probability based on the gamma–gamma and negative exponential turbulence models. While the results of the previous sections are valid for short FSO links characterized by weak irradiance fluctuation only, the performance analysis of this section will capture the performance of short to very long FSO links.

7.6.1 BER AND OUTAGE PROBABILITY OF BPSK-SIM WITH SPATIAL DIVERSITY

Following on from Section 7.2, the general expression for the unconditional BER of an FSO employing BPSK-SIM with an array of photodetectors can be written as

$$P_c = \int_0^\infty Q\left(\sqrt{\gamma(\overline{I})}\right) p(\overline{I}) \, d\overline{I} \tag{7.37}$$

where $\gamma(\overline{I})$ represents the postdetection electrical SNR at the BPSK demodulator input and $p(\overline{I}) = \Pi_{i=1}^{\mathcal{N}} p(I_i)$ is the joint pdf of the uncorrelated irradiance. In evaluating $\gamma(\overline{I})$, the EGC linear combining technique is considered because of its simplicity, but other linear combining schemes can be used as well. The total noise variance is given by $\sigma_{EGC}^2 = \mathcal{N}\sigma_{Th}^2 + \sigma_{Bg}^2$, using a narrow band optical bandpass filter combined with narrow field of view detectors, the background noise can be reduced considerably to make the thermal noise dominant. Therefore, $\sigma_{Bg}^2 < \mathcal{N}\sigma_{Th}^2$ and $\sigma_{EGC}^2 \approx \mathcal{N}\sigma_{Th}^2$. The postdetection $\gamma(\overline{I})$ for the thermal noise-limited performance at the BPSK demodulator input is thus obtained as

$$\gamma(\overline{I}) = \frac{R^2 A^2}{2\mathcal{N}^3 \sigma^2} \left(\sum_{i=1}^{\mathcal{N}} I_i\right)^2 \tag{7.38}$$

The system performance analysis with spatial diversity in weak, moderate and strong turbulence regimes will be based on the gamma–gamma model introduced in Chapter 4. The gamma–gamma parameters representing each of the stated turbulence regimes are as previously presented in Table 6.3. From the numerical solution of Equation 7.37, the BER can thus be obtained. In the limit of strong turbulence, however, the BER expression based on the simple EGC linear combining can be reduced to single integration as follows. Let the random variable Z represent the sum of $\mathcal{N}$ independent negative exponential variables, that is, $Z = \Sigma_{i=1}^{\mathcal{N}} I_i$, then the pdf of Z is obtained by adopting the characteristic function method as

$$p(Z) = \frac{I_0^{-\mathcal{N}} Z^{\mathcal{N}-1} \exp(-Z/I_0)}{\Gamma(\mathcal{N})}, \quad Z \geq 0 \tag{7.39}$$

The unconditional BER with receiver spatial diversity in a negative exponential turbulent atmospheric channel can now be derived as

$$P_e = \frac{1}{\pi\Gamma(\mathcal{N})} \int_0^{\pi/2} \int_0^{\infty} Z^{\mathcal{N}-1} \exp(-K_1(\theta)Z^2 - Z) \, dZ \, d\theta \tag{7.40}$$

where $K_1(\theta) = (RA)^2/4\mathcal{N}^3\sigma^2\sin^2(\theta)$. Expression 7.40 has no closed form but the multiple integral involved can be eliminated by invoking Equation 3.462 reported in Ref. [38]. This leads to the following equation for the BER of BPSK-SIM in a negative exponential fading channel.

$$P_e = \frac{1}{\pi} \int_0^{\pi/2} \frac{1}{\sqrt{(2K_1(\theta))^{\mathcal{N}}}} \exp(1/8K_1(\theta)) D_{-\mathcal{N}}\left(1/\sqrt{2K_1(\theta)}\right) d\theta \tag{7.41}$$

Since Equation 7.41 cannot be further simplified, its upper bound is obtained by maximizing the integrand as

$$P_e \leq \frac{1}{2\sqrt{(2K_1)^{\mathcal{N}}}} \exp(1/8K_1) D_{-\mathcal{N}}\left(1/\sqrt{2K_1}\right) \tag{7.42}$$

where $K_1 = (RA)^2/4\mathcal{N}^3\sigma^2$ and D_p is the parabolic cylinder function whose definition is available in Ref. [38].

The outage probability in the fully developed speckle regime is given by the following

$$P_{\text{Out}} = p\left(\frac{m_{\text{EGC}}}{\mathcal{N}}\left(\sum_{i=1}^{\mathcal{N}} I_i\right) < I_0\right) = p(Z < \mathcal{N} I_0/m_{\text{EGC}}) \tag{7.43}$$

Expression 7.43 is the cumulative distribution function of Z at the specified point and by combining this with Equation 7.39, the outage probability is obtained as

$$P_{\text{Out}} = \int_0^{\mathcal{N}/m_{\text{EGC}}} p(Z) \, dZ = \int_0^{\mathcal{N}/m_{\text{EGC}}} Z^{\mathcal{N}-1} \exp(-Z)/\Gamma(\mathcal{N}) \, dZ \tag{7.44}$$

Equation 7.44 above can be expressed as an implicit function in terms of the incomplete gamma function $\gamma(a, x)$ [38] or as a series as

$$P_{\text{Out}} = \frac{\gamma(\mathcal{N}, \mathcal{N}/m_{\text{EGC}})}{\Gamma(\mathcal{N})} = \frac{1}{\Gamma(\mathcal{N})} \sum_{\ell=0}^{\infty} \frac{(-1)^{\ell}(\mathcal{N}/m_{\text{EGC}})^{\mathcal{N}+\ell}}{\ell!(\mathcal{N}+\ell)} \tag{7.45}$$

where the incomplete gamma function is defined as $\gamma(a, x) = \int_0^x \exp(-t)t^{a-1} \, dt$.

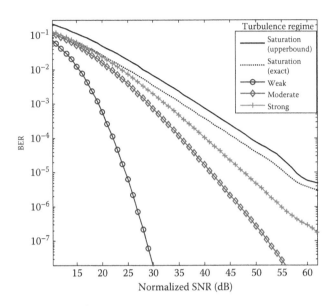

FIGURE 7.11 BPSK-SIM error rate against the normalized SNR in gamma–gamma and negative exponential channels for two photodetectors.

In Figure 7.11, the BER of BPSK-SIM is plotted against the normalized SNR for two photodetectors under different turbulence regimes. When compared with the results obtained with no spatial diversity, the gain of spatial diversity in mitigating the effect of turbulence-induced irradiance fluctuation becomes very clear. A summary of the resulting diversity gains (i.e., reduction in the SNR), with two and three photodetectors at a BER of 10^{-6} is presented in Table 7.1. As expected, the impact of diversity is least in weak turbulence regime resulting in only 2.6 dB reductions in the SNR at a BER of 10^{-6}. Whereas in strong turbulence regime, the gain in SNR is ~12 dB, increasing to ~47 dB in the saturation regime.

As would be expected, the diversity gain is highest in extreme fading conditions since adding more branches will greatly reduce the chance of a catastrophic fading from happening.

TABLE 7.1

Diversity Gain at a BER of 10^{-6} in Gamma–Gamma and Negative Exponential Channels

Number of Photodetectors	Spatial Diversity Gain (dB)		
	Weak	Strong	Saturation
2	2.6	11.6	46.7
3	3.0	21.2	63.6

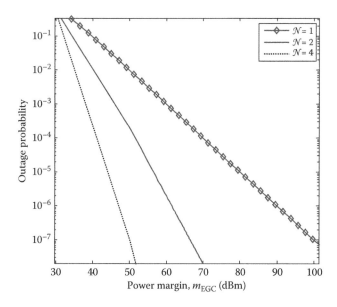

FIGURE 7.12 The outage probability as a function of power margin m_{EGC} (dBm) for $\mathcal{N} = [1, 2, 4]$ in negative exponential channel.

Figure 7.12 shows the extra power margin required to achieve a certain outage probability P_{Out} with and without spatial diversity in negative exponential atmospheric channels. The gain of employing diversity is apparent from the figure just as the case under the BER performance metric. For instance, at a P_{Out} of 10^{-6}, a diversity gain of almost 30 dB is predicted and this increases to ~43 dB with four photodetectors. This significant gain is a fact that with multiple photodetectors, more independent irradiances are received.

For the sake of comparison, the predicted spatial diversity gains based on both metrics (BER and the outage probability) are shown in Figure 7.13 as a function of the number of photodetectors. For both metrics, it is observed that as the number of photodetectors increases beyond four, the diversity gains start to plateau. Although P_{Out} and BER are different by definition, they both result in a similar conclusion. It should, however, be mentioned that using an array of photodetectors adds to the cost and design complexity.

7.6.2 BER and Outage Probability of DPSK-SIM in Negative Exponential Channels

For a DPSK premodulated subcarrier with an array of photodetectors, the most suitable linear combining scheme will be the SelC. This is because in DPSK demodulation, the absolute phase information of the subcarrier is not available and it is only the SelC that does not require the subcarrier phase information. The SNR for a SIM-FSO with SelC is given by Equation 7.46 while the unconditional BER for a DPSK-SIM in a negative exponential channel is given by Equation 7.47

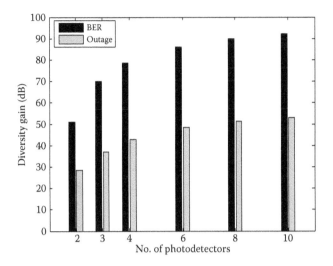

FIGURE 7.13   Diversity gain against number of independent photodetectors at BER and P_{Out} of 10^{-6} in a negative exponential atmospheric channel.

$$\gamma_{\text{SelC}}(I) = \frac{R^2 A^2 I^2}{2\mathcal{N}(\mathcal{N}\sigma_{\text{Th}}^2 + \sigma_{\text{Bg}}^2)} \tag{7.46}$$

$$P_{\text{e(SelC)}} = \int_0^\infty \frac{1}{2}\exp\left(-\frac{\gamma_{\text{SelC}}(I)}{2}\right)p(I_{\text{max}})\,\mathrm{d}I \tag{7.47}$$

The pdf, $p(I_{\text{max}}) = p(\max\{I_i\}_{i=1}^{\mathcal{N}})$ in a negative exponential channel model given by Equation 7.48 is obtained by first finding the cumulative distribution function of I_{max} at an arbitrary point and then differentiating. The resulting pdf is given by Equation 7.48 and the detailed proof presented in Appendix 7.C.

$$p(I_{\text{max}}) = \frac{\mathcal{N}}{I_0}\exp\left(-\frac{I}{I_0}\right)\left(1 - \exp\left(-\frac{I}{I_0}\right)\right)^{\mathcal{N}-1} \tag{7.48}$$

The plot of the pdf of I_{max} is shown in Figure 7.14 for different values of $\mathcal{N}$.

From the combination of Equations 7.46 through 7.48, an expression for the error performance of a DPSK-SIM laser communication system in fully developed speckle atmospheric channel is obtained as

$$P_{\text{e(SelC)}} = \int_0^\infty \frac{\mathcal{N}}{2I_0}\left(1 - \exp\left(-\frac{I}{I_0}\right)\right)^{\mathcal{N}-1}\exp\left(-\frac{I}{I_0} - \frac{R^2 A^2 I^2}{2\mathcal{N}(\mathcal{N}\,\sigma_{\text{Th}}^2 + \sigma_{\text{Bg}}^2)}\right)\mathrm{d}I \tag{7.49}$$

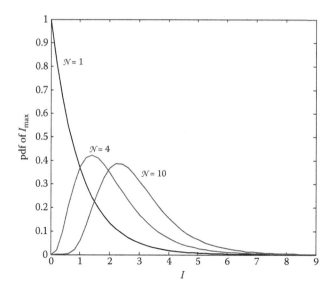

FIGURE 7.14   The pdf, $p(\max\{I_i\}_{i=1}^{\mathcal{N}})$ for $\mathcal{N} = [1, 4, 10]$, and $I_0 = 1$ in a negative exponential channel.

Now the received irradiance I_0 needed to attain an outage probability $P_{\text{Out}} = P(\gamma(I) < \gamma^*)$ in a negative exponential atmospheric channel without diversity is given by Equation 7.49. In arriving at this equation, the threshold SNR has been taken as, $\gamma^* = (RAI^*)^2/2\sigma^2$ where I^* is the receiver sensitivity required to attain the threshold BER*.

$$I_0 = \frac{I^*}{\ln(1 - P_{\text{Out}})^{-1}} \qquad (7.50)$$

With an array of PIN photodetectors employing SelC in a negative exponential atmospheric channel, the received irradiance $I_{0(\text{SelC})}$ needed to attain a given P_{Out} is derived from the combination of Equations 7.46 and 7.48, as

$$I_{0\text{SelC}} = \frac{I^*}{\ln\left(1 - \sqrt[\mathcal{N}]{P_{\text{Out}}}\right)^{-1}} \left(\frac{\mathcal{N}(\mathcal{N}\,\sigma_{\text{Th}}^2 + \sigma_{\text{Bg}}^2)}{\sigma_{\text{Th}}^2 + \sigma_{\text{Bg}}^2}\right)^{1/2} \qquad (7.51)$$

From the foregoing, the diversity gain $I_0/I_{0\text{SelC}}$ can thus be obtained.

The numerical simulations presented in this section are based on the parameters of Table 7.2. In Figure 7.15, the error performance of DPSK-SIM in a negative exponential channel, obtained from Equation 7.49, is shown as a function of the average irradiance. This plot brings to bear the potential gain of SelC in reducing the required sensitivity for a given BER under very strong fading conditions. For example, to achieve a BER of 10^{-6} with no diversity, about 23 dBm of received irradiance is

TABLE 7.2
Numerical Simulation Parameters

Parameter	Value
Symbol rate, R_b	155–625 Mbps
Spectral radiance of the sky, $N(\lambda)$	10^{-3} W/cm^2 μm Sr
Spectral radiant emittance of the sun, $W(\lambda)$	0.055 W/cm^2 μm
Optical band-pass filter bandwidth, $\Delta\lambda$ at $\lambda = 850$ nm	1 nm
PIN photodetector field of view, FOV	0.6 rad
Radiation wavelength, λ	850 nm
Number of photodetectors, $\mathcal{N}$	$1 \leq \mathcal{N} \leq 10$
Load resistance, R_L	50 Ω
PIN photodetector responsivity, R	1
Operating temperature, T_e	300 K

required while with two photodetectors, about −1.7 dBm is needed to achieve the same level of performance. Moreover, as the number of photodetectors increases, the attained diversity gain per additional detector reduces. For instance, for $\mathcal{N} = 2$, the gain per detector at a BER of 10^{-6} is ~12 dB and this reduces to about 5 and 4 dB for $\mathcal{N} = 2$ and 10, respectively. These results are summarized in Table 7.3 for up to 10 photodetectors.

In discussing the outage probability in negative exponential fading channels, Equation 7.51 is plotted in Figure 7.16. In this plot, the threshold average irradiance

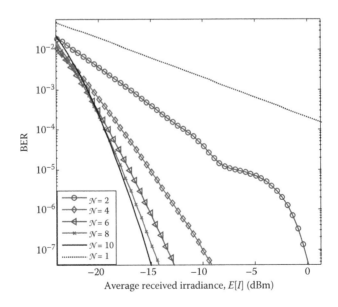

FIGURE 7.15 Error rate of DPSK-SIM against the average received irradiance with spatial diversity in a negative exponential channel for $\mathcal{N} = [2, 4, 6, 8, 10, 1]$.

TABLE 7.3

Gain per Photodetector at a BER of 10^{-6}

$\mathcal{N}$	1	2	4	6	8	10
Average irradiance (dBm)	23.1	−1.7	−12.5	−15.2	116.2	−16.7
Gain (dB) per $\mathcal{N}$	0	12.4	8.9	6.4	4.9	4.0

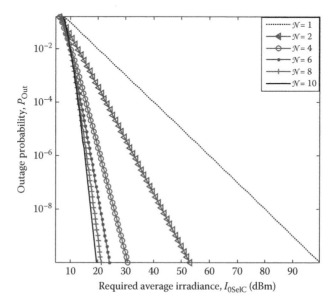

FIGURE 7.16 Outage probability against the average irradiance with SelC spatial diversity in a negative exponential channel for $\Gamma^* = 0$ dBm and $\mathcal{N} = [1, 2, 4, 6, 8, 10]$.

Γ^* is assumed to be 0 dBm. This graph shows that for a very long SIM-FSO link, whose channel fading is modelled by the negative exponential distribution, achieving an outage probability of 10^{-6} or better will require a minimum of 60 dBm received irradiance without SelC.

This power requirement reduces to ~35 and ~23 dBm, respectively, with 2 and 4 photodetectors that are combined using the SelC. To further illustrate the gain of using SelC in the saturation regime, Figures 7.17 and 7.18 show the predicted diversity gain at different values of P_{Out} and $\mathcal{N}$. With two photodetectors and an outage probability of 10^{-6}, the maximum predicted gain per detector is about 14 dB. This predicted gain is even observed to be higher at lower values of P_{Out}. This makes sense, as the use of diversity in a fading channel increases the received signal strength and by extension lowers the outage probability. And in Figure 7.18, it is clearly shown that the gain (dB) per detector peaks at $\mathcal{N} = 2$ and then decreases very rapidly beyond $\mathcal{N} = 4$.

Although up to 10 photodetectors have been shown in the results above, this is mainly for illustration purpose. The use of such a large number of detectors will pose

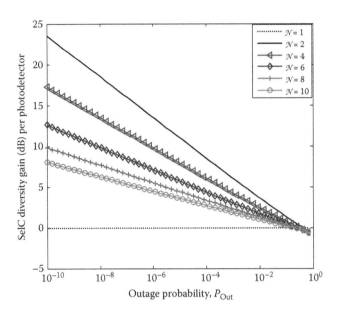

FIGURE 7.17   Predicted SelC diversity gain per photodetector against P_{Out} in the saturation regime for $\mathcal{N} = [1, 2, 4, 6, 8, 10]$.

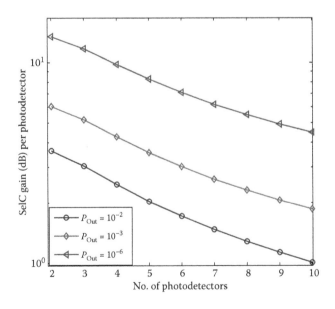

FIGURE 7.18   Predicted SelC diversity gain (dB) per photodetector for $P_{Out} = [10^{-6}, 10^{-3}, 10^{-2}]$ in the saturation regime.

serious implementation difficulties, as they all have to be spaced beyond the spatial coherence distance in order to avoid any signal correlation. An interesting point to note from these results is that, in contrast to the short-range links where the use of SelC is not worthwhile, it is highly recommended in very-long-range links, as it results in a significant reduction in the required receiver sensitivity especially when the photodetector is kept to a maximum of four. This is due to the fact that the irradiance fading experienced in a fully developed speckle regime is dominant over the reduction in the received irradiance due to the reduction in the receiver aperture area. Any scheme that mitigates this dominant irradiance fading will clearly result in improved performance. The predicted average irradiance and diversity gains presented above are valid for as long as the photodetectors received signals are uncorrelated. That is, $\rho_0 < s < \theta_s L$ where θ_s is the divergence angle of the optical source in milliradians and L is the link length in kilometres. For $s < \rho_0$, the received irradiances are correlated and diversity gain is lower as previously discussed in Sections 7.2.4 and 7.5.

7.7   TERRESTRIAL FREE SPACE OPTICAL LINKS WITH SUBCARRIER TIME DIVERSITY

In the previous sections, the use of spatial diversity to lessen the effect of atmospheric turbulence has been discussed. And as highlighted in the spatial diversity results, the best gains are achieved when the detectors are physically separated by a distance greater than the turbulence coherence length. The channel coherence length depends on the turbulence strength and is typically of the order of centimetres. Similarly, the use of aperture averaging requires the receiver aperture to be larger than the turbulence coherence length. This does not only make the system bulky, it is cumbersome and not always feasible. The use of the turbo product code (TPC), as the channel coding scheme, with interleaving has shown to offer good resistance to burst errors and no error floor due to the channel fading in FSO links [42].

This section will be looking at the subcarrier time delay diversity (STDD) as an alternative or complementary means of mitigating the channel fading in SIM-FSO links. The conventional use of multiple subcarriers is to increase throughput/capacity via subcarrier multiplexing. But in this scheme, different subcarriers at different frequencies are used to transmit the delayed copies of the original data. The proposed subcarrier STDD scheme has an advantage of simplicity and low cost to achieve a reasonable diversity gain compared to schemes such as adaptive optics or forward error correction. Moreover, the reduction throughout associated with the temporal diversity can be compensated for through subcarrier multiplexing.

7.7.1   ERROR PERFORMANCE WITH STDD

In the STDD scheme, delayed copies of the data are retransmitted on different subcarriers as shown in Figure 7.19. The STDD scheme relies solely on the statistical temporal variation of the atmospheric turbulence-induced fading. Apart from retransmitting the delayed version of the original data on different subcarriers, other

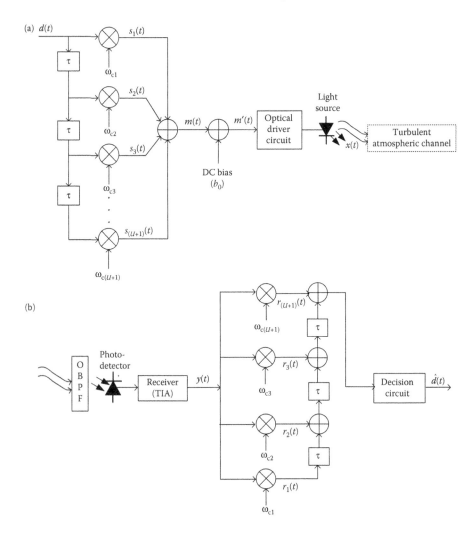

FIGURE 7.19 The subcarrier STDD block diagram: (a) transmitter and (b) receiver TIA-transimpedance amplifier; OBPF, optical bandpass filter.

viable options include using different wavelengths or polarizations for the retransmission. Details of how these two other options are carried out can be found in Refs. [43–45]. Time delay diversity schemes have the advantage of simplicity and low cost to achieve a reasonable diversity gain compared to schemes such as adaptive optics or forward error correction [43,44]. These gains are, however, at a cost of retransmission latency and data rate reduction. The system under consideration must therefore be able to trade low error rate for high latency.

According to the Taylor frozen turbulence hypothesis [43], turbulent eddies responsible for the fading are frozen in space and only move across the beam path by the transverse component of the wind. Thus from the knowledge of the average transverse wind speed, the spatial statistics of turbulence can be translated into temporal statistics. From this transformation, the temporal covariance function of the irradiance fluctuations can be obtained as outlined in Ref. [45]. The temporal covariance is useful in determining the correlation time τ_c. For weak turbulence, τ_c has been experimentally measured to vary between 3 and 15 ms [45]. This suggests the sort of values required for τ so that the TDD systems are efficient in fading channels. In the analysis we assume the following: $\tau \geq \tau_c$, and the minimum required buffer size $R_b\tau$ is set prior to transmission. To illustrate the performance with the proposed STDD, we consider a BPSK-subcarrier intensity-modulated system with a single data carrying subcarrier and U-STDD paths. During symbol duration, the data carrying signal $m(t)$ is given by

$$m(t) = \sum_{i=1}^{U+1} d(t - (i-1)\tau)\cos(\omega_i t + \varphi_i) \qquad (7.52)$$

where $d(.)\in[-1,1]$. The standard RF coherent demodulator employed extracts the reference carrier needed to down convert the received signal to base band. The sum of the demodulator outputs is then fed into the decision circuit as illustrated in Figure 7.19. From this the electrical SNR per bit can then be easily derived as

$$\gamma(\overline{I}) = \frac{(R\xi)^2 P_m}{\sigma^2}\left(\sum_{i=1}^{U+1} I_i\right)^2 \qquad (7.53)$$

The noise variance is given by $\sigma^2 = \sigma_{Bg}^2 + (U+1)\sigma_{Th}^2$, since there are now $(U+1)$ demodulation paths associated with every photodetector. It should be noted that with the addition of U-STDD paths, $\xi = 1/(U+1)A$ becomes the minimum modulation depth needed to avoid any clipping distortions.

7.7.1.1   Error Performance of Short-Range Links

To evaluate the BER with the subcarrier STDD in a log-normal turbulence channel, expression 7.53 is substituted into Equation 6.8 of Chapter 6 and $p(\overline{I})$ replaced with the probability density function (pdf) of the sum of $(U+1)$ independent but identical log-normal variables. Although the sum of independent log-normal variables does not have a closed form [46], it is often approximated as another log-normal variable [39]. There are several approaches to make this approximation; a survey and comparison of these approaches can be found in Ref. [47]. In this work, we will use the moment-matching approach (otherwise called the Wilkinson's method) to make the approximation. This approach is chosen because it is simple, straightforward and has also been reported [48] to work well for small values of σ_1^2 as is the case here. We now make the approximation that the $\sum_{i=1}^{U+1} I_i \cong Z$, where Z is a log-normal variable described as $Z = \exp(v)$ and v is Gaussian. The following are the approximate values of the mean and variance of v obtained by matching the first and second moments of $\sum_{i=1}^{U+1} I_i$ to that of Z; see [48] for detailed proof.

$$\mu_v = \ln(U+1) - \frac{1}{2}\ln\left(1 + \frac{\exp(\sigma_1^2)-1}{U+1}\right) \tag{7.54a}$$

$$\sigma_v^2 = \ln\left(1 + \frac{\exp(\sigma_1^2)-1}{U+1}\right) \tag{7.54b}$$

The resulting BER expression is then simplified using the Gauss–Hermite approximation. The unconditional BER for the BSPK-SIM link with U-STDD thus becomes

$$P_e = \frac{1}{\sqrt{\pi}} \sum_{i=1}^{n} w_i Q\left(\sqrt{K_0'} \exp\left(x_i \sqrt{2}\sigma_v + \mu_v\right)\right) \tag{7.55}$$

where $K_0' = (R\xi I_0)^2 P_m / \sigma_{Bg}^2 + (U+1)\sigma_{Th}^2$. The penalty due to the turbulence fading at a given BER with and without STDD can therefore be obtained from this equation and the ones in Chapter 6.

7.7.1.2 Error Performance of Long-Range Links

For longer-range FSO links (>1 km), the procedure for obtaining the error rate expression is similar to that of the short-range link above except that $p(\bar{I})$ now becomes the pdf of the sum of independent negative exponential variables. This pdf is tractable and is easily derived as [41]

$$p(\bar{I} = Z) = \frac{I_0^{-(U+1)} Z^U \exp(-Z/I_0)}{\Gamma(U+1)} \tag{7.56}$$

where $\Gamma(\cdot)$ is the gamma function. The resulting BER expression given by Equation 7.57, which is obtained by combining Equations 6.8a, 7.53 and 7.56, does not have a closed-form solution. As a result, the BER will have to be evaluated numerically.

$$P_e = \frac{I_0^{-(U+1)}}{\pi\Gamma(U+1)} \int_0^{\pi/2} \int_0^{\infty} Z^U \exp(-K_1'(\theta)Z^2 - Z/I_0)\, dZ\, d\theta \tag{7.57}$$

where $K_0'(\theta) = (R\xi)^2 P_m / 2\sigma^2 \sin^2\theta$.

TABLE 7.4

Simulation Parameters

Parameters	Values
Symbol rate, R_b	155–625 Mbps
Spectral radiance of the sky, $N(\lambda)$	10^{-3} W/cm²µmSr
Spectral radiant emittance of the sun, $W(\lambda)$	0.055 W/cm²µm
Optical band-pass filter bandwidth, $\Delta\lambda$ at $\lambda = 850$ nm	1 nm
PIN photodetector field of view	0.6 rad
Radiation wavelength, λ	850 nm
Link range, L	1 km
Index of refraction structure parameter, C_n^2	0.25×10^{-14}–2.45×10^{-14} m$^{-2/3}$
Load resistance, R_L	50 Ω
PIN photodetector responsivity, R	1
Optical band-pass filter transmissivity	90%
Equivalent temperature, T_e	300 K

From Equation 7.57 it is observed that within the $(0 - \pi/2)$ limit of integration, the function $\sin^2\theta$ is monotonically increasing and an upper bound of Equation 7.57 can then be obtained by maximizing the integrand with $\theta = \pi/2$ to obtain

$$P_e \leq \frac{1}{2\sqrt{(2K_1)^{U+1}}} \exp\left(\frac{\sigma^2}{4(R\xi)^2 P_m}\right) D_{-(U+1)}\left(1/\sqrt{2K_1}\right) \tag{7.58}$$

where D_ρ is the parabolic cylinder function whose definition is available in Ref. [38].

The results presented in this section are based on the simulation parameters of Table 7.4 and the BPSK-SIM scheme. Unless otherwise stated, the background, thermal noise and scintillation are considered as the system limiting factors.

7.7.1.3 Results and Discussion for Short-Range Links

From the combination of Equations 6.7, 6.12 and 7.55, we obtain Figure 7.20 which shows the FSO link's BER with and without STDD at different values of normalized SNR, $\bar{\gamma} = (RI_0)^2 P_m/\sigma^2$. This plot shows $\bar{\gamma}$ to be decreasing as the number of temporal diversity paths increases. To determine the optimum number of diversity paths required, we present in Table 7.5 the following parameters: diversity gain (difference between $\bar{\gamma}$ with and without STDD), fading penalties (difference between $\bar{\gamma}$ in fading channel and under no fading) and the effective data rates. This table explicitly shows that at $\sigma_1^2 = 0.3$, 1-STDD has the highest STDD gain per additional path of ~6.9 dB and the least reduction in data rate. It is interesting to note that the gain per unit path decreases as more paths are added. This makes sense because for a given peak optical power, the modulation index has to be reduced as more paths are added in order to keep the laser within its dynamic range. These findings appear to be independent of how the STDD scheme is implemented. A similar conclusion was previously reported when the delayed data were transmitted on different polarizations/

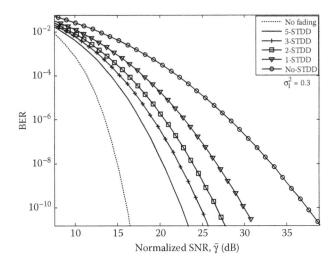

FIGURE 7.20 BER against $\hat{\gamma}$ with and without STDD at 155 Mbps, $\sigma_l^2 = 0.3$, $C_n = 0.75 \times 10^{-14}$ m$^{-2/3}$.

TABLE 7.5
Fading Penalty and STDD Gain at BER = 10^{-9}, $R_b = 155$ Mbps, $\sigma_l^2 = 0.3$ and $\xi = 1/(U - 1)A$

	0-STDD	1-STDD	2-STDD	3-STDD	5-STDD
$\overline{\gamma}$ (dB) (no fading: 15.56 dB)	35.32	28.41	25.34	23.54	21.48
Fading penalty (dB)	19.76	12.85	9.78	7.98	5.92
Diversity gain (dB)	0	6.91	9.89	11.78	13.84
(gain/unit path)	(0)	(6.91)	(4.99)	(3.93)	(2.77)
Effective data rate (Mbps)	R_b	$0.5R_b$	$0.33R_b$	$0.25R_b$	$0.17R_b$

---

wavelengths [43,45]. Hence, single retransmission (1-STDD) will be suggested to mitigate channel fading for short-range links.

Using Equation 7.55 with $\xi = 1/(U + 1)A$, it is shown in Figure 7.21 that the average received irradiance and indeed the diversity gain both increase as the fading strength increases. However, the fading penalties and diversity gains are independent of the data rate. For instance, at $\sigma_l^2 = 0.1$ the estimated gains and fading penalties with 1-STDD stood at 0.36 and ~4.2 dB for $R_b = 155$ and 625 Mbps, respectively. For stronger fading condition with $\sigma_l^2 = 0.5$, these values increase to 3 and ~11 dB, respectively.

7.7.1.4 Results and Discussion for Long-Range Links

The performance of the link in the negative exponential turbulence channel is depicted in Figure 7.22. From the figure it is quite clear that achieving a low-error-rate communication in a long-range FSO link is almost impossible without any diversity

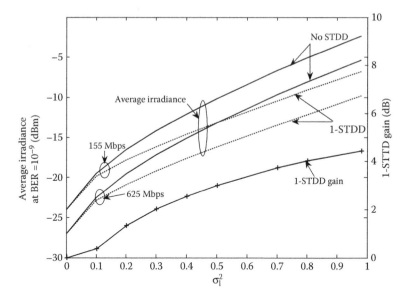

FIGURE 7.21 The average received irradiance at a BER of 10^{-9}, R_b (Mbps) = [155, 625] and 1-STDD gain for different strengths of turbulence.

technique. To achieve a moderate BER of 10^{-6} requires a huge $\overline{\gamma}$ of 112 dB at the receiver if no diversity technique is implemented. And this translates into ~97 dB fading penalty. The implication of this high fading penalty is that at least one fading mitigation technique has to be used in order to establish the link. However with 1-, 2- and 3-STDD the required $\overline{\gamma}$ values are estimated to be 62, 43.6 and 35.8 dB,

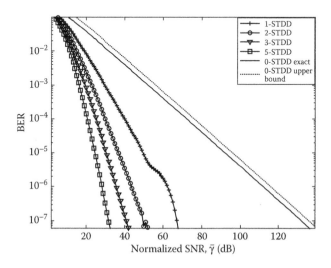

FIGURE 7.22 Link BER performance against the normalized SNR with the subcarrier STDD in the negative exponential turbulence-induced fading channel.

respectively. These gains (reduction in $\overline{\gamma}$ with STDD) are achieved at the price of reduced R_b by a factor of 2, 3 and 4 in that order. The STDD technique can therefore be used for the long-range FSO to achieve a low BER communication.

7.8 APERTURE AVERAGING

This is one of the simplest forms of spatial diversity, where the aperture of receiver lens is larger than the fading correlation length [49,50]. Aperture averaging has been considered to combat scintillation, in particular strong scintillation, by number of researchers [49–51]. With this scheme, averaging is carried out over the relatively fast fluctuations due to the small-size eddies; therefore, by doing so, the frequency content of the irradiance spectrum is shifted to the low spatial frequencies [49,50,52,53]. The aperture averaging factor, which is widely used to quantify the fading reduction by aperture averaging, is given by [49]

$$A = \frac{\sigma_I^2(D)}{\sigma_I^2(0)} \tag{7.59}$$

where $\sigma_I^2(D)$ and $\sigma_I^2(0)$ denote the scintillation index for a receiver lens of diameter D and a 'point receiver' $(D \approx 0)$, respectively.

The strong turbulence is characterized by the way in which the statistical moments of velocity increments increase with the spatial separation L_s. Within the inertial subrange the refractive index structure is defined by the Kolmogorov two-thirds power law given as Equation 3.87b, for full details see Chapter 3. In the spectral domain, the power spectral density of the refractive index fluctuation is related to C_n^2 by

$$\Phi_n(K) = 0.033 C_n^2 K^{-11/3}; \quad 2\pi/L_0 \ll K \ll 2\pi/l_o \tag{7.60}$$

7.8.1 PLANE WAVE

For the plane wave propagation with a smaller inner scale $l_o \ll (L/k)^{0.5}$, the scintillation index is given by

$$\sigma_I^2(D) = \exp\left[\frac{0.49\sigma_I^2}{(1 + 0.65d^2 + 1.11\sigma_I^{12/5})^{7/6}} + \frac{0.51\sigma_I^2(1 + 0.69\sigma_I^{12/5})^{-5/6}}{1 + 0.90d^2 + 0.62d^2\sigma_I^{12/5}}\right] - 1 \tag{7.61}$$

where σ_I is the Rytov variance for the plane wave and parameter d is defined as

$$d = \sqrt{\frac{kD^2}{4L}} \tag{7.62}$$

where L is the propagation distance.

Churnside suggested an interpolation formula to approximate the average aperture averaging for the plane wave given by [51]

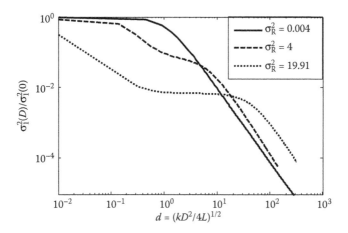

FIGURE 7.23 Aperture averaging factor for a plane wave propagation with the inner scale $l_o = 0$.

$$A \approx \left[1 + 1.07 \left(\frac{D^2 k}{4L} \right)^{7/6} \right]^{-1} \tag{7.63}$$

Under a weak fluctuation irradiance fluctuation, a better approximation (with <7% error) can be achieved by using [49,53,54]

$$A \approx \left[1 + 1.062 \left(\frac{D^2 k}{4L} \right) \right]^{-7/6} \tag{7.64}$$

Figure 7.23 shows the aperture averaging factor for a plane wave for three different regions of turbulence. For a small aperture size, $kD^2/4L \ll 1$, the aperture averaging gain is 1. However, for $kD^2/4L \gg 1$, the variance σ_I^2 decrease with increasing aperture size. Hence, for D being larger than the size of scintillation spots at the receiver $(\lambda L_p)^{0.5}$, the lens can effectively average over the turbulence.

7.8.2 SPHERICAL WAVE

For the spherical wave with zero inner scale and unbounded outer scale, the irradiance flux variance is given by [49]

$$\sigma_I^2(D) = \exp \left[\frac{0.49\sigma_2^2}{(1 + 0.18d^2 + 0.56\sigma_2^{12/5})^{7/6}} + \frac{0.51\sigma_2^2(1 + 0.69\sigma_2^{12/5})^{-5/6}}{1 + 0.90d^2 + 0.62d^2\sigma_2^{12/5}} \right] - 1 \tag{7.65}$$

where σ_2 is the Rytov variance for the spherical wave.

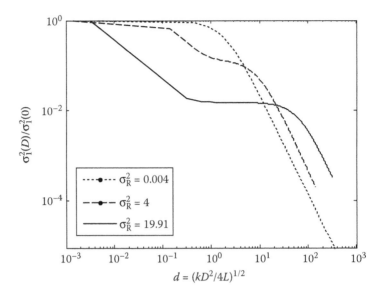

FIGURE 7.24 Aperture averaging factor for a spherical wave propagation with the inner scale $l_o = 0$.

In a weak turbulence region, the aperture averaging for the spherical wave can be approximated as

$$A \approx \left[1 + 0.214 \left(\frac{D^2 k}{4L} \right)^{7/6} \right]^{-1} \tag{7.66}$$

Figure 7.24 shows the aperture averaging factor for a spherical wave for three different regions of turbulence. Comparing Figure 7.24 and Figure 7.23 shows a similar behaviour of aperture averaging for both plane and spherical waves.

7.8.3 Gaussian Beam Wave

For the Gaussian beam wave, the aperture averaging is given by [55,56]

$$A \approx \frac{16}{\pi} \int_0^1 x \, dx \exp \left[-\frac{D^2 x^2}{\rho_0^2} \left(2 + \frac{\rho_0^2}{\omega_0^2 \, \hat{z}^2} - \frac{\rho_0^2 \varnothing^2}{\omega^2(z)} \right) \right] \left[\cos^{-1} x - x\sqrt{1 - x^2} \right] \tag{7.67}$$

where ω_0 is the transmitter beam radius, $\omega(z)$ is the beam size at propagation distance of z and ρ_0 is the coherence length. Figure 7.25 shows the aperture averaging factor for a Gaussian beam for three different regions of turbulence.

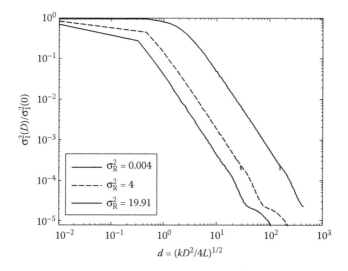

FIGURE 7.25 Aperture averaging factor for a Gaussian beam wave propagation with the inner scale $l_0 = 0$.

APPENDIX 7.A

7.A.1 SUM OF LOG-NORMAL DISTRIBUTION MEAN AND VARIANCE CALCULATION

By definition, a log-normal-distributed variable is given by $I = I_0 \exp(l)$, where l obeys the normal distribution. Let the sum of independent log-normal-distributed irradiance be another log-normal distribution [39], that is

$$\sum_{i=1}^{\mathcal{N}} I_i = \exp(U) = Z \tag{7.A.1}$$

For an identically distributed log-normal random variable [39], $E[Z] = \mathcal{N}E[I]$ and $\sigma_Z^2 = \sigma_I^2$.

From Ref. [57]

$$E[\exp(ar)] = \exp\left(a\mu_r + \frac{a^2 \sigma_r^2}{2} \right) \tag{7.A.2}$$

Applying this to Equation 7.A.1 leads to

$$E[Z] = E[\exp(U)] = \exp\left(\mu_U + \frac{\sigma_U^2}{2} \right) \tag{7.A.3}$$

The expectation of the log-normal random variable I is given by

$$E[I] = I_0 \exp\left(\mu_1 + \frac{\sigma_1^2}{2}\right)$$

(7.A.4)

Note that $E[I] = I_0$, therefore $\exp\left(\mu_1 + \frac{\sigma_1^2}{2}\right) = 1$ from which the log intensity mean, μ_1, is found to be

$$\mu_1 = -\frac{\sigma_1^2}{2}$$

(7.A.5)

It follows from the foregoing that

$$\exp\left(\mu_U + \frac{\sigma_U^2}{2}\right) = \mathcal{N}E[I]$$

(7.A.6)

And having normalized $E[I] = I_0 = 1$ Equation 7.A.6 reduces to

$$\mu_U = \ln(\mathcal{N}) - \frac{\sigma_U^2}{2}$$

(7.A.7)

The second central moment of a log-normal random variable I is given by

$$\sigma_I^2 = \exp\left(\sigma_I^2 + 2\mu_1\right)\left[\exp\left(\sigma_I^2\right) - 1\right]$$

(7.A.8)

Substituting Equation 7.A.5 into Equation 7.A.8 and invoking the relation given by Equation 7.A.2 results in

$$E[\exp(2U)] - (E[\exp(U)])^2 = \mathcal{N}[\exp(\sigma_I^2) - 1]$$

(7.A.9)

$$\exp(2\mu_U + 2\sigma_U^2) - \exp(2\mu_U + \sigma_U^2) = \mathcal{N}[\exp(\sigma_I^2) - 1]$$

(7.A.10)

This implies that

$$\exp(\sigma_U^2) = \left\{\mathcal{N}\exp(-2\ln(\mathcal{N}))[\exp(\sigma_I^2) - 1]\right\} + 1$$

(7.A.11)

Hence

$$\sigma_U^2 = \ln\left(1 + \frac{[\exp(\sigma_I^2) - 1]}{\mathcal{N}}\right)$$

(7.A.12)

And

$$\mu_U = \ln(\mathcal{N}) - \frac{\sigma_U^2}{2}$$

$$= \ln(\mathcal{N}) - \frac{1}{2}\ln\left(1 + \frac{[\exp(\sigma_1^2) - 1]}{\mathcal{N}}\right) \qquad (7.A.13)$$

APPENDIX 7.B

7.B.1 PDF OF $I_{max} = \max\{I_i\}_{i=1}^{\mathcal{N}}$ FOR LOG-NORMAL-DISTRIBUTED VARIABLES

$$I_{max} = \max\{I_i\}_{i=1}^{\mathcal{N}} \qquad (7.B.1)$$

and

$$p(I_{max}) = p(\max\{I_i\}_{i=1}^{\mathcal{N}}) \qquad (7.B.2)$$

Considering an arbitrary received irradiance, I, it follows therefore that: $p(I_{max} < I) = p(I_1 < I, I_2 < I, \ldots I_{\mathcal{N}} < I)$. Since none of the $\{I_i\}_{i=1}^{\mathcal{N}}$ is greater than I_{max}. The following therefore gives the cumulative distribution function (cdf) of I_{max} for $\mathcal{N}$ independent and identically distributed irradiances

$$p(I_{max} < I) = \int_0^I \cdots \int_0^I \int_0^I p(I_1)p(I_2)\, dI_1\, dI_2 \ldots dI_{\mathcal{N}} = \prod_{i=1}^{\mathcal{N}} \int_0^I p(I_i)\, dI_i \qquad (7.B.3)$$

Assuming that the received irradiance is independent and all obey log-normal statistics, then

$$p(I_{max} < I) = \left[\frac{1}{\sqrt{2\pi}\sigma_1} \int_0^I \frac{1}{I}\exp\left\{-\frac{(\ln(I/I_0) + \sigma_1^2/2)^2}{2\sigma_1^2}\right\} dI\right]^{\mathcal{N}}$$

$$= \frac{1}{2^{\mathcal{N}}}\left[1 + \mathrm{erf}\left(\frac{2\ln(I/I_0) + \sigma_1^2}{2\sqrt{2}\sigma_1}\right)\right]^{\mathcal{N}} \qquad (7.B.4)$$

From $(I_{max}) = \dfrac{d(\mathrm{cdf})}{dI}$, then

$$p(I_{max}) = \frac{\mathcal{N}2^{1-\mathcal{N}}\exp(-y^2)}{I\sigma_1\sqrt{2\pi}}[1 + \mathrm{erf}(y)]^{\mathcal{N}-1} \qquad (7.B.5)$$

where $y = \dfrac{\ln(I/I_0) + \sigma_1^2/2}{\sqrt{2}\sigma_1}$. With $\mathcal{N} = 1$ Equation 7.B.5 returns the log-normal distribution as would be expected.

APPENDIX 7.C

7.C.1 PDF OF $I_{\max} = \max\{I_i\}_{i=1}^{\mathcal{N}}$ FOR NEGATIVE EXPONENTIAL DISTRIBUTED VARIABLES

$$I_{\max} = \max\{I_i\}_{i=1}^{\mathcal{N}} \qquad (7.C.1)$$

and

$$p(I_{\max}) = p(\max\{I_i\}_{i=1}^{\mathcal{N}}) \qquad (7.C.2)$$

Considering an arbitrary received irradiance I, it follows therefore that: $p(I_{\max} < I) = p(I_1 < I, I_2 < I, \ldots I_{\mathcal{N}} < I)$. Since none of the $\{I_i\}_{i=1}^{\mathcal{N}}$ is greater than $I_{\max}$. The following therefore gives the cumulative distribution function (cdf) of $I_{\max}$ for $\mathcal{N}$ independent and identically distributed irradiances

$$p(I_{\max} < I) = \int_0^I \ldots \int_0^I \int_0^I p(I_1)p(I_2)\,\mathrm{d}I_1\,\mathrm{d}I_2 \ldots \mathrm{d}I_{\mathcal{N}} = \prod_{i=1}^{\mathcal{N}} \int_0^I p(I_i)\,\mathrm{d}I_i \qquad (7.C.3)$$

Assuming that the received irradiance is independent and all obey negative exponential statistics

$$p(I_{\max} < I) = \left[\frac{1}{I_0} \int_0^I \exp\left(-\frac{I}{I_0}\right) \mathrm{d}I \right]^{\mathcal{N}}$$

$$= \left[1 - \exp\left(-\frac{I}{I_0}\right) \right]^{\mathcal{N}} \qquad (7.C.4)$$

From $(I_{\max}) = \dfrac{\mathrm{d(cdf)}}{\mathrm{d}I}$,

$$p(I_{\max}) = \frac{\mathcal{N}}{I_0} \exp\left(-\frac{I}{I_0}\right)\left(1 - \exp\left(-\frac{I}{I_0}\right)\right)^{\mathcal{N}-1} \qquad (7.C.5)$$

With $\mathcal{N} = 1$ Equation 7.C.5 returns the negative exponential distribution as would be expected.

REFERENCES

1. E. J. Lee and V. W. S. Chan, Optical communications over the clear turbulent atmospheric channel using diversity, *IEEE Journal on Selected Areas in Communications*, 22, 1896–1906, 2004.

2. V. W. S. Chan, Free-space optical communications, *IEEE Journal of Lightwave Technology*, 24, 4750–4762, 2006.
3. E. Jakeman and P. Pusey, A model for non-Rayleigh sea echo, *IEEE Transactions on Antennas and Propagation*, 24, 806–814, 1976.
4. M. Uysal and J. Li, BER performance of coded free-space optical links over strong turbulence channels, *IEEE 59th Vehicular Technology Conference, VTC 2004-Spring*, Milan, Italy, pp. 352–356, 2004.
5. M. Uysal, S. M. Navidpour and J. T. Li, Error rate performance of coded free-space optical links over stong turbulence channels, *IEEE Communication Letters*, 8, 635–637, 2004.
6. K. Kiasaleh, Performance of coherent DPSK free-space optical communication systems in K-distributed turbulence, *IEEE Transactions on Communications*, 54, 604–607, 2006.
7. X. Zhu and J. M. Kahn, Performance bounds for coded free-space optical communications through atmospheric turbulence channels, *IEEE Transaction on Communications*, 51, 1233–1239, 2003.
8. X. Zhu and J. M. Kahn, Markov chain model in maximum likelihood sequence detection for free space optical communication through atmospheric turbulence channels, *IEEE Transaction on Communications*, 51, 509–516, 2003.
9. X. Zhu and J. M. Kahn, Free-space optical communication through atmospheric turbulence channels, *IEEE Transactions on Communications*, 50, 1293–1300, 2002.
10. J. P. Kim, K. Y. Lee, J. H. Kim, and Y. K. Kim, A performance analysis of wireless optical communication with convolutional code in turbulent atmosphere, *International Technical Conference on Circuits Systems, Computers and Communications (ITC-CSCC '97)*, Okinawa, pp.15–18, 1997.
11. T. Ohtsuki, Turbo-coded atmospheric optical communication systems, *IEEE International Conference on Communications (ICC)*, New York, pp. 2938–2942, 2002.
12. I. B. Djordjevic, B. Vasic and M. A. Neifeld, LDPC coded OFDM over the atmospheric turbulence channel, *Optical Express*, 15, 6336–6350, 2007.
13. H. Yamamoto and T. Ohtsuki, Atmospheric optical subcarrier modulation systems using space–time block code, *IEEE Global Telecommunications Conference (GLOBECOM '03)*, New York, pp. 3326–3330, 2003.
14. J. E. Grave and S. Drenker, Advancing free space optical communication with adaptive optics, *Lightwaveonline*, 19, 105–113, 2002.
15. R. K. Tyson, Bit-error rate for free-space adaptive optics laser communications, *Optical Society of America Journal*, 19, 753–758, 2002.
16. W. O. Popoola, Z. Ghassemlooy, J. I. H. Allen, E. Leitgeb and S. Gao, Free-space optical communication employing subcarrier modulation and spatial diversity in atmospheric turbulence channel, *IET Optoelectronics*, 2, 16–23, 2008.
17. W. O. Popoola, Z. Ghassemlooy and E. Leitgeb, Free-space optical communication using subcarrier modulation in gamma–gamma atmospheric turbulence, *9th International Conference on Transparent Optical Networks (ICTON '07)*, Rome, Italy, Vol. 3, pp. 156–160, July 2007.
18. W. O. Popoola, Z. Ghassemlooy and V. Ahmadi, Performance of sub-carrier modulated free-space optical communication link in negative exponential atmospheric turbulence environment, *International Journal of Autonomous and Adaptive Communications Systems (IJAACS)*, 1, 342–355, 2008.
19. M. M. Ibrahim and A. M. Ibrahim, Performance analysis of optical receivers with space diversity reception, *IEEE Proceedings Communications*, 143, 369–372, 1996.
20. B. Braua and D. Barua, Channel capacity of MIMO FSO under strong turbulence conditions, *International Journal of Electrical & Computer Sciences*, 11, 1–5, 2011.
21. C. C. Motlagh, V. Ahmadi and Z. Ghassemlooy, Performance of free space optical communication using M-array receivers at atmospheric condition, *17th ICEE*, Tehran, Iran, May 12–14, 2009.

22. E. Shin and V. W. S. Chan, Optical communication over the turbulent atmospheric channel using spatial diversity, *IEEE Global Telecommunications Conference*, Taipei, Taiwan, Vol. 3, pp. 2055–2060, 2002.
23. E. J. Lee and V. W. S. Chan, Part 1: Optical communication over the clear turbulent atmospheric channel using diversity, *IEEE Journal on Selected Areas in Communications*, 22, 1896–1906, 2004.
24. E. Telatar, Capacity of multi-antenna Gaussian channels, *European Transactions on Telecommunications*, 10, 585–595, 1999.
25. G. J. Foschini, Layered space–time architecture for wireless communication in a fading environment when using multi-element antennas, *Bell Labs Technical Journal*, 1, 41–59, 1996.
26. S. G. Wilson, M. Brandt-Pearce, C. Qianling, and M. Baedke, Optical repetition MIMO transmission with multipulse PPM, *IEEE Journal on Selected Areas in Communications*, 23, 1901–1910, 2005.
27. S. G. Wilson, M. Brandt-Pearce, Q. Cao and J. H. Leveque, Free-space optical MIMO transmission with Q-ary PPM, *IEEE Transactions on Communications*, 53, 1402–1412, 2005.
28. S. M. Navidpour, M. Uysal and M. Kavehrad, BER performance of free-space optical transmission with spatial diversity, *IEEE Transaction on Communications*, 6, 2813–2819, 2007.
29. T. A. Tsiftsis, H. G. Sandalidis, G. K. Karagiannidis and M. Uysal, Optical wireless links with spatial diversity over strong atmospheric turbulence channels, *IEEE Transactions on Wireless Communications*, 8, 951–957, 2009.
30. T. A. Tsiftsis, H. G. Sandalidis, G. K. Karagiannidis and M. Uysal, FSO links with spatial diversity over strong atmospheric turbulence channels, *IEEE International Conference on Communications*, Beijing, China, pp. 5379–5384, 2008.
31. I. Hen, MIMO architecture for wireless communication, *Intel Technology Journal*, 10(2), 157–166, May 2006.
32. F. Xu, A. Khalighi, P. Caussé and S. Bourennane, Channel coding and time-diversity for optical wireless links, *Optics Express*, 17, 872–887, 2009.
33. J. Chen, Y. Ai and Y. Tan, Improved free space optical communications performance by using time diversity, *Chinese Optics Letters*, 6(11), 797–799, 2008.
34. M. Czaputa, T. Javornik, E. Leitgeb, G. Kandus and Z. Ghassemlooy, Investigation of punctured LDPC codes and time-diversity on free-space optical links, *Proceedings of the 2011 11th International Conference on Telecommunications (ConTEL)*, Graz, Austria, pp. 359–362, 2011.
35. W. Zixiong, Z. Wen-De, F. Songnian and L. Chinlon, Performance comparison of different modulation formats over free-space optical (FSO) turbulence links with space diversity reception technique, *Photonics Journal, IEEE*, 1, 277–285, 2009.
36. M. K. Simon and M.-S. Alouini, *Digital Communication over Fading Channels*, 2nd ed. New York: John Wiley & Sons Inc., 2004.
37. H. Moradi, H. H. Refai and P. G. LoPresti, Switched diversity approach for multireceiving optical wireless systems, *Applied Optics*, 50(29), 5606–5614, 2011.
38. I. S. Gradshteyn and I. M. Ryzhik, *Table of Integrals, Series, and Products*, 5th ed. London: Academic Press, Inc., 1994.
39. R. L. Mitchell, Permanence of the log-normal distribution, *Journal of the Optical Society of America*, 58, 1267–1272, 1968.
40. K. Young-Chai, M. S. Alouini and M. K. Simon, Analysis and optimization of switched diversity systems, *IEEE Transactions on Vehicular Technology*, 49, 1813–1831, 2000.
41. G. R. Osche, *Optical Detection Theory for Laser Applications*, New Jersey: Wiley, 2002.

42. Y. Han, A. Dang, Y. Ren, J. Tang and H. Guo, Theoretical and experimental studies of turbo product code with time diversity in free space optical communication, *Optics Express*, 18(26), 26978–26988, 2010.
43. S. Trisno, I. I. Smolyaninov, S. D. Milner and C. C. Davis, Characterization of time delayed diversity to mitigate fading in atmospheric turbulence channels, *Proceedings of SPIE*, 5892, 589215.1–589215.10, 2005.
44. S. Trisno, I. I. Smolyaninov, S. D. Milner and C. C. Davis, Delayed diversity for fade resistance in optical wireless communications through turbulent media, *Proceedings of SPIE*, 5596, 385–394, 2004.
45. C. H. Kwok, R. V. Penty and I. H. White, Link reliability improvement for optical wireless communication systems with temporal domain diversity reception, *IEEE Photonics Technology Letters*, 20, 700–702, 2008.
46. R. Barakat, Sums of independent lognormally distributed random variables, *Journal of the Optical Society of America*, 66(3), 211–216, 1976.
47. N. C. Beaulieu, A. A. Abu-Dayya and P. J. McLane, Estimating the distribution of a sum of independent lognormal random variables, *IEEE Transactions on Communications*, 43, 2869, 1995.
48. S. M. Haas and J. H. Shapiro, Capacity of wireless optical communications, *IEEE Journal on Selected Areas in Communications*, 21, 1346–1357, 2003.
49. L. C. Andrews and R. L. Phillips, *Laser Beam Propagation through Random Media*, 2nd ed. Washington: SPIE Press, 2005.
50. F. S. Vetelino, C. Young, L. Andrews and J. Recolons, Aperture averaging effects on the probability density of irradiance fluctuations in moderate-to-strong turbulence, *Applied Optics*, 46(11), 2099–2108, 2007.
51. J. H. Churnside, Aperture averaging of optical scintillations in the turbulent atmosphere, *Applied Optics*, 30(15), 1982–1994, 1991.
52. L. C. Andrews, R. L. Phillips and C. Y. Hopen, Aperture averaging of optical scintillations: Power fluctuations and the temporal spectrum, *Waves in Random Media*, 10, 53–70, 2000.
53. M. A. Khalighi, N. Aitamer, N. Schwartz and S. Bourennane, Turbulence mitigation by aperture averaging in wireless optical systems, *10th International Conference on Telecommunications, 2009. ConTEL 2009*, Zagreb, Croatia, pp. 59–66, 2009.
54. D. L. Fried, Aperture averaging of scintillation, *Journal of the Optical Society of America*, 57(2), 169–172, 1967.
55. A. K. Majumdar and J. C. Ricklin, *Free-Space Laser Communications: Principles and Advances*, New York: Springer, 2008.
56. J. C. Ricklin, S. M. Hammel, F. D. Eaton and S. L. Lachinova, Atmospheric channel effects on free space laser communication, *Journal of Optical and Fiber Communications Research*, 3, 111–158, 2006.
57. J. W. Goodman, *Statistical Optics*, New York: John Wiley, 1985.

8 Visible Light Communications

In last few years, we have seen a growing research in visible light communications (VLC), and the idea of using LEDs for both illumination and data communications. The main drivers for this technology include the increasing popularity of solid-state lighting, longer lifetime of high-brightness LEDs compared to other sources of artificial light like the incandescent light bulbs, high bandwidth/data rate, data security, no health hazards and low power consumption. The dual functionality provided by VLC (i.e., lighting and data communication from the same high-brightness LEDs) has created a whole range of interesting applications, including but not limited to home networking, high-speed data communication via lighting infrastructures in offices, car-to-car communication, high-speed communication in aeroplane cabins, in-trains data communication, traffic lights management and communications to name a few. The levels of power efficiency and reliability offered by LEDs today are by far superior compared to the traditional incandescent light sources used for lighting. Although the high-brightness white LEDs used in solid-state lighting are still a lot more expensive than the incandescent or compact fluorescent lamp that they are meant to replace, it is expected that the prices will drop considerably with time and wider adoption. In fact, solid-state lighting is anticipated to gain a lot more traction with the recent ban on incandescent light bulbs in Europe, Australia and other countries of the world. Recent research in VLC has successfully demonstrated data transmission at over 500 Mbps over short links in office and home environments. Further research and developments will open up new possibilities to partly resolve some of the issues associated with the present-day infrared, radio/microwave communication systems and lighting technologies. This chapter gives an overview of the visible light communication technology, highlighting the fundamental theoretical background, devices available, modulation and diming techniques and system performance analysis. Multiple-input–multiple-output and cellular visible light communication systems are also covered in this chapter.

8.1 INTRODUCTION

The concept of the visible light as a medium for communication dates back to 1870s when Alexander Graham Bell successfully demonstrated the transmission of audio signal using a mirror made to vibrate by a person's voice. The first practical demonstration of the visible light communication, termed photophone, happened in June 1880 using sunlight as the light source. Bell and his assistant Tainter succeeded in communicating clearly over a distance of some 213 m in the photophone experiment. However, the Bell system had some obvious drawbacks as it depends on sunlight which is intermittent. The phenomenal developments in optoelectronics, particularly solid-state light sources in the last decades, have led to the re-emergence of optical wireless communications.

Solid-state lighting refers to the fact that light is generated through solid-state electroluminescence. It has a relatively short history. In the 1990s, we saw the introduction of high-brightness LEDs for the purpose of general illumination. Within only a few years, LED's luminous efficacy has improved rapidly from less than 0.1 lm/W to over 230 lm/W and with a lifetime as high as 100,000 h [1–3]. We are now seeing the introduction of another upcoming type of solid-state lighting source, known as the organic LED (OLED). OLED has relatively low luminous efficacy (measured value of 100 lm/W [4] and short lifetime compared with the LEDs, thus limiting their application for various colour displays and general illumination for the time being. Nevertheless, OLEDs may also be used as an alternative solution for large area lighting and communications.

Compared with the traditional incandescent (with a luminous efficacy limited to 52 lm/W) and fluorescent (with a luminous efficacy limited to 90 lm/W) lamps used for lighting [2], the peak efficiency of white LEDs exceeds 260 lm/W (this is much lower than the theoretically predicted luminous efficacy reaching as much as 425 lm/W) [1]. In the coming years, we will witness a growing increase in the level of performance offered by illumination LEDs. From the environmental point of view, solid-state lighting (SSL) will be an imperative technology for significant energy savings and consequently a much greener world. This technology has a number of advantages such as [2,5,6]

* A longer life expectancy
* A higher tolerance to humidity
* Are mercury free
* A smaller and compact size
* A much higher energy conversion efficiency (white LEDs with luminous efficacy greater that 200 lm/W are now available)
* Minimum heat generation characteristics compared to all other lighting sources
* Lower power consumption
* Most importantly, fast switching

Due to these advantages, white LEDs are ideal sources for future applications (indoor and outdoor) for dual purpose of lighting and data communications, hence contributing to considerable energy savings at a global level [3,7]. With the availability of highly efficient white LEDs (created by combining the prime colours: red, green and blue) or by using a blue emitter in combination with a phosphor, we are witnessing a surge in research and development in indoor visible light communication (VLC) systems [8,9]. Visible light sources have high optical power output and large emission characteristics to be employed for lighting. Additionally, these devices have specific wireless channel impulse response, which is quite different from that of infrared wireless communications.

The first VLC started at Nakagawa Laboratory in Keio University, Japan in 2003. This was followed by a growing research and development interest at a global scale. By switching phosphorescent white LEDs on and off rapidly, data rates of up to 40 Mbps can be easily achieved. Using the same on-off keying technique, higher data rate in excess of 100 Mbps can be achieved with RGB white LEDs. Resonant cavity LEDs can even go further to achieve data rates up to 500 Mbps. The resonant cavity LEDs use Bragg reflectors, serving as mirrors, to enhance the emitted light. Additionally, it

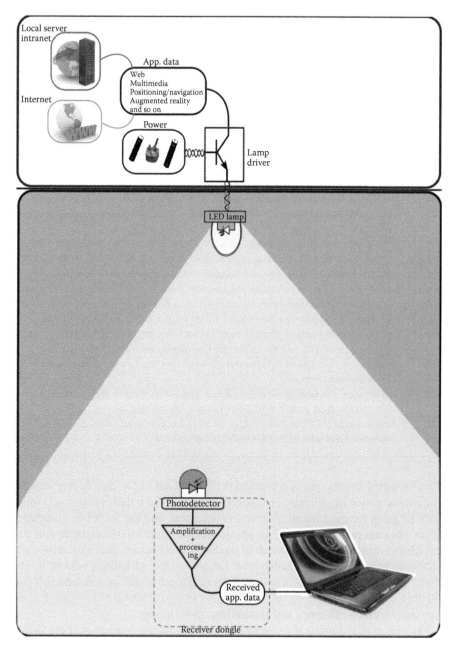

FIGURE 8.1 (**See colour insert.**) An illustration of the VLC concept.

TABLE 8.1
Historical Development Timeline for VLC

2004	Demonstration of LED light systems for high-speed transmission of data to hand-held and vehicle-borne computing devices; Japan.
2005	Land trials of VLC communications to transmit information to mobile phones, with throughput estimated at 10 kbps and several Mbps using fluorescent light and an LED; Japan.
2007	Demonstration of a LED-backlit liquid crystal display (LCD) television while transmitting information to a PDA via light; Fuji Television, Japan.
2007	Visible Light Communications Consortium (VLCC) in Japan proposed two standards: Visible Light Communication System Standard Visible Light ID System Standard JEITA (Japan Electronics and Information Technology Industries Association) accepted these standards as JEITA CP-1221 and JEITA CP-1222.
2008	Development of global standards for home networks, including the use of OWC employing IR and VLC technologies; European Union-funded OMEGA. Demonstration of VLC using five LEDs with a data rate >100 Mbps over longer distances of a few metres using direct LOS. Reduced levels of transmission would have occurred using diffused light from walls outside of the line of sight; European Union-funded OMEGA.
2008	Visible Light Communications Consortium (VLCC) in Japan and the international Infrared Data Association (IrDA) working together on specification standards in the United States.
2009	VLCC issued their first Specification Standard which incorporates and expands upon core IrDA specification and defined spectrum to allow for the use of visible light wavelengths.
2010	Development of VLC technology for communications between a wide range of electronic products, such as high-definition televisions, information kiosks, personal computers, personal digital assistants smartphones and so on; University of California, USA.
2010	Demonstration of VLC with the indoor global positioning system (GPS); Japan.
2010	Transmission of an FM-VLED system at 500 Mbps over 5 m, Siemens and the Heinrich Hertz Institute, Germany.
2010	Development a standard for VLC technologies by the IEEE Wireless Personal Area Networks working group 802.15.7 Task Group 7, which is ongoing.
2011	Demonstration of OFDM-based VLC at 124 Mbps real time, using commercial off-the-shelf phosphorescent white LED, University of Edinburgh, UK.

offers increased spectral purity compared to conventional LEDs, thus further improving communication capabilities. The concept of VLC is illustrated in Figure 8.1, while Table 8.1 gives the most recent historical development timeline for VLC. To achieve higher data rates particularly from the phosphorescent white LEDs, advanced modulation schemes (multilevel and multicarrier modulation techniques), and multiinput/multioutput technique could be readily used. Despite all the advantages offered by the white LEDs, there are a number of challenges (technical as well as nontechnical) that still need to be addressed. One such challenge is in the designing of low-cost devices with high luminous efficiency and outstanding colour quality.

The single-most important factor in VLC is the switching properties of the visible LEDs. They have the ability to be switched on and off very rapidly thereby making it possible to impress data on their radiated optical power/intensity. This process of modulating the optical power/intensity with data is of relatively lower complexity and cost than modulation in RF communication for example. The following gives some distinguishing features and important notes about VLC [10]:

1. We are already seeing the use of LEDs in cars, buses, aircrafts, trains, traffic lights and so on. White LEDs are slowly but surely replacing the inefficient light bulbs in homes, offices and even in street lamps. This trend will continue with high-brightness SSL devices becoming readily available and at a low cost.
2. SSL and indeed VLC can leverage on existing (local) power line and lighting infrastructures in offices, homes and so on.
3. Transmitters and receiver devices, including multiarray/element devices are low cost.
4. Compared with RF technologies, it is safe, does not constitute or suffer from electromagnetic interference, and offers a huge unregulated bandwidth between 400 THz (780 nm) and 800 THz (375 nm) (see Table 8.2).
5. VLC is relatively safe from potential eavesdroppers as the optical radiation is very well confined and does not penetrate walls.
6. Since it does not interfere with the existing RF-based systems and poses no known health problems, it is therefore ideal for hospitals, aircraft cabins, intrinsically safe environments like petrochemical industries and so on.

TABLE 8.2
Comparison of VLC, IR and RF Communication Technologies

Property	VLC	IRB	RFB
Bandwidth	Unlimited, 400–700 nm	Unlimited, 800–1600 nm	Regulated and limited
Electromagnetic interference + hazard	No	No	Yes
Line of sight	Yes	Yes	No
Distance	Short	Short to long (outdoor)	Short to long (outdoor)
Security	Good	Good	Poor
Standards	In progress (IEEE 802.15.7 Task Group)	Well developed for indoor (IrDa), In progress for outdoor	Matured
Services	Illumination + communications	Communications	Communications
Noise sources	Sun light + other ambient lights	Sun light + other ambient lights	All electrical/ electronic appliances
Power consumption	Relatively low	Relatively low	Medium
Mobility	Limited	Limited	Good
Coverage	Narrow and wide	Narrow and wide	Mostly wide

7. Switching at ultrahigh speed (thousands of times per second), well beyond what the human eye can detect. No other lighting technology has this capability.
8. The LEDs used in VLC are energy efficient.

8.2   SYSTEM DESCRIPTION

Figure 8.2 shows a block diagram of a VLC link. Precise dimming appears to be challenging for incandescent and gas-discharge lamps, whereas with LEDs it is quite convenient to accurately control the dimming level. This is because, the LED response time during on- and off-switch operation is very short (a few tens of nano-seconds). Therefore, by modulating the driver current at a relatively high frequency, it is thus possible to switch LEDs on and off without this being perceivable by the human eyes. Thus, the light emitted from an LED is in the form of a repetitive high frequency and a low average power pulse stream. The average luminous flux emitted by an LED is linearly proportional to the relative width of dimming signal. Depending upon the application and safety requirements, the transmitter can be an LED or a semiconductor laser. The LED is preferred over laser if the application is dual pro-pose of illumination and communication as is the case in VLC. The illumination requirement is that the illuminance must be 200–1000 lx for a typical office environ-ment [11]. There are essentially two types of visible LEDs: a single-colour LED (e.g., red, green or blue) and a white LED. There is at present two technologies to generate white light using LEDs (see Figure 8.3). By combining red (~625 nm), green (525 nm)

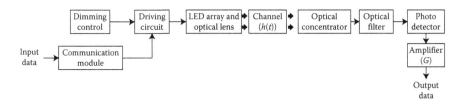

FIGURE 8.2   A block diagram of a VLC system.

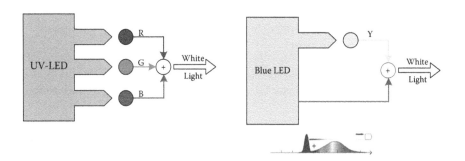

FIGURE 8.3   (**See colour insert.**) Two approaches for generating white emission from LEDs.

and blue (470 nm) (RGB) in a correct proportion (like in a colour television), the white light (or any other colour) can be generated. Typically, these triplet devices consist of a single package with three emitters and combining optics, and they are often used in application where variable colour emission is required. These devices are attractive for VLC as they offer the possibility of wavelength division multiplexing (WDM).

The other technique creates phosphorescent white LEDs. This technique involves the use of blue LED coated with a phosphor layer that emits yellow light. The phosphor layer absorbs a portion of a short wavelength light emitted by the blue LED and then the emitted light from the absorber experiences wavelength shift to a longer wavelength of yellow light. The red-shifted emission mixes additively with the non-absorbed blue component to create the required white colour. At present, the later approach is often favoured due to the lower complexity and cost [12]. However, the slow response of phosphor limits the modulation bandwidth of the phosphorescent white LEDs to a few MHz. A typical VLC link utilizing a white LED is shown in Figure 8.4a, where lighting and the communication link are both provided by the LED. The blue light could be readily extracted from the incoming optical beam by using an optical filter at the receiver. Figure 8.4b and c show the optical spectrum of the emitted white light and the measured modulation bandwidth of Osram Ostar phosphorescent white LED, respectively. For the LED under consideration, the 3 dB cut-off frequency of the white response is ~2.5 MHz at 300 mA driving current compared to ~20 MHz of the blue-only response (see Figure 8.4). The small signal modulation bandwidth of the white LED depends on the LED driving current as shown in Figure 8.5 [12–14].

To achieve high data rate, the bandwidth-limiting effect of the phosphor coating must be circumvented. The following are some techniques to achieve this:

- Blue filtering at the receiver to filter out the slow-response yellowish components [15].
- Pre-equalization at the LED driving module [16].
- Postequalization at the receiver [17].
- Combination of the three above techniques.
- Use of more complex modulation schemes where multiple bits can be carried by each transmitted symbol. This approach involves combining multi-level modulation techniques like quadrature amplitude modulation (QAM) with optical OFDM or discrete multitone (DMT) modulation. When used with blue filtering, the transmission rate can be extended to hundreds of Mbps [13,18–20].

Table 8.3 gives a summary of achievable data rates for different VLC systems for white and blue channels for a number of modulation schemes including non-return to zero (NRZ) on-off keying (OOK) and DMT-QAM modulation schemes. The OOK-NRZ scheme over an equalized channel is simpler compared to the complex modulation schemes since the latter requires a large amount of signal processing at both transmitting and receiving ends. In both cases, the ratio of the achievable transmission data rate over the raw LED modulation bandwidth (~2.5 MHz) is significantly

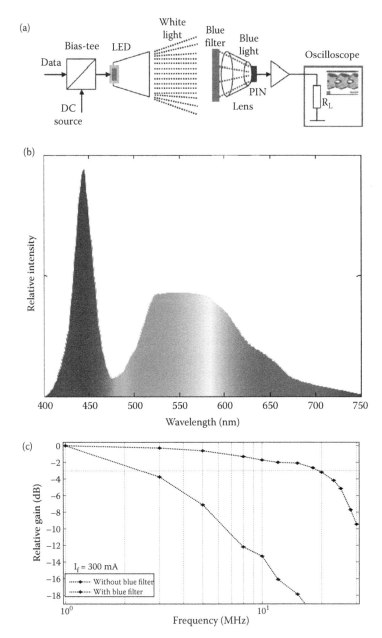

FIGURE 8.4   (**See colour insert.**) (a) VLC link, (b) LED optical spectrum of Osram Ostar white-light LED and (c) modulation bandwidth, with and without blue filtering.

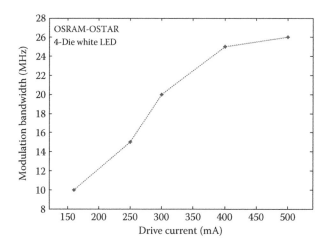

FIGURE 8.5   Measured 3 dB modulation, with blue filtering, as a function of driving current.

TABLE 8.3
Performance of High-Speed VLC Systems

Type of LEDs	Pre-Equalization	Post-equalization	Modulation Scheme	Modulation Bandwidth	Demonstrated Data Rate
White channel			OOK-NRZ	2 MHz	10 Mps (BER < 10⁻⁶)
White channel	x		OOK-NRZ	25 MHz over 2 m	40 Mbps (BER < 10⁻⁶) [16]
Phosphorescent white LEDs			OOK-NRZ	50 MHz	120 Mbps (BER < 10⁻³) [20]
Phosphorescent white LEDs			OOK-NRZ	50 MHz	230 Mbps (BER < 10⁻³) [21]
Phosphorescent white LEDs			DMT-QAM	50 MHz	513 Mbps with APD [22]
Phosphorescent white LEDs	x		OOK-NRZ	45 MHz	80 Mbps (BER < 10⁻⁶) [23]
Phosphorescent white LEDs		x	OOK-NRZ	50 MHz	100 Mbit/s (BER < 10⁻⁹) [17]
Phosphorescent white LEDs			DMT-QAM	25 MHz	100 Mbit/s (BER < 10⁻⁶) [24]
Phosphorescent white LEDs			DMT-QAM	50 MHz	200+ Mbps (BER < 10⁻³) [25]
Phosphorescent white LEDs			DMT-QAM	50 MHz	515 Mbit/s (BER < 10⁻³) [19]

enhanced. However, to achieve a wide modulation bandwidth using equalization or employing multilevel modulation schemes, the system will require a very high signal-to-noise ratio as outlined in Refs. [17,25]. Therefore, it is extremely challenging to achieve Gbps transmission rates using a VLC link.

In fulfilling the lighting requirements, a single high luminous efficiency LED can only provide limited luminous flux and over a limited area. Therefore, to illuminate a much larger environment, spatially distributed LED clusters would be needed, (see, e.g., [26–28]). To enable the capability of providing localized lighting, front-end optics in the form of diffusers, reflectors or collimators will be required to confine/distribute the light appropriately. This approach offers a good degree of freedom when it comes to achieving interesting lighting effects, in terms of colour, location lighting and luminous flux distribution at home and in work places [6,29,30]. In a nutshell, for general lighting, a large number of spatially distributed LEDs would be used for the following reasons: (i) the state-of-the-art LED technology, (ii) lack of sufficient illumination by a single LED and (iii) extremely high-brightness LEDs compromise the eye safety. Spatially distributed LEDs offer localized illumination, more flexibility than conventional light sources, and reduced light leakage in areas where illumination is not required [31,32].

8.2.1 VLC SYSTEM MODEL

Since LEDs are used for the dual propose of illumination and communication, it is necessary to define the luminous intensity and transmitted optical power. The luminous intensity is used for expressing the brightness of an LED; the transmitted optical power indicates the total energy radiated from an LED. Luminous intensity is the luminous flux per solid angle and is given as

$$I = \frac{d\Phi}{d\Omega} \tag{8.1}$$

where Φ is the luminous flux and Ω is the spatial angle.

Φ can be calculated from the energy flux Φ_e as [33]

$$\Phi = K_m \int_{380}^{780} V(\lambda)\Phi_e(\lambda)d\lambda \tag{8.2}$$

where $V(\lambda)$ is the standard luminosity curve, and K_m is the maximum visibility, which is ~683 lm/W at 555 nm wavelength.

The transmitted optical power P_t is given as

$$P_t = K_m \int_{\Lambda_{min}}^{\Lambda_{max}} \int_0^{2\pi} \Phi_e \, d\theta \, d\lambda \tag{8.3}$$

where Λ_{min} and Λ_{max} are determined from the photodiode sensitivity curve.

Figure 8.6 shows a typical office environment. Assuming that an LED lighting has a Lambertian radiation pattern, the radiation intensity at a desk surface is given by [11]

$$I(\varnothing) = I(0)\cos^{m_l}(\varnothing) \tag{8.4}$$

where $\varnothing$ is the angle of irradiance with respect to the axis normal to the transmitter surface, $I(0)$ is the centre luminous intensity and m_l is the order of Lambertian emission defined as

$$m_l = \frac{\ln(2)}{\ln(\cos \Phi_{1/2})} \tag{8.5}$$

where $\Phi_{1/2}$ is the semiangle at half illuminance of an LED.

The horizontal illuminance/intensity at a point (x, y) and the received power at the receiver are given as

$$I_{\text{hor}} = I(0) \cos^{m_l}(\varnothing)/d^2 \cdot \cos(\psi) \tag{8.6}$$

$$P_r = P_t \cdot \frac{(m_l + 1)}{2\pi d^2} \cos^{m_l}(\varnothing) \cdot T_s(\psi) \cdot g(\psi) \cdot \cos(\psi), \quad 0 \le \psi \le \psi_{\text{con}} \tag{8.7}$$

where ψ is the angle of incidence with respect to the axis normal to the receiver surface, $T_s(\psi)$ is the filter transmission, $g(\psi)$ and ψ_{con} are the concentrator gain and FOV, respectively and d is the distance between the VLED and a detector surface.

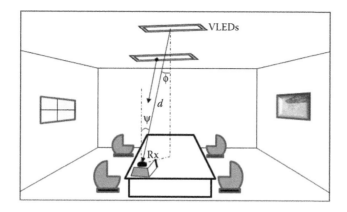

FIGURE 8.6 (**See colour insert.**) Illumination of VLEDs.

The gain of the optical concentrator at the receiver is defined by

$$
g(\psi) = \begin{cases} \dfrac{n^2}{\sin^2 \psi_{con}}, & 0 \le \psi \le \psi_{con} \\ 0, & 0 \ge \psi_{con} \end{cases}
\tag{8.8}
$$

where n is the refractive index.

Currently, most research efforts on VLC are based on single-source systems. Figure 8.7a shows a schematic diagram of distributed LED array (or multisource) for indoor application. Here, each single LED can be viewed as a point light source and therefore the radiation pattern of each LED can be viewed as a function of the solid angle in the three-dimensional space with a well-defined radiation footprint. With all the LEDs fully switched on, the illuminance distribution produced on the floor level is called the *basic illumination pattern*, defined in terms of the solid angle Θ as

$$
f_{ALED}(x, y; d) = \frac{f(\Theta)}{(x^2 + y^2 + d^2)}
\tag{8.9}
$$

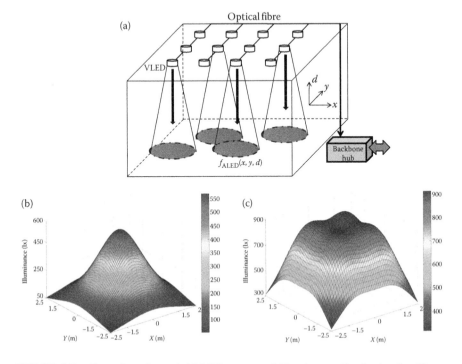

FIGURE 8.7 **(See colour insert.)** (a) LED array, and illuminance distribution for (b) one transmitter and (c) four transmitters.

Distributed multisources LED array systems are more of practical nature for two reasons: (i) most rooms uses multiple light sources to ensure sufficient illumination and (ii) offering spatial diversity, thus avoiding blocking. However, as in diffused OWC link, the multiple sources will suffer from multipath-induced intersymbol interference particularly at higher data rates. There are a number of options to overcome this problem. For example, changing the light sources layouts, reducing the receiver field of view, adopting coding and equalization techniques as well as employing multiplexing schemes.

For VLC systems employing LEDs for illumination and communication, the channel DC gain and the received optical power can be calculated using Equations 3.15 and 8.7, respectively. Table 8.4 shows the parameter for a typical room environment.

Figure 8.7a shows the illuminance using a single transmitter with a semiangle at half power of 70°. The luminous flux maximum value of 568.10 lx is right at the centre. Figure 8.7b shows the illuminance using a single transmitter with a semiangle at half power of 70°. The luminous flux maximum value of 568.10 lx is right at the centre. The illuminance distribution for four transmitters with a semiangle of 70° is depicted in Figure 8.7c, showing a value in the range of 315–910 lx, with an average value of 717 lx.

The optical power distribution for at receiver plane in a LOS path (ignoring the reflection of walls) is shown in Figure 8.8a. The MATLAB® code to simulate the optical power distribution is given in Program 8.1. It can be seen that there is almost uniform distribution of optical power at the centre with a maximum power of 2.3 dBm and a minimum power of −2.3 dBm. However, depending upon the half angle, such uniform power distribution is not possible to achieve. The optical power distribution at a receiver plane with the half angle of 12.5° is shown in Figure 8.8.

TABLE 8.4
System Parameters for a VLC Link

	Parameters	Values
Room	Size	$5 \times 5 \times 3$ m³
	Reflection coefficient	0.8
Source	Location (4 LEDs)	(1.25, 1.25, 3), (1.25, 3.75, 3), (3.75, 1.25, 3), (3.75, 3.75, 3)
	Location (1 LEDs)	(2.5, 2.5, 3)
	Semiangle at half power (FWHM)	70
	Transmitted power (per LED)	20 mW
	Number of LEDs per array	60×60 (3600)
	Centre luminous intensity	300–910 lx
Receive	Receive plane above the floor	0.85 m
	Active area (AR)	1 cm²
	Half-angle FOV	60
	Elevation	90
	Azimuth	0
	Δt	0.5 ns

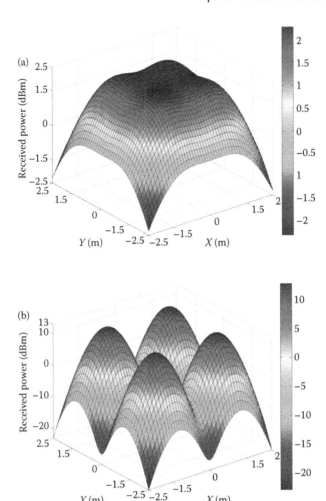

FIGURE 8.8 (**See colour insert.**) Optical power distribution in received optical plane for a FWHM of (a) 70° and (b) 12.5°.

There is more than 35 dB of optical power difference between the minimum and maximum power level, leading to high SNR in some area and dead zone in many areas. In order to make the power distribution uniform, a holographic light shaping diffuse (LSD) can be used.

Program 8.1: MATLAB Codes to Calculate the Optical Power Distribution of LOS Link at Receiving Plane for a Typical Room

```
%%
theta = 70;
% semi-angle at half power
```

```
m1=-log10(2)/log10(cosd(theta));
%Lambertian order of emission
P_LED=20;
%transmitted optical power by individual LED
nLED=60;
% number of LED array nLED*nLED
P_total=nLED*nLED*P_LED;
%Total transmitted power
Adet=1e-4;
%detector physical area of a PD

Ts=1;
%gain of an optical filter; ignore if no filter is used
index=1.5;
%refractive index of a lens at a PD; ignore if no lens is used
FOV=70;
%FOV of a receiver
G_Con=(index^2)/(sind(FOV).^2);
%gain of an optical concentrator; ignore if no lens is used

%%
lx=5; ly=5; lz=3;
% room dimension in meter
h=2.15;
%the distance between source and receiver plane

[XT,YT]=meshgrid([-lx/4 lx/4],[-ly/4 ly/4]);
% position of LED; it is assumed all LEDs are located at same
point for
% faster simulation
% for one LED simulation located at the central of the room,
use XT=0 and YT=0

%%%%%%%%%%%%%%%%%%%%%%%%%%%%%%%%%%%%%%%%%%%%%%%%%%%%%%%
Nx=lx*5; Ny=ly*5;
% number of grid in the receiver plane

x=linspace(-lx/2,lx/2,Nx);
y=linspace(-ly/2,ly/2,Ny);
[XR,YR]=meshgrid(x,y);

D1=sqrt((XR-XT(1,1)).^2+(YR-YT(1,1)).^2+h^2);
% distance vector from source 1
cosphi_A1=h./D1;
% angle vector
receiver_angle=acosd(cosphi_A1);

% alternative methods to calculate angle, more accurate if the
angle are
% negatives
```

```
% nr=[0 0 1];
% RT=[1.25 1.25]; % transmitter location
% for r=1:length(x)
%    for c=1:length(y)
%
%          angleA12=atan(sqrt((x(r)-1.25).^2+(y(c)-1.25).^2)./h);
%          costheta(r,c)=cos(angleA12);
%    end
% end
%


%%
% D2=fliplr(D1);
% % due to symmetry
% D3=flipud(D1);
% D4=fliplr(D3);

H_A1=(ml+1)*Adet.*cosphi_A1.^(ml+1)./(2*pi.*D1.^2);
% channel DC gain for source 1
P_rec_A1=P_total.*H_A1.*Ts.*G_Con;
% received power from source 1;
P_rec_A1(find(abs(receiver_angle)>FOV))=0;
% if the anlge of arrival is greater than FOV, no current is
generated at
% the photodiode.

P_rec_A2=fliplr(P_rec_A1);
% received power from source 2, due to symmetry no need separate
% calculations
P_rec_A3=flipud(P_rec_A1);
P_rec_A4=fliplr(P_rec_A3);

P_rec_total=P_rec_A1+P_rec_A2+P_rec_A3+P_rec_A4;
P_rec_dBm=10*log10(P_rec_total);

%%
Fig.
surfc(x,y,P_rec_dBm);
% contour(x,y,P_rec_dBm);hold on
% mesh(x,y,P_rec_dBm);
```

There had been several attempts to accurately model the channel for indoor visible light communication. Figure 8.9 illustrates the modelling of VLC channels with IM/DD. The visible light channel is modelled as a linear optical AWGN channel and summarized by

$$I_p(t) = RP_t(t) \otimes h(t) + n(t) \tag{8.10}$$

where $P_t(t)$ is the instantaneous transmitted optical power, $h(t)$ is the channel impulse response, $n(t)$ is the signal-independent additive noise and the symbol $\otimes$ denotes convolution.

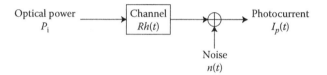

FIGURE 8.9   Modelling of VLC channel with IM/DD.

The time average transmitted optical power P_t is given by

$$P_t = \lim_{T \to \infty} \frac{1}{T} \int_0^T P_i(t)\,dt \quad P_i(t) > 0 \tag{8.11}$$

The average received optical power $P_r = H(0)P_t$, where $H(0)$ is the channel DC gain. Though the channel model in principle is essentially similar to the infrared model used in Chapter 4, the reflectivity of surfaces differ leading to different delay spread and intersymbol interference. The reflectivity of the walls depends on the wavelength and materials used in the wall. Although specular reflections can occur from a mirror or other shiny object, most reflections are typically diffuse in nature, and most are well modelled as Lambertian [11]. A study by Kwonhyung et al. [34] showed that the reflectivity depends on the wavelength as well as the texture. Because of wide linewidth of the optical signal, the reflectivity also varies with wavelength (see Figure 8.10). However, it is observed that the reflectivity of the surface in general

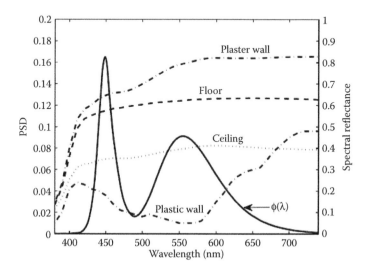

FIGURE 8.10   The PSD (solid line, which corresponds to the left axis) is compared to the measured spectral reflectance (which corresponds to the right axis) of plaster and plastic wall (dash-dot line), floor (dash line) and ceiling (dot line). (Adapted from L. Kwonhyung, P. Hyuncheol and J. R. Barry, *IEEE Communications Letters*, 15, 217–219, 2011.)

is less for visible light compared to infrared. The plaster wall has the highest reflectivity followed by floor and ceiling, respectively. Considering reflection from the wall, the received power is given by the channel DC gain on directed path $H_d(0)$ and reflected path $H_{ref}(0)$ [9].

$$P_r = \sum^{N_{LEDS}} \left\{ P_i H_d(0) + \int_{\text{Reflections}} P_i dH_{ref}(0) \right\} \tag{8.12}$$

The DC channel gain of the first reflection is given by [11]

$$H_{ref}(0) = \begin{cases} \dfrac{A_r(m_l + 1)}{2(\pi d_1 d_2)^2} \rho\, dA_{wall} \cos^{m_l}(\phi_r) \cos(\alpha_{ir}) \\ \times \cos(\beta_{ir}) T_s(\psi) g(\psi) \cos(\psi_r), & 0 \le \psi_r \le \psi_c \\ 0 & \text{elsewhere } \psi_r > \psi_c \end{cases} \tag{8.13}$$

where d_1 and d_2 are the distances between an LED chip and a reflective point, and between a reflective point and a receiver surface, ρ is the reflectance factor, dA_{wall} is a reflective area of small region, ϕ_r is the angle of irradiance to a reflective point, α_{ir} and β_{ir} are the angle of irradiance to a reflective point and the angle of irradiance to a receiver, respectively and ψ_r is the angle of incidence from the reflective surface (Figure 8.11).

The MATLAB code to calculate power distribution due to reflections from the walls is summarized in Program 8.2. Figure 8.12 shows the distribution of P_r, including influence of reflection. As shown the received power is −2.8 to 4.2 dBm for the entire room. The received average power including reflection is about 0.5 dB larger than the directed received average power.

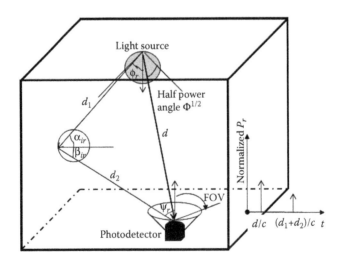

FIGURE 8.11 Propagation model of diffused link.

Program 8.2: MATLAB Codes to Calculate the Optical Power Distribution of First Reflection at the Receiving Plane for a Typical Room

```
%%
theta=70;
% semi-angle at half power
m=-log10(2)/log10(cosd(theta));
%Lambertian order of emission
P_LED=20;
%transmitted optical power by individual LED
nLED=60;
% number of LED array nLED*nLED
P_total=nLED*nLED*P_LED;
%Total transmitted power
Adet=1e-4;
%detector physical area of a PD
rho=0.8;
%reflection coefficient

Ts=1;
%gain of an optical filter; ignore if no filter is used
index=1.5;
%refractive index of a lens at a PD; ignore if no lens is used
FOV=70;
%FOV of a receiver
G_Con=(index^2)/(sind(FOV).^2);
%gain of an optical concentrator; ignore if no lens is used

%%
%%%%%%%%%%%%%%%%%%%%%%%%%%%%%%%%%%%%%%%%%%%%%%%%%%%%%%%%%%%%%%%%%%%%
%%%%%
lx=5; ly=5; lz=2.15;
% room dimension in meter

[XT,YT,ZT]=meshgrid([-lx/4 lx/4],[-ly/4 ly/4],lz/2);
% position of Transmitter (LED);

Nx=lx*5; Ny=ly*5; Nz=round(lz*5);
% number of grid in each surface
dA=lz*ly/(Ny*Nz);
% calculation grid area

x=linspace(-lx/2,lx/2,Nx);
y=linspace(-ly/2,ly/2,Ny);
z=linspace(-lz/2,lz/2,Nz);
[XR,YR,ZR]=meshgrid(x,y,-lz/2);

%%
%first transmitter calculation
TP1=[XT(1,1,1) YT(1,1,1) ZT(1,1,1)];
% transmitter position
TPV=[0 0 -1];
```

```
% transmitter position vector
RPV=[0 0 1];
% receiver position vector

%%
%%%%%%%%%%%%%%%calculation for wall 1%%%%%%%%%%%%%%%%%%%
WPV1=[1 0 0];
% position vector for wall 1

for ii=1:Nx
  for jj=1:Ny
      RP=[x(ii) y(jj) -lz/2];
      % receiver position vector
      h1(ii,jj)=0;
      % reflection from North face
      for kk=1:Ny
          for ll=1:Nz
              WP1=[-lx/2 y(kk) z(ll)];
              D1=sqrt(dot(TP1-WP1,TP1-WP1));
              cos_phi= abs(WP1(3)- TP1(3))/D1;
              cos_alpha=abs(TP1(1)- WP1(1))/D1;

              D2=sqrt(dot(WP1-RP,WP1-RP));
              cos_beta=abs(WP1(1)- RP(1))/D2;
              cos_psi=abs(WP1(3)- RP(3))/D2;

              if abs(acosd(cos_psi))<=FOV
                  h1(ii,jj)=h1(ii,jj)+(m+1)*Adet*rho*dA*...
                 . cos_phi^m*cos_alpha*cos_beta*cos_psi/
                  2*pi^2*D1^2*D2^2);
              end
          end
      end
  end
end
%%
WPV2=[0 1 0];
% position vector for wall 2
%% calculation of the channel gain is similar to wall1
P_rec_A1=(h1+h2+h3+h4)*P_total.*Ts.*G_Con;
% h2, h3 and h4 are channel gain for walls 2,3 and 4, respectively.

P_rec_A2=fliplr(P_rec_A1);
% received power from source 2, due to symmetry no need separate
% calculations
P_rec_A3=flipud(P_rec_A1);
P_rec_A4=fliplr(P_rec_A3);

P_rec_total_1ref=P_rec_A1+P_rec_A2+P_rec_A3+P_rec_A4;
P_rec_1ref_dBm=10*log10(P_rec_total_1ref);

Fig.(1)
surf(x,y,P_rec_1ref_dBm);
```

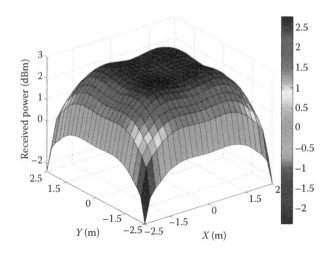

FIGURE 8.12 **(See colour insert.)** The distribution of received power with reflection.

8.2.2 SNR Analysis

The electrical SNR can be expressed in terms of the photodetector responsivity R, received optical power and noise variance as

$$\text{SNR} = \frac{(RP_r)^2}{\sigma_{\text{shot}}^2 + \sigma_{\text{thermal}}^2} \tag{8.14}$$

The shot and thermal noise variances are given by [9]

$$\sigma_{\text{shot}}^2 = 2qRP_rB + 2qI_BI_2B \tag{8.15}$$

$$\sigma_{\text{thermal}}^2 = \frac{8\pi\kappa T_k}{G_{\text{ol}}}C_{\text{pd}}AI_2B^2 + \frac{16\pi^2\kappa T_k\Gamma}{g_{\text{m}}}C_{\text{pd}}^2A^2I_3B^3 \tag{8.16}$$

where the bandwidth of the electrical filter that follows the photodetector is represented by B Hz, κ is the Boltzmann's constant, I_B is the photocurrent due to background radiation, T_k is absolute temperature, G_{ol} is the open-loop voltage gain, C_{pd} is the fixed capacitance of photodetector per unit area, Γ is the FET channel noise factor, g_{m} is the FET transconductance and noise-bandwidth factors $I_2 = 0.562$ and $I_3 = 0.0868$.

Now, considering the multipath case with ISI, the desired received power for the signal and ISI are given by

$$P_{r-sig} = \int_0^T \left\{ \sum_{i=1}^{N_{LEDS}} h_i(t) \otimes x(t) \right\} dt$$

$$P_{r-isi} = \int_T^\infty \left\{ \sum_{i=1}^{N_{LEDS}} h_i(t) \otimes x(t) \right\} dt \qquad (8.17)$$

The shot noise variance is now given as

$$\sigma_{shot-m}^2 = 2qR(P_{r-s} + P_{r-isi})B + 2qI_B I_2 B \qquad (8.18)$$

$$\sigma_{Total-m}^2 = \sigma_{shot-m}^2 + \sigma_{thermal}^2 \qquad (8.19)$$

For the OOK scheme, $BER = Q\left(\sqrt{SNR}\right)$.

Notice that the received optical power is proportional to the square of the photo-detector area (A^2). The shot noise variance is proportional to the detector area. Hence, if shot noise is the dominant noise source, the SNR is proportional to the detector area. The thermal noise is a complicated function of A and hence the noise variance is a complicated function of the photodetector area.

8.2.3 CHANNEL DELAY SPREAD

The received optical power at a point for both the direct and the first-order reflected paths is given in Equation 8.12. For multipath scenario, the total received power is given by

$$P_{rT} = \sum_{i=1}^M P_{d,i} + \sum_{j=1}^N P_{ref,j} \qquad (8.20)$$

where M and N represent the number of direct paths from transmitters to a specific receiver and reflection paths to the same receiver, $P_{d,i}$ is received optical power from the ith direct path and $P_{ref,j}$ is the received optical power from the jth reflected path.

The RMS delay spread is the critical performance criterion for the upper bound of the data transmission rate. The mean excess delay is defined by

$$\mu = \frac{\sum_{i=1}^M P_{d,i} t_{d,i} + \sum_{j=1}^N P_{ref,j} t_{ref,j}}{P_{rT}} \qquad (8.21)$$

And the RMS delay spread is given by

$$D_{RMS} = \sqrt{\mu^2 - (\mu)^2} \qquad (8.22)$$

where

$$\mu^2 = \frac{\sum_{i=1}^{M} P_{di} t_{di}^2 + \sum_{j=1}^{N} P_{ref,j} t_{refi}^2}{P_{rT}} \tag{8.23}$$

The MATLAB code to calculate the RMS delay spread is given in Program 8.3. Figure 8.13a illustrates the RMS delay spread distribution for the case of a single transmitter positioned at (2.5,2.5) and a semiangle of 70°, showing a maximum and average values of 0.4597 and 0.2053 ns, respectively. The RMS delay spread distribution for four transmitters positioned at (1.25,1.25), (1.25,3.75), (3.75,1.25) and (3.75,3.75), and a semiangle of 70° is depicted in Figure 8.13b. The maximum and average values for this case are 1.8284 and 1.096 ns, respectively.

The maximum bit rate that can be transmitted through the channel without the need for an equalizer is given by [35]

$$R_b \leq 1/(10 D_{RMS}) \tag{8.24}$$

Thus the maximum achievable data rate in diffuse channel is 55 Mbps.

Program 8.3: MATLAB Codes to Calculate D_{RMS} Values at Different Receiver Positions

```
%%
C=3e8*1e-9;
%time will be measured in ns in the program
theta=70;
% semi-angle at half power
m=-log10(2)/log10(cosd(theta));
%Lambertian order of emission
% P_LED=20;
% %transmitted optical power by individual LED
% nLED=60;
% number of LED array nLED*nLED
P_total=1;
%Total transmitted power
Adet=1e-4;
%detector physical area of a PD
rho=0.8;
%reflection coefficient

Ts=1;
%gain of an optical filter; ignore if no filter is used
index=1.5;
%refractive index of a lens at a PD; ignore if no lens is used
FOV=60;
%FOV of a receiver
G_Con=(index^2)/(sind(FOV).^2);
%gain of an optical concentrator; ignore if no lens is used
```

```
%%
%%%%%%%%%%%%%%%%%%%%%%%%%%%%%%%%%%%%%%%%%%%%%%%%%%%%%%%%%%%%%%%%%
%%%%%
lx=5; ly=5; lz=3-0.85;
% room dimension in meter

%[XT,YT,ZT]=meshgrid([-lx/4 lx/4],[-ly/4 ly/4],lz/2);
% position of Transmitter (LED);
Nx=lx*10; Ny=ly*10; Nz=round(lz*10);
% number of grid in each surface
dA=lz*ly/((Ny)*(Nz));
% calculation grid area
x=linspace(-lx/2,lx/2,Nx);
y=linspace(-ly/2,ly/2,Ny);
z=linspace(-lz/2,lz/2,Nz);

%%
TP1=[-lx/4 -ly/4 lz/2];
TP2=[lx/4 ly/4 lz/2];
TP3=[lx/4 -ly/4 lz/2];
TP4=[-lx/4 ly/4 lz/2];

% transmitter position
TPV=[0 0 -1];
% transmitter position vector
RPV=[0 0 1];
% receiver position vector

%%
WPV1=[1 0 0];
WPV2=[0 1 0];
WPV3=[-1 0 0];
WPV4=[0 -1 0];

delta_t=1/2;
% time resolution in ns, use in the form of 1/2^m

for ii=1:Nx
  for jj=1:Ny
      RP=[x(ii) y(jj) -lz/2];
      t_vector=0:30/delta_t; % time vector in ns
      h_vector=zeros(1,length(t_vector));

      % receiver position vector
      % LOS channel gain
       D1=sqrt(dot(TP1-RP,TP1-RP));
       cosphi=lz/D1;
       tau0=D1/C;
       index=find(round(tau0/delta_t)==t_vector);
        if abs(acosd(cosphi))<=FOV
        h_vector(index)=h_vector(index)+(m+1)*Adet.
```

```
              *cosphi.^(m+1)./(2*pi.*D1.^2);
        end

   %% reflection from first wall
     count=1;
     for kk=1:Ny
       for ll=1:Nz
         WP1=[-lx/2 y(kk) z(ll)];
         D1=sqrt(dot(TP1-WP1,TP1-WP1));
         cos_phi=abs(WP1(3)-TP1(3))/D1;
         cos_alpha=abs(TP1(1)-WP1(1))/D1;

         D2=sqrt(dot(WP1-RP,WP1-RP));
         cos_beta=abs(WP1(1)-RP(1))/D2;
         cos_psi=abs(WP1(3)-RP(3))/D2;
         tau1=(D1+D2)/C;
         index=find(round(tau1/delta_t)==t_vector);
         if abs(acosd(cos_psi))<=FOV
             h_vector(index)=h_vector(index)+(m+1)*Adet*rho*
                dA*...
             cos_phi^m*cos_alpha*cos_beta*cos_psi/(2*pi^2*
                D1^2*D2^2);
         end
         count=count+1;
       end
     end

   % calculate h_vector from all the walls and all the
     transmitters.

   %%
   t_vector=t_vector*delta_t;
   mean_delay(ii,jj)=sum((h_vector).^2.*t_vector)/sum(h_
     vector.^2);
   Drms(ii,jj)=sqrt(sum((t_vector-mean_delay(ii,jj)).^2.*h_
     vector.^2)/sum(h_vector.^2));

  end
end

surf(x,y, Drms);
%surf(x,y,mean_delay);
axis([-lx/2 lx/2 -ly/2 ly/2 min(min(Drms)) max(max(Drms))]);
```

8.3 SYSTEM IMPLEMENTATIONS

In conventional lighting systems (fluorescent and incandescent lights), a switch/dimmer is normally used to control each individual lamp or a set of lamps. However, such schemes of controlling light may no longer be suitable for LED-based lightings, where a large number of LEDs could be used each requiring different illumination level. Therefore, it would be impractical to assign a switch/dimmer to every LED;

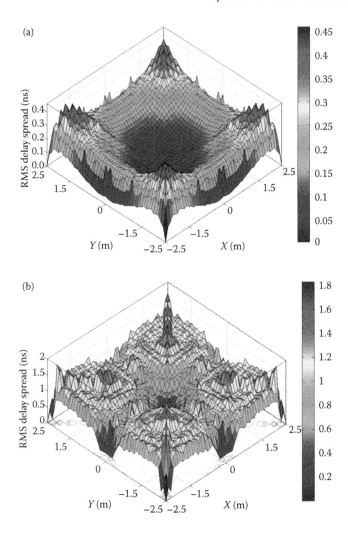

FIGURE 8.13 (**See colour insert.**) RMS delay distribution for (a) a single transmitter positioned at (2.5,2.5), and (b) for four transmitters positioned at (1.25,1.25), (1.25,3.75), (3.75,1.25), (3.75,3.75).

thus the need for different illumination control mechanism. Figure 8.13 shows a system block diagram of a generic control scenario for LEDs. The dimming signal could be based on a number of options, including bit angle modulation, pulse width modulation and so on. With a large number of LEDs, it is possible to specify the desirable illumination level at target locations via the central controller (CC). The LED illumination is characterized by the product of the duty cycle and amplitude of the received signal at a given location. Given the signal duty cycle, one could estimate the amplitude of the light signal. The function of the sensor signal processing (SSP) module is to estimate the illumination level of each LED. Based on the SSP output and the knowledge of the colour of each individual LED, the CC can determine and

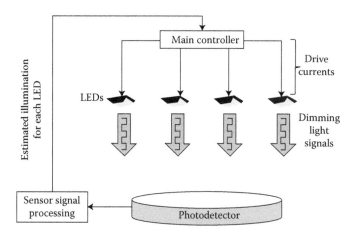

FIGURE 8.14 Schematic diagram of LED control mechanism.

configure the suitable duty cycles for LEDs to accomplish the required illumination level. Therefore, each LED will be intensity modulated differently (Figure 8.14).

In today's buildings, dimming and auto turn on/off are essential features that are widely deployed. For LED lights to be employed as an illumination and as a data communication source simultaneously, the dimming and communication signals must be independent and not interfering with each other. The injected current to the LEDs must therefore carry both the dimming and data communication signals. For VLEDs to control the dimming level, there are a number of dimming control mechanisms that are widely used and new methods have been proposed [36,37]. The amplitude modulation-based dimming is the simplest approach where by intensity modulation of the injected DC current to the LED, the emitted luminous flux is controlled [38]. However, this approach could result in changes of the chromaticity coordinates of the emitted light. Pulse time modulation schemes (PWM, SWFM, etc.) introduced in Chapter 4 are very attractive for both dimming and data communications.

PWM with subcarrier pulse position modulation (SC-PPM) techniques is another technique for controlling the LED brightness [39]. This technique offers simplicity but the dimming frequency is limited by the frequency of SC-PPM to guarantee a reliable data link performance. In addition, the signal format based on PPM has been demonstrated [40]. In this section, a number of diming schemes will be introduced and discussed.

8.3.1 BIT ANGLE MODULATION

BAM, also referred to as the binary code modulation, uses the binary data pattern for encoding the LED dimming levels [41]. Each bit in the BAM pulse train matches to the binary word (see Figure 8.15). For example, in an 8-bit BAM system, the most significant bit b7 matches the pulse with a width of 2^7, bit 6 matches the pulse with the width of 2^6 and so on. The least significant bit b0 matches the pulse width of a unit width. This scheme offers reduced flickering effect and is simple to implement.

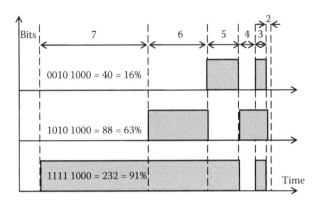

FIGURE 8.15 Time waveforms for BAM signal for VLED dimming control.

8.3.2 PULSE MODULATION SCHEMES

Due to the cost and simplicity, the IM/DD is the preferred method of the signal transmission and reception in OWC. Hence, the baseband modulation techniques using OOK-NRZ is the most popular and most reported modulation techniques for the VLC [9,12,16,17,42,43]. The traditional method of achieving the dimming with data transmission is by combining pulse-width modulation (PWM) [36–38] and pulse amplitude modulation (PAM) [44]. PWM-based dimming schemes, where the pulse duration is used to control the LED drive current, thus adjusting the dimming level (brightness), do not suffer from the wavelength shift due to the current variation in the intensity/amplitude modulation-based dimming mechanism [38]. In PWM-based schemes, a wide range of brightness level (0–100%) can be achieved by simply controlling the modulation index. Additionally, the dimming signal frequency is typically above 100 Hz; therefore, the human eye cannot perceive the current switching [36,37].

PWM is widely used in power engineering as well as communication systems where the leading or trailing edge modulation of a square wave is used to change the brightness of the light, without affecting the colour rendering of the emitted light [45]. As shown in Figure 8.16a and b, the PWM signal can be generated in analogue or digital domains, with the latter giving a much better dimming-level resolution. The duration of the kth pulse (see Figure 8.16c) is given as

$$\tau_k = \tau_0 [1 + M_{\mathrm{pwm}} m(kT_\mathrm{c})] \qquad (8.25)$$

where τ_0 is the unmodulated pulse width representing $M_{\mathrm{pwm}} m(kT_\mathrm{c}) = 0$, and $M_{\mathrm{pwm}} = 2\Delta\tau/T_\mathrm{c}$ is the modulation index $(0 < M_{\mathrm{pwm}} < 1)$, $\Delta\tau$ is the peak modulated time, T_c is the sampling interval and $m(t)$ is the input signal. The frequency spectra of a trailing edge-modulated naturally sampled PWM waveform may be expressed as

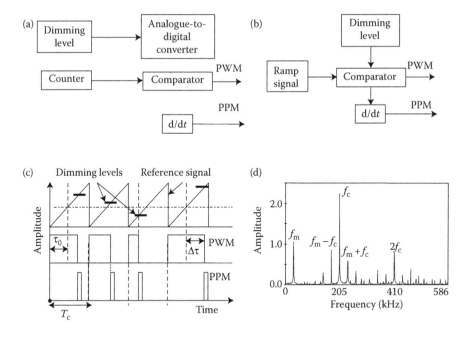

FIGURE 8.16 PWM/PPM block diagrams: (a) digital, (b) analogue waveforms, (c) PWM waveforms and (d) PWM frequency spectrum.

$$V_{ns}(t) = \frac{1}{2} + \frac{M_{pwm}}{2} \sin \omega_m t + \sum_{n=1}^{\infty} \frac{\sin(n\omega_c t)}{n\pi} - \sum_{n=1}^{\infty} \frac{J_0(n\pi M_{pwm})}{n\pi}$$

$$\times \sin(n\omega_c t - n\pi) - \sum_{n=1}^{\infty} \sum_{k=\pm 1}^{\pm\infty} \frac{J_k(n\pi M_{pwm})}{n\pi} \times \sin[(n\omega_c + k\omega_m)t - n\pi] \quad (8.26)$$

and ω_m and ω_c are the input signal and the reference signal frequencies, respectively. $J_k(x)$ is a Bessel function of the first kind, order k. In the absence of an input signal ($M_{pwm} = 0$), this series reduces to the Fourier series of the square carrier wave. Figure 8.16d shows the PWM frequency spectrum, which contains the input signal, the reference frequency and its harmonics and a set of side tones $f_c \pm f_m$ around the reference frequency and its harmonics.

An alternative method of modulation using overlapping PPM (OPPM) has also been suggested, which achieves high-capacity communication and offers flexible dimming [46]. In Ref. [39], a dual-purpose dimming method for illumination as well as data communications has been proposed in the physical layer either by employing PWM or by changing the modulation depth. It is based on the subcarrier PPM where data are separated into groups of log k bits; each and only a single pulse is allocated for each group. The example in Ref. [39] uses a subcarrier frequency of 28.8 kHz with a very low data rate of 4.8 kb/s and a suitable value of the DC level to achieve the desired brightness. In Ref. [47], a tuneable hybrid modulation scheme based on PAM and PPM has been proposed for both lighting and data transmission when the channel condition is not changing. However, it offers both power and bandwidth efficiencies under time-varying channel conditions. PPM-based system, proposed by IEEE 802.15.7, is another scheme with a low data rate capability where the PPM signal is superimposed on top of the PWM dimming control signal; more details in Section 8.3.4 [48]. These schemes offer power efficiency at the cost of bandwidth efficiency. A variable-rate multipulse PPM has also been proposed to achieve brightness control and enhanced bandwidth efficiency [49–52]. This scheme neither requires any high frequency subcarrier for data transmission nor does it need PWM signalling for the brightness control. However, one drawback of this scheme is that the offered transmission data rate is not the same for all levels of brightness. In MPPM, the information-carrying capability is affected by the total number of time slots and the number of pulsed slots per symbol period. For i-slot per symbol, of which j-slots are occupied by the pulse of one slot duration and $i-j$ slots are marked by the empty slots, the number of possible distinct combinations are $k_j^i = i!/j!(i-j)!$.

Selection of $j \in \{1, 2, \ldots, i-1\}$ allows brightness control, where $j=1$ and $j=i-1$, respectively, correspond to the minimum and maximum brightness levels. The illumination level is defined by the brightness index $m_b = j/i$. Note that two possible values of j, that is, $j=0$ (light off) and $j=i$ (light on at all time) are not used since the data transmission is not possible for these two cases.

8.3.3 PWM with Discrete Multitone Modulation

Alternatively combining PWM dimming with DMT-based VLC on the physical layer has been proposed where the control of dimming and VLC transmission on the transmitter side is decoupled [53]. By adopting spectrally efficient modulation schemes, such as quadrature amplitude modulation (QAM) on DMT subcarriers, higher data rates of 513 Mbps have been achieved [54]. DMT is used to compensate for the LED's frequency dependency. With the PWM spectrum containing high-frequency components (see Figure 8.16c), the reference (carrier) signal for PWM must be at least twice the highest subcarrier frequency of DMT to ensure reduced subcarrier interference. For slower PWM rates, there will be significant spectral aliasing, thus resulting in large subcarrier-induced interference.

A block diagram of the basic PWM-DMT VLC system is shown in Figure 8.17. The input data sequence $d(t)$ is first converted into a sequence of symbols using, for example, multilevel QAM, which offers higher spectral efficiency, prior to being applied to a serial-to-parallel (S/P). The modulator and demodulator are implemented by inverse fast Fourier transform (IFFT) and fast Fourier transform (FFT), respectively. The output of the IFFT module is the multicarrier signals given by

$$x(t) = \mathrm{Re}\left\{\sum_{k=1}^{N_{sc}-1} s_k \exp(-j\omega_k t)\right\} \quad 0 \le t \le T \qquad (8.27)$$

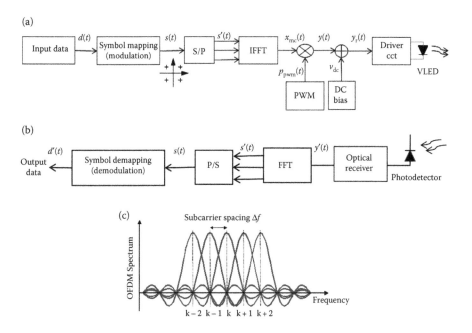

FIGURE 8.17 PWM-DMT block diagram: (a) transmitter, (b) receiver and (c) DMT (OFDM) spectrum.

where s_k is the symbol transmitted in the kth subcarrier channel, $\omega_k = 2\pi k/T$, N_{sc} is the number of subcarriers and T is the duration of the DMT symbol.

The advantages of the DMT systems are robustness in multipath propagation environment; more tolerant to the channel delay spread; use of many subcarriers, increased symbol duration, due to number of subcarrier, relative to the channel delay spread; no or reduced ISI due to the use of guard interval; simplified equalization compared with the single carrier modulation and more resistant to fading. Forward error correction can be used to overcome deep fade experienced by the subcarriers.

For biasing the light source, a DC voltage v_{dc} added to the output of the multiplier prior to intensity modulation of the VLED source. This is to ensure that the signal $x_{mc}(t)$ driving the LED will always be positive. The composite signal is given as

$$y_t(t) = x_{mc}(t) \times p_{pwm}(t) + v_{dc} \qquad (8.28)$$

where $p_{pwm}(t)$ is the periodic PWM signal. The dimming level is the same as the PWM modulation index M_{pwm}.

Figure 8.18 shows PWM waveform with 80% dimming level and PWM-sampled DMT time waveform.

With no PWM signal, the electrical signal applied to the optical drive circuit will be

$$y_t(t) = x_{mc}(t) + v_{dc} \qquad (8.29)$$

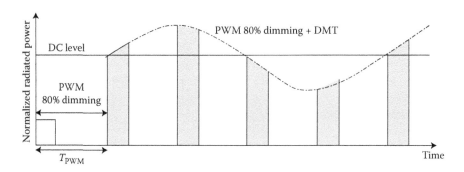

FIGURE 8.18 PWM waveform with 80% dimming level and PWM sampled DMT time waveform.

Assuming the LED is operating in it linear region with f_{3dB} much higher than the signal bandwidth, the emitted optical power is proportional to the driving current $y(t)$. Note that in DMT, the subcarriers may add up constructively, thus leading to a transmitted signal with high peak powers and consequently higher peak-to-average power ratio (PAPR). This may result in nonlinear distortions (harmonics of the subcarrier frequency and the intermodulation products (IMP)) on the emitted signal due to the limited linear range (or nonlinearity) of the LED [55], and needs to be avoided. However, in order to successfully exploit the LED's full dynamic range and improve the system performance and increase the limited resolution of DACs and ADCs, clipping can be employed before the DAC. Clipping will decrease PAPR and increase the average AC power, thus leading to higher SNR [56,57].

At the receiver, the exact opposite of transmitter is carried out as illustrated in Figure 8.17. The photodetector output is passed through the FFT module and the resulting baseband signal is converted back into a serial format via a parallel-to-serial (P/S) converter. Symbol demapping is used to regenerate the transmitted data $d'(t)$.

8.3.4 MULTILEVEL PWM-PPM

A multilevel based on PWM-PPM formats can also be used to support data transmission and illumination simultaneously (see Figure 8.19). In this scheme, a short PPM pulse of duration τ is superimposed on a PWM symbol of length T_f and a pulse duration of T_{on}, with a multilevel pulse amplitude of $A_{ml} = A_{PPM} + A_{PWM}$. At a specific dimming level, the position of the PPM pulse within one PWM frame depends on the input data pattern (see Figure 8.19b). The PWM modulation depth M_{pwm} (or duty cycle) changes with the dimming level and the inclusion of the PPM pulse does not impose any flickering and blinking on the LED brightness. The reason for this is that the bit rate for the PPM is much too high for human eyes to detect it. Both PPM and PWM have the same frame length T_f and are fully synchronized. The PPM frame length $L = 2^M$, where M is the input data bit resolution, where each block of $\log_2 L$ data bits is mapped to one of L possible slots τ. Here, $M = 1$ and $L = 2$. Only two slots in the PPM frame are used to encode the binary data. The PPM slot duration $\tau = T_f/L$. Therefore, $(L - 2)\tau$ is the guard time without PPM data. The average power

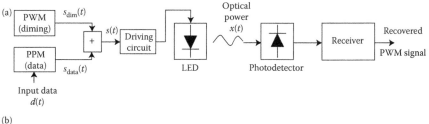

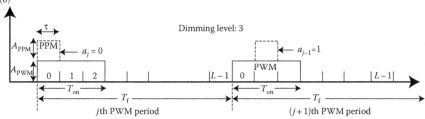

FIGURE 8.19 Multilevel PWM-PMM scheme: (a) system block diagram and (b) timing waveforms.

consumption will be somewhat higher but the interframe dimming is not affected critically since the PPM pulse is added to each PWM frame. For high brightness, the modulation index M_{pwm} of PWM should be high of course.

Figure 8.19b shows the data for the jth and $(j + 1)$th PWM frames with a_j representing a logic zero over the dimming level T_f/N of 3, and a_{j+1} is time shifted by 3. The hybrid signal is given by

$$s(t) = s_{\text{dim}}(t) + s_{\text{data}}(t) \tag{8.30}$$

The dimming and data signal are expressed as

$$s_{\text{dim}}(t) = \sum_{n=-\infty}^{\infty} A_{pwm} \cdot p_1(t - nT_f) \tag{8.31}$$

$$s_{\text{data}}(t) = A_{ppm} \sum_{N=-\infty}^{\infty} \cdot p_2(t - nT_f) \tag{8.32}$$

where A_{pwm} and A_{ppm} are the amplitude of the PWM and PPM signals, $M_{pwm} = T_{on}/T_f$, and $p_1(t)$ and $p_2(t)$ are the square pulses of a unit amplitude given by

$$p_1(t) = \Pi\left(\frac{t - M_{pwm}T_f/2}{M_{pwm}T_f}\right) \tag{8.33}$$

$$p_2(t) = \Pi\left(\frac{t - \tau/2}{\tau}\right) \tag{8.34}$$

Note that T_{on} is proportional to a dimming level, and $T_f = M\tau$.

8.3.5 PWM with NRZ-OOK

In PWM-PPM schemes, the data rate is severely limited and the system is not that flexible. To increase the data rate within the PWM dimming period, the simple NRZ-OOK signal format could be that increases the time slot utilization [40]. Figure 8.20a and b shows a block diagram and time waveforms of PWM-OOK VLC link. The NRZ-OOK data signal is added to the PWM dimming signal in a synchronized manner. In this case, the period of the NRZ-OOK signal is 10 times shorter than the PWM. Note that the data pattern exists even at the off state of the PWM dimming signal. As was outlined in Section 8.3.2, the PWM duty cycle is adjustable to generate the required signal for different environments and conditions. Figure 8.21a shows the measured received PWM-OOK signal with 70% duty cycle for PWM. Using $2^8 - 1$ pseudorandom binary sequence as the data, the eye diagram for PWM-OOK is shown in Figure 8.21b. The recovered NRZ-OOK signal is outlined in Figure 8.21c [40]. In principle, the maximum feasible data rate is limited by driving electronics for LEDs. However, the transition time of the dimming signal must coincide with that of the NRZ-OOK data pulses.

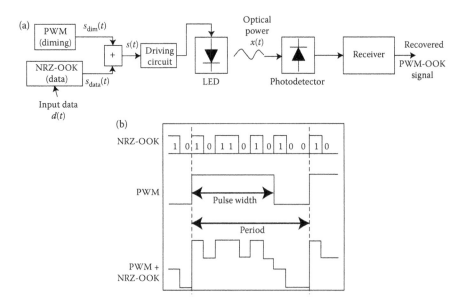

FIGURE 8.20 PWM-OOK system: (a) block diagram and (b) waveforms.

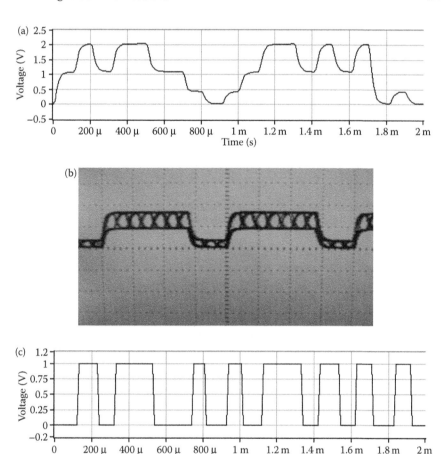

FIGURE 8.21 Measured signal waveforms: (a) received PWM-OOK with 70% duty cycle for PWM, the eye diagram and (b) recovered NRZ-OOK data signal.

8.4 MULTIPLE-INPUT–MULTIPLE-OUTPUT VLC

In MIMO systems, multiple transmitters (i.e., LEDs) and receivers (i.e., photo-detectors) are used to improve the data rates that meet or exceed existing IR LAN as well as offering significantly improved security and ease of deployment [58,59]. MIMO technology offers increased link range and higher data throughput without the need for additional power or bandwidth, by the way of higher spectral efficiency (more bits/s/Hz) and a link reliability and/or diversity. Thus, it is of prime importance in wireless communications. In MIMO systems, the transfer function between a transmitter and a receiver is divided into four components. The first represents the transfer function between a source and surface elements. Figure 8.22a shows a typical VLC MIMO system block diagram. Four LED arrays are used for room lighting as well as for transmitting four independent data streams simultaneously. A receiver array composed of four photodetector elements with

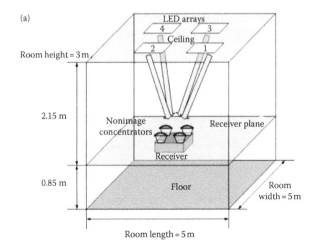

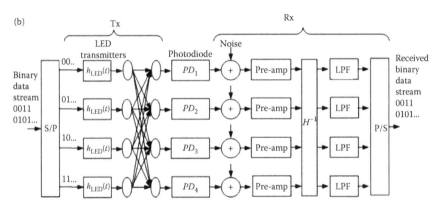

FIGURE 8.22 (a) VLC MIMO system and (b) schematic of VLC MIMO model. LPF, low-pass filter.

nonimaging concentrators are used to collect light emitted from each LED array but with different strengths due to the geometric configuration [58]. Figure 8.22b depicts the schematic of the MIMO model. Serial input data stream is interleaved and used to modulate the individual LED arrays (transmitters) prior to propagating through the channel.

In MIMO systems, the total transmit power is restricted regardless of the number of transmitters $\mathcal{M}$. At the receiver the noise comprises of the background noise, shot noise and thermal noise, for $\mathcal{N}$ optical receivers. The MIMO channel input and output symbols are nonnegative, real intensities (i.e., noncomplex numbers). Since the optical channel does not introduce any nonlinearity, the overall noise components can be considered to be normally distributed. In most cases, it is also reasonable to assume that the background and thermal noise sources are dominant compared with the shot noise. The H-matrix is an $\mathcal{M} \times \mathcal{N}$ matrix of channel attenuations that contains the information relating to the channel DC gain. If there are K-LEDs in the ith array, transmitting to the jth receiver, the DC gain h_{ij} is given by

$$
h_{ij} = \begin{cases} \displaystyle\sum_{k=1}^{K} \frac{A_{\text{pd}-rj}}{d_{ij}^2} R_0(\varphi) \cdot \cos(\beta_{ijk}) & \text{for } 0 \le \beta_{ijk} \le \beta_c \\ 0 & \text{for } \beta_c > \beta_{ijk} \end{cases} \tag{8.35}
$$

where $A_{\text{pd}-rj}$ is the collection area of the jth receiver, d_{ij}^2 is the distance from the ith transmitter to the jth receiver, β_{ijk} is the angle of incidence on the receiver and β_c is the receiver field of view.

The DC gains making up the H-matrix is given as

$$
H = \begin{bmatrix} h_{11} & h_{12} \\ h_{21} & h_{22} \end{bmatrix} \tag{8.36}
$$

The channel output vector in terms of $\mathcal{M}$-dimensional transmitter x and the noise n at the receiver is given by

$$
y = Hx + n \tag{8.37}
$$

Note that the rank of the H-matrix affects the MIMO system performance. If the H-matrix is not a full rank, the speed of the system must decrease in order to compensate for the H-matrix not being a full rank.

The jth received signal is given by

$$
r_j = R \cdot P_t \sum_{i=1}^{i=N_T} h_{ij} \cdot t_i + \sqrt{\sigma_{nj}^2} \tag{8.38}
$$

where R is the photodiode responsivity, P_t is the average transmit optical power and σ_{nj}^2 is the mean square noise current for the jth receiver. The received signals are inserted into the vector y, where $y = [r_1, \ldots, r_j, \ldots, r_{NR}]^T$.

The received data are estimated using the inverse of the H-matrix H' as

$$
d_{\text{est}} = y \cdot H^T \tag{8.39}
$$

The estimated signals are then passed through low-pass filters and a parallel-to-serial converter (P/S) to recover the transmitted serial data. The MATLAB code given in Program 8.4 can be used to simulate the MIMO system and recover the signal. Figure 8.23 shows simulated waveforms observed at the transmitter and receiver.

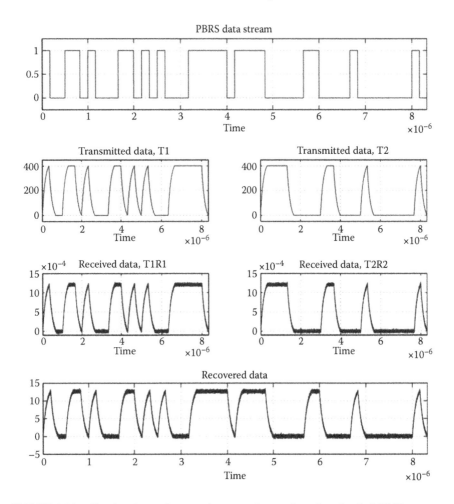

FIGURE 8.23 Simulated waveforms at the transmitter and receiver for the MIMO.

Program 8.4: MATLAB Codes to Simulate MIMO System

```
%% Input Parameters
% LED parameters
LED_halfangle=30;                       %in degrees
p_led=6;

% signal parameters
sig_length=1000;
nsamp=10;
bit_rate_vect=0.5:0.5:5;                % Mbps
bit_rate_index=1;
constellation_vect=[-0.5 0.5]; %OOK NRZ

% Receiver
PD_area=1;
```

```
Responsivity=0.2;
Rx_seperation_xdir=0.05; % seperation of receivers in x direction
Rx_seperation_ydir=0.05; % seperation of receivers in y direction

% room dimensions
room_x=2;                          % room length (m)
room_y=2;                          % Room breadth (m)
c2f=1;                             % Ceiling to floor (m)

% Bits and errors tested
BitsToTest=1e7;
MaxError=30;

% initial index values
x=0;
y=0;
xd=0;
yd=0;

%% Place Transmitters using Cartesien Coordinates
% (0,0) is situated in the top left hand corner of the room (x,y)

% place the transmitters (x,y)
Tx1=[0.5 0.5];
Tx2=[1.5 0.5];
Tx3=[0.5 1.5];
Tx4=[1.5 1.5];

%% Calculate the lambertion order of the transmitting LEDs
lamb_order=-log(2)/log(cosd(LED_halfangle));

%% LED impulse response %%%%%%%%%%%%%%%%%%%%%%%%%%%%%%%%%%%%%%%%%%%%

% Get the LED impulse response
% temp_data=load('impulse response ACULED csv.csv');
% [LED_imp_resp,t_led]=get_led_resp(temp_data);
%
% % Timing vector (time per sample at each bit rate)
% t_s_vect=1./(bit_rate_vect.*nsamp*1e6);
%
% % Resample impulse response to time per sample
% if t_s_vect(bit_rate_index)~=(t_led(2)-t_led(1))
% %    generate new time vector
%      t_vect=[0:floor(max(t_led)/t_s_vect(bit_rate_index))]...
%       *t_s_vect(bit_rate_index);
%      %do a 1d interpolation to get new response
%      LED_imp_resp=interp1(t_led,LED_imp_resp,t_vect);
%      %normalize area under imp_resp to 1
%      LED_imp_resp=LED_imp_resp/sum(LED_imp_resp);
%      %rename as required
%      imp_t=t_vect;
% end
%% calculate total number of Rx moves in all directions
```

```
% calculate no of Rx moves in x direction
Rx_moves_xdir=(room_x-2*Rx_seperation_xdir)/Rx_seperation_
xdir+1;
% calculate no of Rx moves in x direction
Rx_moves_ydir=(room_y-2*Rx_seperation_xdir)/Rx_seperation_
ydir+1;

BER=zeros(Rx_moves_xdir, Rx_moves_ydir);

% total number of moves=Rx_moves_xdir * Rx_moves_ydir
number_of_moves=(Rx_moves_xdir*Rx_moves_ydir);

%% Start main program %%%%%%%%%%%%%%%%%%%%%%%%%%%%%%%%%%%%%%%%%
% start main program loop here for moving the receivers around
the room
Rx_shift_index=1;                % initialize number of Rx moves
while Rx_shift_index<=number_of_moves

  % calculate Rx position
  [Rxp1 Rxp2 Rxp3 Rxp4 x y]=get_Rx_position(Rx_seperation_
  xdir,...
          Rx_seperation_ydir, Rx_shift_index,...
          Rx_moves_xdir, x, y);

  % Calculate the angle of irradiance, illuminance, distance...
  % and H matricies
  [Hmat]=get_H_matricies(Tx1,Tx2,Tx3,Tx4,Rxp1,Rxp2,...
  Rxp3,Rxp4,c2f,lamb_order,PD_area);

% initialize loop variables
% terr=0;
% tbit=0;
  %% Data Transmission & reception
%   while terr<MaxError && tbit< BitsToTest*4
%
%     % Generate OOK NRZ data
%     [data_tx1 OOK1]=OOK_NRZ_Generation(sig_length,nsamp);
%     [data_tx2 OOK2]=OOK_NRZ_Generation(sig_length,nsamp);
%     [data_tx3 OOK3]=OOK_NRZ_Generation(sig_length,nsamp);
%     [data_tx4 OOK4]=OOK_NRZ_Generation(sig_length,nsamp);
%     data_Tx=[OOK1;OOK2;OOK3;OOK4];

  data_Tx=randint(4,10000);
  n=0.0005*rand(4,10000);
  Rx_bit=Hmat*(data_Tx)+n;
  Recover_data=inv(Hmat)*Rx_bit;
  Recover_data(find(Recover_data>0.5))=1;
  Recover_data(find(Recover_data<0.5))=0;
%     detH(x+1,y+1)=det(Hmat);

%
%     % Transmission data                % Scale data
%     T1=conv(data_tx1, LED_imp_resp); T1=p_led*T1/mean(T1);
```

```
%       T2=conv(data_tx2, LED_imp_resp); T2=p_led*T2/mean(T2);
%       T3=conv(data_tx3, LED_imp_resp); T3=p_led*T3/mean(T3);
%       T4=conv(data_tx4, LED_imp_resp); T4=p_led*T4/mean(T4);
%       T=[T1;T2;T3;T4];
%
%       % Data reception %%%%%%%%%%%%%%%%%%%%%%%%%%%%%%%%%%%%%%%%%%%
%       Rvect=Hmat*T;
%
%       % Scale received data vector by P.D.area and responsivity
%       R=Rvect.*PD_area*Responsivity;
%
%       % load measured noise
%       [n]=load_noise_power(R,bit_rate_vect,bit_rate_index);
%
%       % Add noise to the data
%       R=R;
%
%       % Inverse of the H matrix (H^-1)
%       HmatInv=inv(Hmat);
%
%       % Estimate of the received signal vector
%       Rest=HmatInv*R;
%
%       % calculate the delay on the received signal
%       [delay1]=get_delay(Rest(1,:), data_tx1);
%       [delay2]=get_delay(Rest(2,:), data_tx2);
%       [delay3]=get_delay(Rest(3,:), data_tx3);
%       [delay4]=get_delay(Rest(4,:), data_tx4);
%
%       % remove delay
%       Rest_d(1,:)=Rest(1,(delay1:sig_length*nsamp+delay1));
%       Rest_d(2,:)=Rest(2,(delay2:sig_length*nsamp+delay2));
%       Rest_d(3,:)=Rest(3,(delay3:sig_length*nsamp+delay3));
%       Rest_d(4,:)=Rest(4,(delay4:sig_length*nsamp+delay4));
%
%       % measure the BER one channel at a time
%       [sym_error1]=get_sym_error_int(data_tx1,Rest_d(1,:),...
%                       constellation_vect,...
%                       nsamp,sig_length);
%       [sym_error2]=get_sym_error_int(data_tx2,Rest_d(2,:),...
%                       constellation_vect,...
%                       nsamp,sig_length);
%       [sym_error3]=get_sym_error_int(data_tx3,Rest_d(3,:),...
%                       constellation_vect,...
%                       nsamp,sig_length);
%       [sym_error4]=get_sym_error_int(data_tx4,Rest_d(4,:),...
%                       constellation_vect,...
%                       nsamp,sig_length);
%
%       % show progress of simulation
```

```
%      test_pos=[x+1,y+1]*Rx_seperation_xdir;
%      % count the total errors
%      terr=terr+(sym_error1+sym_error2+sym_error3+sym_error4)
%      % count the total number of bits tested
%      tbit=tbit+sig_length*4
%   end

   % Calculate BER

   [n BER(x+1,y+1)]=biterr(Recover_data,data_Tx);

   DET(x+1,y+1)=det(Hmat);

   % move receiver to the next position
   Rx_shift_index=Rx_shift_index+1
end

%% Plotting Options
room_factor_xdir=Rx_moves_xdir/(room_x-Rx_seperation_xdir);
room_factor_ydir=Rx_moves_ydir/(room_y-Rx_seperation_ydir);
x_d=(Rx_seperation_xdir:Rx_seperation_xdir:Rx_moves_xdir/
room_factor_xdir);
y_d=(Rx_seperation_ydir:Rx_seperation_ydir:Rx_moves_ydir/
room_factor_ydir);

Fig.(1)
surfc(BER)
Fig.(2)
surfc(abs(DET))

% Fig.(2)
% contour(x_d,y_d,BER)
%
% Fig.(3)
% meshc(x_d,y_d,BER)

%this function reads the impulse response data
%shortens it and starts the data at time =0

function [imp_resp,t_led]=get_led_resp(temp_data)
temp_imp_resp=temp_data(:,2);
imp_resp=temp_imp_resp(155:250);
temp_t_led=temp_data(:,1);
t_led=temp_t_led(155:250);
t_led=t_led+5.3894e-008; %start data at t=0

% % remove any negative off set and normalise to 1
off_set=min(imp_resp);
imp_resp=imp_resp-off_set;
norm_fact=max(imp_resp);
imp_resp=imp_resp/norm_fact;

end
```

```
function [ Hmat ]=get_H_matricies(Tx1,Tx2,Tx3,Tx4,Rxp1,Rxp2,...
          Rxp3,Rxp4,c2f,lamb_order,PD_area,p_led)

% calculate distance matrix
d11=sqrt(sum((Tx1-Rxp1).^2)+c2f^2);d12=sqrt(sum((Tx1-
Rxp2).^2)+c2f^2);
d13=sqrt(sum((Tx1-Rxp3).^2)+c2f^2);d14=sqrt(sum((Tx1-
Rxp4).^2)+c2f^2);

d21=sqrt(sum((Tx2-Rxp1).^2)+c2f^2);d22=sqrt(sum((Tx2-
Rxp2).^2)+c2f^2);
d23=sqrt(sum((Tx2-Rxp3).^2)+c2f^2);d24=sqrt(sum((Tx2-
Rxp4).^2)+c2f^2);

d31=sqrt(sum((Tx3-Rxp1).^2)+c2f^2);d32=sqrt(sum((Tx3-
Rxp2).^2)+c2f^2);
d33=sqrt(sum((Tx3-Rxp3).^2)+c2f^2);d34=sqrt(sum((Tx3-
Rxp4).^2)+c2f^2);

d41=sqrt(sum((Tx4-Rxp1).^2)+c2f^2);d42=sqrt(sum((Tx4-
Rxp2).^2)+c2f^2);
d43=sqrt(sum((Tx4-Rxp3).^2)+c2f^2);d44=sqrt(sum((Tx4-
Rxp4).^2)+c2f^2);

Dmat=[d11 d12 d13 d14;d21 d22 d23 d24;...
d31 d32 d33 d34;d41 d42 d43 d44];

% calculate theta matrix
Th11=atand(sqrt(sum((Tx1-Rxp1).^2))/c2f);Th12=atand
(sqrt(sum((Tx1-Rxp2).^2))/c2f);
Th13=atand(sqrt(sum((Tx1-Rxp3).^2))/c2f);Th14=atand
(sqrt(sum((Tx1-Rxp4).^2))/c2f);

Th21=atand(sqrt(sum((Tx2-Rxp1).^2))/c2f);Th22=atand
(sqrt(sum((Tx2-Rxp2).^2))/c2f);
Th23=atand(sqrt(sum((Tx2-Rxp3).^2))/c2f);Th24=atand
(sqrt(sum((Tx2-Rxp4).^2))/c2f);

Th31=atand(sqrt(sum((Tx3-Rxp1).^2))/c2f);Th32=atand
(sqrt(sum((Tx3-Rxp2).^2))/c2f);
Th33=atand(sqrt(sum((Tx3-Rxp3).^2))/c2f);Th34=atand
(sqrt(sum((Tx3-Rxp4).^2))/c2f);

Th41=atand(sqrt(sum((Tx4-Rxp1).^2))/c2f);Th42=atand
(sqrt(sum((Tx4-Rxp2).^2))/c2f);
Th43=atand(sqrt(sum((Tx4-Rxp3).^2))/c2f);Th44=atand
(sqrt(sum((Tx4-Rxp4).^2))/c2f);

Thmat=[Th11 Th12 Th13 Th14;Th21 Th22 Th23 Th24;...
   Th31 Th32 Th33 Th34;Th41 Th42 Th43 Th44];
```

```
% calculate the gain of the optical concentrator
% n=1.5;
% OC_gain=n^2./sind(Thmat)^2;
% OC_gain=OC_gain.*PD_area;

% calculate impulse response of each Tx to Rx
Hmat=(lamb_order+1)*(cosd(Thmat).^lamb_order);
Hmat=Hmat./(2*pi*Dmat.^2);
Hmat=Hmat.*cosd(Thmat);
% Hmat=Hmat.*OC_gain;
% Hmat=Hmat.*p_led;
end
```

Table 8.5 illustrates simulated data for a range of VLC MIMO systems [58].

8.5   HOME ACCESS NETWORK

There is a need to deliver multimedia-type services at gigabits per second or higher data rates, in order to address the growing high data traffic at homes, offices and so on. Recently, a European Community project on Home Gigabit Access Network (OMEGA) has experimentally demonstrated a heterogeneous (wireless and wired) approach for proving high-speed home access network [60]. There are a number of data transmission technologies that are used in home access networks, including RF [61], power line communications (PLC) [62], VLC [54,63] and infrared communications (IRC) (see Figure 8.24 and Table 8.6). These technologies are interconnected and controlled by an intelligent MAC layer. The MAC layer selects the most appropriate communication technology in order to attain seamless communications while maintaining the best link performance [54]. RF and PLC technologies are used as backbone links to distribute incoming high data rate information to number locations within the house/office. Both wireless technologies (RF and optical) provide the next level of connectivity between the base stations and the end users.

Optical wireless communication links (LOS on non-LOS) could be based on VLC and the traditional well-established infrared technology. The former could offer illumination as well as data communications in broadcast and bidirectional modes. For very high-speed data rates, of course, the preferred option would be to

TABLE 8.5
Simulated VLC MIMO Systems

Parameters	White Channel			White Channel + Equalization			Blue Channel		
Number of channels	4	16	36	4	16	36	4	16	36
Data rate (Mbit/s)	48	192	432	120	480	1080	160	640	1440
Lens diameter (cm)	0.2	0.44	0.71	0.44	0.8	1.38	1.6	3.6	7.14
Detector size (cm)	0.74 × 0.74	1.68 × 1.68	3.05 × 3.05	1.65 × 1.65	3.08 × 3.08	5.91 × 5.91	6.0 × 6.0	13.7 × 13.7	31.4 × 31.4

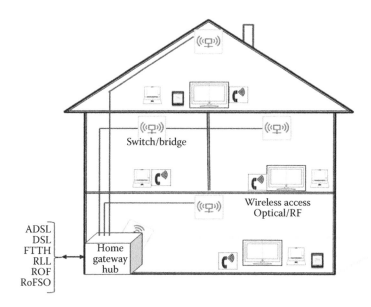

FIGURE 8.24 Home access network.

TABLE 8.6
Technologies for Home Access Networks

Technology	Data Rate
RF—60 GHz LOS	1 Gbps—Bidirectional
PLC	100 Mbps—Bidirectional
VLC	100 Mbps—Broadcast
IRC	1 Gbps—Bidirectional

---

use LOS links. High-speed LOS links with FOVs are relatively simple to implement, and FSO links operating over a few kilometres at gigabit data rates are widely available (see Chapters 1 and 6) [64]. However, in indoor environment a wide coverage area and mobility are is the main issue in LOS configurations. Therefore, to provide a high data rate and a wider coverage area as well as to ensure link availability at all times, a cellular structure or multiple transmitters and multiple receivers using imaging [65] or angle diversity [66] configurations would be the preferred option.

In cellular configuration, a single wavelength is used to cover the entire area (i.e., no wavelength reuse). In cellular systems, there should in principle be minimum overlapping between coverage areas to achieve the optimum power efficiency (see Figure 8.25a). There are a number of cell shapes that could be adopted such as circular, square, equilateral triangle and hexagon. For a given distance between the centre of a polygon and its farthest points, the hexagon has the largest area of the three with no uncovered regions between cells. Each cell has an optical base station

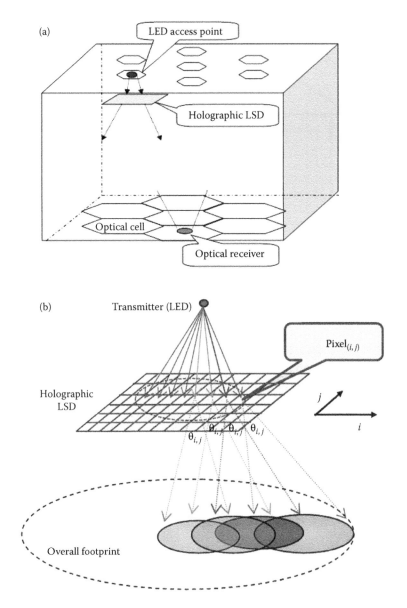

FIGURE 8.25 (**See colour insert.**) Indoor cellular VLC system: (a) block diagram and (b) transmitter with a holographic LSD.

(LED or laser diodes) located at the centre of it, providing lighting as well as illumination. Narrow FOV will ensure LOS path with lower transmit power well below the eye safety level. Using narrow FOV light sources will ensure high speed but reduced transmit power due to the eye safety and reduced coverage area. To overcome these limitations, there are a number of solutions, including (i) MIMO systems with reduced cell sizes and (ii) using holographic light shaping diffuser (LSD) to broaden the FOV, thus ensuring eye safety as well as wider coverage area [67,68].

The hologram is a two-level surface relief diffractive element that affects only the phase of light passing through it. The far-field radiation pattern passing through the hologram is approximately the Fourier transform of the surface relief structure. Using holographic LSD, the effective divergence angle of the transmitter can be extended given by

$$\theta_{output} \approx \sqrt{(\theta_{FOV})^2 + (\theta_{LSD})^2} \tag{8.40}$$

where θ_{FOV} is the divergence angle of light source and θ_{LSD} is the angle of LSD.

The hologram is a two-level surface relief diffractive element that affects only the phase of light passing through it [67]. The far-field radiation pattern passing through the hologram is approximately the Fourier transform of the surface relief structure [69]. As shown Figure 8.25b, LSD is divided into an array of 'pixels' in order to simplify the analysis. For a very small beam profile, the intensity of light can be considered as uniform after passing through a single 'pixel'. The overall coverage area is the sum of individual footprints per pixel. At the receiving end, a wide FOV photoreceiver mounted on to a mobile terminal will ensure seamless connectivity as well as alleviating the need for using pointing and tracking systems.

Following the analysis in Sections 3.1 through 3.3 in Chapter 3, the normalized power distribution and the contour of the power density at the receiving plane for a seven-cell structure with a radius of 12.2 cm (see Figure 8.25a) are is illustrated in Figure 8.26a and b, respectively [70]. It is can be seen that most of the power is concentrated near the centre of each cell, decreasing sharply towards the cell edges. In a seven-cell configuration with a circular footprint, the area within the dotted line circle (see Figure 8.26b) is defined as the 3-dB power attenuation area from the centre of a cell. The rest of the area is defined as the no coverage area or the 'dead zones' with no optical illumination. The over 3-dB coverage area A_{Cov}, where power attenuation is less than 3 dB for four cells is around 4900 cm^2, and the total coverage area of the receiving plane is given by, respectively

$$A_{Cov} = \sum_{i=1}^{N_c} A_{cov_i} \tag{8.41}$$

$$A_{Total} = A_{Cov} + A_{Dz} + A_{Ol} \tag{8.42}$$

where N_c is the number of cells and A_{Dz} and A_{Ol} shown in Figure 8.26b are the 'dead zones' and the overlapping areas, respectively.

To achieve a uniform distribution within each cell, a 30° LSD could be used. Figure 8.26c and d displays predicted power distributions and power density contours for a seven-cell configuration. The over 3-dB coverage area is marked in Figure 8.26c. The total coverage area is now around 8000 cm^2. Note the increase in the power distribution at the boundary regions because of cell overlapping.

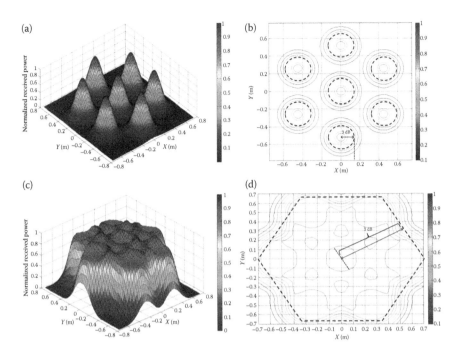

FIGURE 8.26 (**See colour insert.**) Predicted normalized power distribution at the receiving plane: (a) without LSD and (c) with a 30° LSD. Predicted power contour plot at the receiving plane: (b) without LSD and (d) with a 30° LSD.

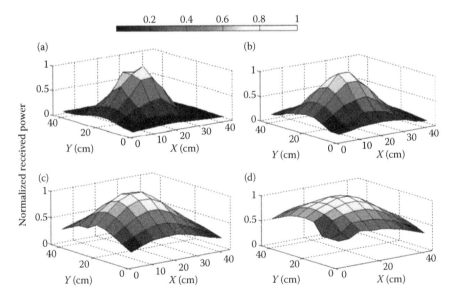

FIGURE 8.27 (**See colour insert.**) Spatial distribution of received power: (a) without LSD, (b) with 10° LSD, (c) with 20° LSD and (d) with 30° LSD.

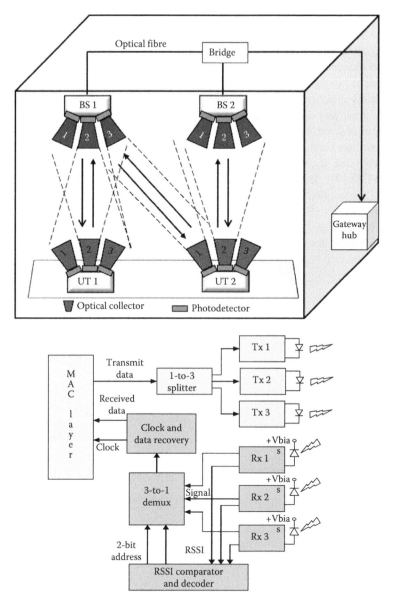

FIGURE 8.28 (**See colour insert.**) A one-dimensional optical wireless cellular system: (a) block diagram and (b) functional block diagram of a BS/UT module.

Figure 8.27 outlines the measured optical power distribution with and without holographic LSDs for a single cell. With LSDs, the power density becomes more uniform when larger-angle LSDs are used. Comparing the power density profiles of a link with no holographic LSD with a 30° holographic LSD (Figure 8.27a and d), it can be seen that a 3-dB transmission boundary has increased from 8 cm to 20 cm (i.e., 625% increasing in the coverage area).

A one-dimensional optical wireless cellular system schematic diagram is shown in Figure 8.28a. The base station (BS) transmitter composed of three transmitters to provide continuous angular coverage of ~25 × 8° are connected via a bridge to each other and the gateway hub. Each user terminal (UT) has three receivers, each with a FOV matched to the transmitter channel. Together, these create an overall reception field of view matching that of the BS transmitter. As UTs move around within the coverage area, they will be within the FOV of one or possibly more BSs. Of course, some form of control scheme would be required to ensure that the correct receiver and transmitter channels are selected with the required link performance metrics.

Figure 8.28b shows the functional block diagram of the physical layer of the BSs and UTs [71]. Transmit data from the MAC is split and used to intensity modulate the light sources. Outputs of receivers are applied to the received signal strength indicator (RSSI) comparator and decoder module the output of which is applied to the 3-to-1 multiplexer in order to select the 'active' receiver. A simple power threshold level is adopted to select the particular receiver. In a situation

TABLE 8.7
Link Budget for a One-Dimensional Optical Wireless Cellular System

Parameters	Cell Centre
Average transmit power	+14d Bm (25 mW)
Received intensity at 3m	−23 dBm/cm²
Estimated collection area (on-axis)	0.63 cm²
Available power	~−25 dBm
Additional receiver loss	~7 dB
Available APD received sensitivity	~35 dBm
Measured system margin	~3 dB

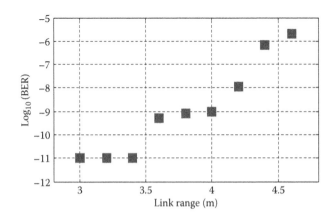

FIGURE 8.29 BER versus the link margin for a one-dimensional optical wireless cellular system.

where RSSI values of two receivers are greater than the threshold level, the receiver with the given lower index number (e.g., Rx1) is selected. This technique called the 'select good enough' approach is inferior compared with the maximal ratio combining or the select best schemes, but it is rather simple to implement. Table 8.7 shows the link budget for the system shown in Figure 8.28. Figure 8.29 shows the BER performance against the on-axis link range, illustrating error-free operation (BER $< 10^{-9}$) up to a 4 m link span, which is ideal for an ideal room environment.

REFERENCES

1. O. Bouchet, M. El Tabach, M. Wolf, D. C. O'Brien, G. E. Faulkner, J. W. Walewski, S. Randel et al., Hybrid wireless optics (HWO): Building the next-generation home network, *6th International Symposium on Communication Systems, Networks and Digital Signal Processing, CNSDSP*, Graz, Austria, 2008, pp. 283–287.
2. M. A. Naboulsi, H. Sizun and F. d. Fornel, Wavelength selection for the free space optical telecommunication technology, *SPIE*, 5465, 168–179, 2004.
3. J. K. Kim and E. F. Schubert, Transcending the replacement paradigm of solid-state lighting, *Optics Express*, 16, 21835–21842, 2008.
4. W. Binbin, B. Marchant and M. Kavehrad, Dispersion analysis of 1.55 μm free-space optical communications through a heavy fog medium, *IEEE Global Telecommunications Conference*, Washington, USA, 2007, pp. 527–531.
5. V. Kvicera, M. Grabner and J. Vasicek, Assessing availability performances of free space optical links from airport visibility data, *7th International Symposium on Communication Systems Networks and Digital Signal Processing (CSNDSP)*, Newcastle upon Tyne, UK, 2010, pp. 562–565.
6. G. J. Foschini, Layered space-time architecture for wireless communication in a fading environment when using multi-element antennas, *Bell Labs Technical Journal*, 1, 41–59, 1996.
7. T. Komine and M. Nakagawa, Fundamental analysis for visible-light communication system using LED lights, *IEEE Transactions on Consumer Electronics*, 50, 100–107, 2004.
8. M. Hoa Le, D. O'Brien, G. Faulkner, Z. Lubin, L. Kyungwoo, J. Daekwang, O. YunJe and W. Eun Tae, 100 Mb/s NRZ visible light communications using a postequalized white LED, *IEEE Photonics Technology Letters*, 21, 1063–1065, 2009.
9. T. Komine and M. Nakagawa, Fundamental analysis for visible-light communication system using LED lights, *IEEE Transactions on Consumer Electronics*, 50, 100–107, 2004.
10. H. Elgala, R. Mesleh, H. Haas and B. Pricope, OFDM visible light wireless communication based on white LEDs, *IEEE 65th Vehicular Technology Conference*, Dublin, Ireland, 2007, pp. 2185–2189.
11. F. R. Gfeller and U. Bapst, Wireless in-house data communication via diffuse infrared radiation, *Proceedings of the IEEE*, 67, 1474–1486, 1979.
12. D. O'Brien, L. Zeng, H. Le-Minh, G. Faulkner, O. Bouchet, S. Randel, J. Walewski et al., Visible light communication, *Short-Range Wireless Communications: Emerging Technologies and Applications*, R. Kraemer and M. Katz, Eds., New Jersey, USA: Wiley Publishing, 2009.
13. M. Hoa Le, D. O'Brien, G. Faulkner, Z. Lubin, L. Kyungwoo, J. Daekwang and O. YunJe, High-speed visible light communications using multiple-resonant equalization, *IEEE Photonics Technology Letters*, 20, 1243–1245, 2008.
14. A. Prokes, Atmospheric effects on availability of free space optics systems, *Optical Engineering*, 48, 066001-10, 2009.

15. S. Randel, F. Breyer, S. C. J. Lee and J. W. Walewski, Advanced modulation schemes for short-range optical communications, *IEEE Journal of Selected Topics in Quantum Electronics*, 16, 1280–1289, 2010.

16. *Guide to Meteorological Instruments and Methods of Observation*, Geneva, Switzerland: World Meteorological Organisation, 2006.

17. M. Grabner and V. Kvicera, On the relation between atmospheric visibility and optical wave attenuation, *16th IST Mobile and Wireless Communications Summit*, Budapest, Hungary, 2007, pp. 1–5.

18. J. A. C. Bingham, Multicarrier modulation for data transmission: An idea whose time has come, *IEEE Communications Magazine*, pp. 5–14, 1990.

19. J. Vucic, C. Kottke, S. Nerreter, A. Buttner, K. D. Langer and J. W. Walewski, White light wireless transmission at 200+ Mb/s net data rate by use of discrete-multitone modulation, *IEEE Photonics Technology Letters*, 21, 1511–1513, 2009.

20. M. Grabner and V. Kvicera, Case study of fog attenuation on 830 and 1550 nm free-space optical links, *Proceedings of the Fourth European Conference on Antennas and Propagation (EuCAP)*, Barcelona, Spain, 2010, pp. 1–4.

21. D. P. Greenwood, Bandwidth specification for adaptive optics systems, *Journal of Optical Soceity of America*, 67, 390–393, 1977.

22. M. A. Al-Habash, L. C. Andrews and R. L. Phillips, Mathematical model for the irradiance probability density function of a laser beam propagating through turbulent media, *Optical Engineering*, 40, 1554–1562, 2001.

23. S. Rajbhandari, Z. Ghassemlooy, J. Perez, H. Le Minh, M. Ijaz, E. Leitgeb, G. Kandus and V. Kvicera, On the study of the FSO link performance under controlled turbulence and fog atmospheric conditions, *Proceedings of the 2011 11th International Conference on Telecommunications (ConTEL)*, 2011, pp. 223–226.

24. S. Rajbhandari, Application of wavelets and artificial neural network for indoor optical wireless communication systems, PhD thesis, Northumbria University, Newcastle upon Tyne, UK, 2010.

25. F. Nadeem, V. Kvicera, M. S. Awan, E. Leitgeb, S. Muhammad and G. Kandus, Weather effects on hybrid FSO/RF communication link, *IEEE Journal on Selected Areas in Communications*, 27, 1687–1697, 2009.

26. M. Abramowitz and I. S. Stegun, *Handbook of Mathematical Functions with Formulas, Graphs and Mathematical Tables*, New York: Dover, 1977.

27. R. Barakat, Sums of independent lognormally distributed random variables, *Journal of the Optical Society of America*, 66, 211–216, 1976.

28. S. M. Haas and J. H. Shapiro, Capacity of wireless optical communications, *IEEE Journal on Selected Areas in Communications*, 21, 1346–1357, 2003.

29. E. J. Lee and V. W. S. Chan, Part 1: Optical communication over the clear turbulent atmospheric channel using diversity, *IEEE Journal on Selected Areas in Communications*, 22, 1896–1906, 2004.

30. E. Telatar, Capacity of multi-antenna Gaussian channels, *European Transactions on Telecommunications*, 10, 585–595, 1999.

31. S. G. Wilson, M. Brandt-Pearce, C. Qianling and M. Baedke, Optical repetition MIMO transmission with multipulse PPM, *IEEE Journal on Selected Areas in Communications*, 23, 1901–1910, 2005.

32. I. Hen, MIMO architecture for wireless communication, *Intel Technology Journal*, 10, 157–166, 2006.

33. T. Komine, *Visible Light Wireless Communications and Its Fundamental Study*, Keio University, PhD thesis, Japan, 2005.

34. L. Kwonhyung, P. Hyuncheol and J. R. Barry, Indoor channel characteristics for visible light communications, *IEEE Communications Letters*, 15, 217–219.

35. F. Xu, A. Khalighi, P. Caussé and S. Bourennane, Channel coding and time-diversity for optical wireless links, *Optics Express*, 17, 872–887, 2009.

36. J. Garcia, M. A. Dalla-Costa, J. Cardesin, J. M. Alonso and M. Rico-Secades, Dimming of high-brightness LEDs by means of luminous flux thermal estimation, *IEEE Transactions on Power Electronics*, 24, 1107–1114, 2009.

37. J. Chen, Y. Ai and Y. Tan, Improved free space optical communications performance by using time diversity, *Chin. Optics Letters*, 6, 797–799, 2008.

38. M. Czaputa, T. Javornik, E. Leitgeb, G. Kandus and Z. Ghassemlooy, Investigation of punctured LDPC codes and time-diversity on free-space optical links, *Proceedings of the 2011 11th International Conference on Telecommunications (ConTEL)*, Zagreb, Croatia, 2011, pp. 359–362.

39. W. Zixiong, Z. Wen-De, F. Songnian and L. Chinlon, Performance comparison of different modulation formats over free-space optical (FSO) turbulence links with space diversity reception technique, *IEEE Photonics Journal*, 1, 277–285, 2009.

40. H. Moradi, H. H. Refai and P. G. LoPresti, Switched diversity approach for multireceiving optical wireless systems, *Applied Optics*, 50, 5606–5614, 2011.

41. K. Young-Chai, M. S. Alouini and M. K. Simon, Analysis and optimization of switched diversity systems, *IEEE Transactions on Vehicular Technology*, 49, 1813–1831, 2000.

42. Z. Lubin, D. O'Brien, M. Hoa, G. Faulkner, L. Kyungwoo, J. Daekwang, O. YunJe and W. Eun Tae, High data rate multiple input multiple output (MIMO) optical wireless communications using white LED lighting, *IEEE Journal on Selected Areas in Communications*, 27, 1654–1662, 2009.

43. Y. Tanaka, T. Komine and S. Haruyam, Indoor visible light data transmission system utilizing white LED lights, *IEICE Transaction of Communication*, E86-B, 2240–2454, 2003.

44. F. Yuan, W. Siu-Hong and L. Hok-Sun Ling, A power converter with pulse-level-modulation control for driving high brightness LEDs, *Twenty-Fourth Annual IEEE Applied Power Electronics Conference and Exposition*, Washington, DC, 2009, pp. 577–581.

45. Y. Han, A. Dang, Y. Ren, J. Tang and H. Guo, Theoretical and experimental studies of turbo product code with time diversity in free space optical communication, *Optics Express*, 18, 26978–26988, 2010.

46. B. Bo, X. Zhengyuan and F. Yangyu, Joint LED dimming and high capacity visible light communication by overlapping PPM, *19th Annual Wireless and Optical Communications Conference (WOCC)*, Shanghai, China, 2010, pp. 1–5.

47. Y. Zengm, R. J. Green, S. Sun and M. S. Leeson, Tunable pulse amplitude and position modulation technique for reliable optical wireless communication channels, *Journal of Communications*, 2, 22–28, 2007.

48. N. C. Beaulieu, A. A. Abu-Dayya and P. J. McLane, Estimating the distribution of a sum of independent lognormal random variables, *IEEE Transactions on Communications*, 43, 2869, 1995.

49. A. B. Siddique and M. Tahir, Joint brightness control and data transmission for visible light communication systems based on white LEDs, *2011 IEEE Consumer Communications and Networking Conference (CCNC)*, Las Vegas, USA, 2011, pp. 1026–1030.

50. F. S. Vetelino, C. Young, L. Andrews and J. Recolons, Aperture averaging effects on the probability density of irradiance fluctuations in moderate-to-strong turbulence, *Applied Optics*, 46, 2099–2108, 2007.

51. J. H. Churnside, Aperture averaging of optical scintillations in the turbulent atmosphere, *Applied Optics*, 30, 1982–1994, 1991.

52. H. Sugiyama and K. Nosu, MPPM: A method for improving the band-utilization efficiency in optical PPM, *Journal of Lightwave Technology*, 7, 465–472, 1989.

53. G. Ntogari, T. Kamalakis, J. Walewski and T. Sphicopoulos, Combining illumination dimming based on pulse-width modulation with visible-light communications based on discrete multitone, *IEEE/OSA Journal of Optical Communications and Networking*, 3, 56–65, 2011.

54. J. Vucic, C. Kottke, S. Nerreter, K. Langer and J. W. Walewski, 513 Mbit/s visible light communications link based on DMT-modulation of a white LED, *Journal of Lightwave Technology*, 28, 3512–3518, 2010.

55. L. C. Andrews, R. L. Phillips and C. Y. Hopen, Aperture averaging of optical scintillations: Power fluctuations and the temporal spectrum, *Waves in Random Media*, 10, 53–70, 2000.

56. M. A. Khalighi, N. Aitamer, N. Schwartz and S. Bourennane, Turbulence mitigation by aperture averaging in wireless optical systems, *10th International Conference on Telecommunications (ConTEL)*, Zagreb, Croatia, 2009, pp. 59–66.

57. E. F. Schubert and J. K. Kim, Solid-state light sources getting smart, *Science*, 308, 1274–1278, 2005.

58. E. Jakeman and P. Pusey, A model for non-Rayleigh sea echo, *IEEE Transactions on Antennas and Propagation*, 24, 806–814, 1976.

59. M. Uysal and J. Li, BER performance of coded free-space optical links over strong turbulence channels, *IEEE 59th Vehicular Technology Conference, VTC 2004-Spring*, Milan, Italy, 2004, pp. 352–356.

60. P. R. Boyce, *Human Factors in Lighting*, 2nd ed. New York, USA: CRC Press, 2003.

61. J. Grubor, S. C. J. Lee, K.-D. Langer, T. Koonen and J. W. Walewski, Wireless high-speed data transmission with phosphorescent white-light LEDs, *33rd European Conference and Exhibition of Optical Communication*, Berlin, Germany, 2007, pp. 1–2.

62. J. Vucic, C. Kottke, S. Nerreter, K. Habel, A. Buttner, K. D. Langer and J. W. Walewski, 125 Mbit/s over 5 m wireless distance by use of OOK-modulated phosphorescent white LEDs, *35th European Conference on Optical Communication*, Vienna, Austria, 2009, pp. 1–2.

63. J. Vucic, C. Kottke, S. Nerreter, K. Habel, A. Buttner, K. D. Langer and J. W. Walewski, 230 Mbit/s via a wireless visible-light link based on OOK modulation of phosphorescent white LEDs, *2010 Conference on Optical Fiber Communication (OFC)*, San Diego, 2010, pp. 1–3.

64. K. Kiasaleh, Performance of coherent DPSK free-space optical communication systems in K-distributed turbulence, *IEEE Transactions on Communications*, 54, 604–607, 2006.

65. B. Braua and D. Barua, Channel capacity of MIMO FSO under strong turbulance conditions, *International Journal of Electrical & Computer Sciences*, 11, 1–5, 2011.

66. C. C. Motlagh, V. Ahmadi and Z. Ghassemlooy, Performance of free space optical communication using M-array receivers at atmospheric condition, *17th ICEE*, Tehran, Iran, May 12–14, 2009.

67. P. L. Eardley, D. R. Wisely, D. Wood and P. McKee, Holograms for optical wireless LANs, *IEE Proceedings Optoelectronics*, 143, 365–369, 1996.

68. M. R. Pakravan, E. Simova and M. Kavehrad, Holographic diffusers for indoor infrared communication systems, *Global Telecommunications Conference*, London, 1996, 1608–1612.

69. I. Fryc, S. W. Brown, G. P. Eppeldauer and Y. Ohno, LED-based spectrally tunable source for radiometric, photometric and colorimetric applications, *Optical Engineering*, 44, 111 309(1–8), 2004.

70. N. A. Galano, Smart lighting: LED implementation and am bient communication applications, Master's thesis, University of California, Berkeley, 2009.

71. Y. Hongming, J. W. M. Bergmans and T. Schenk, Illumination sensing in LED lighting systems based on frequency-division multiplexing, *IEEE Transactions on Signal Processing*, 57, 4269–4281, 2009.

Index

497

.

Printed and bound by CPI Group (UK) Ltd, Croydon, CR0 4YY

18/10/2024

01776267-0011